Protein Crops: Agricultural Sustainability

Protein Crops: Agricultural Sustainability

Editor: William Miller

STATES
ACADEMIC PRESS
www.statesacademicpress.com

States Academic Press,
109 South 5th Street,
Brooklyn, NY 11249, USA

Visit us on the World Wide Web at:
www.statesacademicpress.com

ISBN: 978-1-63989-448-2 (Hardback)

Cataloging-in-Publication Data

Protein crops : agricultural sustainability / edited by William Miller.
p. cm.
Includes bibliographical references and index.
ISBN 978-1-63989-448-2
1. Crops. 2. Proteins. 3. Grain. 4. Oilseeds. 5. Sustainable agriculture. I. Miller, William.
SB91 .P76 2022
631--dc23

Table of Contents

Preface

I am honored to present to you this unique book which encompasses the most up-to-date data in the field. I was extremely pleased to get this opportunity of editing the work of experts from across the globe. I have also written papers in this field and researched the various aspects revolving around the progress of the discipline. I have tried to unify my knowledge along with that of stalwarts from every corner of the world, to produce a text which not only benefits the readers but also facilitates the growth of the field.

Proteins are macromolecules of amino acids which are linked together by peptide bonds. They are an essential nutrient for the human body and form a major structural component of muscles. Crops that have substantial amounts of protein in them are often referred to as protein crops. Common examples of protein crops include peanut, chickpea, soybean, lentil, cowpea and pigeon pea. These crops have a positive impact on the soil as they can fix nitrogen, control erosion and enrich the soil with humus. They also serve as a food source for pollinators. This book includes some of the vital pieces of work being conducted across the world, on various topics related to protein crops. It aims to shed light on some of the unexplored aspects of protein crops and the recent researches in this field. This book is a vital tool for all researching or studying protein crops as it gives incredible insights into emerging trends and concepts.

Finally, I would like to thank all the contributing authors for their valuable time and contributions. This book would not have been possible without their efforts. I would also like to thank my friends and family for their constant support.

Editor

1

Seedling Emergence and Phenotypic Response of Common Bean Germplasm to Different Temperatures under Controlled Conditions and in Open Field

Edited by:
Fred Stoddard,
University of Helsinki, Finland

Reviewed by:
Johann Vollmann,
University of Natural Resources and Life Sciences, Vienna, Austria
Elisa Bellucci,
Marche Polytechnic University, Italy
Robert John French,
Department of Agriculture and Food, Australia

***Correspondence:**
Antonio M. De Ron
amderon@mbg.csic.es

Antonio M. De Ron[1,2], Ana P. Rodiño[1], Marta Santalla[1], Ana M. González[1], María J. Lema[2,3], Isaura Martín[4] and Jaime Kigel[5]*

[1] Biology of Agrosystems, Misión Biológica de Galicia, National Spanish Research Council, Pontevedra, Spain, [2] Sistemas Agroforestales, Unidad Asociada a la Misión Biológica de Galicia (CSIC), Pontevedra, Spain, [3] Phytopathological Station do Areeiro, Provincial Chamber, Pontevedra, Spain, [4] National Center for Plant Genetic Resources, National Institute for Agricultural and Food Research and Technology, Alcalá de Henares, Spain, [5] The Robert H. Smith Faculty of Agriculture, Food and Environment, Hebrew University of Jerusalem, Rehovot, Israel

Rapid and uniform seed germination and seedling emergence under diverse environmental conditions is a desirable characteristic for crops. Common bean genotypes (*Phaseolus vulgaris* L.) differ in their low temperature tolerance regarding growth and yield. Cultivars tolerant to low temperature during the germination and emergence stages and carriers of the grain quality standards demanded by consumers are needed for the success of the bean crop. The objectives of this study were (i) to screen the seedling emergence and the phenotypic response of bean germplasm under a range of temperatures in controlled chamber and field conditions to display stress-tolerant genotypes with good agronomic performances and yield potential, and (ii) to compare the emergence of bean seedlings under controlled environment and in open field conditions to assess the efficiency of genebanks standard germination tests for predicting the performance of the seeds in the field. Three trials were conducted with 28 dry bean genotypes in open field and in growth chamber under low, moderate, and warm temperature. Morpho-agronomic data were used to evaluate the phenotypic performance of the different genotypes. Cool temperatures resulted in a reduction of the rate of emergence in the bean genotypes, however, emergence and early growth of bean could be under different genetic control and these processes need further research to be suitably modeled. Nine groups arose from the Principal Component Analysis (PCA) representing variation in emergence time and proportion of emergence in the controlled chamber and in the open field indicating a trend to lower emergence in large and extra-large seeded genotypes. Screening of seedling emergence and phenotypic response of the bean germplasm under a range of temperatures in controlled growth chambers and under field conditions showed several genotypes, as landraces 272, 501, 593, and the cultivar Borlotto, with stress-tolerance at emergence, and high yield potential that could be valuable genetic material for breeding programs. Additionally, the potential genetic erosion in genebanks was assessed. Regarding bean commercial traits, under low temperature at sowing time seed reached larger size, and crop yield was higher compared to warmer temperatures at the sowing time. Therefore, early sowing of bean is strongly recommended.

Keywords: low temperature tolerance, *Phaseolus vulgaris* L., plant breeding, plant genetic resources, seedling emergence, yield

INTRODUCTION

The common bean (*Phaseolus vulgaris* L.) is native to the Americas where two major domestication centers and gene pools have been described, Andean and Mesoamerican, which differ in their adaptation to different climatic and eco-geographic conditions. Differences in the seed type and size are clear between both genetic pools (Singh et al., 1991; Santalla et al., 2001), having the Andean varieties larger seeds than the Mesoamerican ones. The diffusion of the common bean out of its American domestication centers appears to have been very complex, involving numerous introductions into different continents along a range of agrosystems. Several geographic regions have been proposed as secondary centers of diversification, such as Europe (Santalla et al., 2002; Angioi et al., 2010; Gioia et al., 2013), central-eastern and southern Africa, Brazil, and China (Bellucci et al., 2014). However, once out of the Americas, the spatial isolation between the Mesoamerican and Andean gene pools was not maintained, thus providing increased potential for their hybridization, and introgression. In Europe, this issue is highly relevant for breeding programs. Indeed, their hybridization has led to the recombination of the Mesoamerican and Andean traits resulting in novel and useful genotypes and phenotypes adapted to contrasting environmental conditions (i.e., resistance to biotic and abiotic stress; Rodiño et al., 2006; Angioi et al., 2010; Blair et al., 2010; Santalla et al., 2010). In contrast, various studies suggest that in other regions the introgression between these gene pools appears to be less relevant than in Europe (De Ron et al., 2015).

Early breeding efforts primarily focused on improved disease resistance and adaptation to local environments, with later efforts focused on improved seed quality, improved plant architecture, and breeding for yield (Duc et al., 2015). Yield *per-se*, tolerance to drought, adaptation to poor soils, and nutritional quality are priorities of bean breeders since the 1990s (De Ron et al., 2015). Seed germination and seedling emergence in the small seeded Mesoamerican genotypes is generally faster than that in the Andean ones, and this phenotypic trait has been used to distinguish between the two genetic pools (White and Montes, 1993). Faster emergence may reflect both genetic variation for adaptation to specific environments and effects of seed size in emergence. Seed size has been recognized as a factor affecting bean germination (Hanley et al., 2003; Kaya et al., 2008) and is probably related to water uptake, a key process in seedling emergence (Bewley, 1997). High seed vigor, good germination, and emergence are prerequisites for successful direct sowing in common bean. Thus, a better understanding of the genetics of the processes regulating germination and early growth under different conditions is important not only as a contribution to the knowledge of this species, but also has direct applications in plant breeding, and for germplasm conservation and regeneration.

The bean crop experienced a quick adaptive radiation throughout Europe in the Sixteenth Century (Zeven, 1997), where it was distributed through very different edapho-climatic environments. The microclimate of the cultivated areas, located at different latitudes, and altitudes, could have a strong influence on the recent evolution of this crop (Escribano et al., 1994; Santalla et al., 2002, 2010; Casquero et al., 2006; González et al., 2006; Papa et al., 2006; Rodiño et al., 2006). Several studies showed that a number of varieties with relevance for niche markets still survive on-farm in marginal areas of European countries (Zeven et al., 1999; Eichenberger et al., 2000; Rodiño et al., 2001, 2003, 2009; Negri and Tosti, 2002; Sicard et al., 2005) and in their areas or origin (De Ron et al., 2004; Galván et al., 2006). Common bean is adapted to relatively humid and cool climatic conditions with optimal average daily temperature for reproductive development ranging from 20 to 25°C (Wantanbe, 1953). Temperatures >30°C during the day or >20°C at night result in yield reduction (Hardwick and Andrews, 1980; Rainey and Griffiths, 2005), and seeds germinate poorly below 15°C (Kotowski, 1926; Kigel et al., 2015). Thus, it is necessary to restrict field sowing of beans to warm climates or to delay sowing until the soil is warm enough for satisfactory emergence in cool climates (Hardwick, 1972). Moreover, beans that are slow to germinate are also likely to be slow in growth (Kooistra, 1971). The physiological response of common bean to high temperature stress has been primarily studied through the use of controlled environmental testing in greenhouses and growth chambers (Porch, 2006). However, the long-term goal of breeding for stress tolerance is the development of germplasm with improved field adaptation to different temperatures. Therefore, in order to make maximum use of the available growing period, genotypes must be developed that are tolerant to low temperature during germination and early growth.

Seed germination is the process that commences with uptake of water by the dry seed—i.e., imbibition, and terminates with emergence of the seedling. Thus, the process involves two temporal stages, namely the germination stage and the emergence, or seedling-growth stage (Bewley and Black, 1985; Bewley, 1997). The emergence of the radicle marks the end of the first stage and the onset of the second. The sooner the radicle protrudes through the seed coat, the faster is the emergence. Fast seed germination is considered an important adaptive trait marking a quick transition to the growth phase in the life-cycle of a plant. The time taken for the germination process to be completed is one of the important parameters of seed quality (Copeland and McDonald, 1995; Dutt and Geneve, 2007). Vigorous, rapid, as well as uniform germination and emergence under diverse environmental conditions is a desirable attribute for seedling growth and, ultimately, grain yield in food legumes and cereal crops such as bean, rice, wheat, maize. Crop species vary widely in how fast their seeds germinate, the rate of emergence being the result of the interaction between the seed genotype and specific environmental, or ecological factors (Hernández-Nistal et al., 1989; Alonso-Blanco et al., 2003; Schmuths et al., 2006).

In past years, substantial progress was achieved by plant breeders in adapting crops such as maize, tomato, soybean, and common bean to suboptimal temperatures (Dickson, 1971; Holmberg, 1973; Skrdla and Mock, 1978; Patterson and Payne, 1983). Ideally, the best cold-tolerant genotypes should have successful water imbibition, germination, and emergence at low temperature (Kemp, 1978; Garcia-Huidobro et al., 1982; Dickson and Boettger, 1984a,b; Gummerson, 1986). The interval from seedling to maximum growth and blooming of bean plants should be shortened by selecting lines capable of rapid early growth during periods of low temperature following sowing.

The cultivation of dry bean in South Europe has traditionally taken place with sowings form April–June. But summer cultivation increases production costs by demanding greater irrigation and more tillage because of weed proliferation. Moreover, summer cultivation increases the likelihood of harvest coinciding with the onset of the rainy season, leading to crop failure, and yield losses. Cultivation in the spring period, however, is restricted because low temperatures at sowing delay germination, seedling emergence, and early growth. The alternative, therefore, is to use genotypes tolerant to low temperature at the germination, emergence, and early growth stages. Yet, little research has been done to breed this type of dry bean genotypes. Thus, further identification of cold tolerant genotypes already reported in a few instances (Dickson, 1971; Kooistra, 1971; Bannerot, 1979; Hardwick and Andrews, 1980; Dickson and Boettger, 1984a,b; Scully and Waines, 1987) is necessary.

The purpose of germination testing in genebanks and breeder collections is to provide information on the comparative and foreseeable field planting value of different seed samples. In the case that field testing seed emergence, which can be affected by the field conditions (Ellis et al., 1985), failed more than expected according the tests of germination, there may be a loss of genetic material of gene bank accessions that will imply a process of genetic erosion.

The objectives of this study were (i) to screen the seedling emergence and the phenotypic response of bean germplasm under a range of temperatures in controlled chamber and field conditions to display stress-tolerant genotypes with good agronomic performances and yield potential, and (ii) to compare the emergence of bean seedlings under controlled environment and in open field conditions to assess the efficiency of genebanks standard germination tests for predicting the performance of the seeds in the field.

MATERIALS AND METHODS

Plant Material

Twenty-eight accessions were used in this study—21 landraces, five breeding lines, and two cultivars. The seeds were maintained in the Misión Biológica de Galicia-Spanish National Research Council (MBG–CSIC, Pontevedra, Spain) germplasm facilities at 4°C and 40% RH (relative humidity; **Table 1**).

Controlled Climatic Chamber Experiments

The experiments were carried out at the MBG-CSIC using a climatic chamber set at different day-lengths and temperatures that represent conditions of North-western Spain: tc_1: 14/10 h (day/night) at 14/10°C and 60/80% RH representing field conditions in April (weighted average: 11.5°C, 68.3% RH); tc_2: 15/9 h at 17/12°C and 60/80% RH representing conditions in May (weighted average: 15.1°C, 67.5% RH); and tc_3: 16/8 h at 22/15°C and 60/80% RH representing conditions in June (weighted average: 19.7°C, 66.7% RH). Light was provided by seven very high output (VHO) fluorescent lamps with a photosynthetic photon flux (PPF) of 228 μmol m^{-2} s^{-1}.

A randomized complete block design with three replications was used. Ten seeds of each genotype were sown in sterile peat in plastic containers (30 × 20 × 12 cm), with a plant to plant (seed to seed) distance of 2.5 cm, and row to row distance of 5 cm. Monitoring of emergence was carried out during ca. 30 days, starting 4 days after the beginning of the experiments. Seedlings with a hypocotyl-radicle axis >3 cm were considered as emerged. Proportion of emergence (% of sown seeds) and time to emergence (days from sowing to seedling emergence of all seeds sown) were measured.

Field Experiments

The open field experiments were conducted in the experimental farm of the MBG-CSIC at Pontevedra, Spain (42° 24′ N latitude, 8° 38′ W longitude, 40 masl 14°C average annual temperature, and 1600 mm annual rainfall). The soil, developed from granitic rocks, has a sandy loam texture, and a granular structure. It is classified as Humic Cambisol according to FAO criteria (FAO-ISRIC-ISSS, 1998).

To test temperature effects in the field, seeds were sown by hand at: (a) low temperature conditions (tf_1) on 10 April 2007; (b) moderate temperature conditions (tf_2) on 4 May 2007; and c) warm temperature conditions (tf_3) on 7 June 2007 (Table S1). Field trials were arranged according to a completely randomized block design with 30 plants per sowing treatment and two replications (crop density of 50000 plant ha^{-1}). Single row plots were 3.8 m in length and 0.8 m spaced. Plants were watered as needed, using drip irrigation. The following traits were measured: Emergence time (days) and emergence proportion (%); expansion of the first trifoliate leaf (days); plant height at 10 and 20 days from emergence (centimeters); early plant vigor (from 1 to 9 scale); days from sowing to first open flower, beginning of flowering (50% of plants with at least an open flower), end of flowering (days), and physiological mature pod (days); dry seed weight (g 100 $seed^{-1}$) and seed dimensions (millimeters). Seeds per pod and pods per plant were determined

TABLE 1 | Origin, seed size, genetic pool, market class, and seed age of the common bean genotypes studied.

Genotype[a]	Geographical origin	Seed size (g 100 seed^{-1})	Genetic pool	Market class[b]	Seed age (years)
LANDRACES					
200	43° 07′ N, 8° 55′ W, 247 masl	57	Andean	White kidney	1
272	42° 17′ N, 8° 12′ W, 380 masl	36	Andean	Purple caparron	4
391	42° 13′ N, 8° 16′ W, 545 masl	61	Andean	Red pinto	2
399	43° 29′ N, 6° 05′ W, 37 masl	70	Mesoamerican	Large great northern	2
413	43° 26′ N, 6° 06′ W, 420 masl	79	Andean	Favada pinto	4
419	42° 34′ N, 8° 53′ W, 6 masl	73	Mesoamerican	Large great northern	1
452	43° 31′ N, 7° 06′ W, 298 masl	79	Andean	Favada	1
489	42° 39′ N, 2° 32′ W, 723 masl	40	Andean	Guernikesa	4
501	43° 14′ N, 2° 08′ W, 312 masl	43	Andean	Negro brillante	2
573	40° 30′ N, 5° 45′ W, 959 masl	35	Mesoamerican	Common pinto	1
587	41° 13′ N, 5° 29′ W, 810 masl	37	Mesoamerican	Great northern	1
593	41° 23′ N, 2° 10′ E, 8 masl	30	Mesoamerican	Hook	1
623	41° 23′ N, 2° 10′ E, 8 masl	33	Mesoamerican	Hook	1
837	43° 11′ N, 8° 44′ W, 135 masl	57	Andean	Canario bola	9
838	43° 15′ N, 8° 54′ W, 50 masl	66	Andean	White kidney	1
839	43° 15′ N, 8° 53′ W, 60 masl	81	Andean	Favada	1
842	43° 15′ N, 8° 53′ W, 60 masl	66	Andean	White kidney	9
917	43° 41′ N, 7° 30′ W, 40 masl	87	Andean	Favada	1
921	43° 41′ N, 7° 30′ W, 40 masl	77	Andean	Favada	1
924	43° 41′ N, 7° 30′ W, 40 masl	75	Andean	Favada	1
1058	41° 20′ N, 7° 45′ W, 500 masl	47	Mesoamerican	Great northern	1
BREEDING LINES[c]					
Andecha	43° 29′ N, 5° 26′ W, 16 masl	76	Andean	Favada	4
Bonafema	43° 29′ N, 5° 26′ W, 16 masl	76	Andean	Favada	1
Collacia	43° 29′ N, 5° 26′ W, 16 masl	68	Andean	Favada	1
Montcau	41°23′ N, 2° 07′ E, 63 masl	32	Mesoamerican	Hook	1
Peregrina	43° 23′ N, 8° 08′ W, 178 masl	46	Mesoamerican	Great northern	1
CULTIVARS					
Borlotto	Vilmorin (France)	54	Andean	Cranberry	1
Fukuryu	Agricultural Station (Japan)	56	Andean	Cranberry	1

[a]*All the genotyes are from Spain, with the exception of the last two ones.*
[b]*Santalla et al. (2001).*
[c]*Originated from traditional landraces by individual selection.*

on a plot average basis. Five plants were selected from the center of the plots for the estimation of grain yield (expressed in kg ha^{-1}).

Soil Properties

Five soil samples were collected before sowing in the field. Samples were air dried and passed through a 2 mm sieve prior to analysis. Soil organic matter was estimated by weight loss after ignition (Schulte and Hopkins, 1996). Soil pH in H_2O and 0.1 M KCl were measured (solution ratio 1:2.5). Phosphorus availability was measured by a modification of the Olsen method (Olsen and Dean, 1965). Exchangeable cations were extracted with 1 M NH_4Cl (Peech et al., 1947) and determined by atomic absorption spectrophotometry. Effective cation exchange capacity (ECEC) was calculated as the sum of base cations plus aluminum. The soil had moderate levels of total organic matter (4.2%) due to the long history of cultivation in the experimental plot and had an acidic reaction (pH = 5.7) and significant levels of exchangeable aluminum, as expected in soil derived from granitic like rock. ECEC was relatively low, although the potassium levels were medium to high. High phosphorus availability was due to fertilization with organic manure for many years in the past. Overall, soil analyses showed that the experimental plot was representative of most agricultural soils intensively cultivated in the NW of Spain. The Table S2 summarizes the chemical properties of the soil of the experimental plot.

Data Analyses

Maximum, minimum, average temperatures and RH were daily measured in the controlled chamber and in the open field trials. The statistical analyses were conducted using the general linear model (GLM) procedure of the SAS 2000 statistical package. The least significant difference (LSD) ($P \leq 0.05$) was used to evaluate differences among genotype means. Standard errors and coefficients of variation were also computed (Steel et al., 1997).

Principal component analysis (PCA) was performed by NTSYS-pc v.2.10 (Rohlf, 2000) and the free software R^{mo} modified by García-Pérez (2005) to display the ordination of the genotypes under both environments. The variables used in the PCA were emergence proportion and time to emergence at the three temperature levels (t1, t2, t3) in the controlled chamber and in the open field. According to the scree diagram most of the variation was explained by the first (PC1), second (PC2), and third (PC3) principal components. PC1 and PC2 accounted for 56% of variation and were used for plotting the 28 genotypes.

RESULTS

Screening in Controlled Chamber

Marked differences in emergence proportion and time to emergence were found among genotypes at the different temperatures in the controlled chamber and in the open field sowings, respectively (**Table 2**). In the controlled chamber emergence proportion ranged from 30.0 to 96.7%, with an average of 72.9% at low temperature (tcl = 14/8°C). Under 17/12°C (tc2) and 22/15°C (tc3), some genotypes reached 100% emergence, the averages were 88.2 and 94.4%, respectively. Seedling emergence was delayed under the lower temperature in controlled conditions averaging 27.2 d at 14/8°C (tcl), but emergence time was drastically reduced to 7.8 and 4.6 d when temperature was increased to 17/12°C (tc2) and 22/15°C (tc3), respectively. Effects of genotype, temperature, and genotype × temperature interaction on emergence proportion and time to emergence were significant (**Table 3**), indicating differential and non-uniform responses of genotypes to temperature at the emergence stage.

Screening in Open Field Trials

In terms of average daily temperatures, the high temperature trials (12.7–21.3°C), and the low temperature trials (9.3–20.4°C) overlap. However, the low and high temperature trials were still

TABLE 2 | Emergence proportion and time to emergence of the common bean genotypes studied, grown under three different temperatures in controlled chamber, and open field.

Genotypes	Emergence proportion (%)						Emergence time (d)					
	Controlled chamber			Open field			Controlled chamber			Open field		
	tc1	tc2	tc3	tf1	tf2	tf3	tc1	tc2	tc3	tf1	tf2	tf3
200	86.7	100.0	96.7	50.0	50.0	50.0	28.3	7.3	4.7	11.5	9.5	12.0
272	96.7	86.7	96.7	58.3	58.3	50.0	23.0	8.0	3.7	15.5	9.5	12.0
391	53.3	90.0	83.3	50.0	50.0	50.0	29.7	8.0	4.0	10.0	9.0	13.0
399	63.3	73.3	86.7	75.0	50.0	50.0	30.3	10.0	9.3	13.5	8.5	12.5
413	96.7	100.0	100.0	50.0	50.0	50.0	27.0	7.3	4.7	11.5	8.5	12.5
419	73.3	90.0	90.0	75.0	50.0	40.0	29.0	8.0	4.7	13.5	8.0	13.0
452	80.0	93.3	96.7	50.0	50.0	20.0	26.3	7.3	4.3	15.0	9.0	14.0
489	70.0	70.0	100.0	50.0	50.0	41.7	26.3	7.7	4.0	13.0	10.0	13.0
501	93.3	96.7	100.0	50.0	50.0	35.0	23.7	7.3	4.7	17.0	9.0	14.0
573	33.3	80.0	83.3	50.0	45.0	43.3	29.3	8.0	3.7	12.5	13.5	13.5
587	63.3	100.0	96.7	50.0	50.0	31.7	26.0	8.3	3.7	16.5	9.5	11.0
593	90.0	96.7	100.0	50.0	40.0	50.0	24.3	7.3	4.0	14.5	14.0	13.0
623	30.0	93.3	100.0	35.0	35.0	48.3	27.0	7.3	4.3	16.5	19.0	14.0
837	63.3	80.0	96.7	50.0	50.0	50.0	30.7	8.0	5.0	11.5	8.5	12.5
838	50.0	93.3	86.7	38.3	40.0	38.3	31.3	8.3	5.0	16.0	14.0	13.5
839	33.3	90.0	76.7	31.7	56.7	50.0	30.3	7.3	4.3	17.5	14.5	13.0
842	83.3	90.0	100.0	40.0	50.0	50.0	27.7	7.3	4.3	17.5	10.0	13.0
917	63.3	80.0	86.7	40.0	50.0	20.0	24.7	7.3	4.7	18.0	8.0	14.0
921	70.0	76.7	93.3	31.7	20.0	11.7	26.7	8.0	4.7	19.5	17.5	14.0
924	76.7	76.7	93.3	36.7	21.7	21.7	27.0	7.3	4.3	19.0	23.5	14.0
1058	86.7	90.0	100.0	58.3	58.3	50.0	26.0	7.3	4.7	12.5	10.0	12.5
Andecha	96.7	96.7	100.0	45.0	50.0	31.7	27.3	7.3	4.7	16.0	8.5	13.5
Bonafema	86.7	86.7	96.7	23.3	35.0	35.0	26.7	7.3	5.0	20.5	12.5	13.0
Collacia	53.3	66.7	90.0	16.7	36.7	38.3	29.3	11.0	4.3	20.0	14.0	13.5
Montcau	96.7	100.0	96.7	50.0	58.3	50.0	24.0	7.3	4.0	13.5	8.0	6.0
Peregrina	83.3	93.3	96.7	23.3	33.3	26.7	26.7	8.3	4.0	20.0	13.0	14.0
Borlotto	93.3	100.0	100.0	50.0	50.0	50.0	28.7	7.0	4.5	10.5	8.0	12.0
Fukuryu	73.3	80.0	100.0	33.3	50.0	50.0	25.0	7.3	4.7	17.5	9.0	13.0
Average	72.9	88.2	94.4	45.1	46.0	40.5	27.2	7.8	4.6	15.4	11.3	12.8

TABLE 3 | Mean squares and coefficient of variation (CV) from the analysis of variance for the emergence proportion and time to emergence of the common bean genotypes studied, grown under three different temperatures in controlled chamber.

Source of variation	Df[a]	Emergence proportion (%)	Emergence time (d)
Replications	2	64.0	4.9
Genotype	27	897.0**	10.7**
Temperature	2	10278.4**	12386.3**
Genotype x temperature	54	350.7**	4.8*
Error	110	147.9	3.2
CV (%)		14.3	13.4

[a]*Df, Degrees of freedom.*
, significant at $P \leq 0.05$, $P \leq 0.01$, respectively.*

on the highest end of ideal common bean growing maximum and minimum temperatures (20–25 and 15°C) and thus probably experienced some temperature stress.

Analysis of variance of the 28 genotypes in the three open field experiments with different temperature conditions at sowing time is shown in **Table 4**. All the traits displayed significant differences among genotypes and among sowing times characterized by different temperatures, while significant genotype × sowing time interaction was only observed for dry seed weight, and length, pods per plant and yield. Only six genotypes showed emergence proportion higher than 50% in the earlier sowings at lower temperatures (tf1 and/or tf2), none under the warmer temperatures at the later sowing (tf3), averaging 45.1, 46.0, and 40.5%, respectively, at these sowing dates (**Table 4**). Differences in emergence time were smaller compared to the controlled chamber experiment. Emergence time was reduced from 15.4 d in the earlier, and colder conditions to 11.3 and 12.8 d in the later warmer conditions ($P = 0.05$). Early sowing in April associated with lower temperatures (tf1) significantly delayed by 4–5 d seedling emergence, first trifoliate leaf expansion, and beginning of flowering and by about 14–16 d the end of flowering and physiological maturity, compared to sowing in May (tf2). Sowing at warmer temperatures in June further reduce seed weight and plant yield. These results indicate that early sowing under lower temperatures produced taller plants with larger seeds and higher yield than later sowings under higher temperatures. In fact, the length of the vegetative and reproductive cycle stages was progressively reduced from the colder (tf1) to the warmer (tf2 and tf3) conditions. Thus, under lower temperatures plants had more time for growth, as well as for pod and seed set and maturation, resulting in larger seeds, and higher plant yield. The **Table 5** displays the crop yield of the common bean genotypes studied, grown under three different temperatures in open field. The yield was clearly higher when sowing and growing under tf1 and tf2 than tf3 for all the genotypes, averaging 1188, 849, and 419 kg ha^{-1}, respectively. Best performer was the landrace 419 (3118 kg ha^{-1} at tf1) and the worst was the Japanese cultivar Fukuryu (149 kg ha-1 at tf1), the latter probably due to lack of adaptation to the growing area. Interestingly, the 10 top yielding genotypes corresponded to tf1, while the 10 worst genotypes were sown and grown at tf3, except the aforementioned cultivar Fukuryu (tf1). The performance of some genotypes was consistent through the different temperature conditions, such as 399, 501, 587, and Borlotto.

Comparison of Temperature Effect on Chamber and Field Trials

The analysis of the emergence process was assessed in controlled chamber and in open field under three different temperature ranges by two variables: Time to emergence and proportion of emergence. Maximum, minimum, and the average values of these variables in chamber and field are shown in Table S3. Decrease in number of days to emergence and increase in emergence proportion when temperature increased occurred in the chamber trials but were less clear in the field (**Table 4**). Under favorable conditions (tc2 and/or tc3) 13 genotypes reached 100% emergence in the growth chamber. In the open field, in contrast, only four genotypes showed emergence higher than 50% at tf2 and none at tf3, while emergence level was similar at tf1 and tf2, and higher than at tf3. Emergence time across genotypes was relatively stable at the lower temperatures (tc1) in the growth chamber, ranging from 23.0 to 31.3 d, but in the field the range was 10.0 to 20.5 d in the first sowing (tf1).

Nine groups arose from the PCA ordination (**Table 6, Figure 1**). The x-axis (PC1) represents variation in emergence time and proportion of emergence in the open field. Genotypes located at the left side have earlier emergence and higher emergence proportion than those on the right side. The y-axis represents variation in the same variables in the controlled chamber, with genotypes at the lower side showing earlier emergence, and higher emergence proportion than those at the upper side. Montcau is the only genotype in group 1, five genotypes are included in group 2 (200 272, 413 1058, Borlotto), eight in group 3 (452, 489, 501, 587, 593, 842, Andecha, Fukuryu), four in group 4 (623, 917, Bonafema, Peregrina), two in group 5 (921, 924), three in group 6 (391, 419, 837), three in group 7 (573, 838, 839), only Collacia in group 8 and 399 in group 9.

DISCUSSION

In controlled environments and field conditions, studies have shown that the rate of germination and seedling emergence linearly increases with temperature in several crop species including legumes, such as cowpea, soybean, chickpea, and peanut (Covell et al., 1986; Ellis et al., 1986; Mohamed et al., 1988; Craufurd et al., 1996; Awal and Ikeda, 2002). In this work we found that the variation in emergence in a range of temperatures was greater in controlled chamber than in open field. It can be argued that environmental conditions in the chamber are strictly controlled and the weighted average of temperature and RH had a range of variation wider than in the field. However, seeds in the field are exposed to the natural not controlled environmental variation that could explain the fact that the proportion of emergence is approximately half of the value in the controlled chamber. In field experiments, several factors can affect seed germination, and therefore the proportion

TABLE 4 | Mean squares, coefficient of variation (CV) and genotype means from the analysis of variance of agronomic traits of the common bean genotypes studied, grown under three different conditions in open field.

Trait	Sources of variation						Mean			
	Replications	Genotype (G)	Temperature (T)	G × T	Error	CV (%)	tf1	tf2	tf3	LSD[b]
Emergence time (d)	20.02	31.02**	236.74**	11.14	7.50	20.81	15.36a	11.28c	12.82b	1.26
Emergence proportion (%)	26.46	541.97**	490.51*	146.86	151.19	28.04	45.06ab	46.01a	40.48b	4.59
Leaf expansion (d)	3.72	15.55*	327.25**	10.14	7.88	14.32	22.36a	17.84b	18.61b	1.21
Plant height-10 days (mm)	5.28	58.86**	3333.78**	31.76	25.35	19.89	19.12c	33.95a	22.85b	2.13
Plant height-20 days (mm)	7.86	61.14**	5004.26**	18.87	29.61	12.85	53.30a	37.68b	35.70c	1.66
Early vigor (1-9)	0.53	6.43**	60.06**	2.22	2.54	28.52	6.46a	5.87b	4.42c	0.57
First open flower (d)	25.14	378.28**	1139.59**	22.74	24.69	8.66	61.96a	57.21b	52.95c	1.80
Beginning of flowering (d)	18.34	458.27**	1267.97**	33.01	23.89	7.87	66.79a	62.18b	57.23c	2.18
End of flowering (d)	23.54	77.34**	18036.40**	12.63	9.02	2.77	124.98a	111.14b	88.26c	1.35
Pod physiological maturity (d)	7.04	90.77**	16857.12**	13.36	12.74	3.15	130.20a	114.02b	94.72c	1.39
Dry seed weight (g 100 seed^{-1})	78.38	1863.09**	1825.58**	89.76**	43.11	11.15	63.75a	60.32a	51.99b	3.66
Seed length (mm)	4.74*	65.45**	12.98**	1.78**	0.72	5.14	16.96a	16.62a	16.09b	0.51
Seed width (mm)	0.14	3.69**	3.97**	0.12	0.07	3.31	8.54a	8.33b	7.99c	0.13
Seed thickness (mm)	0.02	4.74**	2.40**	0.10	0.13	5.91	6.28a	6.15b	5.80c	0.12
Seeds pod^{-1}	1.92	4.94**	4.13**	0.61	0.55	17.64	4.35a	4.31a	3.89b	0.30
Pods plant^{-1}	119.76	2108.50**	10732.38**	804.45**	321.42	55.00	31.23b	45.74a	20.14c	0.32
Yield (kg ha^{-1})	67054.50	4086.89**	55260.74**	3145.16**	1841.85	0.53	1188.20b	849.10a	419.00c	2.28
Df[a]	1	27	2	54	56					

*, ** *significant at $P \leq 0.05$, $P \leq 0.01$, respectively.*
[a] *Df, degrees of freedom.*
[b] *Least Significant Difference. Means follow for the same letter are not significant different at $P \leq 0.05$.*

of seedling emergence. One factor may be the presence of soil-borne pathogens (Burke and Kraft, 1974) that affect seeds when the emergence is delayed and the seed remains more time on the soil. Days to emergence in the field varied slightly with increasing temperature in later sowings, but this variation could be due also to other environmental factors such as soil properties and hydrological conditions.

Clear genotypes × temperatures interactions were found for emergence proportion and time to emergence. For instance, genotypes 573 and 623 showed contrasting emergence proportions in response to tf2 and tf3. Furthermore, time to emergence in the field was advanced only by 1 day in 399 compared to 6 days in Collacia.

The bean germplasm from the Iberian Peninsula is characterized by large and extra-large seeds with this trait displaying a broad sense heritability of 0.70 (Escribano et al., 1994; Casquero et al., 2006; González et al., 2006). In our study the seed size varies in average from 30 to 87 g 100 seeds^{-1} being 25% of the studied genotypes large and 43% extra-large seeded. The large seeds were the result of selection by farmers according to market and consumer preferences. Farmers use to germinate the large seeds in nurseries to ensure the process, then to transplant the seedlings to the open field. This practice could have favored selection for earlier emergence, but not for emergence proportion in the field. In fact, the PCA indicates a trend to lower emergence in large and extra-large seeded genotypes: the groups with good emergence (groups 1, 2, and 3) have seven large and extra-large seeded genotypes (50%) and seven medium and small seeded ones (50%). However, the groups with mediocre or bad emergence (groups 4, 5, 6, 7, 8, and 9) include 11 large and extra-large seeded genotypes (79%) and only three medium and small ones (21%).

With regard to market class and seed color, one of the factors that have been suggested as responsible for poor legume emergency in the field is imbibition damage caused by fast water uptake, resulting in cell death, solute leakage, and reduced emergence and growth. However, colored seeds which imbibed more slowly, suffer less damage than the white ones (Powell et al., 1986). Therefore, white seeded market classes are typically more susceptible to this damage. According to the PCA, the groups with good emergence (groups 1, 2, and 3) have eight white seeded genotypes, and six bicolor or colored, while the groups with bad emergence (4, 5, 6, 7, 8, and 9) included 11 white seeded genotypes and only three colored ones. These trends suggest that seed testa color could have some influence in seed germination and seedling emergence, in accordance with the results by other authors (Dickson, 1971; Powell et al., 1986).

Open field trials are the most reliable measure of low temperature tolerance since it is measured in the actual growth environment of the crop. In our research consistent comparable results were not observed between controlled chamber and field environments, in agreement with Kolasinska et al. (2000) and Khajeh-Hosseini and Rezazadeh (2011).

There is evidence that chilling tolerance at juvenile stages of development (germination, emergence, seedling growth) is under independent genetic control from chilling tolerance during

TABLE 5 | Crop yield of the common bean genotypes studied, grown under three different temperatures (tf1, tf2, and tf3) in open field.

Genotype	Yield (kg ha^{-1})		
	tf1	tf2	tf3
200	1025	946	373
272	2236	811	448
391	1463	955	404
399	1594	1152	474
413	468	477	465
419	3118	1092	465
452	818	1128	435
489	367	431	nm*
501	2738	1395	425
573	777	907	355
587	1643	1168	482
593	1072	962	502
623	825	734	382
837	1084	862	393
838	478	544	305
839	1551	1567	445
842	420	602	349
917	1599	1234	356
921	1199	443	339
924	1055	706	526
1058	2861	1006	432
Andecha	1533	785	343
Bonafema	370	461	379
Collacia	884	921	319
Montcau	626	496	428
Peregrina	851	491	350
Borlotto	471	1157	717
Fukuryu	149	325	429
Average	1188	849	419

** nm, not measured.*

reproductive development (Kemp, 1973; Dickson and Petzoldt, 1987; Melo et al., 1997). Coincidence of tolerance to high and low temperature stress has been reported in snap beans (Dickson and Boettger, 1984a,b), oat (Mashiringi and Harahwa, 1985), and maize (Yacoob and Filion, 1986). Porch (2001) found that snap bean lines developed for cold tolerance had significant positive general combining ability (GCA) under high temperatures for traits related to high temperature tolerance.

It is generally accepted that germination rate is affected by seed shape or size and, therefore, the quantity of nutrients stored in the seed (Cui et al., 2002; Hanley et al., 2003; Nonogaki, 2006; Kaya et al., 2008). In the experiments here reported, large seeded genotypes needed more days to seedling emergence than small seeded ones, both in the growth chamber and in the open field experiments, and showed lower emergence in the field under real growing conditions. This fact may be related to the history of the bean crop in southwest Europe, since its introduction in the early Sixteenth Century. Farmers probably selected large seeded common bean genotypes that resembled the Old World faba bean (*Vicia faba* L.) that was cultivated at that time, thus introducing the new crop as a novelty, which explains the name "faba" or "haba" used often in Spain for the common bean. In spite of the pleiotropic effect that seed size probably had in delaying germination and reducing emergency, famers continued to select large seeded genotypes of common bean such as the currently grown in the northwest of the Iberian Peninsula (Escribano et al., 1994; Rodiño et al., 2001, 2009).

Nine groups arose from the PCA representing variation in emergence time and proportion of emergence in the controlled

TABLE 6 | Characteristics of the groups arising from the PCA of the common bean genotypes studied under different environments.

Group	Genotype	Genetic pool[a]	Seed color[b]	Seed size[c]	Emergence score[d]
1	Montcau	M	w	m	g
2	200	A	w	m	g
	272	A	b	m	g
	413	A	b	x	g
	1058	M	w	l	g
	Borlotto	A	c	l	g
3	452	A	w	x	g
	489	A	c	s	g
	501	A	c	m	g
	587	M	w	m	g
	593	M	w	m	g
	842	A	w	x	g
	Andecha	A	w	x	g
	Fukuryu	A	c	l	g
4	623	M	w	m	m
	917	A	w	x	m
	Bonafema	A	w	x	m
	Peregrina	M	w	l	m
5	921	A	w	x	b
	924	A	w	x	b
6	391	A	c	l	m
	419	M	w	l	m
	837	A	c	l	m
7	573	M	c	m	m
	838	A	w	x	m
	839	A	w	x	m
8	Collacia	A	w	x	b
9	399	M	w	x	b

[a]A, Andean; M, Mesoamerican.
[b]w, white; b, bicoloured; c, colored.
[c]x, extra-large (> 65 g 100 seeds^{-1}); l, large (> 50 g 100 seeds^{-1}); m, medium (> 35 g 100 seeds^{-1}); s, small (< 35 g 100 seeds^{-1}).
*[d]According to the PCA ordination in **Figure 1**: g, good; m, mediocre; b, bad.*

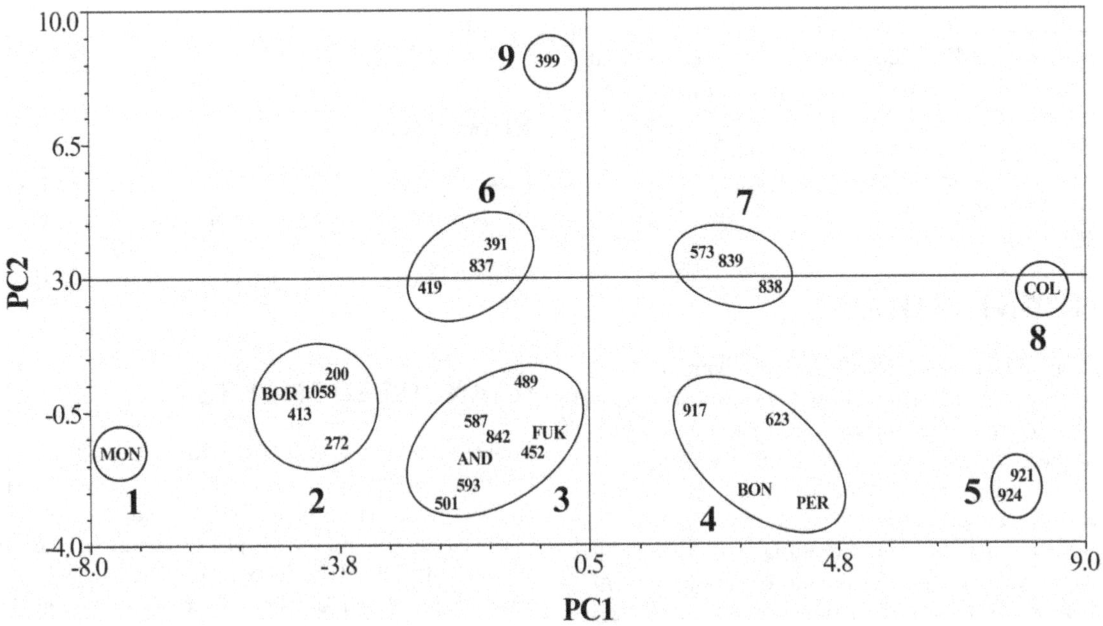

FIGURE 1 | Ordination the common bean genotypes studied according to the Principal Component Analysis (PCA). PC1, first principal component; PC2, second principal component.

chamber and in the open field. PCA indicates a trend to lower emergence in large and extra-large seeded genotypes. The genotypes 399 and 419 (large great northern market class), 921, 924, and Collacia (favada market class) are examples of white large seeded genotypes with poor germination in the field. In particular, according to the PCA, the favada market class genotypes (two in group 3 and six in groups 4, 5, 7, and 8), with very high market value, have lower proportion of emergence, probably due to the fact that farmers practice was to germinate the seed in the nursery, and afterwards transplant the seedling into the open field. Since then neither natural selection nor breeding had taken place to improve the germination and emergence of the favada market class genotypes or the large great northern ones whose seeds have similar market and uses by consumers. Nevertheless, in the present study some of these genotypes (e.g., 399, 419, 917, 921, and 924) showed high emergence (>80%) when germinated at low temperature in a controlled environment chamber. This was probably due to the fact that these genotypes were selected in the past to be germinated in stable environments, not in unstable open field conditions.

In the experiments here reported, the emergence process was assessed by time to emergence and the proportion of seeds producing emerged seedlings, with relevance in the conservation of germplasm in gene banks. In order to conserve the genetic structure of the original accessions, regeneration of germplasm in the field should be performed only when the results of the germination tests display low germination. Emergence in controlled conditions in climatic chamber resembles the viability test used in gene banks, whereas sowing in the field is used to regenerate accessions. According to our results the viability tests should be complemented with vigor tests that can provide better estimation of field emergence (Hampton and Tekrony, 1995). For the same reason, in regeneration processes of bean germplasm it would be advisable to germinate seeds under favorable conditions (e.g., in nurseries) before transplanting the seedlings into the field, in order to prevent (or minimize) genetic erosion due to the mortality of viable seeds.

According to our results, temperature conditions had a relevant role in the crop yield, together with the environmental factors. The plant productivity, which is the major expression of the genotypes fitness, hade the higher values when sowing and growing the plants under low temperature, decreasing when the temperature was increased. It is relevant also to link the yield with the process of emergence of seedlings at different temperatures in the open field and in the controlled chamber. Taken into account these results, the genotypes 272, 501, 593, and 1058, and the cultivar Borlotto had assembled the best conditions for early sowing achieving good yield performance.

In conclusion, seed germination, seedling emergence, plant growth, and crop yield under different temperatures are of relevance for the selection of common bean genotypes with better performance under stress temperatures, but also may lead to potential genetic erosion in germplasm collections. As a result of the screen of seedling emergence and phenotypic response of bean germplasm under a range of temperatures in controlled chamber and field conditions, some genotypes such as landraces 272, 501, 593, and the cultivar Borlotto were identified as temperature stress-tolerant at sowing time and seedling emergence with good agronomic performance and yield potential and they could be a valuable genetic material for breeding programs. Additionally, the efficiency of bean genebanks standard germination tests for predicting the performance of the seeds in the field was assessed comparing the emergence of bean seedlings under controlled environment and in open field. Regarding bean commercial traits, under low temperature at sowing time bean seeds reached larger size, and the crop yield was higher compared to warmer temperatures at this stage. Therefore, early sowing of bean is strongly recommended.

AUTHOR CONTRIBUTIONS

AD: conception and experimental design of the work, including chamber, and field experiments; revising the work and approval of the version to be published. AR: experimental design of the work; acquisition and analysis of chamber and field data for the work; drafting and revising the work and approval of the version to be published. MS: interpretation of field and chamber data for the work; drafting and revising the work and approval of the version to be published. AG: statistical analysis of field and chamber experimental data; revising the work and approval of the version to be published. ML: interpretation of soil data for the work; revising the work and approval of the version to be published. IM: interpretation of germplasm data for the work; revising the work and approval of the version to be published. JK: analysis and interpretation of physiological data for the work; revising the work and approval of the version to be published.

FUNDING

Research was supported by the projects AGL2014-51809-R and RFP2013-00001 from the Spanish Government (Ministerio de Economía y Competitividad) and AGI/CSIC I+D+I 2014 OTR00114 from the Galician Government-CSIC (Spain).

ACKNOWLEDGMENTS

The authors are grateful to CRF-INIA (Alcalá de Henares, Spain), Polytechnic University of Cataluña (Barcelona, Spain), Agricultural Station (Hokkaido, Japan), and SERIDA (Villaviciosa, Spain) for supplying common bean germplasm. We thank to Albert Vandenberg and Pedro Revilla for reviewing the manuscript and for laboratory facilities, Josefina Hernández-Nistal for scientific assistance, Salvador Rodríguez for technical assistance, and Diputación de Pontevedra (Pontevedra, Spain) for farm facilities.

REFERENCES

Alonso-Blanco, C., Bentsink, L., Hanhart, C. J., Blankestijn-De Vries, H., and Koornneef, M. (2003). Analysis of natural allelic variation at seed dormancy loci of *Arabidopsis thaliana*. *Genetics* 164, 711–729.

Angioi, S. A., Rau, D., Attene, G., Nanni, L., Bellucci, E., Logozzo, G., et al. (2010). Beans in Europe: origin and structure of the European landraces of *Phaseolus vulgaris* L. *Theor. Appl. Genet.* 121, 829–843. doi: 10.1007/s00122-010-1353-2

Awal, M. A., and Ikeda, T. (2002). Effects of changes in soil temperature on seedling emergence and phenological development in field-grown stands of peanut (*Arachis hypogaea*). *Environ. Exp. Bot.* 47, 101–113. doi: 10.1016/S0098-8472(01)00113-7

Bannerot, H. (1979). Cold tolerance in beans. *Ann. Rep. Bean Improv. Coop.* 22, 81–84.

Bellucci, E., Bitocchi, E., Rau, D., Rodriguez, M., Biagetti, E., Giardino, A., et al. (2014). "Genomics of origin, domestication and evolution of *Phaseolus vulgaris*," in *Genomics of Plant Genetic Resources*, eds R. Tuberosa, A. Graner, and E. Frison (Dordrecht: Springer), 483–507.

Bewley, J. D. (1997). Seed germination and plant dormancy. *Plant Cell* 9, 1055–1066. doi: 10.1105/tpc.9.7.1055

Bewley, J. D., and Black, M. (1985). *Seeds: Physiology of Development and Germination. New York, NY: Plenum Press.*

Blair, M. W., Knewston, S. J. B., Astudillo, C., Li, C. M., Fernandez, A. C., and Grusak, M. A. (2010). Variation and inheritance of iron reductase activity in the roots of common bean (*Phaseolus vulgaris* L.) and association with seed iron accumulation QTL. *BMC Plant Biol.* 10:215. doi: 10.1186/1471-2229-10-215

Burke, D. W., and Kraft, J. M. (1974). Responses of beans and peas to root pathogens accumulated during monoculture of each crop species. *Phytopathology* 64:4. doi: 10.1094/phyto-64-546

Casquero, P. A., Lema, M., Santalla, M., and De Ron, A. M. (2006). Performance of common bean landraces from Spain in the Atlantic and Mediterranean environments. *Genet. Resour. Crop Evol.* 53, 1021–1032. doi: 10.1007/s10722-004-7794-1

Copeland, O. L., and McDonald, B. M. (1995). *Principles of Seed Science and Technology. 3rd Edn.* New York, NY: Chapman and Hall.

Covell, S., Ellis, R. H., Roberts, E. H., and Summerfield, R. J. (1986). The influence of temperature on seed germination rate in grain legumes. I. Acomparison of chickpea, lentil, soybean and cowpea at constant temperatures. *J. Exp. Bot.* 37, 705–715. doi: 10.1093/jxb/37.5.705

Craufurd, P. Q., Ellis, R. H., Summerfield, R. J., and Menin, L. (1996). Development in cowpea (*Vigna unguiculata*). 1. The influence of temperature on seed germination and seedling emergence. *Exp. Agric.* 32, 1–12. doi: 10.1017/S0014479700025801

Cui, K., Peng, S., Xing, Y., Xu, C., Yu, S., and Zhang, Q. (2002). Molecular dissection of seedling-vigor and associated physiological traits in rice. *Theor. Appl. Genet.* 105, 745–753. doi: 10.1007/s00122-002-0908-2

De Ron, A. M., Menéndez-Sevillano, M. C., and Santalla, M. (2004). Variation in primitive landraces of common bean (*Phaseolus vulgaris* L.) from Argentina. *Genet. Resour. Crop Evol.* 51, 883–894. doi: 10.1007/s10722-005-1934-0

De Ron, A. M., Papa, R., Bitocchi, E., González, A. M., Debouck, D. G., Brick, M. A., et al. (2015). "Common bean," in *Grain Legumes. Series: Handbook of Plant Breeding*, ed A. M. De Ron (New York, NY; Dordrecht; Heidelberg; London: Springer), 1–36.

Dickson, M. H. (1971). Breeding beans, *Phaseolus vulgaris* L., for improved germination under unfavorable low temperature conditions. *Crop Sci.* 11, 848–850. doi: 10.2135/cropsci1971.0011183X001100060024x

Dickson, M. H., and Boettger, M. A. (1984a). Emergence, growth and blossoming of bean (*Phaseolus vulgaris*) at sub-optimal temperatures. *J. Am. Soc. Hortic. Sci.* 109, 257–260.

Dickson, M. H., and Boettger, M. A. (1984b). Effect of high and low temperatures on pollen germination and seed set in green beans. *J. Am. Soc. Hortic. Sci.* 109, 372–374.

Dickson, M. H., and Petzoldt, R. (1987). Inheritance of low temperature tolerance in beans at several growth stages. *Hortic. Sci.* 22, 481–483.

Duc, G., Agrama, H., Bao, S., Berger, J., Bourion, V., Burstin, J., et al. (2015). Breeding annual legumes for adaptation to low input cropping systems and new areas: methods to approach more complex traits and target new variety ideotypes. *Crit. Rev. Plant Sci.* 34, 381–411. doi: 10.1080/07352689.2014.898469

Dutt, M., and Geneve, R. L. (2007). Time to radicle protrusion does not correlate with early seedling growth in individual seeds of impatiens and petunia. *J. Am. Soc. Hortic. Sci.* 132, 423–428.

Eichenberger, K., Gugerli, F., and Schneller, J. J. (2000). Morphological and molecular diversity of Swiss common bean cultivars (*Phaseolus vulgaris* L. Fabaceae) and their origin. *Bot. Helv.* 110, 61–77.

Ellis, R. H., Covell, S., Roberts, E. H., and Summerfield, R. J. (1986). The influence of temperature on seed germination rate in grain legumes. II. Interspecific variation in chickpea (*Cicer arietinum* L.) at constant temperature. *J. Exp. Bot.* 37, 1503–1515. doi: 10.1093/jxb/37.10.1503

Ellis, R. H., Hong, T. D., and Roberts, E. H. (1985). "*Handbooks for Genebanks: no,*" in *Handbook of Seed Technology for Genebanks, Vol. 2. Compendium of Specific Germination Information and Test Recommendations* (Rome: International Board for Plant Genetic Resources (IBPGR)).

Escribano, M. R., De Ron, A. M., and Amurrio, J. M. (1994). Diversity in agronomical traits in common bean populations from Northwestern Spain. *Euphytica* 76, 1–6. doi: 10.1007/B. F.00024014

FAO-ISRIC-ISSS (1998). *World Reference Base for Soil Resources.* 84th World Soil Resources Reports, Food and Agriculture Organization of the United Nations, Rome.

Galván, M. Z., Menéndez-Sevillano, M. C., De Ron, A. M., Santalla, M., and Balatti, P. A. (2006). Genetic diversity among wild common beans from Northwestern Argentina based on morphoagronomic and RAPD data. *Genet. Res. Crop Evol.* 53, 891–900. doi: 10.1007/s10722-004-0981-20

Garcia-Huidobro, J., Monteith, J. L., and Squire, G. R. (1982). Time-temperature and germination of pearl millet (*Pennisetum typhoides* S & H). *J. Exper. Bot.* 33, 288–296.

García-Pérez, A. (2005). *Métodos Avanzados de Estadística Aplicada. Técnicas Avanzadas.* Madrid: UNED.

Gioia, T., Logozzo, G., Attene, G., Bellucci, E., Benedettelli, S., Negri, V., et al. (2013). Evidence for introduction bottleneck and extensive inter-gene pool (Mesoamerica x Andes) hybridization in the European common bean (*Phaseolus vulgaris* L.) germplasm. *PLoS ONE* 8:e75974 doi: 10.1371/journal.pone.0075974

González, A. M., Monteagudo, A. B., Casquero, P. A., De Ron, A. M., and Santalla, M. (2006). Genetic variation and environmental effects on agronomical and commercial quality traits in the main European market classes of dry bean. *Field Crop Res.* 95, 336–347. doi: 10.1016/j.fcr.2005.04.004

Gummerson, R. J. (1986). The effect of constant temperatures and osmotic potentials on the germination of sugar Beet. *J. Exp. Bot.* 37, 729–741. doi: 10.1093/jxb/37.6.729

Hampton, J. G., and Tekrony, D. M. (1995). *Handbook of Vigour Test Methods.* Zurich: International Seed Testing Association (ISTA).

Hanley, M. E., Unna, J. E., and Darvill, B. (2003). Seed size and germination response: a relationship for fire-following plant species exposed to thermal shock. *Oecologia* 134, 18–22. doi: 10.1007/s00442-002-1094-2

Hardwick, R. C. (1972). The emergence and early growth of french and runner beans (*Phaseolus vulgaris* L., and *Phaseolus coccineus* L.) sown at different dates. *J. Hortic. Sci.* 47, 395–410. doi: 10.1080/00221589.1972.11514483

Hardwick, R. C., and Andrews, D. J. (1980). Selection for cold tolerance in *Phaseolus vulgaris*, yield of selected lines grown in warm and cool environments. *Ann. App. Biol.* 95, 249–259. doi: 10.1111/j.1744-7348.1980.tb04744.x

Hernández-Nistal, J., Rodríguez, D., Nicolás, G., and Aldasoro, J. J. (1989). Abscisic acid and temperature modify the levels of calmodulin in embryonic axes of *Cicer arietinum*. *Physiol. Plant.* 75, 255–260 doi: 10.1111/j.1399-3054.1989.tb06177.x

Holmberg, S. A. (1973). Soybeans for cool temperature climates. *Agric. Hortic. Gen.* 31, 1–20.

Kaya, M., Kaya, G., Kaya, M. D., Atak, M., Saglam, S., Khawar, K. M., et al. (2008). Interaction between seed size and NaCl on germination and early seedling growth of some Torkish cultivars of chickpea (*Cicer arietinum* L.). *J. Zhejiang Univ. Sci. B* 9, 371–377. doi: 10.1631/jzus.B0720268

Kemp, G. A. (1973). Initiation and development of flowers in beans under suboptimal temperature conditions. *Can. J. Plant Sci.* 53, 623–627. doi: 10.4141/cjps73-122

Kemp, G. A. (1978). Growth of primary leaves of beans (*Phaseolus vulgaris* L.) under sub-optimal temperatures. *Can. J. Plant Sci.* 58, 169–174. doi: 10.4141/cjps78-025

Khajeh-Hosseini, M., and Rezazadeh. (2011). The electrical conductivity of soak-water of chickpea seeds provides a quick test indicative of field emergence. *Seed Sci. Technol.* 39, 692–696 doi: 10.15258/sst.2011.39.3.18

Kigel, J., Rosental, L., and Fait, A. (2015). "Seed physiology and germination of grain legumes," in *Grain Legumes. Series: Handbook of Plant Breeding*, ed A. M. De Ron (New York, NY; Dordrecht; Heidelberg; London: Springer), 327–363.

Kolasinska, K., Szyrmer, J., and Dul, S. (2000). Relationship between laboratory seed quality tests and field emergence of common bean seed. *Crop Sci.* 40, 470–475. doi: 10.2135/cropsci2000.402470x

Kooistra, E. (1971). Germination of beans (*Phaseolus vulgaris* L.) at low temperatures. *Euphytica* 20, 208–213. doi: 10.1007/BF00056080

Kotowski, F. (1926). Temperature relation to germination of vegetable seed. *Proc. Am. Soc. Hortic. Sci.* 23, 176–184.

Mashiringi, N. A., and Harahwa, G. (1985). The response of twenty wheat (*Triticum aestivum*) varieties to growing conditions in Zimbabwe. *Zimbabwe Agric. J.* 82, 11–15.

Melo, L. C., Dos Santos, J. B., and Ramalho, M. A. P. (1997). Choice of parents to obtain common bean (*Phaseolus vulgaris*) cultivars tolerant to low temperatures at the adult stage. *Braz. J. Genet.* 20, 283–292.

Mohamed, H. A., Clark, J. A., and Ong, C. K. (1988). Genotypic differences in the temperature responses of tropical crops. I. Germination characteristics of groundnut (*Arachis hypogaea* L.) and pearl-millet (*Pennisetum typhoides* S & L.). *J. Exp. Bot.* 39, 1121–1128. doi: 10.1093/jxb/39.8.1121

Negri, V., and Tosti, N. (2002). *Phaseolus* genetic diversity maintained on-farm in central Italy. *Genet. Res. Crop Evol.* 49, 511–520. doi: 10.1023/A:1020902508166

Nonogaki, H. (2006). Seed germination - The biochemical and molecular mechanisms. *Breed. Sci.* 56, 93–105. doi: 10.1270/jsbbs.56.93

Olsen, S. R., and Dean, L. A. (1965). "Phosphorous" in *Methods of Soil Science*, ed C. A. Black (*Madison, WI:* American Society of Agronomy*)*, 403–430.

Papa, R., Nanni, L., Sicard, D., Rau, D., and Attene, G. (2006). "*The evolution of genetic diversity in Phaseolus vulgaris L.*," in *Darwin's Harvest: New Approaches to the Origins, Evolution and Conservation of Crops*, eds T. J. Motley, N. Zerega, and H. Cross (New York, NY: Columbia University Press), 121–142.

Patterson, B. D., and Payne, L. A. (1983). Screening for chilling resistance in tomato seedlings. *Hortic. Sci.* 18, 340–341.

Peech, M., Alexander, L. T., Dean, L., and Reed, J. F. (1947). *Methods of Soil Analysis for Soil Fertility Investigations.* US Dept. of Agriculture Circ.

Porch, T. G. (2001). *Genetics and Applications of Heat Tolerance in Common Bean.* (Dissertation Ph.D.) thesis, Cornell University, Ithaca.

Porch, T. G. (2006). Application of stress indices for heat tolerance screening of common bean (*Phaseolus vulgaris*). *J. Agron. Crop Sci.* 192, 390–394. doi: 10.1111/j.1439-037X.2006.00229.x

Powell, A. A., Oliveira, D.e, M., A., and Matthews, S. (1986). The role of imbibition damage in determining the vigour of white and coloured seed lots of dwarf French beans. *J. Exp. Bot.* 37, 716–722. doi: 10.1093/jxb/37.5.716

Rainey, K. M., and Griffiths, P. D. (2005). Inheritance of heat tolerance during reproductive development in snap bean (*Phaseolus vulgaris* L.). *J. Am. Soc. Hortic. Sci.* 130, 700–706.

Rodiño, A. P., Gonzalez, A. M., Santalla, M., De Ron, A. M., and Singh, S. P. (2006). Novel genetic variation in common bean from the Iberian Peninsula. *Crop Sci.* 46, 2540–2546. doi: 10.2135/cropsci2006.02.0104

Rodiño, A. P., Monteagudo, A. B., De Ron, A. M., and Santalla, M. (2009). Ancestral landraces of common bean from the South of Europe and their agronomical value for breeding programmes. *Crop Sci.* 49, 2087–2099. doi: 10.2135/cropsci2008.07.0427

Rodiño, A. P., Santalla, M., De Ron, A. M., and Singh, S. P. (2003). A core collection of common bean from the Iberian Peninsula. *Euphytica* 131, 165–175. doi: 10.1023/A:1023973309788

Rodiño, A. P., Santalla, M., Montero, I., Casquero, P., and De Ron, A. M. (2001). Diversity in common bean (*Phaseolus vulgaris* L.) germplasm from Portugal. *Genet. Res. Crop Evol.* 48, 409–417. doi: 10.1023/A:1012248002436

Rohlf, F. J. (2000). *NTSYS-pc. Numerical Taxonomy and Multivariate Analysis. Exeter Software.* New York, NY: Exeter Software.

Santalla, M., De Ron, A. M., and De La Fuente, M. (2010). Integration of genome and phenotypic scanning gives evidence of genetic structure in Mesoamerican common bean (*Phaseolus vulgaris* L.) landraces from the southwest of Europe. *Theor. Appl. Genet.* 120, 1635–1651. doi: 10.1007/s00122-010-1282-0

Santalla, M., De Ron, A. M., and Voysest, O. (2001). "European bean market classes," in *Catalogue of Bean Genetic Resources,* eds M. Amurrio, M. Santalla, and A. M De Ron (Pontevedra: PHASELIEU FAIR 3463-MBG-CSIC. Fundación Pedro Barrié de la Maza), 77–94.

Santalla, M., Rodiño, A. P., and De Ron, A. M. (2002). Allozyme evidence supporting South-western Europe as a secondary center of genetic diversity for common bean. *Theor. App. Genet.* 104, 934–944. doi: 10.1007/s00122-001-0844-6

Schmuths, H., Bachmann, K., Weber, W. E., Horres, R., and Hoffmann, M. H. (2006). Effects of preconditioning and temperature during germination of 73 natural accessions of *Arabidopsis thaliana. Ann. Bot.* 97, 623–634. doi: 10.1093/aob/mcl012

Schulte, E. E., and Hopkins, B. G. (1996). "Estimation of organic matter by weight loss- on ignition," in *Soil Organic Matter Analysis and Interpretation*, eds F. R. Magdoff, M. A.*Tatabai, and E. A. Hanlon (Madison, WI: Soil Science Society of America),* 21–31.

Scully, B., and Waines, J. G. (1987). Germination and emergence response of common and tepary beans to controlled temperature. *Agron. J.* 79, 287–291. doi: 10.2134/agronj1987.00021962007900020022x

Sicard, D., Nanni, L., Porfiri, O., Bulfon, D., and Papa, R. (2005). Genetic diversity of *Phaseolus vulgaris* L and *P. coccineus* L. landraces in central Italy. *Plant Breed.* 124, 464–472. doi: 10.1111/j.1439-0523.2005.01137.x

Singh, S. P., Gepts, P., and Debouck, D. G. (1991). Races of common bean (*Phaseolus vulgaris* Fabaceae). *Econ. Bot.* 45, 379–396. doi: 10.1007/BF02887079

Skrdla, W. H., and Mock, J. S. (1978). Evaluation of maize plant introductions for cold tolerance. *Euphytica* 27, 27–32. doi: 10.1007/BF00039116

Steel, R. G. D., Torrie, J. H., and Dickey, D. A. (1997). *Principles and Procedures of Statistics. A Biometrical Approach.* New York, NY: McGraw-Hill.

Wantanbe, H. (1953). Studies on the unfruitfulness of beans. 3. Influences of temperature on blooming and of relative humidity on the pollen activities of beans. *J. Hortic. Ass. Jpn.* 22, 172–176.

White, J. W., and Montes, C. (1993). The influence of temperature on seed germination in cultivars of common bean. *J. Exp. Bot.* 44, 1795–1800. doi: 10.1093/jxb/44.12.1795

Yacoob, R. K., and Filion, W. G. (1986). Temperature-stress response in maize: a comparison of several cultivars. *Can. J. Genet. Cytol.* 28, 1125–1131. doi: 10.1139/g86-156

Zeven, A. C. (1997). The introduction of the common bean (*Phaseolus vulgaris* L.) into Western Europe and the phenotypic variation of dry bean collected in the Netherlands in 1946. *Euphytica* 94, 319–328. doi: 10.1023/A:10029402 20241

Zeven, A. C., Waninge, J., Van Hintum, T., and Singh, S. P. (1999). Phenotypic variation in a core collection of common bean (*Phaseolus vulgaris* L.) in the Netherlands. *Euphytica* 109, 93–106. doi: 10.1023/A:10036654 08567

Conflict of Interest Statement: The authors declare that the research was conducted in the absence of any commercial or financial relationships that could be construed as a potential conflict of interest.

2

Enhancing Legume Ecosystem Services through an Understanding of Plant–Pollinator Interplay

Edited by:
Antonio M. De Ron,
Spanish National Research Council –
Misión Biológica de Galicia, Spain

Reviewed by:
Fred Stoddard,
University of Helsinki, Finland
Pedro A. Casquero,
University of Leon, Spain

***Correspondence:**
María J. Suso
ge1susom@uco.es

María J. Suso[1], Penelope J. Bebeli[2], Stefanie Christmann[3], Célia Mateus[4], Valeria Negri[5], Miguel A. A. Pinheiro de Carvalho[6,7], Renzo Torricelli[5] and Maria M. Veloso[8]*

[1] *Department of Plant Breeding, Institute for Sustainable Agriculture, Spanish National Research Council, Córdoba, Spain,* [2] *Laboratory of Plant Breeding and Biometry, Department of Crop Science, Agricultural University of Athens, Athens, Greece,* [3] *International Center for Agricultural Research in the Dry Areas, Rabat, Morocco,* [4] *Agrarian and Forestry Systems and Plant Protection Unit, National Institute for Agrarian and Veterinarian Research/Linking Landscape, Environment, Agriculture and Food, Oeiras, Portugal,* [5] *Department of Agricultural, Food and Environmental Sciences, University of Perugia, Perugia, Italy,* [6] *ISOPlexis Genebank, University of Madeira, Funchal, Portugal,* [7] *Mediterranean Institute of Environmental and Agricultural Sciences, University of Evora, Evora, Portugal,* [8] *Biotechnology and Genetic Resources Unit, National Institute for Agrarian and Veterinarian Research/Linking Landscape, Environment, Agriculture and Food, Oeiras, Portugal*

Legumes are bee-pollinated, but to a different extent. The importance of the plant–pollinator interplay (PPI), in flowering crops such as legumes lies in a combination of the importance of pollination for the production service and breeding strategies, plus the increasing urgency in mitigating the decline of pollinators through the development and implementation of conservation measures. To realize the full potential of the PPI, a multidisciplinary approach is required. This article assembles an international team of genebank managers, geneticists, plant breeders, experts on environmental governance and agro-ecology, and comprises several sections. The contributions in these sections outline both the state of the art of knowledge in the field and the novel aspects under development, and encompass a range of reviews, opinions and perspectives. The first three sections explore the role of PPI in legume breeding strategies. PPI based approaches to crop improvement can make it possible to adapt and re-design breeding strategies to meet both goals of: (1) optimal productivity, based on an efficient use of pollinators, and (2) biodiversity conservation. The next section deals with entomological aspects and focuses on the protection of the "pest control service" and pollinators in legume crops. The final section addresses general approaches to encourage the synergy between food production and pollination services at farmer field level. Two basic approaches are proposed: (a) Farming with Alternative Pollinators and (b) Crop Design System.

Keywords: heterosis breeding, legume landraces, breeding for seed yield, pest control, Farming with Alternative Pollinators and Crop Design System

Agricultural land serves first and foremost to produce food and, to an increasing extent, it should contribute to the "green economy" by enhancing diversity in agro-ecosystems. It is therefore often necessary to support farmers in their efforts to improve the biodiversity level on their farms. Legumes are bee pollinated. To realize the full potential of the plant–pollinator interplay (PPI), a multidisciplinary approach is required. The article assembles an international team of genebank managers, geneticists, plant breeders, experts on environmental governance and agro-ecology and comprises several sections. Section "Basic Application-Oriented Questions in Heterosis Breeding: The Potential of Insect-Mediated Outcrossing in Grain Legumes" deals with heterosis breeding strategies. Regarding hybrid and heterotic population breeding, current strategies for using heterosis center on finding ways of reducing the cost and increasing the efficiency of producing hybrid seed. Optimum pollination should be focused on seed production technology based on insect-aided outcrossing. Section "Exploring the Role of Plant–Pollinator Interplay in Grain Legume Landrace Germplasm Conservation and Management" describes landrace pollination aspects from the perspective of *ex situ* and on-farm multiplication. The role played by pollinators can have positive or detrimental effects. While gene flow between different multiplication seed fields can be problematic, insufficient outcrossing within a single population can also create troubles. Data should be available to genebank curators, breeders and farmers to allow the selection of the multiplication strategy in the presence of pollinators and to predict or assess its impact on landrace genetic structure and seed production. Section "Developing Strategies for Seed Production and Seed Producers in Forage Legumes" focuses on strategies for seed production and seed producers in forage legumes. In forage legumes, the potential seed yield (PSY) is very high; many ovules are present per unit area at anthesis, but only a small part of them turns into mature seeds. Seed production is influenced by several factors including the presence of pollinators around seed crop. The section describes general strategies for seed production in forage legumes. Section "Protecting the Pollination Service from Pesticides" brings entomological aspects and focuses on the protection of the "pest control service" and pollinators in legume ecosystems. The use of pesticides to control pests not only affects negatively the pollinators but also the pests' natural enemies. Enhancing the pest control service of legume ecosystems, by avoiding pesticides and adopting adequate alternative control methods, and by promoting ecological infrastructures, has a positive impact on pollinators. Section "General Approaches to Encourage the Synergy between Food Production and Pollination Services" addresses general approaches to encourage the synergy between food production and pollination services. It is increasingly recognized that pollination has multiple aspects: farmers may use both pollinator-friendly crops and implement practices in the overall production scheme and management of their farms to increase the occurrence, health and visitation of pollinators, whether these are wild or managed. Designing the right crop and identifying the suite of practices, appropriate and effective in a particular site, is where management of pollination becomes key. Two basic approaches are proposed to manage pollination by farmers at field level: (a) Farming with Alternative Pollinators (FAP) and (b) taking into account that flowering crops could be effective determinants of bee diversity and density, the Crop Design System (CDS).

BASIC APPLICATION-ORIENTED QUESTIONS IN HETEROSIS BREEDING: THE POTENTIAL OF INSECT-MEDIATED OUTCROSSING IN GRAIN LEGUMES

Background

Heterosis (or hybrid vigor) is a natural phenomenon whereby hybrid offspring of genetically diverse individuals display improved physical and functional characteristics relative to their parents (Fu et al., 2014). Heterosis has been increasingly applied in breeding for nearly a century, with the aim of developing more vigorous, higher- yielding, and better-performing cultivars. Hybrid breeding is a success story in several allogamous species such as maize (*Zea mays* L.), sunflower (*Helianthus annuus* L.), and other crops, where, greater yield has been achieved (Coors and Pandey, 1999). Although it is technically more challenging, hybrid breeding is also being approached in partially outcrossing crops such as canola (*Brassica napus* L.) (Longin et al., 2012). Heterosis results in from 20% to over 50% yield increase in inbred species (Tester and Langridge, 2010). For many self-pollinating crop plants, hybrid breeding has been established only recently. The reason for this may lie in the limitations of methods available for controlling pollination and hybrid performance prediction, in particular when established heterotic groups are absent. Strategies for using heterosis more widely to increase yields in inbred crops center on finding ways of reducing the cost and increasing the efficiency of producing hybrid seed (Zhao et al., 2015). In grain legumes, very limited progress has been made in the use of heterosis. So far, hybrid pigeonpea [*Cajanus cajan* (L.) Millsp.] is the only success story to tell in legumes (Palmer et al., 2011).

Goal and Key Constraints

This section deals with the hybrid breeding technology for the exploitation of heterosis in legumes; it does not review the *status quo* of hybrid breeding, but highlights the main problems currently limiting our capacity to utilize heterotic potential by using the knowledge from the PPI.

The biological constraints of legumes (for more details, see Crop Design System) hamper the implementation of a cost-efficient hybrid seed production technology. The details have been discussed recently (for review, see Palmer et al., 2011; Suso et al., 2015). Briefly, Palmer et al. (2001), listed four components that are crucial for the successful development of hybrids in food legumes: parental combinations that produce heterotic progenies with yield higher than the best pure-line cultivars, a stable male-sterile and female-fertile system, an efficient pollen transfer mechanism from pollen parent to pod parent, and a profitable level of seed increase for the seedsman and grower. However, in legumes, most of the above mentioned components are lacking, making hybrid research an uphill task. Legume hybrid breeding has been less successful, than in other crops because of the lack of knowledge about high-yielding heterotic groups coupled with difficulties in implementing a cost-effective system for hybrid seed production as result of an efficient pollen transfer mechanism from pollen parent to pod parent (Palmer et al., 2011).

Heterotic groups – genotypes or populations – that produce high-yielding progeny after crossing need to be identified. The determination and identification of heterotic combinations or associations require a large number of parental recombinations to be evaluated over a number of years, in many locations, and with an adequate number of replications per location. In legumes, a limiting factor in the study of heterosis is the lack of hybrid seed available for the agronomic performance tests. Most studies were based on a low number of environments with small-sized plots and low seed density (Ortiz-Perez et al., 2007; Palmer et al., 2012).

Key Examples

Despite these drawbacks, major attempts have been undertaken in the public and private domain to develop hybrid breeding programs in some legumes. Moreover, these attempts at crop improvement programs were recently stimulated by the demand for increased agricultural productivity per area in the face of dynamic environmental and biotic threats which the rapid global environmental changes will make increasingly challenging (Tester and Langridge, 2010). We summarized current knowledge on the extent of hybrid breeding technology through selected examples: a successful story, pigeonpea [*Cajanus cajan* (L.) Millsp], a predominantly selfing crop, soybean [*Glycine max* (L.) Merr.] and a partially allogamous crop, faba bean (*Vicia faba* L.).

Pigeonpea

The research by ICRISAT and its collaborators into sterility systems and agronomic performance studies was the catalyst that ensured the success of the pigeonpea hybrid. Saxena et al. (2013) reported that ICRISAT scientists, working with their Indian counterparts, developed the Hybrid ICPH 2671, which showed around 47% superiority for grain yield over the popular cv. control cultivar 'Maruti' in multilocation on-station testing and in the on-farm trials. The cytoplasmic-nuclear male sterility, combined with natural outcrossing of the crop (outcrossing was facilitated by a prolonged flowering period) was used to develop viable hybrid breeding technology. Saxena et al. (2013) reported that good amounts of seed set were obtained even with 25–30% outcrossing. The pollen vectors may visit the male-sterile plants several times, and on each visit, a certain proportion of the flowers are pollinated to set the pods, while the un-pollinated flowers should drop. This is followed by the emergence of new flowers on the same plant, some of which again set pods through open pollination. The outstanding performance of this hybrid has led to its release for cultivation in India by both a private seed company and a public university.

Soybean

A number of male-sterility systems with appropriate maintainers and restorers are available for soybean hybrid technology. The identification of cytoplasmic-nuclear male-sterile lines along with their maintainers and restorers has been achieved by intraspecific and interspecific hybridizations. Heterosis studies have shown that yield levels above the best parent are possible (Palmer et al., 2011).

Once a stable male-sterility system is obtained, it is necessary to transfer the pollen from the male parent to the female parent. For soybean, a major limitation is the movement of pollen from male parents to female parents. Most of cross-pollinations have been conducted with hand-emasculated plants. Manual cross-pollination to produce large quantities of hybrid seed is difficult and time-consuming. Ortiz-Perez et al. (2006a,b) and Palmer et al. (2009) recommended insect-aided technology for hybrid seed production and considered that the exploitation of the heterosis depends on basic information available on different aspects of the PPI. The role of pollen vectors in soybean production has been generally disregarded, with producers relying on self-pollination and pesticides to maintain yield levels (Milfont et al., 2013). However, growing evidence links improved soybean production to the activity of flower visitors. The studies by Martins (2013), Milfont et al. (2013), and Monasterolo et al. (2015) showed that pollinator activity is important for this crop, leading to increased yield.

Insect-aided crossing potential depends on the functional floral pollinator-related traits of the female and male parents. Studies in cultivated soybean revealed significant differences for floral discovery, attraction and reward traits of the seed parent (Ortiz-Perez et al., 2006a). Ortiz-Perez et al. (2008) indicated that the bees' preference for certain parental lines was the key factor in reducing hybrid seed production. Among various traits that influence outcrossing, volatiles are crucial. Differences in the emission of flower volatile compounds among soybean cultivars mediate the attractiveness of high and low seed set male-sterility lines. Suso et al. (2010) showed that higher seed setting of male-sterile lines was associated with increases in the length of the standard and changes in its lobe shape. Wild soybean species have a greater ability for outcrossing. *G. syndetika* Pfeil and Craven is a perennial species noted for its relatively intense perfume. In this species, several volatile organic compounds were identified (Palmer et al., 2012; Pappas et al., 2012). Altogether, these results can be exploited to genetically enhance annual soybean cultivars for increased outcrossing.

Faba bean

Although good hybrid combinations have been found, none of the many published CMS (cytoplasmic male sterility) systems are employed in practical breeding, mostly due to instability and spontaneous reversion to pollen fertility (for an in-depth review, see Palmer et al., 2011). Hybrid cultivars are not available commercially yet, and are unlikely, due to the high cost of producing and growing very large seeds (Gnanasambandam et al., 2012).

Partially allogamous crops are bred as either self- or cross-fertilizers. Breeding as a self-fertilized crop (line cultivar) leaves heterosis unexploited; moreover, there is a risk of lower yield stability (Maalouf et al., 1999). Genetic uniformity is now being questioned in sustainable agriculture, thus open-pollinated highly heterogeneous and heterozygous populations (mixture or synthetic cultivars) should be an objective in itself (Gasim and Link, 2007; Goldringer et al., 2010).

The heterozygosity level of a population is a function of its outcrossing rate. The expression of heterotic effects in synthetics is particularly dependent on the degree of outcrossing (Maalouf et al., 1999). A prerequisite of breeding synthetic cultivars is the breeding of lines with high cross-fertilization rate (Gasim and Link, 2007).

Conclusions and Future Perspectives

Hybrids and heterotic populations can contribute to food security through higher yield and resilience. The development of hybrid technology is severely limited, primarily because of its high seed cost due to low outcrossing. Optimization of the insect-aided outcrossing technology is required and should be coupled with the development of a clear understanding of the PPI.

In the context of pollinator decline (Klein et al., 2007), the efficient use of pollinators emerges as a key target (Christmann and Aw-Hassan, 2012; IPBES, 2013; Cost Action FA1307, 2014; Garibaldi et al., 2014). Besides, legume breeding for sustainable agriculture is linked to the development of non-food services such as environmental services (Helenius and Stoddard, 2007; Duc et al., 2014). As it has been the case in other problem-solving matters in plant breeding, the solution lies in the interdisciplinary domain rather than in unique expertise. Structured dialog, as suggested in the framework of the European Innovation Partnerships - Focus Group on Genetic Resources[1], among pollination biologists, entomologists, as well as breeders, farmers and, beekeepers, offers promising possibilities to overcome the constraint. Significant breakthroughs can be made by mixing disciplines: combining expertise in the core disciplines of agro-ecology, apiculture/entomology, and breeding could result in the development of insect-mediated outcrossing technology.

EXPLORING THE ROLE OF PLANT–POLLINATOR INTERPLAY IN GRAIN LEGUME LANDRACE GERMPLASM CONSERVATION AND MANAGEMENT

General Introduction to Grain Legume Pollination

In modern agriculture, 19 grain legume crops provide food for mankind, worldwide, or locally (Broughton et al., 2003; Smýkal et al., 2015). Eleven of these crops have records of naturally occurring cross-pollination mediated by PPI (Purseglove, 1968; Klein et al., 2007; Suso et al., 2015). Among these crops are soybean (*G. max*), common bean (*Phaseolus vulgaris* L.), runner bean (*P. coccineus* L.), lima bean (*P. lunatus* Benth.), faba bean (*V. faba* L.), adzuki bean (*Vigna angularis* (Wild.) Ohwi & H. Ohashi), mungo bean [*V. radiata* (L.) R. Wilczek], and cowpea [*V. unguiculata* (L.) Walp.], etc. Many legume crops are considered predominantly self-pollinated (Suso et al., 2011) and their domestication process has favored this pollination behavior (Allard, 1999) with the corresponding adaptation of flower structure and traits (further information is presented in see Crop Design System). Self-pollination has contributed to crop segregation against wild crop relatives (McCouch et al., 2012) and other related legume crops. However, self-pollination can limit the crop diversity and adaptation to the environment, and many crops have kept alternative pollination and reproduction mechanisms to overcome these constrains. Thus, legume self-pollination cannot be seen as an unvarying process (Lloyd and Schoen, 1992), and can be compatible with a variable amount of pollen transfer within the inflorescence or between plants within populations or between populations.

Assessing Grain Legume Landrace Diversity and Structure

Many grain legumes are traditionally cultivated and maintained worldwide resulting in the evolution of wide genetic diversity represented by landraces (Negri et al., 2013). Genesys (2015) allow us to have an idea of the number of these landraces conserved *ex situ*. There are as many as 75,796 bean landraces, 69.11% of bean germplasm held in genebanks, 19,702 cowpea landraces (63.13%), 31,880 chickpea landraces (57.88%), and 5,754 faba bean landraces (27.95%).

Crop landraces are locally adapted heterogeneous populations and a potential source of genotypes with functional traits needed for PPI and breeding improvement. Landraces have been the matter of widespread analysis but not always of consensual understanding (Zeven, 1998; Pinheiro de Carvalho et al., 2013). However, the importance of landraces, including legume landraces, for crop evolution, food production and agriculture sustainability (Upadhyaya et al., 2011) and eventually for the implementation of the IPM, FAP, and CDS approaches, is widely recognized (see Reducing Pesticide Application by an Integrated Pest Management Approach, Farming with Alternative Pollinators - An Approach to Gain a Mass-Basis for Pollinator Protection, and Crop Design System). Legume landraces are the result of evolutionary adaptation to agro-ecological conditions, and their structure includes the commitment to genetic singularity (assembly of genes and alleles defining the landrace), ecological complexity (defined by environmental adaptation) and socio-cultural value (Brush, 1999; Negri et al., 2009; Freitas et al., 2011).

Legume landraces are submitted to a general trend in modern agriculture characterized by its fast replacement by modern cultivars, with increasing danger of erosion and extinction (Almekinders and de Boef, 1999; Negri et al., 2009; Pinheiro de Carvalho et al., 2016). This trend determines an overall concern for their *ex situ* conservation in germplasm collections, while almost 150 thousands landraces of six major grain legumes are actually stored (see Genesys, 2015).

An important factor for understanding the role of legume landraces in PPI is the knowledge of landrace population structure, in order to comprehend their genetic diversity and the presence of genotypes with required functional traits, and to optimize their *ex situ* and *in situ* (on-farm) conservation and management and/or use in breeding programs. According to our knowledge, the research into this issue has been insufficiently developed. Legume landraces are, by nature, heterogeneous and

[1] http://ec.europa.eu/eip/agriculture/en/content/genetic-resources-cooperation-models

have a complex structure, whereas traits and genes are evolving in dynamic equilibrium with environmental conditions. The heterogeneity of the landrace can be promoted by farmers in the field through seed exchange, natural or accidental selection and/or gene flow (including that mediated by pollinators through the PPI) within or between closely related landrace populations.

The sampling methodology of landrace material is aimed at gathering as much genetic diversity as possible, while keeping useful landraces traits. If the landrace is strictly self-pollinated, then the seeds harvested will change in terms of the increase of genotypes, leading to a greater fitness in population. In partially cross-pollinated landraces, which allow limited pollen exchange between flowers mediated by pollinators, even small natural outcrossing rates can create many population crosses, since they are composed of several genotypes cultivated in a proximity which allows PPI. In such conditions PPI pollen transfer and allele shuffling will increase the heterozygosity of the landrace. Recent data shows that natural outcrossing can contribute to increasing the genetic diversity of crop populations (Papa and Gepts, 2004; Barnaud et al., 2008).

Impact of Pollinators on the Structure of Grain Legume Landraces: Risks and Opportunities

The impact of pollinators on the heterogeneous structure of landraces is still not well understood. Pollinators can promote self-pollination or outcrossing within each landrace and among landraces or cultivars. Thus, landraces can be vulnerable to changes determined by crosses with other cultivars. Outcrossing with other homogenous cultivars can result in genetic loss of one or several landraces' useful genotypes and their functional traits. On the other hand, inbreeding depression appears to be related, with genetic erosion and reduction of heterozygosis, when the natural outcross or environmental adaptation are limited. The evolution of the landrace in the field and outcrossing among its genotypes should be facilitated by PPI to avoid inbreeding depression (Street et al., 2008c). Additionally, selfing was found to alter the performance of an open-pollinated population in a progressive way, leading to the expression change of those characters associated with pollinator discovery and attraction (Nadal et al., 2003; Suso and Río, 2014). Martin and Adams (1987), studying the extent of outcrossing in 15 bean landraces in Malawi, identified three different categories regarding their structure. One category was a group of four seed types without evidence of outcrossing among genotypes as result of recent landrace cultivation in the studied region, differences in their flowering times, absence of pollinators activity or pollen incompatibility. In the second category, intermediate outcrossing events between landraces were observed, while in the third category landrace structure had been largely determined by many outcrossing events (Martin and Adams, 1987). These observations point to a temporal change of landrace structure and diversity, promoted by PPI, and need to be taken into consideration, during landrace management.

The global strategy of *ex situ* conservation of legume landraces has resulted in the interruption of evolutionary processes (Brush, 1999). The process can restart periodically, during gene banks management actions, with regeneration or multiplication of accessions. Under cultivation or in on-farm conservation, landraces continue evolving and adapting to environmental conditions. However, this evolution can be threatened by genetic erosion resulting in the extinction of landraces when farmers abandon them or in their genetic pollution by bred cultivars. These cultivars are characterized by genetic gains in yield, but are homogeneous and may not hold the useful traits required for PPI improvement and hybrid seed production (Palmer et al., 2011; see also Key Examples). There are benefits derived from natural outcrossing within landraces for creating new variability and selecting highly adapted yielding cultivars (Saxena and Kumar, 2010). Cross-pollination can also be hazardous for the breeder wishing to maintain the genetic identity and purity of the crop cultivar (Gómez, 2004). The natural outcrossing occurring in several grain legume crops can affect the reliability of the breeding programs when improved cultivars outcross with unimproved plants in nearby fields (Rahman et al., 2001).

The gene flow occurring between landraces and related species depends on outcrossing rates, in addition to the socioeconomic conditions of their cultivation. Oaxacan farmers' practice multiple landrace cropping systems using common bean, runner bean and *Phaseolus dumosus* across an altitudinal gradient as a strategy to mitigate risks across a broad range of agronomic and household needs that cannot be addressed with single species or races (Worthington et al., 2012). De Mooy et al. (1990) showed that under traditional intercropping systems, outcrossing plays a positive role in maintaining genetic variability and stability of cowpea landraces. However, the outcrossing between landraces and bred cultivars increases the risks of genetic pollution (De Mooy et al., 1990). Additionally, the natural outcrossing in pigeon pea can be considered a constraint for the maintenance of genetic purity of the bred cultivars and genetic seed stocks. New cultivars have been contaminated in a short period of 3–4 years (Gupta et al., 1981). However, Singh et al. (2005) reviewed the long-term Indian pigeon pea breeding program and showed that cultivars were selected from landrace variations created by natural outcrossing, hybridization and segregation events (Singh et al., 1990 cited by Saxena, 2005; Singh et al., 2013).

Ex situ and On-Farm Approaches and the Plant–Pollinator Interplay. A Complex Issue

An important issue in genebank management is to maintain the necessary level of the heterozygosity and heterogeneity of landraces, which can be achieved through the management of pollinator-crop interplay (Suso et al., 2015). Trying to identify gaps in the management of grain legume landraces, Maggioni et al. (2002) analyzed the maintenance and regeneration protocols of 11 germplasm collections and realized that about 50% of them are multiplied under open-pollination conditions. An average contamination rate of 12% for these landraces was estimated, with a high variation between genotype frequency (Maggioni et al., 2002). These data point out the need to focus on collection management protocols, with particular attention to the regeneration procedures and conditions. Recently, recommendations have been drawn up for cowpea, chickpea, faba bean and lentil landrace management (Dumet et al., 2008; Street et al., 2008a,b,c) aiming at supporting the positive results of outcrossing within landraces and to avoid the detrimental effects of cross-pollination with different cultivars.

The Global Crop Diversity Trust has also published regeneration guidelines for many grain legumes, including cultivated chickpea landraces. Even though chickpea (*Cicer arietinum* L.) is mainly self-pollinated, an outcrossing rate of 5.6% has been detected (Rubio et al., 2010), mediated by insects (Purseglove, 1968). An isolation distance of 90 cm between plots has been recommended to avoid unwanted outcrosses (Street et al., 2008c). In pigeonpea, covering the flowering plants with muslin bags or insect-proof cages has been recommended to avoid cross-pollination (Upadhyaya et al., 2008). Regeneration of faba bean landrace germplasm (Street et al., 2008a) can take place in screen houses to avoid insect-facilitated cross-pollination with foreign landraces or cultivars, and it allows within-accession crossing to prevent inbreeding depression. The regeneration in open-pollinated conditions should keep isolation distances to facilitate pollination by natural out crossing within accession.

Self-pollinated *Lens culinaris* Medik. shows a degree of outcrossing, ranging from 0.06 to 5.12%, depending on the cultivar, location, and year (Horneburg, 2006). Additionally, the outcrossing rate between individual flowers in the inflorescence can reach 22.2% within cultivars. Its regeneration will therefore require specific management aimed at controlling pollination, either by allowing insect pollination inside isolation cages or tunnels or in-field free open pollination, keeping the minimum distance between different landraces or cultivars to avoid undesirable outcrosses.

It is vital to maintain landrace integrity, because some genotypes have kept their floral structure, promoting natural outcrosses, which have almost disappeared through crop domestication. At the same time, the genetic mechanisms facilitating the maintenance of outcrossing are relevant in breeding (for both hybrids and populations). Therefore, these flower functional traits will be a useful tool for crop improvement as well as for *in situ* and on-farm conservation (see also Key Examples).

Thus, when regeneration actions are implemented, it is equally important to take notice of pollinators [particularly honey bees (*A. mellifera*), bumble bees (*Bombus* sp.) and hover flies (Syrphidae)] visiting crop flowers, and evaluate the accessions for additional phenological and morphological data. Street et al. (2008b) recommended: (1) coordinating periodic field inspections with entomologists during the growing season, to identify the most efficient visitor/pollinator and determine the relationship between pollinator and flower morphology; (2) evaluating the agronomic data in relation to phenology and flower morphology and structure. Besides, all off-type plants resulting from crossing should be kept and mixed to build a basis population to provide breeders with high-value material (for its outcrossing potential).

DEVELOPING STRATEGIES FOR SEED PRODUCTION AND SEED PRODUCERS IN FORAGE LEGUMES

Seed Production in Forage Legumes

Genetic and Environmental Factors in Seed Production

Forage legumes show a wide variation in traits influencing breeding programs, cultivar maintenance and seed production. Perennial species are generally cross-pollinated by insects and, consequently, predominantly allogamous. Cultivars of these species (synthetic and improved populations from mass selection or recurrent selection) are populations of different, highly heterozygous genotypes assembled by the breeder in a delicate equilibrium, conditioned by the cross-mating system, that must be maintained during the lifetime of a cultivar, since eventual inbreeding has deleterious effects on forage and seed yields (Ceccarelli, 1986).

Annual medics and subterranean clover are generally self-pollinating species, with cleistogamous flowers and a self-tripping mechanism (Katznelson and Morley, 1965; Lesins and Lesins, 1979). Because of their reproductive system (predominant autogamy) and the huge germplasm variation existing in nature, the breeding of these species has mostly been based on a pure-line selection from natural populations (Piano and Pecetti, 2010).

Seed yield, which is an important factor in determining the success of a cultivar, is determined by genetic and environmental (*sensu lato*) factors. The latter include the choice of the best-suited environment (for soil and climate characteristics) and the choice of an appropriate agronomic technique, where seed sowing rate, care of the establishment (fertilization, irrigation, grazing, and/or cutting management, weeds and pest control) and number of pollinators should be carefully set up. Sub-optimal environmental conditions, cultural and harvesting practices and absence of pollinators negatively condition forage seed yield (Lorenzetti, 1981, 1993; Lorenzetti and Negri, 1995). While research has already clarified the best environmental conditions for seed production (for the main forage species at least), there is still a need to carry out further research on genetic factors affecting seed yield, on which we will focus the following section, with particular reference to minor legumes.

Potential Seed Yield vs. Realized Seed Yield

In most forage legumes, the realized seed yield (RSY) is very low compared with the potential seed yield (PSY). The PSY can be estimated multiplying the Potential Seed Number (PSN) per unit area by the average thousand seed weight (TSW) (**Table 1**). The former depends on inflorescence number per unit area, flowers per inflorescence, ovules-per flower (i.e., on the reproductive system size). It is noteworthy that each of these components interacts with the others. With the exception of ovule number

TABLE 1 | Component seed production, potential and realized seed yield of some perennial forage legumes (modified from Lorenzetti, 1993).

Species	Inflorescence ($N°/m^2$)	Flowers per inflorescence	Ovules per flower	Potential Seed Number/m^2	Thousand seed weight (TSW) (g)	Potential seed yield (PSY) tha^{-1}	Seed harvest realized (RSY)	
							tha^{-1}	% of PSY
M. sativa	3750	16	10	600,000	2.2	12.0	0.5	4
T. pratense	750	110	2	165,000	1.6	2.6	0.5	19
T. repens	600	100	6	240,000	0.6	1.8	0.4	22
L. corniculatus	400	6	40	96,000	1.2	1.2	0.2	17
H. coronarium	2915*,**	30***	4-5	393,525	4.5	17.7	0.2	1
O. viciifolia	2223*	60	1	133,380	18.0	24.0	0.3	1

**Martiniello (1998); **Satta et al. (2000); ***Yagoubi and Chriki (2000).*

per flower, the less variable component, the other components show a high variability in all species (Lorenzetti, 1981; Rosellini et al., 1994), but few data on their hereditability exist. PSY is very high in forages since many ovules are present per unit area at anthesis; however, only a small part of these ovules turns into mature seed and only a part of this seed is collected (**Table 1**). RSY, in fact not only depends on the PSY, but also on the efficiency of the reproductive system (seed setting and maturation) and the percentage of harvested seed.

Pollination and fertilization are essential steps in the formation of mature seeds (Fairey and Hampton, 1998) and mostly depend on pollination. Pollinators like honey bees (*A. mellifera*), leafcutter bees (*Megachile rotundata* Fab.), alkali bees (*Nomia melanderi* Cockerell), and bumble bees (*Bombus* sp.), while foraging, operate flower tripping and fertilization and seed formation occur. Consequently, an adequate number of insects is a key factor in the production of forage seeds. Furthermore, crossing operated by pollinators allows the maintenance of a high level of heterozygosity and heterosis which, as mentioned above, condition important agronomic traits, such as forage and seed yields (Veronesi and Lorenzetti, 1983; Strickler, 1999; Strickler and Vinson, 2000; Pecetti and Proietti, 2003).

The environment strongly influences the success of pollination and fertilization through its effects on flowering pattern, flowering duration, pollen production and vitality, pollen tube growth, stigma receptivity, and ovule viability (Clifford and Scott, 1989; Hampton, 1990; Thomas and Pasumarty, 1996). However, the interplay plant-pollinator also plays a role. For example, in alfalfa, the bee foraging behavior is influenced by flower color (Steiner et al., 1992), flower aroma (Loper et al., 1974), and nectar availability (Kauffeld and Sorensen, 1971; Pecetti and Tava, 2000; Palmer et al., 2009). The latter are partially under genetic control (Loper and Waller, 1970; Loper and Lapioli, 1971; Loper and Berdel, 1978; Pecetti and Tava, 2000; Pecetti et al., 2002) while flower color is totally under genetic control (Barnes, 1966).

Moreover, fertilization may be reduced by genetically determined reproductive barriers such as incompatibility, plant sterility, or unreduced-gametes (Falcinelli, 1999). In addition, a high percentage (50%) of embryo sac degeneration and/or fertilized egg abortion is observed in outcrossing forage legumes (Thomas, 1996) as a possible consequence of the genetic load resulting from deleterious recessive alleles (Wiens, 1984; Burbidge and James, 1991; Pasumarty et al., 1993; Rosellini et al., 1998). Breeding for traits like flower attractiveness to pollinators or low genetic load, factors that indeed condition reproductive efficiency, has been limited up to now.

Poor seed retention of forage legumes is another constraint in achieving PSY, since it causes considerable seed loss before and during harvesting. Forage legumes have not been fully domesticated yet and often show pod dispersal and/or dehiscence of pod prior to harvest. For example, seed shattering coupled with dehiscence of pod is high in *Lotus corniculatus* L., while pod dispersion is high in annual medics, *Trifolium incarnatum* L. and *T. repens*, where losses up to 40% of the total seed yield are reported (Clifford and McCartin, 1985).

Strategies for Increasing Seed Production in Forage Legumes

Generally, the breeding programs of forage legumes are mainly designed to achieve a balance between persistence, quality, yield, and animal safety (Sewell et al., 2011), while they should also be addressed to increase seed production, since the commercial success of a forage cultivar also depends on its ability to produce seed. High seed yield facilitates the spread of the cultivar because low-cost seed can be easily placed on the market (Lorenzetti and Negri, 1995). In addition, several studies have shown that in forage legumes, seed yield is positively correlated with biomass production (Bolaños-Aguilar et al., 2002; Torricelli et al., 2007).

There are still ample margins for genetic improvement in seed production in forage legumes relying on existing genetic variation for traits that condition SYP (such as inflorescence number and seed number per plant, seed number per inflorescence and seed weight per inflorescence) and there is a high correlation with seed yield per plant (Bolaños-Aguilar et al., 2002; Sengul, 2006), attractiveness for pollinators and seed shattering.

Breeding in these crops is still at an early stage compared with other species. Both improved cultivars and populations are characterized by a wide genetic base, and thus, there is sufficient variation for seed yield to be substantially increased by selection and breeding. Improvements in seed yield will confer substantial benefits to breeding companies and farmers producing forages. It is noteworthy in the context of this paper that cultivars specifically bred for increased attractiveness to pollinators will also help in maintaining the wild pollinator fauna and increase honey production by honeybees (with a consequent increase in economic returns for farmers).

PROTECTING THE POLLINATION SERVICE FROM PESTICIDES

The Pest Control Service in Legume Crops

Legume crops are regularly attacked by several pests, pathogens and weeds, at different times of their growth cycles (Stevenson et al., 2007; El-Wakeil and El-Sebai, 2009; Pérez-De-Luque et al., 2010; Sillero et al., 2010; Stoddard et al., 2010; Kharrat and Ouchari, 2011; Stoddard, 2013; Arlauskiene et al., 2015), which strongly affect crop yields worldwide. This frequently compromises the choice of farmers for legume crops (Jensen et al., 2010). These attacks not only affect the crops' health but also limit the rotation and intercropping schemes that use legumes (Jensen et al., 2010).

Among the biotic constraints of legume crops, insect pests play a major role in yield reduction and several species are reported to be present in legume crops (Stevenson et al., 2007; Stoddard et al., 2010). More than 70 arthropod species have been reported attacking and damaging faba bean (Stoddard et al., 2010). However, not all these enemies are economically important (Nuessly et al., 2004; Stevenson et al., 2007).

The control of pests is crucial to optimizing production; however, frequently, natural enemies are not able to control pest outbreaks efficiently just by themselves in agricultural ecosystems. Presently, the use of pesticides still dominates the pest control programs of legume crops (Stevenson et al., 2007) and natural enemies or introduced (released) ones are chronically exposed to cocktails of pesticides. Predicted changes in global climate are likely to further exacerbate such problems in the future (Goulson et al., 2015).

Pesticides have a negative impact on the pests' enemies (and also on pollinators, as explained below) by direct mortality and by physiological and behavioral sub-lethal effects. Pesticide effects on their development, adult longevity, immunology, fecundity, sex ratio, mobility, navigation/orientation, feeding behavior, oviposition behavior and learning performance have a great influence in the abundance and pest control performance of those enemies (see details in Desneux et al., 2007). Pesticides disrupt/ weaken the natural pest control service in agricultural ecosystems.

The Effect of Pesticides on the Pollination Service

Pollination service is another ecological service essential in agricultural ecosystems, for both crops and wild plants. Many insect species contribute to pollination, including bees, hoverflies, butterflies, moths, and some beetles, by visiting a wide or narrow range of flower species to collect food (nectar and/or pollen), thus acting as generalist or specialist pollinators. Bees rely on either nectar or pollen at different stages of their life cycle with a total dependence on flowers, and are considered the most efficient pollinators. European bee species comprise the managed western honey bee (*A. mellifera*), 68 bumble bee species and over 1000 semi-social and solitary bee species (EASAC-The European Academies' Science Advisory Council, 2015).

There has been a steady decline in many pollinator groups, which has been attributed to several factors, many related to agricultural intensification, such as habitat loss and fragmentation, decreased resource diversity and pesticides application (Steffan-Dewenter et al., 2002; Kremen et al., 2007; Carré et al., 2009; Potts et al., 2010; Brown et al., 2012; Nieto et al., 2014).

Insecticides are the group of pesticides that pose the most direct risk to insect pollinators, especially the neonicotinoids (Woodcock, 2012; Sanchez-Bayo and Goka, 2014; Chagnon et al., 2015; Rundlöf et al., 2015). However, herbicide application also has the indirect effect of damaging the floral resources on which pollinators depend and delaying the flowering, so the timing between the period when food is most needed by pollinators and food availability is disrupted (Boutin et al., 2014; Nieto et al., 2014). Additionally, exposure to some fungicides can greatly increase the toxicity of insecticides, such as neonicotinoids and pyrethroids (Desneux et al., 2007; Goulson et al., 2015).

Pollinators can be exposed to insecticides during their application, by contact with residues, or from the ingestion of contaminated pollen, nectar, or guttation fluid. Exposure occurs in agricultural crops, flowering weedy crop boundaries and also in the neighboring semi-natural habitats where pollinators nest and forage, owing to pesticide drift (Potts et al., 2010). Numerous deaths of pollinators can result from insecticide application to a crop containing blooming weeds, even if the crop itself is not in bloom (Riedl et al., 2006; Gregorc and Ellis, 2011).

Exposure to insecticides may have a lethal effect (e.g., Gradish et al., 2010; Brittain and Potts, 2011; Tapparo et al., 2012; Pisa et al., 2015) or sub-lethal behavioral or physiological effects, such as a reduction in orientation abilities, reduction, or absence of foraging behavior, reduced feeding and life expectancy or compromised pollinators detoxification mechanisms and immune responses, rendering them more susceptible to diseases and parasites (Desneux et al., 2007; Mommaerts et al., 2010; Potts et al., 2010; Brittain and Potts, 2011; Henry et al., 2012; Pettis et al., 2012; Woodcock, 2012; Feltham et al., 2014; Godfray et al., 2014; Goulson et al., 2015).

The life-history traits of pollinator species affect their sensitivity to insecticides, such as body size (contact absorption rate and forage ranges), sociality (colonies essentially bioaccumulate pesticides and social pollinators experience exposure at greater doses for longer periods than solitary ones), flight season (overlapping with pesticide application periods), voltinism (univoltine species foraging when pesticides are applied are at a particular high risk), floral specialization (for polylectic pollinators, sources containing pesticides may be diluted by foraging in other plants or habitats), nesting behavior (ground or above ground nesting may be differently exposed to pesticides), and sex (different intrinsic susceptibility and body size) (see details in Williams et al., 2010; Brittain and Potts, 2011). Although synergistic interactions among traits remain to be explored, individual traits can be useful in predicting and understanding responses of related species to environmental disturbances (Williams et al., 2010).

It must be taken into account that the effects of pesticides on honey bees are poorly representative of the effects on other pollinators, including other bees. Indeed, bees (Apoidea) constitute a highly diverse group, and bees from different taxonomic groups differ widely in their vulnerability to pesticide exposure (Desneux et al., 2007).

In conclusion pollination service benefits from a healthy, robust pest control service in the respective agricultural ecosystem, which reduces the need for artificial pest control such as pesticide application.

Reducing Pesticide Application by an Integrated Pest Management Approach

The development of pests' resistance to pesticides, and the economic, health and environmental risks they pose have highlighted the need to reduce pesticide applications (see the European Union Sustainable Use of Pesticides Directive 2009/128/EC) and to adopt alternative pest management methods. To date, no single method provides efficient pest control, and an integrated approach (integrated pest management - IPM) is needed, in which methods are combined in order to maintain enemy populations below their respective economic threshold levels.

The use of pesticides is allowed in the IPM approach, but only when absolutely necessary and after a careful selection (Riedl et al., 2006; Stevenson et al., 2007; Macfadyen et al., 2015). Pest monitoring, together with the concept of economic threshold, allows the perception on when and where to apply pesticides and avoids unnecessary spraying. Removal and protection of bee colonies is also advised (Riedl et al., 2006).

Integrated pest management in legume crops has been widely discussed by several authors (e.g., Holt and Wratten, 1986; Redden et al., 2005; Sharma et al., 2007; Stevenson et al., 2007; Duc et al., 2010; Pérez-De-Luque et al., 2010; Rispail et al., 2010; Sillero et al., 2010; Stoddard, 2013). The wide range of regions where legumes are grown determines a high diversity of crop enemies (Stevenson et al., 2007), and their presence and abundance varies from region to region. Therefore, it is not feasible to define a generalized IPM strategy- it has to be adapted locally or at a regional level.

The biological control method may include releases of artificially reared beneficial organisms; however, a more sustainable way to implement this is to give the natural enemies occurring there the best conditions to increase their density and to enhance their impact on the pests - conservation biological control (CBC). This enhancement of the pest control service is achieved by managing the habitat and manipulating environmental factors (Rebek et al., 2005; Stevenson et al., 2007; Macfadyen et al., 2015).

The success of CBC, and also of the pollination service, is strongly linked to the temporal and spatial availability and quality of ecological infrastructures, inside and outside the farm limits (within a radius of approximately 100–200 m) (Boller et al., 2004), such as shrubs, herbaceous plants (ground covers, inter-row strips, or crop boundaries), wooded field edges, stones, and bare undisturbed soil areas, inside the crop land or in the vicinity (Landis et al., 2000; Westerkamp and Gottsberger, 2000; Richards, 2001; Rebek et al., 2005; Schellhorn et al., 2008; Van Driesche et al., 2008; Köpke and Nemecek, 2010; Shackelford et al., 2013; Sigsgaard et al., 2013). Diverse floral resources (pollen and nectar), with phenological complementarity, apart from their evident importance for pollinators, are of great importance for both predators and parasitoids (Landis et al., 2000; Brown et al., 2012; Döring et al., 2012). When managing existing or introduced floral resources, implementing or preserving different types of shelters, it is essential to know a lot about the ecosystem being manipulated, including the morphology and bio-ecology of the target organisms present there (Landis et al., 2000).

Generally, faba bean offers a rich foraging habitat for several beneficial insects, with mass flowering and substantial amounts of nectar and pollen (Westphal et al., 2003; Nuessly et al., 2004; Köpke and Nemecek, 2010; Stoddard, 2013). For many natural enemies and pollinators, faba bean is considered a substitute for the loss of habitat diversity that results from agricultural intensification (Richards, 2001). Diverse legume leys provide a variety of flowers which bloom sequentially throughout the season to extend the availability of pollen and nectar for pollinators above that of standard leys (Brown et al., 2012).

When adopting IPM and selecting the most suitable pest management methods, it must be taken into account that not only pesticides may be dangerous to natural enemies and pollinators.

Cultural practices such as removing weeds, vegetal debris and dead wood with insect galleries, as well as tillage procedures, may have a very negative impact on the beneficial fauna by the destruction of food and refuge sources (Fuchs and Hirnyck, 2000; Riedl et al., 2006; Stevenson et al., 2007; Williams et al., 2010), and so a trade-off of benefits- disadvantages must be considered to reach a correct decision (Saunders et al., 2015).

GENERAL APPROACHES TO ENCOURAGE THE SYNERGY BETWEEN FOOD PRODUCTION AND POLLINATION SERVICES

Farming with Alternative Pollinators – An Approach to Gain a Mass-Basis for Pollinator Protection

The Urgent Need to Get a Mass-Basis for the Protection of Wild Pollinators

The global decline of pollinator diversity is mainly a consequence of human mismanagement (agricultural chemicals, monocultures, landscape fragmentation, habitat loss) as referred to before, and increasingly of different climate change effects and the interaction of these stress factors (Potts et al., 2010). Climate change ultimately requires us to reflect on the PPI from a broader perspective to gain sufficient support for the protection of wild pollinators. Even in the case of agricultural crops, honey bees provide only 8–15% of services (Nabhan and Buchmann, 1997; see also MA, 2005, p. 759).

Pollinators have four key functions within ecosystems and for mankind, which make them indispensable and highlights the need for humans to enhance farming practices and the use of landscapes:

- 87 of the 115 most important food crops depend or benefit from pollination (Klein et al., 2007); this list of high-value crops includes some food legumes such as faba bean, pigeon pea, common bean etc. The total yield, e.g., from faba bean with open access to pollinators is 185% higher compared to autonomous self-pollination (Nayak et al., 2015). This example demonstrates the importance of pollinators for food security and sustainable intensification of agricultural production. Some forage legumes such as alfalfa, clover and vetch depend mainly on wild pollinator species (alfalfa/Megachilidae; clover/*Bombus*);
- 85% of all terrestrial plants require pollinators (Ollerton et al., 2011), and so a loss of pollinators could cause a cascade of extinctions (Biesmeijer et al., 2006) and even pollinator population collapse if stress factors reached a tipping point (Lever et al., 2014). Pollinator protection by sustainable legume production can therefore contribute to the protection of the biodiversity of natural plants;
- Cross-pollination can enhance genetic diversity and thus adaptation to climate change (Christmann and Aw-Hassan, 2012), which is of importance for (1) legume-farmers, e.g., in remote mountainous areas with limited or no access to hybrids, and (2) the much higher number of natural plant species having no human breeders;
- Most other ecosystem services rely on pollination to a certain extent, and the partial, but simultaneous loss of these ecosystem services too could cause a spiral of poverty (Christmann et al., in review).

Cultivated legumes do not have exclusive mutualism between crop and pollinator as, e.g., fig, vanilla, or cocoa, but they are mainly pollinated by wild bees and also by honeybees. Legumes can play an important role in sustaining the diversity of wild bees and some wild bee species are indispensable to sustain some legumes, as the flower type of Fabaceae neither allows all insect pollinators to provide the service nor permits hand-pollination by humans. The extinction of wild bees (leaf cutting bees, solitary bees, or bumblebees) could significantly endanger legume production, in particular faba bean production in temperate or high altitude (above 2500 m) regions and seed production of alfalfa. Faba bean flowers early and faces increasing risks of seasonal abnormalities in the course of climate change, which can prevent honeybees from providing service. Honeybees require 15°C and sunny days without rain and strong wind, whereas the availability of, e.g., bumblebees around the field can safeguard farmers' revenues despite unfavorable weather conditions.

Wild pollinators mostly work in a small radius (50–2000 m) around their nests (Kohler et al., 2008). Thus, they are more affected by monocultures, ploghing, chemicals, loss of diverse field edges and landscape fragmentation than honeybees, which fly up to 5 km and are supported by beekeepers (transport, food, hives). So the protection of wild pollinators requires a mass-basis of local stakeholders ready to engage themselves in long term pollinator protection. Raising knowledge and awareness of these issues can increase the readiness of smallholders and some benevolent producers to engage in pollinator protection, but the number of commercial producers ready to reduce chemicals and to enhance habitats might be limited, if this action is based on good-will alone (Christmann et al., in review).

The Methodology of FAP

Hence, the leading research question for FAP (Christmann and Aw-Hassan, 2012) is: how to get a mass-basis of intrinsically motivated local actors? Such a high number of protagonists cannot be paid; even in industrialized countries, the incentive must be grounded on the economic value of ecosystem services (The Economics of Ecosystems and Biodiversity (TEEB)-approach; TEEB, 2010).

Based on broad evidence that a high diversity of pollinators can significantly enhance the harvest of many crops in quantity and quality and that distance between nest and crop is the limiting factor (see literature in Ricketts et al., 2008; Christmann and Aw-Hassan, 2012; Garibaldi et al., 2014), the hypothesis is that by specific crop-based habitat enhancement the total net income from many horticulture crops can be significantly increased; the difference in income between FAP- and control-fields will be sufficient to motivate farmers in favor of pollinator protection by habitat enhancement.

Farming with Alternative Pollinators compares the net revenue of FAP-fields and control fields: whereas control fields have the main crop in the entire field, FAP-fields use only 75% of the field for the main crop and 25% for habitat enhancement, consisting of three-season-forage buffets for pollinators and shelter (both based on further crops, spices, herbs, and medicinal plants), nesting support made out of local materials and a water source, e.g., a puddle or pond. Twelve or more different crops with different flower types and colors nourish pollinators for three seasons. The habitat enhancement in the field itself can optimally reduce the distance between nest and foraging crop. The different flower types and colors of petals attract a higher diversity of pollinators to FAP-fields during the flowering of the main crop. FAP uses a wide range of marketable plants instead of wildflower strips (e.g., (Garibaldi et al., 2014 recommend wildflower strips) to gain the highest possible income gain as an incentive. As the habitat zone and higher insect abundance also attract predators, FAP does not measure the economic value of pollination services, but of habitat enhancement. The methodology for fields and orchards is described in detail in Christmann et al., in review. The options of a purposeful combination of FAP and CBC should be explored and the potential of FAP for enhancing genetic diversity of crops and productivity of seeds should be analyzed. Such supplementary positive impacts on crop productivity by enhanced cross-pollination might additionally motivate farmers to enhance pollinator habitats.

Farming with Alternative Pollinators is a new approach with a limited number of field experiments in Uzbekistan and currently Morocco. The pilot project in Uzbekistan (main crops: cucumber and sour cherry; partly faba bean and partly sainfoin in the habitat zone; Christmann et al., 2014, in review) did on-farm trials mainly with smallholders, because in general they use less or no agricultural chemicals as they prefer low production costs and healthy food for their own families. Trials with smallholders can develop first data to convince commercial farmers too. Starting with large-scale producers can cause discussions about the use of chemicals. Also, due to their high number, smallholders can be important protagonists for pollinator protection.

Results of First Field Experiments

The FAP-pilot project confirmed: the FAP-fields attracted a clearly higher number of different pollinator and predator species, including effective wild bee species; for sour cherry, the harvest was double on FAP-sites, for cucumber even higher (Christmann et al., 2014, in review). Solitary bees (important pollinators for cucumber) used the nesting support. Due to the project duration (18 months), it was not possible to assess the adoption rate in the long term, but the readiness to protect pollinators and to take effective action based on data concerning yield increase was much higher in both project regions (Christmann et al., in review) than described in case studies recommending information campaigns (e.g., Kasina et al., 2009a).

Potential of FAP with Special Regard to Legumes

Pollinator-friendly legumes can significantly contribute to the habitat zone. Due to their specific flower type and the (for crops) unusual color of the petals, they can attract, e.g., wild bees, which are also effective for many other crops. According to the experience in Uzbekistan it is not difficult to convince farmers to seed faba bean or sainfoin in the habitat zone, as they know they also will benefit directly from these plants. They suggested using crops themselves, and would not consent to plants such as *Laburnum anagyroides* Medik or seed *Indigofera tinctoria* L., ornamental plants or wildflowers (weeds) to attract pollinators (Christmann et al., in review).

Pollinator-dependent legumes might benefit also substantially from FAP as a main crop. Current research on faba bean supports this assumption, since pollinators from nearby natural habitats can increase the harvest significantly (Musallam et al., 2004; Aouar-Sadli et al., 2008; Andersson et al., 2014). Also the production of string bean, soybean, kidney bean, and various peas etc. as main crops, as well as alfalfa-seed production, might benefit from FAP (Klein et al., 2007; referred also in the earlier paragraphs of this article).

As soon as the additional income gain by FAP-measures for these crops as a main crop has been assessed and planting instructions are spread, the intrinsic motivation of farmers for pollinator protection might rise significantly, as the approach does not require investment or long-term training. Including more marketable plants like other crops, spices and medicinal plants, rather than just pollinator-friendly legumes, can provide the necessary mass-basis for pollinator protection. Enhancing the environment might be a better option than the suggestion to breed for pollinator independent crops (Chaplin-Kramer et al., 2014), which has several negative effects: it does not take advantage of heterosis effect and bears the risk of lower yields (see "Basic Application-Oriented Questions in Heterosis Breeding: The Potential of Insect-Mediated Outcrossing in Grain Legumes"), creates the need for continuous breeding to adapt to the changing climate and runs the risk of local pollinator populations starving even more.

Crop Design System

In the CDS approach, breeders incorporate the potential benefits of pollen vectors into practices to increase the efficiency of hybrid seed technology and, in parallel, increase the occurrence, health and visitation of pollinators, whether these are wild, or managed by developing pollinator-friendly crops.

Understanding the CDS Approach

The rationale is the following:

(a) Animal-pollinated crops grown today are often suboptimal in attracting and sustaining their pollinator populations (Bailes et al., 2015). Cultivars are generally bred by the artificial selection of agronomic traits that are of commercial interest but with little regard to pollinator-related traits and preferences. Consequently, insect-pollinated cultivars may not be attractive to pollinators (Kobayashi et al., 2010).

Legumes are known to have flowers with moveable parts that have to be actively handled by insects for pollination to take place. A flower has to be actively moved by a visitor to access the flower reward and, consequently, to perform pollination (Cordoba and Cocucci, 2011). Due to the morphological specialization in floral phenotype, effective pollen transfer requires several traits of the floral mechanism of plants to be adjusted to the morphology and behavior of animal pollen vectors (Garibaldi et al., 2015). Simple morphological mismatches appearing in the flowers are an obvious problem. For instance, hybrid seed production depends on the production and use of male-sterility lines. These lines are difficult to work with naturally in the fields, as the foraging behaviors of bees often change when confronted with such anomalies (Ortiz-Perez et al., 2006b). Information show that some traits may respond to pollinator-driven natural selection (Suso and Río, 2014). Suso and Río (2015) identified some architectural and floral patterns in faba bean that are the result of natural selection and reflect local adaptation to differences in pollination environments. Volatile compounds emitted by crop flowers mediate plant/pollinator interactions (Klatt et al., 2013).

In the CDS approach, breeders and farmers develop cultivars with enhanced heterozygosity and heterogeneity as result of appropriate functional floral traits (discovery, attraction, and reward) within the crop for supporting the bee pollinator populations used as agents of crossings. One idea proposed is to support native pollinators for insect-mediated outcrossing by designing a crop with appropriate functional flower traits (Suso et al., 2005; Palmer et al., 2009; Suso and Maalouf, 2010; Suso et al., 2015).

Legumes have great potential to be served by bee pollinators as crossing agents because they are visited by a great number of bees, such as bumble bees, honey bees, and wild bees (Free, 1993; Delaplane and Mayer, 2000). Non Apis bees, such as solitary bees, are an essential group of visitors and of incalculable value for pollination (e.g., Pierre et al., 1999; Kasina et al., 2009a,b; Suso et al., 2015), even in self-pollinated species such as soybean (Martins, 2013).

However, there is a gap in our knowledge about which particular insects visit and cross-pollinate a particular legume crop, and why. For many legumes, there is little specific data available and even if information on pollinators is available, the findings are often ambiguous (Suso et al., 2015). Besides, the pollinator species can vary from country to country, e.g., faba beans are mainly pollinated by wild bees in Spain and England, but by different species of *Bombus* in France (Bond and Kirby, 1999, 2001; Pierre et al., 1999).

(b) Declines in bee populations have increased interest in the ecological services of legumes (Köpke and Nemecek, 2010). Legume cultivation provides some key environmental benefits such as floral resources. Although the yield produced by legumes remains a major objective for breeders, other issues such as environmentally friendly, resilient production in the context of climate change, as well as new ecological services such as pollinator protection, are required (Duc et al., 2014). Improved value for the conservation of bee pollinators should be a key breeding objective.

Concerns and Constrains

Over many centuries, farmers have often selected many legume crops for their autogamous abilities or self-fertility. According to Hammer, (1984 cited by Smýkal et al., 2015), the "domestication syndrome" for legumes includes the transition from outcrossing to selfing. The shift from outcrossing or facultative selfing to strict inbreeding has been described as the single most common trend in legume domestication (Rick, 1988). One common view on the mating system of grain legumes, among germplasm curator and breeders, is that they display autogamous or mostly autogamous mating systems (Suso et al., 2011).

However, for the development of insect-aided outcrossing technology, selfing has to be bred out again and functional floral traits should be managed toward outcrossing. The transition from outbreeding ancestors to autogamously selfing lineages has been derived many times during plant evolution and, in most cases, it is accompanied by a characteristic set of morphological and functional changes to the flowers, together termed the selfing syndrome (Sicard and Lenhard, 2011). The ability of plants to evolve new mating behaviors is a notable feature of plant evolution (Chen et al., 2007).

Legumes possess flowers perfectly capable of outcrossing (Suso et al., 2015). According to Palmer et al. (2009), considering that pollinators respond to differences in discovery, attractiveness and reward, outcrossing could be increased through selection for floral characteristics that enhance its frequency and efficiency. A basic understanding of the variety of mechanisms by which flower morphology and structure influence pollinators' behavior and consequently pollen transfer, should allow plant breeders to manipulate the PPI in ways that increase insect-mediated outcrossing. Bailes et al. (2015), reviewing key topics in plant-pollinator research, also remark that floral traits are a promising and little-explored avenue for the improvement of crop yields.

Regarding the pollination crisis, breeders should confront the decline of pollinators. One tendency in plant breeding is to opt for plants that do not need pollinators (Partap and Ya, 2012; Chaplin-Kramer et al., 2014). However, yield barriers must be broken. Breaking yield barriers could be achieved by increasing potential yield by improvements in the efficiency of heterosis exploitation (Tester and Langridge, 2010). The clear message is that insect-aided cross-pollination is a key factor in the development of hybrid seed technology and pollinators are essential in providing this service. Changes from a crop that does not require cross-pollination, to crops that respond optimally to the presence of pollen vectors is the key. The proposed breeding strategy is to combine highly self-fertile genotypes which respond optimally to the presence of pollinators to produce high-yielding cultivars. Thus, insect-aided outcrossing allows the exploitation of heterosis potential but, in the absence of pollinators, a minimum yield is achieved (Suso et al., 1996). A delayed selfing mechanism provides reproductive assurance while allowing a high level of outcrossing when pollinators are not a limiting factor (Shirk and Hamrick, 2014).

Making the Most of the PPI

In order to maximize pollination success, the employment of new agronomic practices, and the maintenance of semi-natural habitats in the agricultural landscape are essential (Andersson et al., 2014; Nayak et al., 2015; referred to also in section Farming with Alternative Pollinators - An Approach to Gain a Mass-Basis for Pollinator Protection) and should be coupled with the breeding procedure. The CDS approach, considering the management of natural resources, pollen vectors and improving the genetic characteristics of crops for functional floral traits, may prove faster, and more effective. The increasing demands for environmentally friendly agricultural practices and the decline in pollinators have established a favorable context for the development of pollinator-friendly crops, while placing demands on plant breeders to develop cultivars that meet more complex objectives (Duc et al., 2014). But, unfortunately, the analysis of Duc et al. (2014), reflects a relatively small representation of scientific articles related to the hosting of pollinators. Additionally, in spite of the fact that hybrid breeding is considered a breeding technology to increase crop production in a changing world (Tester and Langridge, 2010), the option of considering hybrids and heterotic populations has received minor attention among the breeding approaches. Thus, substantial research is needed to understand better how plants with appropriate functional floral traits could enhance insect-mediated outcrossing and the conservation of pollinator populations, which would help solve practical issues in the development of hybrids and heterotic populations and mitigate pollination declines.

AUTHOR CONTRIBUTIONS

All authors contributed to manuscript writing and read and approved the final manuscript. However, regarding that the article include quite different contributions, aspects and disciplines, each author has led and participated mainly in one section, and is going to take responsibility of the particular section. Thus, the individual contributions of each author are being specified. MJS conceived and design the paper and carried out the Section "Basic Application-Oriented Questions in Heterosis Breeding: The Potential of Insect-Mediated Outcrossing in Grain Legumes". and Crop Design System. PB and MC carried out the Section "Exploring the Role of Plant–Pollinator Interplay in Grain Legume Landrace Germplasm Conservation and Management". VN and RT carried out the Section "Developing Strategies for Seed Production and Seed Producers in Forage Legumes". CM and MV carried out Section "Protecting the Pollination Service from Pesticides". SC carried out section Farming with Alternative Pollinators (FAP) - An Approach to Gain a Mass-Basis for Pollinator Protection.

ACKNOWLEDGMENTS

MJS gratefully acknowledge the EU-FP7-KBBE-2009-3 project SOLIBAM (Strategies for Organic and Low-input Integrated Breeding and Management) and the H2020-SFS-2014-2 project DIVERSIFOOD (Embedding crop diversity and networking for local high quality food systems- GA number: 633571) for financial support. MJS, PB, MC, and RT wish to acknowledge the support from the COST Action FA1307 Super-B (www.superb-project.eu), supported by COST (European Cooperation in Science and Technology). SC wishes to acknowledge the support from BMUB, funding the pilot research in Uzbekistan within the International Climate Change Initiative.

REFERENCES

Allard, R. W. (1999). History of plant population genetics. *Annu. Rev. Genet.* 33, 1–27. doi: 10.1146/annurev.genet.33.1.1

Almekinders, C., and de Boef, W. (1999). The challenge of collaboration in the management of crop genetic diversity. *ILEIA Newsl.* 3-4, 5–7.

Andersson, G. K. S., Ekroos, J., Stjernman, M., Rundlöf, M., and Smith, H. G. (2014). Effects of farming intensity, crop rotation and landscape heterogeneity on field bean pollination. *Agric. Ecosytst. Environ.* 184, 145–148. doi: 10.1016/j.agee.2013.12.002

Aouar-Sadli, M., Louadi, K., and Doumandji, S. E. (2008). Pollination of the broad bean (*Vicia faba* L. var. major) (Fabaceae) by wild bees and honey bees (Hymenoptera: Apoidea) and its impact on the seed production in the Tizi-Ouzou area (Algeria). *Afr. J. Agric. Res.* 3, 266–272.

Arlauskiene, A., Kadziuliene, Z., Sarunaite, L., Deveikyte, I., and Semaskiene, R. (2015). "A contribution of legumes to diversity of crops in organic farming," in *Proceedings of the Eucarpia International Symposium on Protein Crops*, Pontevedra, 130.

Bailes, E. J., Ollerton, J., Pattrick, J. G., and Glover, B. J. (2015). How can an understanding of plant–pollinator interactions contribute to global food security? *Curr. Opin. Plant Biol.* 26, 72–79. doi: 10.1016/j.pbi.2015.06.002

Barnaud, A., Trigueros, G., McKey, D., and Joly, H. I. (2008). High outcrossing rates in fields with mixed sorghum landraces: how are landraces maintained? *Heredity* 101, 445–452. doi: 10.1038/hdy.2008.77

Barnes, D. K. (1966). Flower color inheritance in diploid and tetraploid alfalfa: a reevaluation, in U.S. *Dept. Agric. Tech. Bul.* 1353, 3–28.

Biesmeijer, J. C., Roberts, S. P., M., Reemer, M., Ohlemüller, R., Edwards, M., et al. (2006). Parallel declines in pollinators and insect-pollinated plants in Britain and the Netherlands. *Science* 313, 351–354. doi: 10.1126/science.112786

Bolaños-Aguilar, E. D., Huyghe, C., Ecalle, C., Hacquet, J., and Julier, B. (2002). Effect of cultivar and environment on seed yield in alfalfa. *Crop Sci.* 42, 45–50. doi: 10.2135/cropsci2002.0045

Boller, E. F., Häni, F., and Poehling, H.-M. (2004). *Ecological Infrastructures: Ideabook on Functional Biodiversity at the Farm Level - Temperate Zones of Europe. IOBCwprs Commision Integrated Production Guidelines Endorsment*, Eschikon: Swiss Centre for Agricultural Extension and Rural Development, 212.

Bond, D., and Kirby, E. J. M. (1999). Anthophora plumipes (Hymenoptera: Anthophoridae) as a pollinator of broad bean (*Vicia faba* major). *J. Apicul. Res.* 38, 199–203. doi: 10.1080/00218839.1999.11101010

Bond, D. A., and Kirby, E. J. M. (2001). Further observations of Anthophora plumipes visiting autumn-sown broad bean (*Vicia faba* major) in the United Kingdom. *J. Apicult. Res.* 40, 113–114.

Boutin, C., Strandberg, B., Carpenter, D., Mathiassen, S. K., and Thomas, P. J. (2014). Herbicide impact on non-target plant reproduction: what are the toxicological and ecological implications? *Environ. Pollut.* 185, 295–306. doi: 10.1016/j.envpol.2013.10.009

Brittain, C., and Potts, S. G. (2011). The potential impacts of insecticides on the life-history traits of bees and the consequences for pollination. *Basic Appl. Ecol.* 12, 321–331. doi: 10.1016/j.baae.2010.12.004

Brown, R. J., Döring, T. F., Storkey, J., Smith, J., Jones, H. E., and Potts, S. G. (2012). "Legume based plant mixtures for delivery of ecosystem services: pollinators," in *Proceedings of the SAC and SEPA Biennial Conference, Agriculture and the Environment IX, Valuing Ecosystems: Policy, Economic and Management Interactions*, ed. K. McCracken (Edinburgh: SAC Environmental), 156–162.

Broughton, W. J., Hernández, G., Blair, M., Beebe, S., Gepts, P., and Vanderleyden, J. (2003). Beans (*Phaseolus* spp.) – model food legumes. *Plant Soil* 252, 55–128. doi: 10.1023/A:1024146710611

Brush, S. B. (1999). *Genes in the Field. On-Farm Conservation of Crop Diversity.* Boca Raton, FL: Lewis Publishers. doi: 10.1201/9781420049824

Burbidge, A. H., and James, S. H. (1991). Post-zygotic seed abortion in the genetic system of Stylidium (Angiospermae, Stylidiaceae). *J. Hered.* 82, 319–328.

Carré, G., Roche, P., Chifflet, R., Morison, N., Bommarco, R., Harrison-Cripps, J., et al. (2009). Landscape context and habitat type as drivers of bee diversity in European annual crops. *Agric. Ecosyst. Environ.* 133, 40–47. doi: 10.1016/j.agee.2009.05.001

Ceccarelli, S. (1986). "Problems of seed production in forage crops," in *Seed Production Technologyed*, eds J. P. Srivastava and L.T. Simarski (Aleppo: International Center for Agricultural Research in the Dry Areas).

Chagnon, M., Kreutzweiser, D., Mitchell, E. A. D., Morrissey, C. A., Noome, D. A., and Van der Sluijs, J. P. (2015). Risks of large-scale use of systemic insecticides to ecosystem functioning and services. *Environ. Sci. Pollut. Res.* 22, 119–134. doi: 10.1007/s11356-014-3277-x

Chaplin-Kramer, R., Dombeck, E., Gerber, J., Knuth, K. A., Mueller, N. D., and Ziv, G., et al. (2014). Global malnutrition overlaps with pollinator-dependent micronutrient production. *Proc. R. Soc. B* 281:20141799. doi: 10.1098/rspb.2014.1799

Chen, K. Y., Cong, B., Wing, R., Vrebalov, J., and Tanksley, S. D. (2007). Changes in regulation of a transcription factor lead to autogamy in cultivated tomatoes. *Science* 318, 643–645. doi: 10.1126/science.1148428

Christmann, S., and Aw-Hassan, A. A. (2012). Farming with alternative pollinators (FAP) –an overlooked win-win-strategy for climate change adaptation. *Agric. Ecosyt. Environ.* 161, 161–164. doi: 10.1016/j.agee.2012.07.030

Christmann, S., Aw-Hassan, A., Rajabov, T., Khamraev, A. S., Symank, A., Ahmedov, A., et al. (2014). *Farming with Alternative Pollinators (FAP) – A Practical Guide for Smallholders,* Technical Report, Tashkent.

Clifford, P. T. P., and McCartin, S. J. M. (1985). Effects of pre-harvest treatment and mower and header types on seed loss and hard seed content at mowing, recovery and separation when harvesting a white clover seed crop. *N. Z. J. Exp. Agric.* 13, 307–316. doi: 10.1080/03015521.1985.10426099

Clifford, P. T. P., and Scott, D. (1989). Inflorescence, bumble bee, and climate interactions in seed crops of a tetraploid red clover (*Trifolium pratense* L.). *J. Appl. Seed Prod.* 7, 38–45.

Coors, J. G., and Pandey, S. (1999). *The Genetics and Exploitation of Heterosis in Crops.* Mexico: CIMMYT.

Cordoba, S. A., and Cocucci, A. A. (2011). Flower power: its association with bee power and floral functional morphology in papilionate legumes. *Ann. Bot.*108, 919–931. doi: 10.1093/aob/mcr196

Cost Action FA1307 (2014). *Sustainable Pollination in Europe: Joint Research on Bees and Other Pollinators (SUPER-B).* Available at: http://www.cost.eu/domains_actions/fa/Actions/FA1307 [Accessed August 11, 2015].

DeMooy, B. E., deMooy, C. J., and Burke, D. W. (1990). "Estimation of percentage natural outcrossing in cowpea (*Vigna unguiculata* L.) Walp in Botswana," in *Proceedings of the Reports of Bean Improvement Cooperative and National Dry Bean Council Research Conference, Annual Report*, Botswana, 122. Available at: http://nald c.nal.usda.gov/catalog/IND92013799

Delaplane, K. S., and Mayer, D. F. (2000). *Crop Pollination By Bees.* New York, NY: CABI Publishing. doi: 10.1079/9780851994482.0000

Desneux, N., Decourtye, A., and Delpuech, J. M. (2007). The sublethal effects of pesticides on beneficial arthropods. *Annu. Rev. Entomol.* 52, 81–106. doi: 10.1146/annurev.ento.52.110405.091440

Döring, T. F., Storkey, J., Baddeley, J. A., Crowley, O., Howlett, S. A., McCalman, H., et al. (2012). "Legume based plant mixtures for delivery of multiple ecosystem services: weed diversity and weed control," in *Proceedings of the SAC and SEPA Biennial Conference, Agriculture and the Environment IX, Valuing Ecosystems: Policy, Economic and Management Interactions*, ed. K. McCracken (Edinburgh: SAC Environmental), 163–168.

Duc, G., Agrama, H., Bao, S., Berger, J., Bourion, V., De Ron, A. M., et al. (2014). Breeding annual grain legumes for sustainable agriculture: new methods to approach complex traits and target new cultivar ideotypes. *Crit. Rev. Plant Sci.* 34, 381–411. doi: 10.1080/07352689.2014.898469

Duc, G., Bao, S., Baumc, M., Redden, B., Sadiki, M., Suso, M. J., et al. (2010). Diversity maintenance and use of *Vicia faba* L. genetic resources. *Field Crop. Res.* 115, 270–278. doi: 10.1016/j.fcr.2008.10.003

Dumet, D., Adeleke, R., and Faloye, B. (2008). "Regeneration guidelines: cowpea," in *Crop specific Regeneration Guidelines [CDROM]*, eds M. E. Dulloo, I. Thormann, M. A. Jorge, and Hanson, J. (Rome: Hanson Rome in CGIAR System-wide Genetic Resource Programme), 8.

EASAC-The European Academies' Science Advisory Council (2015). *Ecosystem Services, Agriculture and Neonicotinoids.* EASAC policy report 26. Available at: www.easac.eu

El-Wakeil, N. E., and El-Sebai, T. N. E. (2009). Role of biofertilizer on faba bean growth, yield, and its effect on bean aphid and the associated predators. *Arch. Phytopathol. Plant Protect.* 42, 1144–1153. doi: 10.1080/032354007016 50882

Fairey, D. T., and Hampton, J. G. (1998). *Forage Deed Production: Temperate Species*, Vol. 1. New York, NY: CAB International.

Falcinelli, M. (1999). Temperate forage seed production: conventional and potential breeding strategies. *J. New Seed* 1, 37–66. doi: 10.1300/J153v01n01_04

Feltham, H., Park, K., and Goulson, D. (2014). Field realistic doses of pesticide imidacloprid reduce bumblebee pollen foraging efficiency. *Ecotoxicology* 23, 317–323. doi: 10.1007/s10646-014-1189-7

Free, J. B. (1993). *Insect Pollination of Crops.* New York, NY: Academic Press.

Freitas, G., Ganança, J. F. T., Nóbrega, H., Nunes, E., Costa, G., Slaski, J. J., et al. (2011). Morphological evaluation of common bean (*Phaseolus vulgaris* L.) diversity on the Island of Madeira. *Genet. Resour. Crop Evol.* 58, 861–874. doi: 10.1007/s10722-010-9624-y

Fu, D., Xiao, M., Hayward, A., Fu, Y., Liu, G., Jiang, G., et al. (2014). Utilization of crop heterosis: a review. *Euphytica* 197, 161–173. doi: 10.1007/s10681-014-1103-7

Fuchs, S. J., and Hirnyck, R. E. (2000). *Crop Profile for Lentil in Idaho State.* Available at: http://www.ipmcenters.org/cropprofiles/docs/IDlentils.pdf [accessed May 2015].

Garibaldi, L. A., Bartomeus, I., Bommarco, R., Klein, A., Cunningham, S. A., and Aizen, M. A., et al. (2015). Trait matching of flower visitors and crops predicts fruit set better than trait diversity. *J. Appl. Ecol.* 52, 1436–1444. doi: 10.1111/1365-2664.12530

Garibaldi, L. A., Carvalheiro, L. G., Leonhardt, S. D., Aizen, M. A., Blaauw, B. R., Isaacs, R., et al. (2014). From research to action: enhancing crop yield through wild pollinators. *Front. Ecol. Environ.* 12, 439–447. doi: 10.1890/ 130330

Gasim, S., and Link, W. (2007). Agronomic performance and the effect of self-fertilization on German winter faba beans. *J. Cent. Eur. Agric.* 8, 121–128.

Genesys (2015). *Gateway to Plant Genetic Resources.* Available at: https://www.genesys-pgr.org/ (accessed July, 2015).

Gnanasambandam, A., Paull, J., Torres, A., Kaur, S., Leonforte, T., Li, H., et al. (2012). Impact of molecular technologies on faba bean (*Vicia faba* L.). *Breed. Strategies Agron.* 2, 132–166.

Godfray, H. C. J., Blacquière, T., Field, L. M., Hails, R. S., Petrokofsky, G., Potts, S. G., et al. (2014). A restatement of the natural science evidence base concerning neonicotinoid insecticides and insect pollinators. *Proc. R. Soc. B* 281:20140558. doi: 10.1098/rspb.2014.0558

Goldringer, I., Dawson, J., Rey, F., Vettoretti, A., Chable, V., Lammerts van Bueren, E., et al. (2010). "Breeding for resilience: a strategy for organic and low-input farming systems?," in *Proceedings of the EUCARPIA 2nd Conference of the Organic and Low- Input Agriculture Section*, Paris.

Gómez, O. (2004). *Evaluation of Nicaraguan Common Bean (Phaseolus vulgaris L.) landraces. Doctor's dissertation.* Acta Universitatis agriculturae Sueciae, Agraria.

Goulson, D., Nicholls, E., Botías, C., and Rotheray, E. L. (2015). Bee declines driven by combined stress from parasites, pesticides, and Lack of flowers. *Science* 347:6229. doi: 10.1126/science.1255957

Gradish, A. E., Scott-Dupree, C. D., Shipp, L., Harris, C. R., and Ferguson, G. (2010). Effect of reduced risk pesticides for use in greenhouse vegetable production on *Bombus* impatiens (Hymenoptera: Apidae). *Pest Manag. Sci.* 66, 142–146. doi: 10.1002/ps.1846

Gregorc, A., and Ellis, J. D. (2011). Cell death localization in situ in laboratory reared honey bee (*Apis mellifera* L.) larvae treated with pesticides. *Pestic Biochem. Phys.* 99, 200–207. doi: 10.1016/j.pestbp.2010.12.005

Gupta, S. C., Reddy, L. J., Sharma, D., Green, J. M., Murthi, A. N., and Saxena, K. B. (1981). "Maintenance of pigeonpea cultivars", in *Proceedings of the International Workshop on Pigeonpeas*, Vol. 1 (Patancheru: ICRISAT), 295–302.

Hammer, K. (1984). Das domestications syndrom. *Kultupfanze* 11, 11–34.

Hampton, J. G. (1990). Genetic variability and climatic factors affecting herbage legume seed production: an introduction. *J. Appl. Seed Prod.* 8, 45–51.

Helenius, J., and Stoddard, F. L. (2007). "Agro-ecosystem services from increased usage and novel applications of legumes. Integrating legume biology for sustainable agriculture," *Sixth European Conference on Grain Legumes*, Lisbon: Book of Abstracts.

Henry, M., Béguin, M., Requier, F., Rollin, O., Odoux, J. F., Aupinel, P., et al. (2012). A common pesticide decreases foraging success and survival in honey bees. *Science* 336, 348–350. doi: 10.1126/science.1215039

Holt, J., and Wratten, S. D. (1986). Components of resistance to *Aphis fabae* in faba bean. *Entomol. Exp. Appl.* 40, 35–40. doi: 10.1111/j.1570-7458.1986.tb02152.x

Horneburg, B. (2006). Outcrossing in lentil (*Lens culinaris*) depends on cultivar, location and year, and varies within cultivars. *Plant Breed.* 125, 638–640. doi: 10.1111/j.1439-0523.2006.01290.x

IPBES (2013). *Report of the Second Session of the Plenary of the Intergovernmental Science-Policy Platform on Biodiversity and Ecosystem Services.* Nairobi: United Nations Environment Program. Available at: http://www.ipbes.net [accessed August 11, 2015].

Jensen, E. S., Peoples, M. B., and Hauggaard-Nielsen, H. (2010). Faba bean in cropping systems. *Field Crop. Res.* 115, 203–216. doi: 10.1016/j.fcr.2009.10.008

Kasina, J. M., Hagen, M. L., Kraemer, M., Nderitu, J., Martius, C., and Wittmann, D. (2009a). Bee pollination enhances crop yield and fruit quality in Kakamega, western Kenya. *E. Afr. Agric. For. J.* 75, 1–11.

Kasina, J. M., Mburu, J., Kraemer, M., and Holm-Mueller, K. (2009b). Economic benefit of crop pollination by bees: a case of kakamega small-holder farming in western Kenya. *J. Eco. Entom.* 102, 467–473. doi: 10.1603/029.102.0201

Katznelson, J., and Morley, F. H. W. (1965). A taxonomic revison of sect. *Calycomorphum* of the genus *Trifolium*. I. The geocarpic species. *Isr. J. Bot.* 14, 112–134.

Kauffeld, N. M., and Sorensen, E. I. (1971). Interrelation of honey bee preference of alfalfa clones and flower color, aroma, nectar volume, and sugar concentration. *Kansas Agric. Exp. Stn. Res. Publ. No* 163, 1–14.

Kharrat, M., and Ouchari, H. (2011). Faba bean status and prospects in Tunisia. *Grain Legumes* 56, 11–12.

Klatt, B. K., Burmeister, C., Westphal, C., Tscharntke, T., and von Fragstein, M. (2013). Flower volatiles, crop varieties and bee responses. *PLoS ONE* 8:e72724. doi: 10.1371/journal.pone.0072724

Klein, A.-M., Vaissiere, B. E., Cane, J. H., Steffan-Dewenter, I., Cunningham, S. A., Kremen, C., et al. (2007). Importance of pollinators in changing landscapes for world crops. *Proc. R. Soc. B.* 274, 303–313. doi: 10.1098/rspb.2006.3721

Kobayashi, K., Tsukamoto, S., Tanaka, A., Nikura, S., and Ohsawa, R. (2010). Selective flower visitation behavior by pollinators in a radish F1 seed production field. *Breed. Sci.* 60, 203–211. doi: 10.1270/jsbbs.60.203

Kohler, F., Verhulst, J., Klink, R., and Kleijn, D. (2008). At what spatial scale do high-quality habitats enhance the diversity of forbs and pollinators in intensively farmed landscapes? *J. Appl. Ecol* 45, 753–762. doi: 10.1111/j.1365-2664.2007.01394.x

Köpke, U., and Nemecek, T. (2010). Ecological services of faba bean. *Field Crop. Res.* 115, 217–233. doi: 10.1016/j.fcr.2009.10.012

Kremen, C., Williams, N. M., Aizen, M. A., Gemmill-Herren, B., LeBuhn, G., and Minckley, R. et al. (2007). Pollination and other ecosystem services produced by mobile organisms: a conceptual framework for the effects of land-use-change. *Ecol. Lett.* 10, 299–314. doi: 10.1111/j.1461-0248.2007.01018.x

Landis, D. A., Wratten, S. D., and Gurr, G. M. (2000). Habitat management to conserve natural enemies of arthropod pests in agriculture. *Annu. Rev. Entomol.* 45, 175–201. doi: 10.1146/annurev.ento.45.1.175

Lesins, K. A., and Lesins, I. (1979). *Genus Medicago (Leguminosae): A Taxogenetic Study*. Dordrecht: Kluwer, 228. doi: 10.1007/978-94-009-9634-2

Lever, J. J., van Nees, E. H., Scheffer, M., and Bascompte, J. (2014). The sudden collapse of pollinator communities. *Ecol. Lett.* 17, 350–359. doi: 10.1111/ele.12236

Lloyd, D. G., and Schoen, D. J. (1992). Self- and cross-fertilization in plants. I. Functional dimensions. *Int. J. Plant Sci.* 153, 358–369. doi: 10.1086/297040

Longin, C. F. H., Mühleisen, J., Maurer, H. P., Zhang, H., Gowda, M., and Reif, J. C. (2012). Hybrid breeding in autogamous cereals. *Theor. Appl. Genet.* 125, 1087–1096. doi: 10.1007/s00122-012-1967-7

Loper, G. M., and Lapioli, A. M. (1971). Photoperiodic effects on the emanation of volatiles from alfalfa (*Medicago sativa* L.) florets. *Plant Physiol.* 49, 729–732. doi: 10.1104/pp.49.5.729

Loper, G. M., and Waller, G. D. (1970). Alfalfa flower aroma and flower selection by honey bees. *Crop Sci.* 10, 66–68. doi: 10.2135/cropsci1970.0011183X001000010026x

Loper, G. M., Waller, G. D., and Berdel, R. L. (1974). Olfactory screening of alfalfa clones for uniform honey bee selection. *Crop Sci.* 14, 120–122. doi: 10.(2135)/cropsci1974.0011183X001400010037x

Loper, G. M., and Berdel, R. L. (1978). Seasonal emanation of ocimene from alfalfa flowers with three irrigation treatments. *Crop Sci.* 18, 447–452. doi: 10.2135/cropsci1978.0011183X001800030022x

Lorenzetti, F. (1981). "Relationship between dry matter and seed yield in Leguminous forage plants," in *Proceedings of the Rep. Eucarpia Fodder Crops Section Meeting*, ed. Van Bogaert (Merebelke: Rijksstation voor Plantenvereding), 57–74.

Lorenzetti, F. (1993). "Achieving potential herbage seed yields in species of temperate regions," in *Proceedings of the XVII International Grassland Congress*, eds M. J. Baker, J. R. Crush, and L. R. Humphreys (Palmerston North: Keeling & Mundy LTD), 1621–1628.

Lorenzetti, F., and Negri V. (1995). Produzione potenziale e produzione reale delle colture da seme di leguminose foraggere. *Riv. Agro. Anno* 29, 242–259.

MA (2005). *Ecosystems and Human Well-Being: Current State and Trends*, Vol. 1, eds R. Hassan, R. Scholes, and N. Ash. Washington, DC: Island Press.

Maalouf, F. S., Suso, M. J., and Moreno, M. T. (1999). Choice of methods and indices for identifying the best parentals for synthetic varieties in faba bean. *Agronomie* 19, 705–712. doi: 10.1051/agro:19990805

Macfadyen, S., Davies, A. P., and Zalucki, M. P. (2015). Assessing the impact of arthropod natural enemies on crop pests at the field scale. *Insect Sci.* 22, 20–34. doi: 10.1111/1744-7917.12174

Maggioni, L., Ambrose, M., Schachl, R., Duc, G., and Lipman, E. (2002). *Report of a Working Group on Grain Legumes.* Rome: International Plant Genetic Resources Institute, 149.

Martin, B., and Adams M. W. (1987). Landraces of *Phaseolus vulgaris* (Fabaceae) in Northern Malawi a: legional variation. *Econ. Bot.* 41, 190–203. doi: 10.1007/BF02858965

Martiniello, P. (1998). Influence of agronomic factors on the relationship between forage production and seed yield in perennial forage grasses and legumes in a Mediterranean environment. *Agronomie* 18, 591–601. doi: 10.1051/agro:19980809

Martins, D. J. (2013). "People, plants and pollinators: uniting conservation, food security, and sustainable agriculture in east of Africa," in *Conservation Biology: Voices from the Tropics*, eds N. S. Sodhi, L. Gibson, and P. H. Raven (Hoboken, NJ: John Wiley & Sons, Ltd.), 232–238.

McCouch, S. R., Kovach, M. J., Sweeney, M., Jiang, H., and Semon, M. (2012). "The dynamics of rice domestication: a balance between gene flow and genetic isolation," in *Biodiversity in Agriculture: Domestication, Evolution, and Sustainability*, eds P. Gepts, T. R. Famula, R. L. Bettinger, S. B. Brush, A. B. Damania, P. E. McGuire, et al. (Cambridge: Cambridge University Press), 311-329. doi: 10.1017/CBO9781139019514.018

Milfont, M., Rocha, E. E. M., Lima, A. O. N., and Freitas, B. M. (2013). Higher soybean production using honeybee and wild pollinators, a sustainable alternative to pesticides and autopollination. *Environ. Chem. Lett.* 11, 335–341. doi: 10.1007/s10311-013-0412-8

Mommaerts, V., Reynders, S., Boulet, J., Besard, L., Sterk, G., and Smagghe, G. (2010). Risk assessment for side-effects of neonicotinoids against bumblebees with and without impairing foraging behavior. *Ecotoxicology* 19, 207–215. doi: 10.1007/s10646-009-0406-2

Monasterolo, M., Musicante, M. L., Valladares, G. R., and Salvo, A. (2015). Soybean crops may benefit from forest pollinators. *Agric. Ecosyst. Environ.* 202, 217–222. doi: 10.1016/j.agee.2015.01.012

Musallam, I. W., Haddad, N. J., Tawaha, A.-R. M., and Migdadi, O. S. (2004). The importance of bee-pollination in four genotypes of faba bean (*Vicia faba* L.). *Int. J. Agricult. Biol.* 6, 9–12.

Nabhan, G. P., and Buchmann, S. L. (1997). "Services provided by pollinators," in *Nature's Services. Societal Dependence on Natural Ecosystems*, ed. G. C. Daily (Washington, DC: Island Press), 133–150.

Nadal, S., Suso, M. J., and Moreno, M. (2003). Management of *Vicia faba* genetic resources: changes associated to the selfi ng process in the major, equina and minor groups. *Gen. Resour. Crop Evol.* 50, 183–192. doi: 10.1023/A:1022944017530

Nayak, G. K., Roberts, S. P. M., Garratt, M., Breeze, T. D., Tscheulin, T., Harrison-Cripps, J., et al. (2015). Interactive effect of floral abundance and semi-natural habitats on pollinators in field beans (*Vicia faba*). *Agric. Ecosyst. Environ.* 199, 58–66. doi: 10.1016/j.agee.2014.08.016

Negri, V., Maxted, N., and Veterlainen, M. (2009). "European landrace conservation: an introduction," in *'European Landraces: On-farm Conservation, Management and Use'. Bioversity Technical Bulletin No. 15, Bioversity International*, eds M. Vetelainen, V. Negri, and N. Maxted (Italy: Bioversity International publisher), 1–22.

Negri, V., Pacicco, L., Bodesmo, M., and Torricelli, R. (2013). *The First Italian Inventory of In Situ Maintained Landraces*. Perugia: Morlacchi Editore.

Nieto, A., Roberts, S. P. M., Kemp, J., Rasmont, P., Kuhlmann, M., García Criado, M., et al. (2014). *European Red List of bees*. Luxembourg: Publication Office of the European Union.

Nuessly, G. S., Hentz, M. G., Beiriger, R., and Scully, B. T. (2004). Insects associated with Faba bean, *Vicia faba* (Fabales: Fabaceae), in southern Florida. *Fla. Entomol.* 87, 204–211. doi: 10.1653/0015-4040(2004)087[0204:IAWFBV]2.0.CO;2

Ollerton, J., Winfree, R., and Tarrant, S. (2011). How many flowering plants are pollinated by animals? *Oikos* 120, 321–326. doi: 10.1111/j. 1600-0706.2010.18644x

Ortiz-Perez, E., Horner, H. T., Hanlin, S. J., and Palmer, R. G. (2006a). Evaluation of insect-mediated seed set among soybean lines segregating for male sterility at the ms6 locus. *Field Crop. Res.* 97, 353–362. doi: 10.1016/j.fcr.2005.11.007

Ortiz-Perez, E., Horner, H. T., Hanlin, S. J., and Palmer, R. G. (2006b). Insect-mediated seed-set evaluation of 21 soybean lines segregating for male sterility at 10 different loci. *Euphytica* 152, 351–360. doi: 10.1007/s10681-006-9222-4

Ortiz-Perez, E., Cianzio, S. R., Wiley, H., Horner, H. T., Davis, W. H., and Palmer, R. G. (2007). Insect-mediated cross-pollination in soybean [*Glycine max* (L.) Merrill] I. Agronomic performance. *Field Crop. Res.* 101, 259–268. doi: 10.1016/j.fcr.2006.12.003

Ortiz-Perez, E., Wiley, H., Horner, H. T., Davis, W. H., and Palmer, R. G. (2008). Insect-mediated cross-pollination in soybean [*Glycine max* (L.) Merrill]: II. Phenotypic recurrent selection. *Euphytica* 162, 269–280. doi: 10.1007/s10681-007-9612-2

Palmer, R. G., Gai, J., Sun, H., and Burton, J. W. (2001). Production and evaluation of Hybrid Soybean. *Plant Breed. Rev.* 21, 263–307. doi: 10.1002/9780470650196.ch7

Palmer, R. G., Perez, P. T., Ortiz-Perez, E., Maalouf, F., and Suso, M. J. (2009). The role of crop-pollinator relationships in breeding for pollinator-friendly legumes: from a breeding perspective. *Euphytica* 170, 35–52. doi: 10.1007/s10681-009-9953-0

Palmer, R. G., Gai, J., Dalvi, V. A., and Suso, M. J. (2011). "Male Sterility and Hybrid Production Technology," in *Biology and Breeding of Food Legumes*, eds A. Pratap and J. Kumar (Oxfordshire: CABI Publishers), 193–207. doi: 10.1079/9781845937669.0193

Palmer, R. G., Horner, H. T., Suso, M. J., Vangorp, M. L., and Pappas, A. L. (2012). "Soybean heterosis: the first steps," in *Proceedings of the International Conference on Utilization of Heterosis in Crops (ICUHC). Toward a New Era of Crop Heterosis* (Xi'an: Abstracts).

Papa, R., and Gepts, P. (2004). *Gene Flow Between Crops and Their Wild Progenitors. Encyclopedia of Plant and Crop Science*. New York, NY: Marcel Dekker, Inc. 488–491.

Pappas, A. L., Koziel, J. A., Healy, R. A., Horner, H. T., Bailey, T. B., Brown, A. H. D., et al. (2012). "Floral and aromatic analysis of the soybean perennial Glycine syndetika for future agronomic exploitation," in *Proceedings of the XXII International Congress on Sexual Plant Reproduction, Melbourne*, 13.

Partap, U., and Ya, T. (2012). The human pollinators of fruit crops in Maoxian County, Sichuan, China. *Mt. Res. Dev.* 32, 176–186. doi: 10.1659/MRD-JOURNAL-D-11-00108.1

Pasumarty, S. V., Matsumura, T., Higuchi, S., and Yamada, T. (1993). Causes of low seed set in white clover (*Trifolium repens* L.). *Grass Forage Sci.* 48, 79–83. doi: 10.1111/j.1365-2494.1993.tb01839.x

Pecetti, L., and Tava, A. (2000). Effect of flower color and sampling time on volatile emanation in alfalfa flowers. *Crop Sci.* 40, 126–130. doi: 10.2135/cropsci2000.401126x

Pecetti, L., Tava, A., Felicioni, A., Pinzauti, M., and Piano, E. (2002). Effect of three volatile compounds from lucerne flowers on their attractiveness towards pollinators. *Bull. Insectol.* 55, 21–27.

Pecetti, L., and Proietti, S. (2003). Estimation of cross-pollination by leafcutter bees in two lucerne genotypes with different flower colour. *J. Genet. Breed.* 57, 219–224.

Pérez-De-Luque, A., Eizenberg, H., Grenz, J. H., Sillero, J. C., Ávila, C., Sauerborn, J., et al. (2010). Broomrape management in faba bean. *Field Crop. Res.* 115, 319–328. doi: 10.1016/j.fcr.2009.02.013

Pettis, J. S., Engelsdorp, D., van, Johnson, J., and Dively, G. (2012). Pesticide exposure in honey bees results in increased levels of the gut pathogen Nosema. *Naturwissenschaften* 99, 153–158. doi: 10.1007/s00114-011-0881-1

Piano, E., and Pecetti, L. (2010). "Minor legume species," in *Fodder Crops and Amenity Grasses, Handbook of Plant Breeding*, Vol. 5, eds B. Boller, U. K. Posselt, and F. Veronesi (New York, NY: Springer), 477–500. doi: 10.1007/978-1-4419-0760-8_20

Pierre, J., Suso, M. J., Moreno, M. T., Esnault, R., and Le Guen, J. (1999). Diversite et efficacite de l'entomofaune pollinisatrice (Hymenoptera: Apidae) de la feverole (*Vicia faba* L.) sur deux sites, en France et en Espagne. *Ann. Soc. Entomol.Fr.(N.S.)* 35(suppl.), 312–318.

Pinheiro de Carvalho, M. A. A., Bebeli, P. J., Barata da Silva, A. M., Bettencourt, E., Slaski, J. J., and Dias, S. (2016). "Agro-biodiversity. the importance of inventories in the assessment of crop diversity and its time and spatial changes," in *Genetic Erosion and Biodiversity*, eds M. R. Ahuja and S. M. Jain (Berlin: Springer), 130–147.

Pinheiro de Carvalho, M. A. A., Bebeli, P. J., Bettencourt, E., Costa, G., Dias, S., dos Santos, T. M. M., et al. (2013). Cereal landraces genetic resources in worldwide GeneBanks. *Agron. Sustain. Dev.* 33, 177–203. doi: 10.1007/s13593-012-0090-0

Pisa, L. W., Amaral-Rogers, V., Belzunces, L. P., Bonmatin, J. M., Downs, C. A., Goulson, D., et al. (2015). Effects of neonicotinoids and fipronil on nontarget invertebrates. *Environ. Sci. Pollut. Res.* 22, 68–102. doi: 10.1007/s11356-014-3471-x

Potts, S. G., Biesmeijer, J. C., Kremen, C., Neumann, P., Schweiger, O., and Kunin, W. E. (2010). Global pollinator declines: trends, impacts and drivers. *Trends Ecol. Evol.* 25, 345–353. doi: 10.1016/j.tree.2010.01.007

Purseglove (1968). *Tropical Crops: Dicotyledons*. London: Longman Group Limited, 246–250.

Rahman, M. A., Rahman, M. M., and Sarkar, M. A. (2001). Progress in isolation and purification of Lathyrus sativus breeding lines *Lathyrus. Lathyrism Newsl.* 2, 39–40.

Rebek, E. J., Sadof, C. S., and Hanks, L. M. (2005). Manipulating the abundance of natural enemies in ornamental landscapes with floral resource plants. *Biol. Control* 33, 203–216. doi: 10.1016/j.biocontrol.2005.02.011

Redden, B., Leonforte, T., Ford, R., Croser, J., and Slattery, J. (2005). "Pea (*Pisum sativum*)," in *Genetic Resources, Chromosome Engineering, and Crop Improvement*, eds R. Singh and P. Juahar (Boca Raton, FL: CRC Press), 49–83.

Richards, A. J. (2001). Does low biodiversity resulting from modern agricultural practice affect crop pollination and yield? *Ann. Bot.* 88, 165–172. doi: 10.1006/anbo.2001.1463

Rick, C. M. (1988). "Evolution of mating systems in cultivated plants," in *Plant Evolutionary Biology*, eds L. D. Gottlieb and S. Jain (London: Chapman & Hall), 133–147.

Ricketts, T. H., Regetz, J., Steffan-Dewenter, I., Cunningham, S. A., Kremen, C., Bogdanski, A., et al. (2008). Landscape effects on crop pollination services:

are there general patterns? *Ecol. Lett.* 11, 499–515. doi: 10.1111/j.1461-0248.2008.01157.x

Riedl, H., Johansen, E., Brewer, L., and Barbour, J. (2006). *How to Reduce Bee Poisoning From Pesticides, PNW 591.* Moscow, ID: University of Idaho, 24.

Rispail, N., Kalo, P., Kiss, G. B., Ellis, T. H. N., Gallardo, K., Richard, D., et al. (2010). Model legumes contribute to faba bean breeding. *Field Crop. Res.* 115, 253–269. doi: 10.1016/j.fcr.2009.03.014

Rosellini, D., Lorenzetti, F., and Bingham, E. T. (1998). Quantitative ovule sterility in *Medicago sativa. Theor. Appl. Genet.* 97, 1289–1295. doi: 10.1007/s001220051021

Rosellini, D., Veronesi, F., and Falcinelli, M. (1994). Recurrent selection of microgametophytic vigour in alfa-alfa and correlated response at the sporophytic level. *Crop Sci.* 34, 933–936. doi: 10.2135/cropsci1994.0011183X003400040017x

Rubio, J., Fernandez-Romero, M. D., Millán, T., Gil, J., and Suso, M. J. (2010). "Outcrossing rate and genetic structure on an open-flowering population of *Cicer arietinum* based on microsatellite markers," in *Proceedings of the 5th Lnternaional Food Legumes Research Conference on and 7the European Conference on Grain Legumes for Global Health Legume Crops and Products for Food, Feed and Environmental Benefits* (Antalyap: Book of Abstracts), 220.

Rundlöf, M., Andersson, G. K. S., Bommarco, R., Fries, I., Hederström, V., Herbertsson, L., et al. (2015). Seed coating with a neonicotinoid insecticide negatively affects wild bees. *Nature* 521, 77–80. doi: 10.1038/nature14420

Sanchez-Bayo, F., and Goka, K. (2014). Pesticide residues and bees—a risk assessment. *PLOS ONE* 9:e94482. doi: 10.1371/journal.pone.0094482

Satta, A., Acciaro, M., Floris, I., LentinI, A., and Sulas, L. (2000). "Insect pollination of sulla (*Hedysarum coronarium* L.) and its effect on seed production," in *Legumes for Mediterranean Forage Crops, Pastures and Alternative Uses*, ed. S. L. Zaragoza (Paris: CIHEAM), 373–377.

Saunders, M. E., Peisley, R. K., Rader, R., and Luck, G. W. (2015). Pollinators, pests, and predators: recognizing ecological trade-offs in agroecosystems. *Ambio* 45, 1–11. doi: 10.1007/s13280-015-0696-y

Saxena, K. B. (2005). "Pigeonpea [*Cajanus cajan* (L.) Millsp.]," in *Genetic Resources, Chromosome Engineering, and Crop Improvement: Grain Legumes*, Vol. 1, eds R. J. Singh and P. P. Jauhar (Boca Raton, CA: CRC Press), 85–110. doi: 10.1201/9780203489284.ch4

Saxena, K. B., and Kumar, R. V. (2010). Insect-aided natural out-crossing in four wild relativesof pigeonpea. *Euphytica* 173, 329–335. doi: 10.1007/s10681-009-0088-0

Saxena, K. B., Kumar, R. V., Tikle, A. N., Saxena, M. K., Gautam, V. S., Rao, S. K., et al. (2013). ICPH 2671 – the world's first commercial food legume hybrid. *Plant Breed.* 132, 479–485. doi: 10.1111/pbr.12045

Schellhorn, N. A., Bellati, J., Paull, C. A., and Maratos, L. (2008). Parasitoid and moth movement from refuge to crop. *Basic Appl. Ecol.* 9, 691–700. doi: 10.1016/j.baae.2007.12.010

Sengul, S. (2006). Using path analysis to determine Lucerne (*Medicago sativa* L.) seed yield and its components. *N. Z. J. Agric. Res.* 49, 107–115. doi: 10.1080/00288233.2006.9513700

Sewell, J., Hill, R., and Reich, J. (2011). Persistence of Grazing Tolerant Lucernes under Australian, Conditions. Pasture Persistence. *Grassland Res. Pract. Ser.* 15, 187–190.

Shackelford, G., Steward, P. R., Benton, T. G., Kunin, W. E., Potts, S. G., Biesmeijer, J. C., et al. (2013). Comparison of pollinators and natural enemies: a meta-analysis of landscape and local effects on abundance and richness in crops. *Biol. Rev. Camb. Philos. Soc.* 88, 1002–1021. doi: 10.1111/brv.12040

Sharma, H. C., Gowda, C. L. L., Stevenson, P. C., Ridsill-Smith, T., Clement, S. L., Rao, R., et al. (2007). "Host plant resistance and insect pest management in chickpea," in *Chickpea Breeding & Management*, eds S. S. Yadav, R. Redden, W. Chen, and B. Sharma (Oxfordshire: CABI International), 520–537.

Shirk, R. Y., and Hamrick, J. L. (2014). High but variable outcrossing rates in the invasive Geranium carolinianum (*Geraniaceae*). *Am. J. Bot.* 101, 1200–1206. doi: 10.3732/ajb.1400224

Sicard, A., and Lenhard, M. (2011). The selfing syndrome: a model for studying the genetic and evolutionary basis of morphological adaptation in plants. *Ann. Bot.* 107, 1433–1443. doi: 10.1093/aob/mcr023

Sigsgaard, L., Betzer, C., Naulin, C., Eilenberg, J., Enkegaard, A., and Kristensen, K. (2013). The effect of floral resources on parasitoid and host longevity: prospects for conservation biological control in strawberries. *J. Insect Sci.* 13, 1–7. doi: 10.1673/031.013.10401

Sillero, J. C.,Villegas-Fernández, A. M., Thomas, J., Rojas-Molina, M. M., Emerand, A. A., Fernández-Aparicio, M., et al. (2010). Faba bean breeding for disease resistance. *Field Crop. Res.* 115, 297–307. doi: 10.1016/j.fcr.2009.09.012

Singh, L., Gupta, S. C., and Farris, D. G. (1990). "Pigeonpea: breeding," in *The Pigeonpea*, ed. Y. L. Nene (Wallingford: CAB International), 375–399.

Singh, N. B., Singh, I. P., and Singh, B. B. (2005). "Pigeonpea breeding," in *Advances in Pigeonpea Research*, eds M. Ali and S. Kumar (Kanpur: Indian Institute of Pulses Research), 67–95.

Singh, N., Tyagi, R. K., and Pandey, C. (2013). *Genetic Resources of Pigeonpea (Cajanus cajan): Conservation for Use.* New Delhi: National Bureau of Plant Genetic Resources, 49p.

Smýkal, P., Coyne, C. J., Ambrose, M. J., Maxted, N., Schaefer, H., and Matthew, W., et al. (2015). Legume crops phylogeny and genetic diversity for science and breeding. *Crit. Rev. Plant Sci.* 34, 43–104. doi: 10.1080/07352689.2014.897904

Steffan-Dewenter, I., Munzenberg, U., Burger, C., Thies, C., and Tscharntke, T. (2002). Scale-dependent effects of landscape context on three pollinator guilds. *Ecology* 83, 1421–1432. doi: 10.1890/0012-9658(2002)083[1421:SDEOLC]2.0.CO;2

Steiner, J. J., Beuselinck, P. R., Peaden, R. N., Kojis, W. P., and Bingham, E. T. (1992). Pollinator effects on crossing and genetic shift in a three-flower-color alfalfa population. *Crop Sci.* 32, 73–77. doi: 10.2135/cropsci1992.0011183X003200010016x

Stevenson, P. C., Dhillon, M. K., Sharma, H. C., and El Bouhssini, M. (2007). "Insect pests of lentil and their management," in *Lentil: An Ancient Crop for Modern Times*, eds S. S. Yadav, D. L. McNeil, and P. C. Stevenson (Berlin: Springer), 331–348. doi: 10.1007/978-1-4020-6313-8_20

Stoddard, F. L., Nicholas, A. H., Rubiales, D., Thomas, J., and Villegas-Fernández, A. M. (2010). Integrated pest management in faba bean. *Field Crop. Res.* 115, 308–318. doi: 10.1016/j.fcr.2009.07.002

Stoddard, F. L. (2013). *Agronomic case studies in Legume Futures. Legume Futures Report 1.2.* Available at: http://www.legumefutures.de/images/Legume_Futures_Report_1.2.pdf

Street, K., Ismail, A., and Rukhkyan, N. (2008a). "Regeneration guidelines: faba bean," in *Crop Specific Regeneration Guidelines [CD-ROM]*, eds M. E. Dulloo, I. Thormann, M. A. Jorge, and J. Hanson (Rome: CGIAR System-wide Genetic Resource Programme), 9.

Street, K., Ismail, A. and Rukhkyan, N. (2008b). "Regeneration guidelines: lentil," in *Crop specific Regeneration Guidelines [CD-ROM]*, eds M. E. Dulloo, I. Thormann, M. A. Jorge, and J. Hanson (Rome: CGIAR System-wide Genetic Resource Programme), 10.

Street, K., Rukhkyan, N., and Ismail, A. (2008c). "Regeneration guidelines: chickpea," in *Crop Specific Regeneration Guidelines [CD-ROM]*, eds M. E. Dulloo, I. Thormann, M. A. Jorge, and J. Hanson (Rome: CGIAR System-wide Genetic Resource Programme), 10.

Strickler, K. (1999). Impact of flower standing crop and pollinator movement on alfalfa seed yield. *Environ. Entomol.* 28, 178–187. doi: 10.1093/ee/28.2.178

Strickler, K., and Vinson, J. (2000). A simulation of the effect of pollinator movement on alfalfa seed production. *Environ. Entomol.* 29, 907–918. doi: 10.1603/0046-225X-29.5.907

Suso, M. J., Bebeli, P. J., and Palmer, R. (2015). "Reproductive biology of grain legumes," in *Grain Legumes, Handbook of Plant Breeding*, ed. A. de Ron (Berlin: Springer Science Business Media), 364–399.

Suso, M. J., Harder, L., Moreno, M. T., and Maalouf, F. (2005). New strategies for increasing heterozygosity in crops: *Vicia faba* mating system as a study case. *Euphytica* 143, 51–65. doi: 10.1007/s10681-005-2526-y

Suso, M. J., and Maalouf, F. (2010). Direct and correlated responses to upward and downward selection for outcrossing in *Vicia faba. Field Crop. Res.* 116, 116–126. doi: 10.1016/j.fcr.2009.12.001

Suso, M. J., Moreno, M. T., Mondragao Rodrigues, F., and Cubero, J. I. (1996). Reproductive biology of *Vicia faba*: role of pollination conditions. *Field Crop. Res.* 46, 81–91. doi: 10.1016/0378-4290(95)00089-5

Suso, M. J., Nadal, S., and Palmer, R. G. (2010). "Potential power of the plant-pollinator relationship as a tool to enhance both environmental and production services of grain legumes in the context of low-input agriculture: what do we know?," in *Proceedings of the Breeding for resilience: a strategy for organic and low-input farming systems. Eucarpia 2nd Conference of the Organic and Low-Input Agriculture Section; 2010 Dec 1-3*, eds I. Goldringer, J. Dawson, F. Rey, and A. Vettoretti (Paris: INRA), 23–26.

Suso, M. J., and Río, R. (2014). Faba bean gene-pools development for low-input agriculture: understanding early stages of natural selection. *Euphytica* 196, 77–93. doi: 10.1007/s10681-013-1015-y

Suso, M. J., and Río, R. (2015). A crop–pollinator inter-play approach to assessing seed production patterns in faba bean under two pollination environments. *Euphytica* 201, 231–251. doi: 10.1007/s10681-014-1200-7

Suso, M. J., Vishnyakova, M., Ramos, A., Duc, G., and Ambrose, M. (2011). An international survey on state of the art of grain legume management in gene banks. *J. Agr. Sci. Tech. B* 1, 975–981.

Tapparo, A., Marton, D., Giorio, C., Zanella, A., Soldà, L., Marzaro, M., et al. (2012). Assessment of the environmental exposure of honey bees to particulate matter containing neonicotinoid insecticides coming from corn coated seeds. *Environ. Sci. Technol.* 46, 2592–2599. doi: 10.1021/es2035152

TEEB (2010). *The Economics of Ecosystems and Biodiversity. Ecological and Economic Foundations*, ed. P. Kumar. London: Earthscan.

Tester, M., and Langridge, P. (2010). Breeding technologies to increase crop production in a changing world. *Science* 327, 818–822. doi: 10.1126/science.1183700

Thomas, R. G. (1996). "Maximising seed production potential in white clover: factors influencing seed set per floret," in *Proceedings of the White Clover: New Zealand's Competitive Edge, A Joint Symposium Between Agronomy Society of New Zealand and New Zealand Grassland Association Held at Lincoln University* (Christchurch: Agronomy Society of New Zealand), 41–44.

Thomas, R. G., and Pasumarty, S. V. (1996). Limitations to seed set in white clover (*Trifolium repens* L.). I. Preliminary observations. *J. App. Seed Prod.* 14, 59–66.

Torricelli, R., Colesanti, N., and Falcinelli, M. (2007). "Improved seed production in a new Italian cultivar of lucerne (*Medicago sativa* L.)," in *Proceedings of the Sixth International Herbage Seed Conference, Gjennestad, Norway - 18-20 June 2007*, Bioforsk, 100–106.

Upadhyaya, H. D., Dwivedi, S. L., Ambrose, M., Ellis, N., Berger, J., Sm'ykal, P., et al. (2011). Legume genetic resources: management, diversity assessment, and utilization in crop improvement. *Euphytica* 180, 27–47. doi: 10.1007/s10681-011-0449-3

Upadhyaya, H. D., Reddy, K. N., and Sastry, D. V. S. S. R. (2008). "Regeneration guidelines: pigeon pea," in *Crop Specific Regeneration Guidelines [CD-ROM]*, eds M. EDulloo, I. Thormann, M. A. Jorge, and J. Hanson (Rome: CGIAR System-wide Genetic Resource Programme), 8.

Van Driesche, R. G., Lyon, S., Sanderson, J. P., Bennett, K. C., Stanek, E. J., III, and Zhang, R. (2008). Greenhouse trials of *Aphidius* colemani (Hymenoptera: Braconidae) bankerplants for control of aphids (Hemiptera: Aphididae) in greenhouse spring floral crops. *Fl. Entomol.* 91, 583–591.

Veronesi, F., and Lorenzetti, F. (1983). Productivity and survival of alfalfa hybrid and inbred plants under competitive conditions. *Crop Sci.* 23, 557–580. doi: 10.2135/cropsci1983.0011183X002300030031x

Westerkamp, C., and Gottsberger, G. (2000). Diversity pays in crop pollination. *Crop Sci.* 40, 1209–1222. doi: 10.2135/cropsci2000.4051209x

Westphal, C., Steffan-Dewenter, I., and Tscharntke, T. (2003). Mass flowering crops enhance pollinator densities at a landscape scale. *Ecol. Lett.* 6, 961–965. doi: 10.1046/j.1461-0248.2003.00523.x

Wiens, D. (1984). Ovule survivorship, brood size, life history, breeding systems, and reproductive success in plants. *Oecologia* 64, 47–53. doi: 10.1007/BF00377542

Williams, N. M., Crone, E. E., Roulston, T. H., Minckley, R. L., Packer, L., and Potts, S. G. (2010). Ecological and life-history traits predict bee species responses toenvironmental disturbances. *Biol. Conserv.* 143, 2280–2291. doi: 10.1016/j.baae.2010.11.008.

Woodcock, T. S. (2012). *Pollination in the Agricultural Landscape: Best Management Practices for Crop Pollination.* Guelph, ON: University of Guelph.

Worthington, M., Soleri, D., Aragón-Cuevas, F., and Gepts, P. (2012). Genetic Composition and Spatial Distribution of Farmer-managed Phaseolus Bean Plantings: an Example from a Village in Oaxaca, Mexico. *Crop Sci.* 52, 1721–1735. doi: 10.2135/cropsci2011.09.0518

Yagoubi, N., and Chrki, A. (2000). Estimation of mating system parameters in *Hedysarum coronarium* L. (Leguminoseae, Fabaceae). Agronomie. *EDP Sciences* 20, 933–942. doi: 10.1051/agro:2000103

Zeven, A. C. (1998). Landraces: a review of definitions and classifications. *Euphytica* 104, 127–139. doi: 10.1023/A:1018683119237

Zhao, Y., Mette, M. F., and Reif, J. C. (2015). Genomic selection in hybrid breeding. *Plant Breed.* 134, 1–10. doi: 10.1111/pbr.12231

Conflict of Interest Statement: The authors declare that the research was conducted in the absence of any commercial or financial relationships that could be construed as a potential conflict of interest.

3

The Future of Lupin as a Protein Crop in Europe

Edited by:
Soren K. Rasmussen,
University of Copenhagen, Denmark

Reviewed by:
Ethalinda K. S. Cannon,
Iowa State University, USA
Jens Stougaard,
Aarhus University, Denmark

***Correspondence:**
José J. Pueyo,
Institute of Agricultural Sciences,
ICA-CSIC,
Serrano, 115-bis,
28006 Madrid, Spain
jj.pueyo@csic.es

*M. Mercedes Lucas [1], Frederick L. Stoddard [2], Paolo Annicchiarico [3], Juana Frías [4], Cristina Martínez-Villaluenga [4], Daniela Sussmann [5], Marcello Duranti [6], Alice Seger [7], Peter M. Zander [8] and José J. Pueyo [1]**

[1] Institute of Agricultural Sciences, ICA-CSIC, Madrid, Spain, [2] Department of Food and Environmental Sciences, University of Helsinki, Helsinki, Finland, [3] Research Centre for Fodder Crops and Dairy Productions, CRA, Lodi, Italy, [4] Institute of Food Science Technology and Nutrition, ICTAN-CSIC, Madrid, Spain, [5] Fraunhofer-Institute for Process Engineering and Packaging, Freising, Germany, [6] Department of Food Environment and Nutritional Sciences, Università degli Studi di Milano, Milan, Italy, [7] Terrena Lup'Ingredients, Martigne-Ferchaud, France, [8] Leibniz Centre for Agricultural Landscape Research, ZALF, Müncheberg, Germany

Europe has become heavily dependent on soya bean imports, entailing trade agreements and quality standards that do not satisfy the European citizen's expectations. White, yellow, and narrow-leafed lupins are native European legumes that can become true alternatives to soya bean, given their elevated and high-quality protein content, potential health benefits, suitability for sustainable production, and acceptability to consumers. Nevertheless, lupin cultivation in Europe remains largely insufficient to guarantee a steady supply to the food industry, which in turn must innovate to produce attractive lupin-based protein-rich foods. Here, we address different aspects of the food supply chain that should be considered for lupin exploitation as a high-value protein source. Advanced breeding techniques are needed to provide new lupin varieties for socio-economically and environmentally sustainable cultivation. Novel processes should be optimized to obtain high-quality, safe lupin protein ingredients, and marketable foods need to be developed and offered to consumers. With such an integrated strategy, lupins can be established as an alternative protein crop, capable of promoting socio-economic growth and environmental benefits in Europe.

Keywords: lupin, protein crop, plant protein, breeding, protein ingredients, protein foods, food supply chain

The demand of the ever-growing world population for dietary protein is no longer sustainable through animal products alone. Soya bean has become the prevalent source of plant proteins for food and feed, and Europe depends on soya bean imports for 70% of its plant protein requirements. White lupin (*Lupinus albus* L.), yellow lupin (*L. luteus* L.) and narrow-leafed lupin (*L. angustifolius* L.), are native European legumes that represent a significant alternative to soya bean. Their seed protein content is high (up to 44%) and its quality is good, they offer potential health benefits, and they contribute to the sustainability of cropping systems. Lupins are successful protein crops in Australia, where an important industry has developed to use lupin protein and other fractions, yet lupin production in Europe is insufficient to guarantee the stable and sufficient supply required for its use by the food and feed industry. Lupin is grown in several European countries, and although its grain yield is the world's highest in some parts of Europe, its cropping area remains modest and yields are highly variable. A slight rise of European cultivated area and production has occurred during the period 2000–2013 (**Figure 1**), representing 17.6% of the world's production during that period (**Figure 2**). The Andean lupin (*L. mutabilis* Sweet) has been brought into cultivation in some parts of South America, but is not present in Europe on a commercial scale.

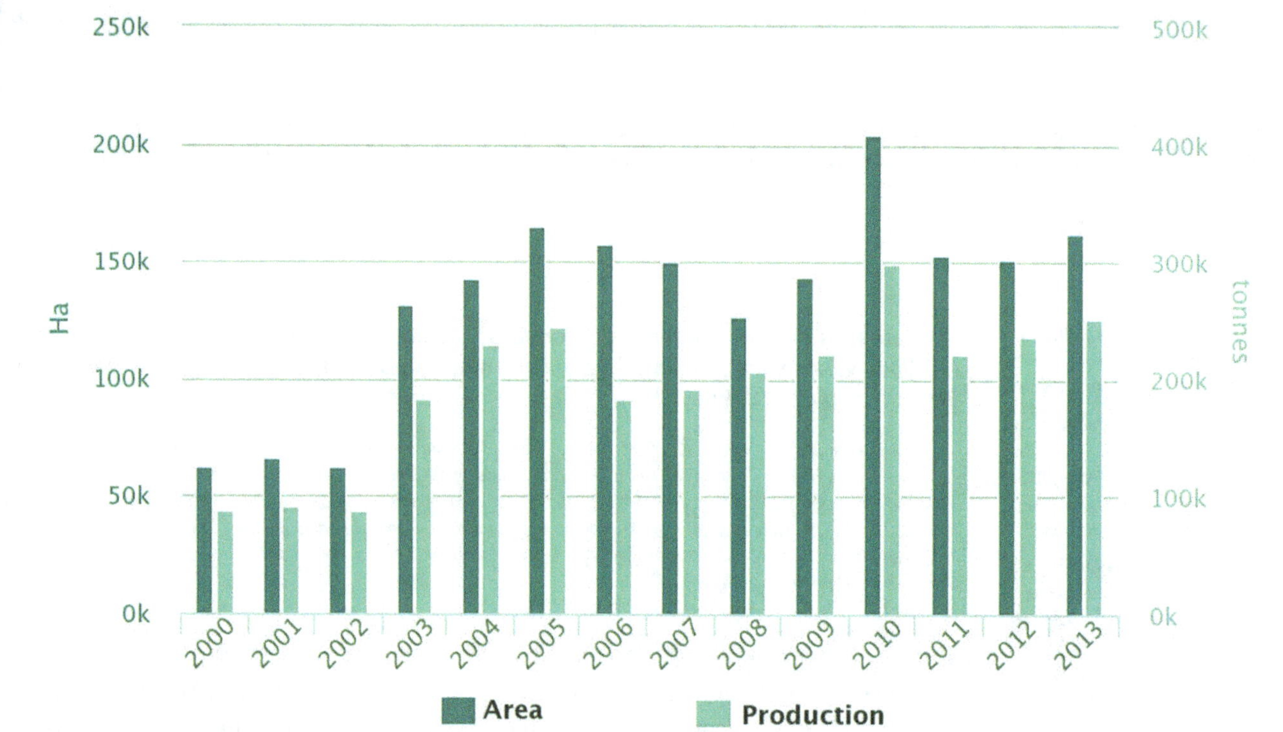

FIGURE 1 | Lupin cultivated areas and production in Europe. Source: FAOSTAT 2015.

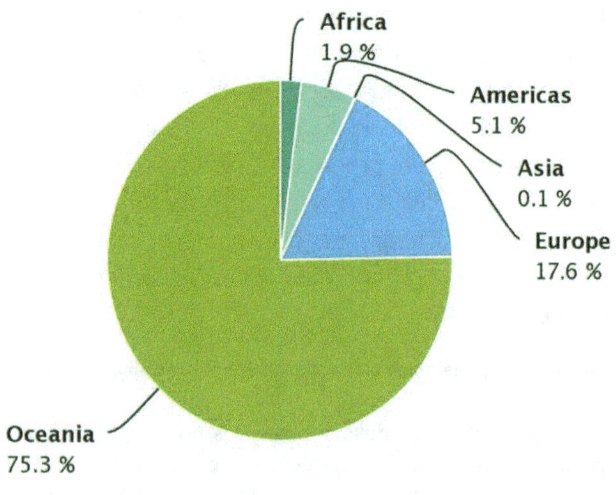

FIGURE 2 | Worldwide distribution of lupin production. Source: FAOSTAT 2015.

Insufficient grain yield in certain areas, largely due to limited breeding, constitutes a significant barrier to the expansion and economic sustainability of lupin cultivation in Europe. Lupin breeding has exploited little of the gene pool (Berger et al., 2012), partly due to the need to maintain a low alkaloid content (sweet seed). Landrace germplasm offers tremendous opportunities for increasing white lupin yield (Annicchiarico et al., 2010). Lupins are relatively more tolerant to several abiotic stresses than other legumes, and have a proven potential for the recovery of poor and contaminated soils (Fernández-Pascual et al., 2007; Coba de la Peña and Pueyo, 2012). Identifying germplasm with tolerance to a range of abiotic stresses may allow lupin cultivation to expand into a wider range of agro-climatic conditions across Europe. In areas subjected to frequent or occasional frosts, high vernalization requirement along with intrinsic cold tolerance is required, as sudden winter frost may lead to high mortality even in regions with relatively mild winters (Annicchiarico and Iannucci, 2007). Summer drought and progressive soil salinization are major climatic stresses in Mediterranean areas, yet information is lacking on the intrinsic genetic variation for drought tolerance, particularly in white lupin. Poor adaptation to calcareous soils is probably the main limit to the crop expansion, but white lupin landraces from Egypt or Italy that display tolerance to free calcium have been identified (Annicchiarico and Thami-Alami, 2012). The primary source of phosphate fertilizer (mined rock phosphate) is considered likely to be depleted within 40 years, emphasizing the importance of sustainable P management. When available P is limiting, lupins form specialized cluster root structures and/or release P-mobilizing carboxylates that free it from insoluble forms (Lambers et al., 2012), which could be exploited more widely to reduce the need for P supplementation.

To combine high protein quality, adaptation to environmental stresses and high yield, a suite of modern techniques is required. Extensive research in Australia has provided valuable information on the genetics of narrow-leafed lupin, and its genome sequence is nearly complete. Genomic selection is a particularly promising tool for improving complex, quantitatively inherited traits (Heffner et al., 2009). Combining phenotyping and genotyping data of the population into a model enables breeding values for each marker to be estimated. Genotyping-by-sequencing (GBS) can provide thousands of single nucleotide polymorphism (SNP) markers at much lower cost than earlier techniques (Elshire et al., 2011), but should be optimized for individual crops. GBS can help to identify and establish valuable genetic markers needed for genomic selection of each lupin species, and ultimately, to select cultivars adapted to the wide variety of European soils and climates, and to allow productive cultivation on poor and marginal lands.

Lupins form root nodules in which biological nitrogen fixation (BNF) takes place in symbiosis with compatible soil bacteria of the genus *Bradyrhizobium*. In general, European lupins that grow in acid soils can establish effective symbiosis with bradyrhizobia that are usually present in these soil, although they may be absent from neutral or alkaline soils. Where indigenous rhizobia are inadequate, inoculants boost BNF, improving grain yield and often its protein quality. BNF is strongly related to the physiological state of the host plant, so that environmental stresses not only have a detrimental effect on metabolism, growth and development of the plant, but also affect symbiosis. The *Lupinus-Bradyrhizobium* symbiosis has been described as being relatively tolerant to abiotic stresses (Fernández-Pascual et al., 2007). Rhizobial strain and lupin genotype interactions influence nodulation score, nitrogen fixation and plant growth. Specific *Bradyrhizobium* strains should be developed together with stress-tolerant lupin genotypes, to promote lupin as a successful component of sustainable crop production systems.

To use lupin as a source of high quality and healthy proteins for food, several quality traits are worth extra attention: in particular, alkaloid content, which must be maintained as low as possible, and the conglutin-γ protein fraction, which is of interest to control insulin resistance and diabetes (Terruzzi et al., 2011). The oil content of white lupin is 8–14% and its nutritional quality is excellent (Boschin et al., 2008). If this trait is further improved by breeding, it would increase the economic sustainability of the crop by making it dual-purpose for protein and oil, like soya bean. Iron deficiency is one of the most common and widespread nutritional disorder worldwide. Traditional iron supplementation methods can have negative effects on the consumer, and are unavailable to many people. The Fe-rich protein ferritin is abundant in some legume seeds, including lupin (Strozycki et al., 2007), and may represent a safe way to increase dietary iron intake. Lupin proteins influence lipid and glucose metabolism, as well as blood pressure levels (Duranti and Morazzoni, 2011). Amino acid sequences that may be related to hypotensive and lipid-lowering activities have been detected in lupin proteins (Bettzieche et al., 2009), although the components responsible for these effects are still undetermined and the underlying mechanisms remain ill-defined. Another interesting aspect of lupin protein functionality is its possible effects on inflammatory processes and changes in the gut microbiome, which has a significant influence on several physiological parameters, including metabolism, nutrient absorption and immune function (Walsh et al., 2014).

Food processing affects structural features of proteins and peptides, exerting a strong impact on their techno-functional and bioactive properties. Some proteins display their activity through peptides that must be liberated upon digestion to achieve functionality, while other proteins, such as conglutin-γ, seem to be required in an intact form. The production of protein ingredients with improved nutritive, technological and health benefits presents the food industry with unique challenges. Technological approaches remain to be optimized to improve production, cost effectiveness, and sustainable and environmental feasibility of high-quality protein ingredients. The selection of pre-treatments, such as germination in mineral-fortified conditions (Frias et al., 2009), and the optimization of processing conditions to facilitate protein extraction and isolation, represent important technological targets, oriented toward sustainable and non-invasive techniques. Off-flavor production is an important constraint in legume flour milling, storage and use (Azarnia et al., 2011). Pre-treatment with electromagnetic microwave radiation provides potential advantages over conventional hydro-thermal pre-treatment, inactivating enzymes that cause off-flavor formation and enhancing seed brittleness, thereby assisting milling and fractionation with little effect on protein functionality, as has been demonstrated in oat (Keying et al., 2009). These techniques need to be properly adapted to lupin seeds. Different fractionation, isolation and purification techniques combined with pre-processing are used to obtain protein concentrates/isolates with targeted techno-functional properties. Wet fractionation techniques are conventionally used to obtain relatively pure protein isolates, although they are energy-intensive procedures requiring large amounts of water and they may affect protein functionality (Muranyi et al., 2013). Many processing conditions for protein isolation are well established, but they are amenable to improvement, as shown by the influence of various parameters on target properties of the protein preparation and on overall protein yield. Dry milling combined with air classification is an alternative strategy to wet extraction, and may provide ideal disentanglement with lower energy and water demands (Schutyser and van der Goot, 2011).

Protein isolates prepared by isoelectric precipitation exhibit excellent water absorption, emulsification and foam-forming activities (Rodríguez-Ambriz et al., 2005). In contrast, legume protein isolates prepared by salt-induced extraction and dilution-precipitation with water have outstanding oil-binding capacity, emulsion stability and specific viscoelasticity, which affect their rheological properties, e.g., for bakery products (Marco and Rosell, 2008). Hence, the techno-functional properties of lupin proteins must be examined to optimize processing conditions and select suitable processing schemes, such as protein modification by microstructuring, to obtain ingredients for different food applications. Protein hydrolysates containing bioactive peptides are of great interest for the design of functional foods and nutraceuticals. Proteolysis of lupin storage proteins confers added value and targets functional properties as well as decreasing allergenicity (Clemente, 2000), and is commonly achieved by enzymatic treatments and fermentation. Recently developed high hydrostatic pressure and ultrasound-assisted methods increase the hydrolytic efficiency of enzymes (Zhang et al., 2012; Garcia-Mora et al., 2015). Sensory qualities and storage stability are key parameters to be taken into account when introducing lupin ingredients into the food industry. For example, neutral taste should be assured and maintained over a certain period in storage. Therefore, endogenous enzyme activities must be characterized to ensure the quality of the protein produced and to monitor storage stability. Lupin fiber could help to stabilize lupin protein based structures, as shown for other plant fiber materials (Laine et al., 2011). Accompanying components affect the techno-functional properties of the final protein products, so this has to be considered.

Technological challenges to optimize the production and processing of lupin protein relate to the integration of the process into a holistic concept that considers oil and fiber recovery, use of co-products for animal feed, bio-fuel (Pakarinen et al., 2011), or soil amendments, or even water recycling. After protein extraction, the large amount of dietary fiber (up to 40% of seed mass) has a potential role in functional foods. Innovative, cost-effective and environmentally friendly technologies need to be explored for fiber purification and fractionation. Lupin oil is an attractive product because of its balanced fatty acid composition and its content of bioactive lipids (Kalogeropoulos et al., 2010). Within the context of co-product utilization, animal feed is one of the main contributors to overall profitability.

The safety aspects of lupin ingredients include the formation of biogenic amines and the presence of allergens. Lupin allergy is still quite rare, but cross-reactivity between peanut and lupin exists (Fiocchi et al., 2009), which has led to the inclusion of lupins in the allergen list of the EU directive on labeling (EU Directive 2006/142/EC). Conglutins-α, -β, -γ, and -δ are candidate allergens in lupin, of which conglutin-α has strong allergenicity (Holden et al., 2008). Considering the severity of allergic reaction to peanut, the cross-reactivity of new lupin derivatives needs to be carefully assessed, and commercially exploited lupin products properly labeled, to minimize the danger for potential allergic consumers.

Lupin ingredients are already used in foods in Europe but their use is much less common than that of soya bean or pea ingredients, despite their beneficial properties. Lupin ingredients (flour and protein isolates and concentrates) are mainly used in bakery and gluten-free products, albeit as minor components (<5% of the ingredients). Nevertheless, new products containing lupin ingredients enter the European market each year (Mintel GNPD, http://www.mintel.com/global-new-products-database). Given its high protein content, lupin flour is considered an excellent raw material to supplement different food products (Pollard et al., 2002) and can be used as an egg substitute in cakes, pancakes, biscuits, pasta, or bread (Dervas et al., 1999). The rich yellow color of lupin flour and some protein concentrates has considerable appeal and can be of value in pasta or noodle dishes (Doxastakis et al., 2002), yet it can be easily eliminated when necessary (Guémes-Vera et al., 2008). Lupin flour can be incorporated into wheat flour to improve the nutritional value of the final products with little or no detriment in product sensory quality, and lupin fiber can also be used as a source of dietary fiber (Clark and Johnson, 2002). In general, the addition of up to 10% lupin flour improves water binding, texture, shelf-life and aroma of bread, although the mixing time and dough stability decrease as the proportion of lupin flour is increased (Doxastakis et al., 2002; Pollard et al., 2002).

While European consumers have a positive opinion on plant protein consumption, many do not know that lupin is a plant protein source comparable to soya bean. Among the main trends in the current European food market are naturalness; environmentally friendly; vegetarian alternative to meat, milk or eggs; and health properties (Mintel GNPD). The development of novel lupin-based foods responding to these desiderata should probably focus first on replacement of animal products (meat alternative, vegetarian spreads, dessert creams, ice-cream, and vegetable drinks). Gluten-free products are often lower in protein than standard goods, and lupin-based goods (bakery products, pasta, breakfast cereals) would restore or even increase the protein content. Further targets include high-protein food products with excellent sensory properties (sausages, pasta, snacks, drinks, bread). Specific consumer groups can be addressed with suitable nutritional values, taste and texture. The major international nutraceutical markets (EU, USA, Japan) are expected to grow in the near future, with major areas of interest being chronic degenerative cardiovascular diseases, type-2 diabetes and neuroprotection. Despite their health-promoting effects (Duranti and Morazzoni, 2011; Terruzzi et al., 2011), lupin ingredients are scarcely used in probiotics and nutraceuticals, as only a few lupin-containing food supplements and products with health claims are present in Europe (Mintel GNPD). The European Food Safety Authority (EFSA) currently allows no health claim about lupin ingredients. If proper clinical trials demonstrate a preventive role of lupin-based foods toward ailments such diabetes, cardiovascular diseases, metabolic syndrome, and obesity, these health-promoting features may confer an important advantage over other protein crops.

Lupin production in Europe declined steadily during the second half of the 20th century (Preissel et al., 2015), mainly because of low productivity driven by seasonal variability (Cernay et al., 2015), the low price of lupin grain (de Visser et al., 2014), and EU policies favoring the importation of soya bean. Concern about the sustainability and dependence of non-European protein sources, as well as potential environmental advantages, have re-established interest in lupins and other legume crops, and production has increased since 2003 (**Figure 1**). To establish lupin-based products as serious alternatives, the supply chain of these products must be assessed in terms of production costs, yields and quality, as well as factors influencing the acceptance of lupin-based products by European consumers (de Visser et al., 2014). The complete supply chain has to be addressed, analyzing the factors contributing to low lupin production by European farmers, and the conditions necessary to increase supply and make lupin a viable alternative to existing crops. The potential benefits of lupins at the farm level include N fixation with positive direct and indirect effects for subsequent crop yields, lower phosphorus requirements, reduced soil loss, and biodiversity benefits (Nemecek et al., 2008). Value-chain characteristics to be considered comprise agricultural production in terms of cost factors and yield effects, agricultural policies driving farmers' decisions, production quantities and qualities demanded by processing companies, sustainability impacts, and factors influencing the acceptance of lupin products by European consumers. Life-cycle assessment methods (LCA) and participatory impact assessment techniques should be combined in a complementary manner. An in-depth investigation of the causes and driving forces leading to the current production and consumption situation of lupins in Europe would indicate which characteristics lupins have to fulfill and how framework situations need to be changed (e.g., agricultural production conditions, level of substitution in end-products) to make lupins into profitable protein crops.

The rising global demand for meat, dairy and fish products for human consumption is recognized as unsustainable, owing to the high environmental impact of animal production. Furthermore, diseases associated with excessive consumption of animal products are increasing in frequency. The world population continues to grow, so a trend toward diets containing more plant protein seems not just strongly recommended, but inevitable, and it would be wise to anticipate and direct this trend by adopting knowledge-based strategies. To best achieve this objective, advances in virtually all phases of the supply chain are needed. Lupins should be established as fundamental crops in the various agro-climatic zones and marginal lands of Europe, and their yields and adaptation genetically improved to ensure a continuous supply of quality grain. The combination of their low needs for fertilizer N and P with their exploitation of poor, degraded, stress-affected or contaminated soils to safely produce protein-rich crops could contribute to socio-economic development and environmental sustainability. Sustainable, innovative, and cost-efficient processing methods to produce high-protein ingredients should be devised to guarantee the socio-economic value of these crops. Reformulating traditional foods, rather than developing totally novel ones, may be a more effective strategy to capture the imagination of the European consumer, and lead to desirable changes in overall diet. There is huge potential market demand for lupin-based products, with niches in growing sectors, such as vegetarians, vegans, and people with intolerance or allergy to gluten, soya, milk, or egg. The incorporation of lupin ingredients as a source of protein for human consumption depends largely upon their nutritional quality, but also on their ability to be used as, or incorporated into, foods that will be readily consumed. There are European enterprises with a strong background and wealth of expertise on market analysis, consumer acceptance issues, legal issues and marketing tools, which are required to develop new and successful lupin products that constitute a serious European alternative to the prevailing soya bean-derived food monopoly.

Funding

This work was supported by a grant from MINECO (AGL2013-40758-R).

Acknowledgments

The authors wish to thank all participants in the LUSH Proteins Consortium for their valuable contribution.

References

Annicchiarico, P., Harzic, N., and Carroni, A. M. (2010). Adaptation, diversity, and exploitation of global white lupin (*Lupinus albus* L.) landrace genetic resources. *Field Crop Res.* 119, 114–124. doi: 10.1016/j.fcr.2010.06.022

Annicchiarico, P., and Iannucci, A. (2007). Winter survival of pea, faba bean and white lupin cultivars in contrasting Italian locations and sowing times, and implications for selection. *J. Agr. Sci.* 145, 611–622. doi: 10.1017/S0021859607007289

Annicchiarico, P., and Thami-Alami, I. (2012). Enhancing white lupin (*Lupinus albus* L.) adaptation to calcareous soils through selection of lime-tolerant plant germplasm and *Bradyrhizobium* strains. *Plant Soil* 350, 131–144. doi: 10.1007/s11104-011-0889-5

Azarnia, S., Boye, J. I., Warkentin, T., and Malcolmson, L. (2011). Changes in volatile flavour compounds in field pea cultivars as affected by storage conditions. *Int. J. Food Sci. Technol.* 46, 2408–2419. doi: 10.1111/j.1365-2621.2011.02764.x

Berger, J. D., Buirchell, B. J., Luckett, D. J., and Nelson, M. N. (2012). Domestication bottlenecks limit genetic diversity and constrain adaptation in narrow-leafed lupin (*Lupinus angustifolius* L.). *Theor. Appl. Genet.* 124, 637–652. doi: 10.1007/s00122-011-1736-z

Bettzieche, A., Brandsch, C., Eder, K., and Stangl, G. I. (2009). Lupin protein acts hypocholesterolemic and increases milk fat content in lactating rats by influencing the expression of genes involved in cholesterol homeostasis and triglyceride synthesis. *Mol. Nutr. Food Res.* 53, 1134–1142. doi: 10.1002/mnfr.200800393

Boschin, G., D'Agostina, A., Annicchiarico, P., and Arnoldi, A. (2008). Effect of genotype and environment on fatty acid composition of *Lupinus albus* L. seed. *Food Chem.* 108, 600–606. doi: 10.1016/j.foodchem.2007.11.016

Cernay, C., Ben-Ari, T., Pelzer, E., Meynard, J.-M., and Makowski, D. (2015). Estimating variability in grain legume yields across Europe and the Americas. *Sci. Rep.* 5, 11171. doi: 10.1038/srep11171

Clark, R., and Johnson, S. (2002). Sensory acceptability of foods with added lupin (*Lupinus angustifolius*) kernel fiber using pre-set criteria. *J. Food Sci.* 67, 356–361. doi: 10.1111/j.1365-2621.2002.tb11410.x

Clemente, A. (2000). Enzymatic protein hydrolysates in human nutrition. *Trends Food Sci. Tech.* 11, 254–262. doi: 10.1016/S0924-2244(01)00007-3

Coba de la Peña, T., and Pueyo, J. J. (2012). Legumes in the reclamation of marginal soils, from cultivar and inoculant selection to transgenic approaches. *Agron. Sustain. Dev.* 32, 65–91. doi: 10.1007/s13593-011-0024-2

de Visser, C. L. M., Schreuder, R., and Stoddard, F. L. (2014). The EU's dependency on soya bean import for the animal feed industry and potential for EU produced alternatives. *OCL* 21, D407. doi: 10.1051/ocl/2014021

Dervas, G., Doxastakis, G., Hadjisavva-Zinoviadi, S., and Triantafillakos, N. (1999). Lupin flour addition to wheat flour doughs and effect on rheological properties. *Food Chem.* 66, 67–73. doi: 10.1016/S0308-8146(98)00234-9

Doxastakis, G., Zafiriadis, I., Irakli, M., Marlani, H., and Tananaki, C. (2002). Lupin, soya and triticale addition to wheat flour doughs and their effect on rheological properties. *Food Chem.* 77, 219–227. doi: 10.1016/S0308-8146(01)00362-4

Duranti, M., and Morazzoni, P. (2011). Nutraceutical properties of lupin seed proteins. A great potential still waiting for full exploitation. *Agro Food Ind. Hi Technol.* 22, 20–23.

Elshire, R. J., Glaubitz, J. C., Sun, Q., Poland, J. A., Kawamoto, K., Buckler, E. S., et al. (2011). A robust, simple genotyping-by-sequencing (GBS) approach for high diversity species. *PLoS ONE* 6:e19379. doi: 10.1371/journal.pone.0019379

Fernández-Pascual, M., Pueyo, J. J., de Felipe, M. R., Golvano, M. P., and Lucas, M. M. (2007). Singular features of the *Bradyrhizobium-Lupinus* symbiosis. *Dyn. Soil Dyn. Plant* 1, 1–16.

Fiocchi, A., Sarratud, P., Terracciano, L., Vacca, E., Bernardini, R., Fuggetta, D., et al. (2009). Assessment of the tolerance to lupine-enriched pasta in peanut-allergic children. *Clin. Exp. Allergy* 39, 1045–1051. doi: 10.1111/j.1365-2222.2009.03199.x

Frias, J., Gulewicz, P., Martinez-Villaluenga, C., Pilarski, R., Blazquez, E., Jimenez, B., et al. (2009). Influence of germination with different selenium solutions on nutritional value and cytotoxicity of lupin seeds. *J. Agric. Food Chem.* 57, 1319–1325. doi: 10.1021/jf8028368

Garcia-Mora, P., Penas, E., Frias, J., Gomez, R., and Martinez-Villaluenga, C. (2015). High-pressure improves enzymatic proteolysis and the release of peptides with angiotensin I converting enzyme inhibitory and antioxidant activities from lentil proteins. *Food Chem.* 171, 224–232. doi: 10.1016/j.foodchem.2014.08.116

Guémes-Vera, N., Pena-Bautista, R. J., Jimenez-Martinez, C., Davila-Ortiz, G., and Calderon-Dominguez, G. (2008). Effective detoxification and decoloration of *Lupinus mutabilis* seed derivatives, and effect of these derivatives on bread quality and acceptance. *J. Sci. Food Agric.* 88, 1135–1143. doi: 10.1002/jsfa.3152

Heffner, E. L., Sorrells, M. E., and Jannink, J. L. (2009). Genomic selection for crop improvement. *Crop Sci.* 49, 1–12. doi: 10.2135/cropsci2008.08.0512

Holden, L., Sletten, G. B. G., Lindvik, H., Faeste, C. K., and Dooper, M. M. (2008). Characterization of IgE binding to lupin, peanut and almond with sera from lupin-allergic patients. *Int. Arch. Allergy Immunol.* 146, 267–276. doi: 10.1159/000121461

Kalogeropoulos, N., Chiou, A., Ioannou, M., Karathanos, V. T., Hassapidou, M., and Andrikopoulos, N. K. (2010). Nutritional evaluation and bioactive microconstituents (phytosterols, tocopherols, polyphenols, triterpenic acids) in cooked dry legumes usually consumed in the Mediterranean countries. *Food Chem.* 121, 682–690. doi: 10.1016/j.foodchem.2010.01.005

Keying, Q., Changzhonga, R., and Zaigui, L. (2009). An investigation on pretreatments for inactivation of lipase in naked oat kernels using microwave heating. *J. Food Eng.* 95, 280–284. doi: 10.1016/j.jfoodeng.2009.05.002

Lambers, H., Bishop, J. G., Hopper, S. D., Etienne Laliberté, E., and Zúñiga-Feest, A. (2012). Phosphorus-mobilization ecosystem engineering: the roles of cluster roots and carboxylate exudation in young P-limited ecosystems. *Ann. Bot.* 110, 329–348. doi: 10.1093/aob/mcs130

Laine, P., Toppinen, E., Kivela, R., Taavitsainen, V. M., Knuutila, O., Sontag-Strohm, T., et al. (2011). Emulsion preparation with modified oat bran: optimization of the emulsification process for microencapsulation purposes. *J. Food. Eng.* 104, 538–547. doi: 10.1016/j.jfoodeng.2011.01.014

Marco, C., and Rosell, C. M. (2008). Functional and rheological properties of protein enriched gluten free composite flours. *J. Food Eng.* 88, 94–103. doi: 10.1016/j.jfoodeng.2008.01.018

Muranyi, S., Otto, C., Pickardt, C., Koehler, P., and Schweiggert-Weisz, U. (2013). Microscopic characterisation and composition of proteins from lupin seed (*Lupinus angustifolius* L.) as affected by the isolation procedure. *Food Res. Int.* 54, 1419–1429. doi: 10.1016/j.foodres.2013.10.004

Nemecek, T., von Richthofen, J. S., Dubois, G., Casta, P., Charles, R., and Pahl, H. (2008). Environmental impacts of introducing grain legumes into European crop rotation. *Eur. J. Agron.* 28, 380–393. doi: 10.1016/j.eja.2007.11.004

Pakarinen, A., Maijala, P., Stoddard, F. L., Santanen, A., Tuomainen, P., Kymalainen, M., et al. (2011). Evaluation of annual bioenergy crops in the boreal zone for biogas and ethanol production. *Biomass Bioenerg.* 35, 3071–3078. doi: 10.1016/j.biombioe.2011.04.022

Pollard, N. J., Stoddard, F. L., Popineau, Y., Wrigley, C. W., and MacRitchie, F. (2002). Lupin flours as additives: dough mixing, breadmaking, emulsifying, and foaming. *Cereal Chem.* 79, 662–669. doi: 10.1094/CCHEM.2002.79.5.662

Preissel, S., Reckling, M., Schläfke, N., and Zander, P. (2015). Magnitude and farm-economic value of grain legume pre-crop benefits in Europe: a review. *Field Crops Res.* 175, 64–79. doi: 10.1016/j.fcr.2015.01.012

Rodríguez-Ambriz, S. L., Martinez-Ayala, A. L., Millan, F., and Davila-Ortiz, G. (2005). Composition and functional properties of *Lupinus campestris* protein isolates. *Plant Food Hum. Nutr.* 60, 99–107. doi: 10.1007/s11130-005-6835-z

Schutyser, M. A. I., and van der Goot, A. J. (2011). The potential of dry fractionation processes for sustainable plant protein production. *Trends Food Sci. Technol.* 22, 154–164. doi: 10.1016/j.tifs.2010.11.006

Strozycki, P. M., Szczurek, A., Lotocka, B., Figlerowicz, M., and Legocki, A. B. (2007). Ferritins and nodulation in *Lupinus luteus*: iron management in indeterminate type nodules. *J. Exp. Bot.* 58, 3145–3153. doi: 10.1093/jxb/erm152

Terruzzi, I., Senesi, P., Magni, C., Montesano, A., Scarafoni, A., Luzi, L., et al. (2011). Insulin-mimetic action of conglutin-gamma, a lupin seed protein, in mouse myoblasts. *Nutr. Metab. Cardiovas.* 21, 197–205. doi: 10.1016/j.numecd.2009.09.004

Walsh, C. J., Guinane, C. M., O'Toole, P. W., and Cotter, P. D. (2014). Beneficial modulation of the gut microbiota. *FEBS Lett.* 588, 4120–4130. doi: 10.1016/j.febslet.2014.03.035

Zhang, T., Jiang, B., Miao, M., Mu, W., and Li, Y. (2012). Combined effects of high-pressure and enzymatic treatments on the hydrolysis of chickpea protein isolates and antioxidant activity of the hydrolysates. *Food Chem.* 135, 904–912. doi: 10.1016/j.foodchem.2012.05.097

Conflict of Interest Statement: The authors declare that the research was conducted in the absence of any commercial or financial relationships that could be construed as a potential conflict of interest.

4

The Potential for Engineering Enhanced Functional-Feed Soybeans for Sustainable Aquaculture Feed

Edited by:
José J. Pueyo,
Consejo Superior de Investigaciones Científicas, Spain

Reviewed by:
Abu Hena Mostafa Kamal,
University of Texas at Arlington, USA
Marty Riche,
Florida Atlantic University, USA
Alfonso Clemente,
Spanish National Research Council, Spain

***Correspondence:**
Eliot M. Herman
emherman@email.arizona.edu

Eliot M. Herman and Monica A. Schmidt*

School of Plant Sciences, University of Arizona, Tucson, AZ, USA

Aquaculture is the most rapidly growing segment of global animal production that now surpasses wild-capture fisheries production and is continuing to grow 10% annually. Sustainable aquaculture needs to diminish, and progressively eliminate, its dependence on fishmeal-sourced feed from over-harvested fisheries. Sustainable aquafeed sources will need to be primarily of plant-origin. Soybean is currently the primary global vegetable-origin protein source for aquaculture. Direct exchange of soybean meal for fishmeal in aquafeed has resulted in reduced growth rates due in part to soybean's anti-nutritional proteins. To produce soybeans for use in aquaculture feeds a new conventional line has been bred termed *Triple Null* by stacking null alleles for the feed-relevant proteins Kunitz Trypsin Inhibitor, lectin, and P34 allergen. *Triple Null* is now being further enhanced as a platform to build additional transgene traits for vaccines, altered protein composition, and to produce high levels of β-carotene an intrinsic orange-colored aquafeed marker to distinguish the seeds from commodity beans and as the metabolic feedstock precursor of highly valued astaxanthin.

Keywords: aquaculture, aquafeed, soybean, carotenoid, vaccines, Kunitz trypsin inhibitor, soybean agglutinin, P34 allergen

AQUACULTURE IS THE FASTEST GROWING SEGMENT OF GLOBAL ANIMAL PRODUCTION THAT WILL REQUIRE NEW SUSTAINABLE SOURCES OF FEED

The global human population is projected to grow to 9.6 billion individuals by 2050 and will continue to 12 billion individuals or more by 2100 (Gerland et al., 2014). This population growth will require an increase of 70% more animal production that translates to a need for 235% more animal feed. Soybean is the largest plant-source input of protein and reduced nitrogen for formulated animal feed that drives a global industry of production and consumption. Aquaculture is most rapidly growing sector of the animal food-production sectors, increasing at nearly 10% per year (Food and Agriculture Organization [FAO], 2012). Aquaculture-sourced fish surpasses wild capture fisheries in annual production and this differential is projected to rapidly increase due to overfishing and habitat destruction in parallel to increased production in the aquaculture sector. As aquaculture displaces wild capture fisheries providing feed sources for cultured animals presents an immense challenge. Many commercially raised fish are carnivorous (e.g., trout and salmon), while others are omnivores and herbivores (e.g., catfish, carp, and tilapia); therefore, different fish species vary in their capacity to effectively use different kinds of feed. Aquaculture currently over-relies on fishmeal to provide high-quality feed protein. Annual fishmeal production has been constant for the last 15 years at approximately 6.5 million metric tons. Current or projected increase of fishmeal production to support aquaculture growth cannot be sustained (Watson et al., 2015).

A sustainable feed solution for aquaculture would be for farmed fish to be fed renewable plant-sourced protein and oil products harvested from terrestrial farms (Gatlin et al., 2007; Naylor et al., 2009; Rust et al., 2011). While substituting plant protein for fishmeal is now routine at low levels, in most cases increasing the proportion of plant protein in fish feed limits fish growth rates and feed efficiency. There is an opportunity to engineer seed crops as functional feed ingredient that specifically address the challenges of increasing the proportion of plant-based protein in commercial fish aquafeeds. Soybean is currently used as a primary source of vegetable protein supplement for aquafeed (Tacon et al., 2012) and is a large US export (Hardy, 1996). Soybean, particularly at high inclusion levels, in aquaculture can result in reduced growth rate dues at least in part from soybean's anti-nutritional proteins (Krogdahl et al., 1994, 2010; Baeverfjord and Krogdahl, 1996; Burrells et al., 1999; Bakke-McKellep et al., 2000; Buttle et al., 2001; Lilleeng et al., 2007; Sahlmann et al., 2013). One approach is to use biotechnology to suppress anti-nutritional proteins or to alter seed protein composition (see Herman, 2003, for review). To some degree conventional breeding and stacking of traits derived from non-biotech sources such as collections and mutation to enhance seed composition can also meet trait goals. Other enhancements that are not encoded within the soybean genome will require adding foreign genes to produce stacked trait soybeans optimized for high-performance feed-ingredient (Herman, 2009).

BREEDING LOW ANTI-NUTRITIONAL BIOACTIVITY SOYBEANS AND ITS DEVELOPMENT AS A PLATFORM TO EXPRESS ADDITIONAL TRANSGENE TRAITS

Developing a Conventional Low Bioactivity Soybean Seed; the Creation of Triple Null

Soybeans null for bioactive seed proteins that have been isolated from the USDA soybean collection include Kunitz trypsin inhibitor (KTI) null (Orf and Hymowitz, 1979; Hymowitz, 1986), soybean agglutinin (SBA) null (Orf et al., 1978; Goldberg et al., 1983) and immunodominant soybean allergen P34 protein nulls (Josephs et al., 2006). Each of these nulls has the potential to partially address concerns of soybean feed/food consumption and stacked together these traits can form a platform for engineering enhanced soybean varieties. A *Triple Null* stack of the recessive nulls of KTI, SBA, and P34 in the standard cultivar *Williams* (the archetype used to elucidate the soybean genome) has been produced (Schmidt et al., 2015a). The parental components of the *Triple Null (p34,kti,le)* stack include frame shift mutants of P34 (Josephs et al., 2006; Bilyeu et al., 2009) and KTI (Jofuku et al., 1989) that still accumulate a small amount of authentic protein due to a shift of the start methionine inward in the open reading frame Schmidt et al. (2015a). The *Lectin* (LE) null results from a large transposon and does not produce a functional transcript or lectin protein accumulation (Orf et al., 1978; Goldberg et al., 1983). Proteomic analysis of *Triple Null* shows that the line lacks this trio of bioactive proteins while retaining the full complement of other proteome constituents without any other collateral bioactive protein alterations (Schmidt et al., 2015a).

Triple Null Soybeans Can Be Transformed to Stack Additional Traits

To enable further enhancement of *Triple Null* its capacity for biolistic transformation was assessed. Somatic embryos were induced from *Triple Null* and used in test biolistic transformation studies using a hygromycin-selection cassette (see Schmidt and Herman, 2008; Schmidt et al., 2011, 2015b for methods). The resulting transformed lines were selected and regenerated into somatic embryos that were then germinated to produce a T_0 population of plants. The resulting T_1 seeds are a segregating population typically requiring two or three generations of recurrent selection to produce homozygote lines. **Figure 1** shows PCR analysis of a segregating population assessed by the presence of the *Hygromycin* marker demonstrating that *Triple Null* that is bred into a *Williams 82* background is capable of being transformed by the same protocols as cv *Jack*. This indicates that the traits of *Triple Null* can be used as a platform to stack additional transgene traits and can be exchanged for other standard transformation soybean lines.

POTENTIAL BIOTECHNOLOGY TRAITS TO ENHANCE *TRIPLE NULL'S* PERFORMANCE AS AN INGREDIENT IN AQUAFEED

The capacity to transform *Triple Null* is enabling for further biotechnological modifications using gene expression and silencing to further optimize performance of the seeds in aquafeeds. As an aquafeed source there are potential targets of expressing bioactive proteins such as vaccines and to further remodel the seed proteome such as by optimizing its protein content for digestibility. There are additional modifications that could be added to soybean that produces products that are not feasible to produce from soybean's genome. These traits could reproduce lipid-soluble carotenoids and omega fatty acids that would otherwise be obtained from fish oil with the same sustainability issues as fishmeal or from chemical synthesis and/or modification.

Protein Enhancement

The soybean seed proteome consists of two dominant seed storage proteins (SPs) 7S conglycinin and 11S glycinin and a number of additional moderately abundant proteins that are often bioactive including Kunitz and Bowman Birk trypsin inhibitors, lectin, P34 allergen, urease and lipoxygenases. *Triple*

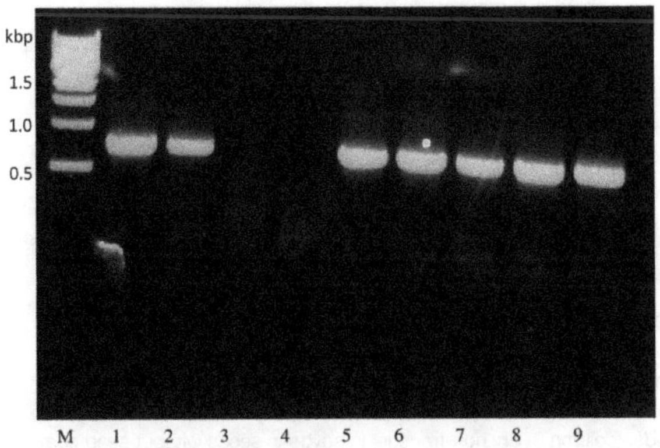

FIGURE 1 | *Triple Null* soybean is transformable via biolistic transformation of somatic embryos. Genomic PCR was performed on T_1 segregating *Triple Null* plants transformed with the hygromycin resistance marker gene expression cassette. Presence of the PCR amplicon ~ 700 bp in all samples except 3 and 4 indicate the presence of the hygromycin gene in those plants DNA, indicating a stable transformation event that is segregating in an expected Mendelian ratio.

Null stacking alleles from the USDA soybean collection addresses some of the most problematic bioactive proteins for aquafeed (Krogdahl et al., 1994; Buttle et al., 2001; Lilleeng et al., 2007). The SPs comprise about 65% of the seed's protein content and is regulated by genotype as a breeding trait. Most of the major soybean proteins are members of small gene families (for example Nielsen et al., 1989) making it difficult to identify conventional alleles that that silence all of the individual genes. Biotechnology approaches using co-suppression and RNAi has been used to knockout soybean genes and gene families of β-conglycinin (Kinney et al., 2001; Kim et al., 2014), P34 allergen (Herman et al., 2003), and both β-conglycinin and glycinin (Schmidt et al., 2011). Other lines have reduced Bowman Birk trypsin inhibitor (BBI) content by co-expressing an inactive variant (Livingstone et al., 2007) as not to further reduce the sulfur amino acid content of soybeans.

Silencing major soybean SPs results in compensatory remodeling of the proteome but not the seed protein content. This proteome compensation is termed protein rebalancing (Herman, 2014) that in soybean results when a major SP is silenced by either mutation or targeted genetic modification that should result in a significant shortfall of accumulated seed protein. The soybean remodels its proteome resulting in other intrinsic seed proteins increasing in accumulation to compensate for loss of major proteins that maintains both protein content (Schmidt et al., 2011) and total amino acid composition (Schmidt et al., 2011; Kim et al., 2014). From the biotechnologist's perspective proteome rebalancing can be exploited as a technology to replace major soybean SPs with other transgene products (Herman and Schmidt, 2014). Suppressing β-conglycinin that comprises 20% of the total seed protein results in enhanced accumulation of glycinin that maintains the standard protein content. Glycinin consists of a small five gene family to test the potential of exploiting the proteome rebalancing process a model GFP gene was constructed as a glycinin mimic and introgressed into β-conglycinin silenced seeds as an additional glycinin allele. This resulted in increasing GFP accumulation eight-fold from 1 to 8% of the total protein (Schmidt and Herman, 2008). Although this appears to be a potential engineering strategy to increase foreign protein content in seeds the proteome rebalancing processes by maintaining protein content and relative total amino acid this strategy does not appear to be effective to increase the sulfur amino acid content of soybean by expressing zein in a β-conglycinin suppressed background (Kim et al., 2014).

Feed based Vaccines Are Economic Solution to Disease

In animal production animal density is a key economic factor. High animal density is an economic necessity of production and this aggravates the problems of animal competition, waste management, and controlling the potential for disease. For fish production a number of bacterial and viral diseases have emerged that impact production. For many fish species especially the omnivorous and herbivorous species the low individual value of each fish limits the potential expenditure for antibiotic treatment or preventive vaccination. Fish can be immunized orally (Companjen et al., 2005; Plant and LaPatra, 2011; Tobar et al., 2011, 2015) or by bath exposure and if the vaccine antigens are inexpensive and effective this can enable disease prevention. There are problems to be resolved to make these protocols as effective as individual injection immunization particularly with primary sensitization (Munang'andu et al., 2015; Mutoloki et al., 2015). There is a large literature of prototype vaccines engineered by plant biotechnology for inclusion in food or feed especially for situations requiring mass immunization at minimal cost (Daniell et al., 2009). Plant biopharma is well matched to the needs and economic limitations of aquaculture (Clarke et al., 2013). Soybean feed-based vaccines can meet this need as a stacked trait. Soybean can be induced to produce up to 1% of its total protein shown as a prototype vaccine (Piller et al., 2005; Moravec et al., 2006). To produce greater levels of vaccine proteins technology has been developed to produce high-levels of heterologous proteins in soybean (Schmidt and Herman, 2008). An engineering strategy has been developed where transgenes encoding heterologous proteins can be produced at high levels, >8% of the total protein, by mimicking the gene of a compensating glycinin. Using this type of technology it should be feasible to design feed vaccines that separately or simultaneously produce several antigens potentially stacked with a strong adjuvant such as the enterotoxin LTB (Moravec et al., 2006) that could be produced in different formulations to vary and stagger the sensitization. This approach may improve the performance of oral vaccines for aquaculture.

Carotenoid Enhancement

Astaxanthin (3,3′-dihydroxy 4,4′-diketo-β-carotene) is the carotenoid responsible for giving many crustaceans and some bird species their signature pink/red color. Its principle commercial use is a flesh colorant for fish. Farm-raised salmonoid fish are deprived of phytoplankton and/or algae that would naturally provide this colorful compound, making it necessary to supplement fish diet with additive pigments in order to ensure that the resultant filets appeal to consumers. In the U.S. alone, astaxanthin sales per year are about $200 million for just the salmon aquaculture industry (Lorenz and Cysewski, 2000). In order for plants to be an effective sustainable supply of this colorant it would have be both biosynthesized and accumulated at relevant levels and produced in a suitable delivery plant organ system. Proof-of-concept in the successful production of astaxanthin has been achieved in tobacco nectar (Mann et al., 2000; Ralley et al., 2004) and leaves (Hasunuma et al., 2008) and tubers of both potato (Gerjets and Sandmann, 2006) and carrot (Jayaraj et al., 2008). The highest level obtained was 64 μg astaxantin/g of the nectar of transgenic tobacco plants. Production of this carotenoid in seeds has been shown to be possible but with marginal accumulation levels: transgenic canola (*Brassica napus*) seeds contained 0.2 μg astaxanthin/g (Fujisawa et al., 2009) and transgenic soybean seeds 7 μg astaxanthin/g dry seed (Pierce et al., 2015).

Variations in the amount of astaxanthin produced in different plant systems can be attributed to the inherent ability of certain plant organs to accumulate carotenoids, leaf and flower tissue more than seeds, and the catalytic attributes of the carotenoid biosynthetic enzymes. Enzymes in the carotenoid pathway have been reported to act upon an assortment of various substrates, often resulting in a mixture of carotenoids produced rather than a majority of a specific sought metabolite. Following on research showing that enhanced β-carotene levels could be achieved by the overexpression of a bacterial *Pantoea* phytoene synthase gene in soybean seeds (**Figure 2**) (Schmidt et al., 2015b), this construct was then combined individually with two ketolase genes from different sources: one from bacteria *Brevundimonas* spp and the other algae known for its high concentration of astaxanthin *Haematococcus pluvialis*. A mixture of carotenoids resulted with only one line using the bacterial ketolase producing detectable levels of astaxanthin (Pierce et al., 2015). Further emphasizing the importance of enzyme choice, three β-carotene ketolase enzymes were used from different sources in *Arabidopsis* leaves and found a wide variety of successful production of the desired astaxanthin carotenoid (Zhong et al., 2011). Up to 2 mg/g was produced when the ketolase gene from *Chlamydomonas*, compared to 0.24 mg/g when *Chlorella* was the gene source and non-detectable levels when *Haematococcus* ketolase gene was used. These findings emphasis the importance of enzyme selection in the biosynthesis of carotenoids giving promise to the future production of sustainable and commercially viable levels of this carotenoid pigment in a suitable plant delivery system.

Feed-based Omega 3 Fatty Acids

Desired oil modifications in seeds for aquaculture use would include the omega 3 (ω3) fatty acids. These fatty acids have been shown to enhance fish growth rates (Tocher, 2015) and are one of the human health promoting constituents in fish (for review, Calder and Yaqoob, 2009). Many plant seed oils are abundant in the ω-3 fatty acid α linolenic acid ($18{:}3^{\Delta 9,12,15}$) yet the health promoting long chain ω3 fatty acids are more limiting in the human diet with their principle source being marine fish. Although a vital component to both human and marine life nutrition, long chain ω-3 fatty acids eicosapentaenoic (EPA 20: $5^{\Delta 5,8,11,14,17}$) and docosahexaenoic acid (DHA $22{:}6^{\Delta 4,7,10,13,16,19}$) must be consumed in diets due to human and fish being very limited in their ability to produce them (Williams and Burdge, 2006). The most successful research to produce DHA and EPA in oil seeds crops was reported in *Brassica* seeds and used nine introduced gene expression cassettes using fatty acid modifying enzymes from various marine sources to achieve 0.2% DHA and up to 15% EPA of total fatty acids (Wu et al., 2005). Research to produce these lipid compounds in plants has moved toward soon being able to make commercially viable levels for the aquaculture community.

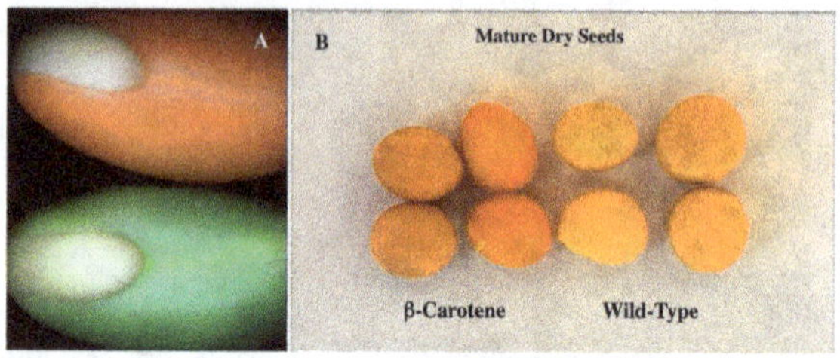

FIGURE 2 | β-carotene biofortified transgenic soybean seeds. (A) developing soybean cotyledons from (bottom) non-transgenic soybean compared to (top) chloroplast localized seed specific bacterial phytoene synthase transgenic soybean. (**B**; column 4) dry non-transgenic soybean seeds, (column 3) dry non-transgenic soybean seeds without seed coat, (column 2) dry transgenic soybean seeds without seed coat and (column 1) dry transgenic β-carotene soybean seeds. Enhanced β-carotene accumulation in transgenic soybean seeds provides a visually obvious orange color in both developing cotyledons and dry seeds.

SOYBEANS IN FUNCTIONAL FEED WILL MEET AN URGENT NEED TO SUPPORT THE GROWING AQUACULTURE INDUSTRY

Triple Null (*p34, kti, le*) is a potential platform in a path to stack additional transgene traits for functional aquafeed. The growing aquaculture industry and global demands for high quality animal protein will require the optimization of feed performance. Over the past two decades biotechnology has had a remarkable role in increasing yield and productivity. The next frontier is to enhance the performance of crops for specific end uses. The rapidly growing aquaculture industry over-dependent on unsustainable fishmeal sources is an ideal circumstance to explore the potential to design high performance functional feed ingredients.

AUTHOR CONTRIBUTIONS

This data contained and production of this manuscript is a collaborative effort of both MS and EH who contributed to all aspects of the manuscript's production.

REFERENCES

Baeverfjord, G., and Krogdahl, A. (1996). Development and regression of soybean meal induced enteritis in Atlantic salmon, *Salmo salar* L., distal intestine: a comparison with the intestines of fasted fish. *J. Fish Dis.* 19, 375–387. doi: 10.1046/j.1365-2761.1996.d01-92.x

Bakke-McKellep, A. M., Press, C., Baeverfjord, G., Krogdahl, Å., and Landsverk, T. (2000). Changes in immune and enzyme histochemical phenotypes of cells in the intestinal mucosa of Atlantic salmon, *Salmo salar* L., with soybean meal-induced enteritis. *J. Fish Dis.* 23, 115–127. doi: 10.1111/j.1365-2095.2007.00519x

Bilyeu, K., Ren, C., Nguyen, H. T., Herman, E. M., and Sleper, D. A. (2009). Association of a four basepair insertion in the P34 gene with the low allergen trait in soybean. *Plant Genome* 2, 141–148. doi: 10.3835/plantgenome2009.01.0006

Burrells, C., Williams, P. D., Southgate, P. J., and Crampton, V. O. (1999). Immunological, physiological and pathological responses of rainbow trout (*Oncorhynchus mykiss*) to increasing dietary concentration of soybean proteins. *Vet. Immunol. Immunopathol.* 72, 277–288. doi: 10.1016/S0165-2427(99)00143-9

Buttle, L. G., Burrells, A. C., Good, J. E., Williams, P. D., Southgate, P. J., and Burrells, C. (2001). The binding of soybean agglutin (SBA) to the intestinal epithelium of Atlantic salmon, *Salmo salar* and rainbow trout, *Oncorhynchus mykiss*, fed high levels of soybean meal. *Vet. Immunol. Immunopathol.* 80, 237–244. doi: 10.1016/S0165-2427(01)00269-0

Calder, P. C., and Yaqoob, P. (2009). Omega 3 polyunsaturated fatty acids and human health outcomes. *Biofactors* 35, 266–272. doi: 10.1002/biof.42

Clarke, J. L., Waheed, M. T., Lossel, A. G., Martinussen, I., and Daniell, H. (2013). How can plant genetic engineering contribute to cost effective fish vaccine development for promoting sustainable aquaculture? *Plant Mol. Biol.* 83, 33–40. doi: 10.1007/s11103-013-0081-9

Companjen, A. R., Florack, D. E., Bastiaans, J. H., Matos, C. I., Bosch, D., and Rombout, J. H. (2005). Development of a cost-effective oral vaccination method against viral disease in fish. *Dev. Biol. (Basel)* 121, 143–150.

Daniell, H., Singh, N. D., Mason, H., and Streatfield, S. J. (2009). Plant-made vaccine antigens and biopharmaceuticals. *Trends Plant Sci.* 14, 669–679. doi: 10.1016/j.plants.2009.09.009

Food and Agriculture Organization [FAO] (2012) *The State of World Fisheries and Aquaculture*. Rome: FAO. Available at: http://www.fao.org/docrep/016/i2727e/i2727e.pdf?

Fujisawa, M., Takita, E., Harada, H., Sakurai, N., Suzuki, H., Ohyama, K., et al. (2009). Pathway engineering of *Brassica napus* seeds using multiple key enzyme genes involved in ketocarotenoid formation. *J. Exp. Bot.* 60, 1319–1332. doi: 10.1093/jxb/erp006

Gatlin, D. M., Barrows, F. T., Brown, P., Dabrowski, K., Gaylord, T. G., Hardy, R. W., et al. (2007). Expanding the utilization of sustainable plant products in aquafeeds: a review. *Aquacult. Res.* 38, 551–579. doi: 10.1111/j.1365-2109.2007.01704.x

Gerjets, T., and Sandmann, G. (2006). Ketocartoenoid formation in transgenic potato. *J. Exp. Bot.* 57, 3639–3645. doi: 10.1093/jxb/erl103

Gerland, P., Raftery, A. E., Ševčíková, H., Li, N., Gu, D., Spoorenberg, T., et al. (2014). World population stabilization unlikely this century. *Science* 346, 234–237. doi: 10.1126/science.1257469

Goldberg, R. B., Hoschek, G., and Vodkin, L. O. (1983). An insertion sequence blocks the expression of a soybean lectin gene. *Cell* 33, 465–475. doi: 10.1016/0092-8674(83)90428-2

Hardy, R. W. (1996). Alternate protein sources for salmon and trout diets. *Anim. Feed Sci. Technol.* 59, 71–80. doi: 10.1016/0377-8401(95)00888-8

Hasunuma, T., Miyazawa, S., Yoshimura, S., Shinzaki, Y., Tomizawa, K., Shindo, K., et al. (2008). Biosynthesis of astaxanthin in tobacco leaves by transplastomic engineering. *Plant J.* 55, 857–868. doi: 10.1111/j.1365-313X.2008.03559.x

Herman, E. M. (2003). Genetically modified soybeans and food allergies. *J. Exp. Bot.* 54, 1317–1319. doi: 10.1093/jxb/erg164

Herman, E. M. (2009). *Biotechnology Could Combine Positive Traits in Seed-Based Fish Feed.* St. Louis, MO: Global Aquaculture Advocate, 88–90.

Herman, E. M. (2014). Soybean seed proteome rebalancing. *Front. Plant Sci.* 5:437. doi: 10.3389/fpls.2014.00437

Herman, E. M., Helm, R. M., Jung, R., and Kinney, A. J. (2003). Genetic modification removes an immunodominant allergen from soybean. *Plant Physiol.* 132, 36–43. doi: 10.1104/pp.103.021865

Herman, E. M., and Schmidt, M. A. (2014). "Towards using biotechnology to modify soybean seeds as protein biofactories," in *Recent Advancements in Plant Expression in Crop Plants*, eds K. Azhakanandam, A. Silverstone, H. Daniell, and M. R. Davey (Berlin: Springer), 193–212. doi: 10.1007/978-1-4939-2202-4_5

Hymowitz, T. (1986). Genetics and breeding of soybeans lacking the Kunitz trypsin inhibitor. *Adv. Exp. Med. Biol.* 199, 291–298. doi: 10.1007/978-1-4757-0022-0_18

Jayaraj, J., Devlin, R., and Punja, Z. (2008). Metabolic engineering of novel ketocarotenoid production in carrot plants. *Trans Res.* 17, 489–501. doi: 10.1007/s11248-007-9120-0

Jofuku, K. D., Schipper, R. D., and Goldberg, R. B. (1989). A frameshift mutation prevents Kunitz trypsin inhibitor mRNA accumulation in soybean embryos. *Plant Cell* 1, 427–435. doi: 10.1105/tpc.1.4.427

Josephs, L. M., Hymowitz, T., Schmidt, M. A., and Herman, E. M. (2006). Evaluation of Glycine germplasm for nulls of the immunodominant allergen P34/Gly m Bd 30k. *Crop Sci.* 46, 1755–1763. doi: 10.2135/cropsci2005.12-0500

Kim, W.-S., Jez, J. J., and Krishnan, H. B. (2014). Effects of proteome rebalancing and sulfur nutrition on the accumulation of methionine rich δ-zein in transgenic soybeans. *Front. Plant Sci.* 5:633. doi: 10.3389/fpls.2014.00633

Kinney, A. J., Jung, R., and Herman, E. M. (2001). Cosuppression of the α-subunits of β-conglycinin in transgenic soybean seeds induces the formation of endoplasmic reticulum-derived protein bodies. *Plant Cell* 13, 1165–1178. doi: 10.1105/tpc.13.5.1165

Krogdahl, Å., Lea, T. B., and Olli, J. L. (1994). Soybean proteinase-inhibitors affect intestinal trypsin activities and amino acid digestibilities in rainbow trout (*Oncorhynchus mykiss*). *Comp. Biochem. Physiol.* 107, 215–219. doi: 10.1016/0300-9629(94)90296-8

Krogdahl, Å., Penn, M. H., Thorsen, J., Refstie, S., and Bakke, A. M. (2010). Important antinutrients in plant feedstuffs for aquaculture: an update on recent findings regarding responses in salmonids. *Aquac. Res.* 41, 333–344. doi: 10.1111/j.1365-2109.2009.02426.x

Lilleeng, E., Froystad, M. K., and Krogdahl, A. (2007). Effects of diets containing soybean meal on trypsin mRNA expression and activity in Atlantic salmon (*Salmo salar* L). *Comp. Biochem. Physiol. A. Mol. Integr. Physiol.* 147, 25–36. doi: 10.106/j.cbpa.2006.10.043

Livingstone, D., Beilinson, V., Kalyaeva, M., Schmidt, M. A., Herman, E. M., and Nielsen, N. C. (2007). Reduction of protease inhibitor activity by expression of a mutant Bowman-birk gene in soybean seed. *Plant Mol. Biol.* 64, 397–408. doi: 10.1007/s11103-007-9163-x.

Lorenz, R. T., and Cysewski, G. R. (2000). Commercial potential for *Haematococcus* microalgae as a neutral source of astaxantin. *Trends Biotechnol.* 18, 160–167. doi: 10.1016/S0167-7799(00)01433-5

Mann, V., Harker, M., Pecker, I., and Hirschberg, J. (2000). Metabolic engineering of astaxanthin production in tobacco flowers. *Nat. Biotechnol.* 18, 888–892. doi: 10.1038/78515

Moravec, T., Schmidt, M. A., Herman, E. M., and Woodford-Thomas, T. (2006). Production of *Escherichia coli* heat labile toxin (LT) B subunit in soybean seed and its analysis of its immunogenicity as an oral vaccine. *Vaccine* 25, 1647–1657. doi: 10.1016/j.vaccine.2006.11.010

Munang'andu, H. M., Mutoloki, S., and Evensen, O. (2015). An overview of challenges limiting the design of protective mucosal vaccines for finfish. *Front. Immunol.* 6:542. doi: 10.3389/fimmu.2015.00542

Mutoloki, S., Munang'Andu, H. M., and Evensen, O. (2015). Oral vaccination of fish- antigen preparations, uptake, and immune induction. *Front. Immunol.* 6:519. doi: 10.3389/fimmu.2015.00519

Naylor, R. L., Hardy, R. W., Bureau, D. P., Chiu A., Elliott, M., Farrell, A. P., et al. (2009). Feeding aquaculture in an era of finite resources. *Proc. Natl. Acad. Sci. U.S.A.* 106, 15103–15110. doi: 10.1073/pnas.0905235106

Nielsen, N. C., Dickinson, C. D., Cho, T. J., Thanh, V. H., Scallon, B. J., Fischer, R. L., et al. (1989). Characterization of the glycinin gene family in soybean. *Plant Cell* 1, 313–328. doi: 10.1105/tpc.1.3.313

Orf, J. H., and Hymowitz, T. (1979). The inheritance of the absence of the Kunitz trypsin inhibitor in seed protein of soybeans. *Crop Sci.* 19, 107–109. doi: 10.2135/cropsci1979.0011183X001900010026x

Orf, J. H., Hymowitz, T., Pull, S. P., and Pueppke, S. G. (1978). Inheritance of soybean seed lectin. *Crop Sci.* 18, 899–900. doi: 10.2135/cropsci1978.0011183X001800050058x

Pierce, E. C., LaFayette, P. R., Ortega, M. A., Joyce, B. L., Kopsell, D. A., and Parrott, W. A. (2015). Ketocarotenoid production in soybean seeds through metabolic engineering. *PLoS ONE* 10:e0138196. doi: 10.1371/journal.pone.0138196

Piller, K. J., Clemente, T. E., Jun, S. M., Petty, C. C., Sato, S., Pascual, D. W., et al. (2005). Expression and immunogenicity of an *Escherichia coli* K99 fimbriae subunit antigen in soybean. *Planta* 222, 6–18. doi: 10.1007/s00425-004-1445-9

Plant, K. P., and LaPatra, S. E. (2011). Advances in fish vaccine delivery. *Dev. Comp. Immunol.* 35, 1253–1259. doi: 10.1016/j.dci.2011.03.007

Ralley, L., Enfissi, E. M. A., Misawa, N., Schuch, W., Bramley, P. M., and Fraser, P. D. (2004). Metabolic engineering of ketocarotenoid formation in higher plants. *Plant J.* 39, 477–486. doi: 10.1111/j.1365-313X.2004.02151.x

Rust, M. B., Barrows F. T., Hardy, R. W., Lazur, A., Naughten, K., and Silverstein J. (2011). *The Future of Aquafeeds. NOAA Technical Memorandum NMFS F/SPO-124. NOAA National Marine Fisheries Service.* Available at: http://www.nmfs.noaa.gov/aquaculture/docs/feeds/the_future_of_aguafeeds_final.pdf

Sahlmann, C., Sutherland, B. J., Kortner, T. M., Koop, B. F., Krogdahl, A., and Bakke, A. M. (2013). Early response of gene expression in the distal intestine of Atlantic salmon (*Salmo salar* L.) during the development of soybean meal induced enteritis. *Fish Shellfish Immunol.* 34, 599–609. doi: 10.1016/j.fsi.2012.11.031

Schmidt, M. A., Barbazuk, W. B., Stanford, M., May, G., Song, Z., Hong, W., et al. (2011). Silencing of soybean seed storage proteins results in a rebalanced protein composition preserving seed protein content without major collateral changes in the metabolome and transcriptome. *Plant Physiol.* 156, 330–345. doi: 10.1104/pp.111.173807

Schmidt, M. A., and Herman, E. M. (2008). Proteome rebalancing in soybean seeds can be exploited to enhance foreign protein accumulation. *Plant Biotechnol. J.* 6, 832–842. doi: 10.1111/j.1467-7652.2008.00364.x

Schmidt, M. A., Hymowitz, T., and Herman, E. M. (2015a). Breeding and characterization of soybean Triple Null; a stack of recessive alleles of kunitz trypsin inhibitor, soybean agglutinin, and P34 allergen nulls. *Plant Breed.* 134, 310–315. doi: 10.1111/pbr.12265

Schmidt, M. A., Parrott, W. A., Hildebrand, D. F., Berg, R. H., Cooksey, A., Pendarvis, K., et al. (2015b). Transgenic soya bean seeds accumulating β-carotene exhibit the collateral enhancements of oleate and protein content traits. *Plant Biotechnol. J.* 13, 590–600. doi: 10.1111/pbi.12286

Tacon, A. G. J., Hasan, M. R., Allan, G., El-Sayed, A. F. M., Jackson, J., Kaushik, S. J., et al. (2012). "Aquaculture feeds : addressing the long-term sustainability of the sector," in *Proceedings of the Global Conference on Aquaculture 2010. Farming the Waters for People and Food*, eds R. P. Subasinghe, J. R. Arthur, D. M. Bartley, S. S. De Silva, M. Halwat, N. Hishamunda, et al. (Rome: FAO), 896.

Tobar, I., Aranciba, S., Torres, C., Vera, V., Soto, P., Carrasco, C., et al. (2015). Successive oral immunizations against Piscirickettsia salmonis and infectious salmon anemia virus are required to maintain a jlong-term protection in farmed salmonids. *Front. Immunol.* 6:244. doi: 10.3389/fimmu.2015.00244

Tobar, J., Jerez, S. A., Caruffo, M., Bravo, C., Contreras, F., Bucarey, S., et al. (2011). Oral vaccination of Atlantic salmon (*Salmo salar*) against salmonid rickettsial septicaemia. *Vaccine* 29, 2336–2340. doi: 10.1016/j.vaccine.2010.12.107

Tocher, D. R. (2015). Omega 3 long chain polyunsaturated fatty acids and aquaculture in prespective. *Aquaculture* 449, 94–107. doi: 10.1016/j.aquaculture.2015.01.010

Watson, R. A., Nowara, G. B., Hartmann, K., Green, B. S., Tracey, S. R., and Carter, C. G. (2015). Marine foods sourced from farther as their use of global ocean primary production increases. *Nat. Commun.* 6:7365. doi: 10.1038/ncomms8365

Williams, C. M., and Burdge, G. (2006). Long chain n-3 PUFA: plant v marine sources. *Proc. Nutr. Soc.* 65, 42–50. doi: 10.1079/PNS2005473

Wu, G., Truksa, M., Datla, N., Vrinten, P., Bauer, J., Zank, T., et al. (2005). Stepwise engineering to produce high yields of very long-chain polyunsaturated fatty acids in plants. *Nat. Biotechnol.* 23, 1013–1017. doi: 10.1038/nbt1107

Zhong, J.-J., Huang, J.-C., Liu, J., Li, Y., Jiang, Y., Xu, Z.-F., et al. (2011). Functional characterization of various algal carotenoid ketolases reveals that ketolating zeaxanthin efficiently is essential for high production of astaxanthin in transgenic *Arabidopsis*. *J. Exp. Bot.* 62, 3659–3669. doi: 10.1093/jxb/err070

Conflict of Interest Statement: The authors declare that the research was conducted in the absence of any commercial or financial relationships that could be construed as a potential conflict of interest.

The reviewer AC and handling Editor declared their shared affiliation, and the handling Editor states that the process nevertheless met the standards of a fair and objective review.

5

The Global Expansion of Quinoa: Trends and Limits

Edited by:
Paul Christiaan Struik,
Wageningen University, Netherlands

Reviewed by:
Colin Kahlil Khoury,
International Center for Tropical Agriculture, Colombia
Eibertus Nicolaas van Loo,
Wageningen University and Research Centre, Netherlands

***Correspondence:**
Didier Bazile
didier.bazile@cirad.fr

Didier Bazile[1], Sven-Erik Jacobsen[2] and Alexis Verniau[3]*

[1] UPR Green, Department of Environment and Societies, French Agricultural Research and International Cooperation Organization, CIRAD, Montpellier, France, [2] Department of Plant and Environmental Sciences, Faculty of Science, University of Copenhagen, Taastrup, Denmark, [3] Ecole Supérieure d'Agricultures, Angers, France

Quinoa (*Chenopodium quinoa* Willd.) was first domesticated in Andean countries over 7000 years ago. Following the Spanish conquest, quinoa was rejected as "Indian food." After centuries of neglect, the potential of quinoa was rediscovered during the second half of the 20th century. Since then, the number of countries importing quinoa increased, with new producers appearing on the map and quinoa now being cultivated in areas outside the Andean countries. The geographical increase in distribution of quinoa has highlighted the difficulty of access to quality seed, which is a key factor for testing the crop outside the Andes. In this context, research partnerships have helped promote the exchange of quinoa germplasm and have allowed trials to be undertaken in non-traditional areas of cultivation. The number of countries growing the crop has increased rapidly from eight in 1980, to 40 in 2010, and to 75 in 2014. A further 20 countries have sown quinoa for the first time in 2015. In this paper, we analyze this trend and discuss the limits of quinoa's expansion. As commercial production of quinoa is expected to develop, changes in international regulatory frameworks on genetic resources are needed in order to facilitate plant breeding for the most adaptive varieties for each region.

Keywords: *Chenopodium quinoa* Willd., plant genetic resources, adaptation, climate change, benefit-sharing seed regulations, farmers' rights

INTRODUCTION

Quinoa (*Chenopodium quinoa* Willd.), is an annual species that originates from South America. Its domestication is thought to have begun in Andean region around 7000 years ago. Generations of farmers have been involved in quinoa selection, which explains the high levels of genetic diversity found today.

Quinoa has remained a staple food for indigenous people of the Andes over the centuries. Following the Spanish conquest, quinoa was rejected as "Indian food" but it has never disappeared despite the introduction of Old World species. After centuries of neglect, the nutritional status of quinoa was rediscovered during the second half of the 20th century, leading to a renaissance of its production (Repo-Carrasco et al., 2003).

Thanks to the high levels of genetic diversity, the crop is highly resilient to agro-ecological extremes (soils, rainfall, temperature, and altitude) and is tolerant to frost, drought, and salinity (Ruiz et al., 2014, 2015). Quinoa can be divided into different groups or ecotypes, reflecting its diffusion from the center of origin around Lake Titicaca. Each of these ecotypes is associated with sub-centers of diversity (Risi and Galwey, 1984), and highly adapted to specific environments. For example, the sea level ecotype from the central and southern part of Chile is the most adapted to temperate environments (Galwey, 1993; Jacobsen, 1997, 2003; Christiansen et al., 2010; Bendevis et al., 2014a,b), useful in developing new varieties for northern latitudes (Jacobsen, 2015). Quinoa diversity is divided into five main ecotypes (Bazile et al., 2013):

- Highlands in Peru and Bolivia;
- Inter-Andean valleys in Colombia, Ecuador, and Peru;
- Salares in Bolivia, Chile, and Argentina;
- Yungas in Bolivia;
- Sea level in Chile.

Research partnerships have often facilitated the exchange of germplasm and have had a powerful impact on this development by strengthened collaborations (Izquierdo et al., 2003). However, partnerships between research institutions for germplasm exchanges need to consider legal and ethical aspects related to the access to genetic resources for experimentation and fair commercial development (Bazile, 2015).

Today, quinoa is still considered a minor crop for global food and agriculture and often classified as a Neglected and Underutilized Species (NUS) with a high potential of development. However, its current distribution is now changing due to changes in consumption patterns. In this article, we analyze this trend, and present a case for the increase in production of quinoa around the world.

SYNTHESIS OF EXISTING DATA

A survey was conducted of resource persons who have been involved in quinoa research over the global level during the last 30 years. The survey identified those countries involved in quinoa research and cultivation. In addition, an analysis was conducted on the issues of access to quality seed of quinoa.

Quinoa project reports were studied to complement the information obtained through surveys. The genotypes used were described, together with their agromorphological characteristics and their origin was determined. The data generated was used to identify the key elements in the expansion of quinoa's cultivation.

A NEGLECTED AND UNDERUTILIZED SPECIES (NUS) WITH A RAPID EXPANSION

From at least 7000 years ago until the beginning of the 1980s, quinoa has only been connected to the Andes. However, when researchers in other countries understood the potential and benefits of quinoa, experimentation has not stopped growing. Studies have been performed in an increasing number of countries. The number of countries growing quinoa has risen rapidly from 8 in 1980 to 75 in 2014, with a further 20 countries which sowed quinoa for the first time in 2015 (Bazile and Baudron, 2015).

The first known experiment conducted outside the Andes took place in 1935 in Kenya (Elmer, 1942) using the cream-colored seed variety, obtained from the Royal Botanical Gardens, Kew, UK. Trials were conducted the response of quinoa to various nutrient deficiencies in 1948, on the tolerance of quinoa to salinity in 1950, and quinoa's growth response to temperature in 1968. Research was then undertaken using Chilean germplasm in the 1980s led by Colorado State University, USA. The same time period saw the beginning of commercial quinoa cultivation in Canada. Other countries followed and quinoa was introduced to UK (1983), Denmark (1984), Tibet (1984), India (1985), The Netherlands (1986), China (1988), Brazil and Cuba (1989) (Bazile et al., 2015).

The project "American and European Test of Quinoa" (Mujica et al., 2001) overseen by Angel Mujica and Sven-Erik Jacobsen together with Juan Izquierdo from the Food and Agriculture Organization of the United Nations (FAO) that began in 1996, was an important effort in the 1990s. The testing followed two other important activities, which were focused on cultivation in Europe: the EU project under the AIR[1] program "Quinoa – a multiple crop for EC's Agriculture diversification, 1993–1997," coordinated by University of Cambridge, UK, and the Food and Agriculture COST Action 814 on "Crop Adaptation to Cool and Wet Climates, 1991–1996," coordinated by the Ministry of Agriculture, Belgium. Through international cooperation initiatives for promoting quinoa, field trials were set up in other countries such as Sweden, Poland, Czech Republic, Austria, Germany, Italy, and Greece.

The General Assembly of the United Nations declared 2013 as the International Year of Quinoa (IYQ). Quinoa was afforded a high profile as a crop with the potential to grow in importance in world agriculture (Jacobsen, 2003; Ruiz et al., 2015). The rapid expansion of the harvested area, with a doubling of countries from 2013, is rapidly changing the perception and representation of quinoa from a minor to a potential major crop (**Figure 1**).

The main quinoa producers in the world, however, are still Bolivia and Peru. In 2013, the area under quinoa cultivation in Bolivia was 75000 ha and in Peru 45000 ha. These two countries produce more than 80% of quinoa in the world, followed by Ecuador, USA, China, Chile, Argentina, France, and Canada, which together represent 15–20% of the world production (Bazile, 2015; Bazile et al., 2015). The area under quinoa cultivation in Europe has increased from 0 in 2008 to 5000 ha in 2015, mainly in France, Spain, and UK (Abbott, personal communication). China has first experimented with cultivating quinoa in Tibet in 1984, while today quinoa is cultivated in nine Chinese regions and over 2500 ha.

The activities carried out due to the IYQ have directly contributed to awareness raising for quinoa cultivation. FAO is actively involved in testing quinoa in 27 countries outside the

[1]FP3-AIR – Specific research and technological development and demonstration programme (EEC) in the field of agriculture and agro-industry, including fisheries, 1990–1994.

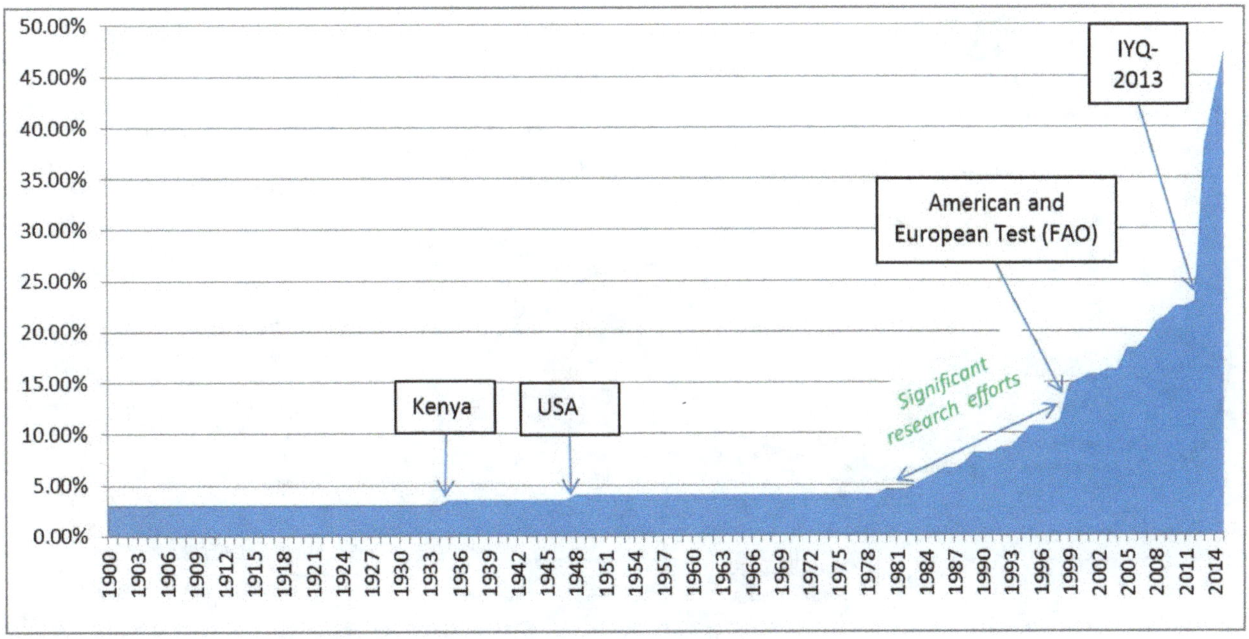

FIGURE 1 | Percentage of UN countries with quinoa experimentation or cultivation (1900–2015).

Andean region, with the aim to promoting food security, due to the nutritional qualities of quinoa and its resistance to abiotic stresses. During 2013–2015, evaluation of quinoa varieties were conducted in:

- Central and Southern Asia (Kyrgyzstan, Tajikistan, Sri Lanka, and Bhutan);
- Western Asia and North Africa (Algeria, Egypt, Iraq, Iran, Lebanon, Mauritania, Sudan, and Yemen);
- Africa (Djibouti, Kenya, Somalia, South Sudan, Ethiopia, Uganda, Zambia, Burkina Faso, Cameroon, Chad, Niger, Senegal, Togo, Ghana, and Guinea).

Size, shape, and compactness of the inflorescence are considered to be important for the rate of maturation. A large open inflorescence will dry quicker after rain and morning dew than a small, compact one, but it may also be prone to seed loss, as quinoa is relatively little domesticated, with few, modern varieties available (Jacobsen, 1997).

The breeding of quinoa, especially Europe and North America, concentrated on gains in time to maturity (around 150 days), high yield and uniformity (Zurita-Silva et al., 2014; Peterson et al., 2015). Another aim has been to reduce the content of saponins in the seed hull. Saponin is a bitter-tasting compound acting as a general defense against biotic stresses. For commercial production, however, the removal of saponins is an additional cost. As a result, sweet cultivars have been developed for use in Europe, especially in the Netherlands (Mastebroek et al., 2000, 2002) and in Denmark (Jacobsen, 2015).

Today quinoa is presently cultivated or tested in 95 countries of the world (Bazile, 2015). This global expansion of quinoa looks set to continue as increasing numbers of countries are testing quinoa.

GLOBAL REGULATORY FRAMEWORKS FOR QUINOA CULTIVATION

Considering the evolution of global regulations on seeds, two periods must be distinguished for their importance on biodiversity sovereignty: the signing of the Convention of Biological Biodiversity (CBD in Rio de Janeiro, Brazil in 1992) marks a particularly break and contrasts the before and the after for plant genetic resources' access and sharing. The CBD stipulates national sovereignty over biodiversity and obliges parties to form bilateral agreements for accessing their genetic resources (United Nations, 1992). In order to facilitate innovation in plant breeding, the International Treaty on Plant Genetic Resources for Food and Agriculture of 2004 provides a specific instrument for exchanging germplasm of the major food and forage crops (FAO, 2009). It established a global and Multilateral System (MLS) to allow farmers, plant breeders, and scientists to exchange plant genetic materials (Louafi et al., 2013).

Quinoa, unfortunately, is not one of the species pertaining to Annex 1 of the Treaty, which is a list of those species included the MLS of exchanges. The Declaration of Cordoba (2012) from the International Seminar "Crops for the XXI Century," the first international event celebrating the UN IYQ 2013, proposed the addition of minor crops to be included in the Annex I to the Treaty. To date, however, there has been no consensus reached.

The different regulations on plant genetic resources are usually applied at different levels (local, national, and international) and for different purposes (genetic resources, varieties and seeds,

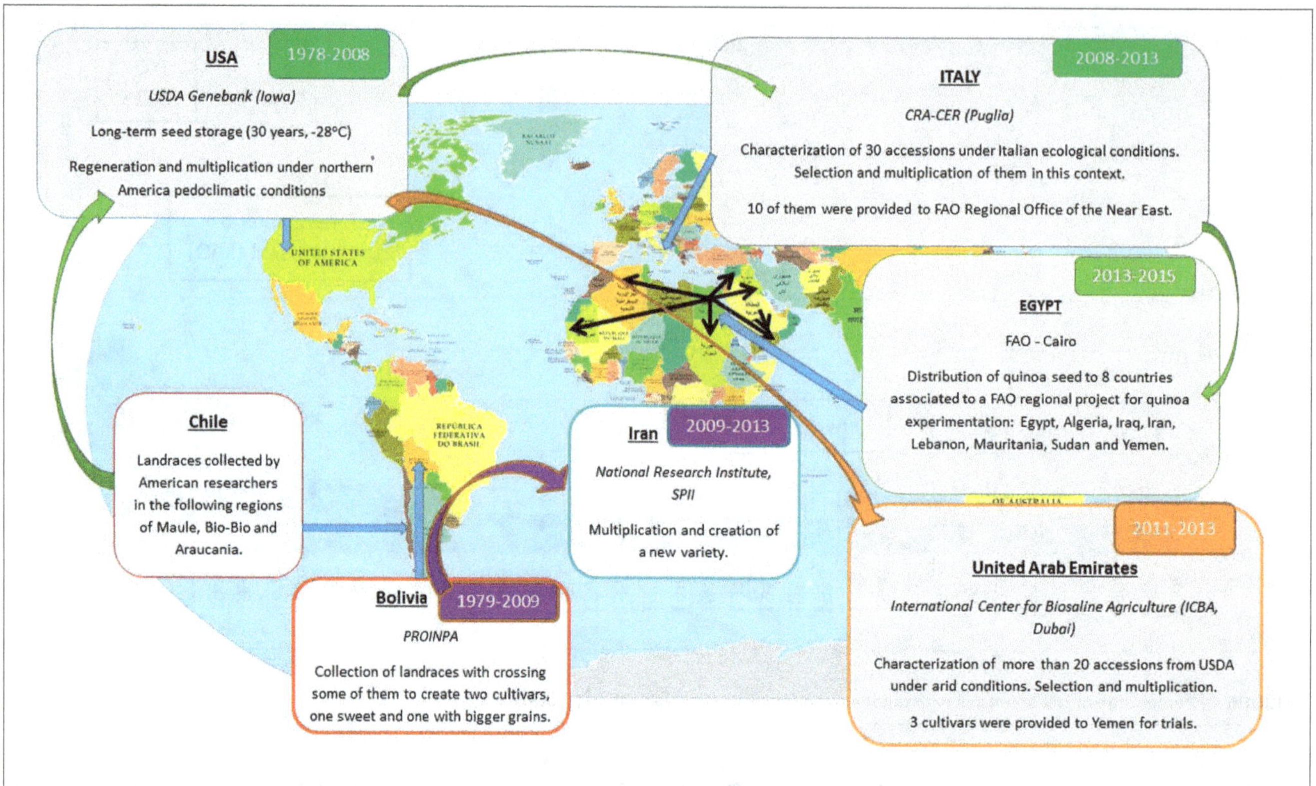

FIGURE 2 | An example of quinoa seed exchanges from Andean countries to new producers' countries.

agricultural by-products, etc.). For quinoa, however, there is no single existing legal framework providing a comprehensive coverage of all the issues related to the genetic resources and their sustainable management (Chevarria-Lazo et al., 2015). The CBD does regulate bilateral access and benefit sharing, but this is difficult to apply to quinoa as the crop is now planted internationally, not restricted to the Andean region, and this has been the case for decades.

Prior to 1992, 25 countries had developed *ex situ* collections of quinoa accessions, which are now distributed without legal necessity of prior agreement from Andeans countries (Rojas et al., 2015). As many countries had established collections prior to coming into force of the CBD, this now means that these countries may develop new varieties from this germplasm without having to refer to the country of origin. Nevertheless, individuals and institutions do not have the right to share quinoa germplasm outside their country as per the terms of the CBD, but this pertains only to those countries that have ratified the CBD. A country a signatory to the CBD, however, has only a moral obligation not to go against it. To date, the CBD has been ratified by 196 countries (the Parties), and 168 have signed it before. However, there are a few countries, such as the USA, have not yet ratified this convention.

Overall, these legal regulations or restrictions at the global level to exchange germplasm across countries, only a small part of quinoa's genetic diversity is available to serve quinoa's adaptation in new environments for expansion of its cultivation.

Exchange of quinoa germplasm has been undertaken both formally via legal provisions (Standard Material Transfer Agreements – SMTAs), or informally via research networks (**Figure 2**). Only 25% of genetic material exchanges correspond to individual exchanges, while 75% of the material has been exchanged among research networks. Considering personal links into existing partnerships, more than 85% of the respondents of the survey declared not signing MTAs.

GENEFLOWS OF QUINOA

Early stages of expansion highlighted the interest among importing countries and consumers for adapting quinoa to new environments, and it facilitated exchange of genetic resources. Now we are entering another phase of quinoa development, and a turning point with many countries interested in quinoa cultivation but with limitations for access to genetic resources.

Figure 2 illustrates the seed exchange among different partners involved in quinoa trials. Collections of quinoa accessions were created during the 1980s by Wilson (1988a,b, 1990) and taken back to the USA. The United States Department of Agriculture (USDA), Iowa, USA, maintains over 229 accessions, which are sent to those who request them. In Italy, the Cereal Research Centre (CER-CRA) obtained and evaluated 100 accessions for cultivation and development of varieties adapted to Mediterranean conditions. After undertaking germplasm characterization, low performing accessions were eliminated and some of the selected accessions were then distributed to Egypt (**Figure 2**). The FAO Regional Office for Near East in Cairo, Egypt, assisted in distributing quinoa accessions among National Research Institutions from eight countries (Algeria, Egypt, Iraq, Iran, Lebanon, Mauritania, Sudan, and Yemen) to conduct an evaluation of these genotypes in semi-arid and arid conditions. This example of seed flow demonstrates the importance of research networks in distributing quinoa seed at the global level. It also has limitations because of change of codes (unique identifier assigned to the accession in a genebank) from one genebank to another, and from Andean countries to other countries. Avoiding code change is one main challenge for data and germplasm sharing with transparency, but sometimes code change serves as a form of protection for plant breeders to mask the origin of the material they are developing.

The current situation provides a specific legal framework for access and exchange of genetic resources that have a strong impact on use and innovation. The Nagoya Protocol on Access and Benefit-sharing was adopted in Japan in 2010, entering in force in 2014 (70 Ratifications and 92 Signatures). It provides an international agreement, which aims at sharing the benefits from the utilization of genetic resources in a fair and equitable way, and contributing to the conservation of biological diversity and the sustainable use of its components. Considering the limits of the CBD for innovation in plant breeding, the Nagoya Protocol provides a new pathway for enhancing legal transparency on procedures for access and benefit-sharing, and for monitoring the utilization of genetic resources. Andean countries need to be active in this international agenda if they want to protect their interests on quinoa. Solutions need to be developed to better harmonize the different existing legal frameworks and to create new complementary ones. The rapid spread of quinoa at the global level provides an opportunity to consider more in depth the implications on seed exchange of the current regulatory instruments for genetic resources, so that they can be improved and implemented. In addition, it also offers an opportunity for plant breeders to reconsider models of varietal innovations in plant breeding to manage genetic resources for recombination without registration of intellectual property rights (IPR) for new varieties. In addition, adaptation of varieties to global climate changes needs to cope with an increasing unpredictability. Genetic structure of future varieties must be able to evolve to face those changes (Murphy et al., 2016). Landraces (crop populations with high level of intra-varietal diversity) from Central and Southern Chile, and in some cases from the highlands of Peru and Bolivia, are those most frequently often tested outside the Andes. In addition, researchers are mostly using only three varieties (Puno, Regalona, and Titicaca) registered through Plant Variety Protection (PVP; Bendevis et al., 2014a,b; Jacobsen, 2015). The geographical expansion of quinoa cultivation has led to an increase in the number of PVP into action worldwide. Some countries (Netherlands, Chile, Canada, Denmark, Peru, UK, and Israel) have developed new varieties, and have set up a PVP certificate in the UPOV system. These commercial varieties facilitate access to seeds for quinoa testing. However, due to their physiological similarity being less day length sensitive than original material from Peru and Bolivia, the genetic diversity for quinoa experimentation and adaptation in new environments is narrow. Genebanks in the Andean region conserve more than 88% of the accessions available worldwide, which is considerable when taking into account the 16,422 accessions of quinoa and its wild relatives are conserved in 59 genebanks in 30 countries (Rojas et al., 2015). Despite the high levels of quinoa diversity conserved, this is currently not exchanged globally due to highly political issues of national sovereignty (Bazile, 2015).

Intellectual Property Rights should be addressed, recognizing the work of the Andean people in the selection and conservation of local quinoa landraces (Fuentes et al., 2012), and maintaining and adding value to quinoa's biodiversity for the benefit of world food security and poverty reduction. The future global seed system for quinoa cultivation needs to focus on increasing linkages among farmers, researchers, plant breeders, and politicians. This requires an iterative dialog at the national, regional, and international levels among all the stakeholders involved in managing and using quinoa diversity.

AUTHOR CONTRIBUTIONS

DB made the survey and networks analysis on quinoa global expansion and has written the first draft of the manuscript for revision by the co-authors. He is currently developing a Global Collaborative Network on Quinoa. S-EJ participated in many introduction of quinoa around the world. He shared his experience to integrate his global view on quinoa expansion to increase the quality of the mini-review. AV participated in various national projects on quinoa trials in the Near East and North Africa during his studies. He mapped the seed exchanges with the example of (**Figure 2**) in this manuscript.

ACKNOWLEDGMENTS

We would like to acknowledge the contributions of all colleagues, specifically from research institutions, that aided the authors in collecting the data for analyses and in understanding the complex relationships among them.

REFERENCES

Bazile, D. (2015). *Le Quinoa: Les Enjeux d'une Conquête*. Paris: Éditions Quæ, 112.

Bazile, D., and Baudron, F. (2015). "The dynamics of the global expansion of quinoa growing in view of its high biodiversity," in *State of the Art Report on Quinoa Around the World in 2013*, eds D. Bazile, H. D. Bertero, and C. Nieto (Roma: FAO & CIRAD), 42–55.

Bazile, D., Bertero, H. D., and Nieto, C. (eds). (2015). *State of the Art Report on Quinoa Around the World in 2013*. Roma: FAO & CIRAD, 589.

Bazile, D., Fuentes, F., and Mujica, A. (2013). "Historical perspectives and domestication," in *Quinoa: Botany, Production and Uses*, eds A. Bhargava and S. Srivastava (Wallingford: CABI Publishing), 16–35.

Bendevis, M. A., Sun, Y., Rosenqvist, E., Shabala, S., Liu, F., and Jacobsen, S.-E. (2014a). Photoperiodic effects on short-pulse14C assimilation and overall carbon and nitrogen allocation patterns in contrasting quinoa cultivars. *Environ. Exp. Bot.* 104, 9–15. doi: 10.1016/j.envexpbot.2014.03.002

Bendevis, M. A., Sun, Y., Shabala, S., Rosenqvist, E., Liu, F., and Jacobsen, S.-E. (2014b). Differentiation of photoperiod-induced ABA and soluble sugar responses of two quinoa (*Chenopodium quinoa* Willd.) Cultivars. *J. Plant Growth Regul.* 33, 562–570. doi: 10.1007/s00344-013-9406-9

Chevarria-Lazo, M., Bazile, D., Dessauw, D., Louafi, S., Trommetter, M., and Hocdé, H. (2015). "Quinoa and the exchange of genetic resources: improving the regulation systems," in *State of the Art Report on Quinoa Around the World in 2013*, eds D. Bazile, H. D. Bertero, and C. Nieto (Roma: FAO & CIRAD), 83–105.

Christiansen, J. L., Jacobsen, S.-E., and Jørgensen, S. T. (2010). Photoperiodic effect on flowering and seed development in quinoa (*Chenopodium quinoa* Willd.). *Act Agric. Scand.* 60, 539–544.

Elmer, L. A. (1942). Quinoa (*Chenopodium quinoa*). *East Afr. Agric. J.* 8, 21–23. doi: 10.1080/03670074.1942.11664212

FAO (2009). *International Treaty on Plant Genetic Resources for Food and Agriculture.* Rome: FAO, 68.

Fuentes, F., Bazile, D., Bhargava, A., and Martinez, E. A. (2012). Implications of farmers' seed exchanges for on-farm conservation of quinoa, as revealed by its genetic diversity in Chile. *J. Agric. Sci.* 150, 702–716. doi: 10.1017/S0021859612000056

Galwey, N. W. (1993). The potential of quinoa as a multipurpose crop for agricultural diversification: a review. *Ind. Crops Prod.* 1, 101–106. doi: 10.1016/0926-6690(92)90006-H

Izquierdo, J., Mujica, A., Marathee, J. P., and Jacobsen, S.-E. (2003). Horizontal, technical cooperation in research on quinoa (*Chenopodium quinoa* Willd.). *Food Rev. Int.* 19, 25–29. doi: 10.1081/FRI-120018865

Jacobsen, S.-E. (1997). Adaptation of quinoa (*Chenopodium quinoa*) to Northern European agriculture: studies on developmental pattern. *Euphytica* 96, 41–48. doi: 10.1023/A:1002992718009

Jacobsen, S. E. (2003). The worldwide potential of quinoa (*Chenopodium quinoa* Willd.). *Food Rev. Int.* 19, 167–177. doi: 10.1081/FRI-120018883

Jacobsen, S. E. (2015). "Adaptation and scope for quinoa in northern latitudes of Europe," in *State of the Art Report on Quinoa Around the World in 2013*, eds D. Bazile, H. D. Bertero, and C. Nieto (Roma: FAO & CIRAD), 436–446.

Louafi, S., Bazile, D., and Noyer, J. L. (2013). "Conserving and cultivating agricultural genetic diversity: transcending established divides," in *Cultivating Biodiversity to Transform Agriculture*, ed. E. Hainzelin (Heidelberg: Springer), 181–230. doi: 10.1007/978-94-007-7984-6_6

Mastebroek, H. D., Limburg, H., Gilles, T., and Marvin, H. J. P. (2000). Occurrence of sapogenins in leaves and seeds of Quinoa (*Chenopodium quinoa* Willd.). *J. Sci. Food Agric.* 80, 152–156. doi: 10.1002/(SICI)1097-0010(20000101)80:1<152::AID-JSFA503>3.0.CO;2-P

Mastebroek, H. D., Van Loo, E. N., and Dolstra, O. (2002). Combining ability for seed yield traits of *Chenopodium quinoa* breeding lines. *Euphytica* 125, 427–432. doi: 10.1023/A:1016030129541

Mujica, A., Jacobsen, S. E., Izquierdo, J., and Marathee, J. P. (2001). *Resultados de la Prueba Americana y Europea de la Quinua.* Puno: FAO, UNA & CIP.

Murphy, K. M., Bazile, D., Kellogg, J., and Rahmanian, M. (2016). Seeds without Borders: development of a worldwide consortium on evolutionary participatory breeding in quinoa. *Front. Plant Sci.* 7:608. doi: 10.3389/fpls.2016.00608

Peterson, A., Jacobsen, S.-E., Bonifacio, A., and Murphy, K. (2015). A crossing method for quinoa. *Sustainability* 7, 3230–3243. doi: 10.3390/su7033230

Repo-Carrasco, R., Espinoza, C., and Jacobsen, S.-E. (2003). Nutritional value and use of the Andean crops quinoa (*Chenopodium quinoa*) and kañiwa (*Chenopodium pallidicaule*). *Food Rev. Int.* 19, 179–189. doi: 10.1081/FRI-120018884

Risi, J. C., and Galwey, N. W. (1984). The *Chenopodium* grains of the Andes: inca crops for modern agriculture. *Adv. Appl. Biol.* 10, 145–216.

Rojas, W., Pinto, M., Alanoca, C., Gomez Pando, L., Leon-Lobos, P., Alercia, A., et al. (2015). "Quinoa genetic resources and ex situ conservation," in *State of the Art Report on Quinoa Around the World in 2013*, eds D. Bazile, H. D. Bertero, and C. Nieto (Roma: FAO & CIRAD), 56–82.

Ruiz, K. B., Biondi, S., Martínez, E. A., Orsini, F., Antognoni, F., and Jacobsen, S.-E. (2015). Quinoa – a model crop for understanding salt tolerance mechanisms in halophytes. *Plant Biosyst.* 150, 357–371. doi: 10.1080/11263504.2015.1027317

Ruiz, K. B., Biondi, S., Oses, R., Acuña-Rodríguez, I. S., Antognoni, F., Martinez-Mosqueira, E. A., et al. (2014). Quinoa biodiversity and sustainability for food security under climate change. A review. *Agron. Sustain. Dev.* 34, 349–359. doi: 10.1007/s13593-013-0195-0

United Nations (1992). *Convention on Biological Diversity.* New York, NY: UNEP, 30.

Wilson, H. D. (1988a). Quinoa biosystematics I: domesticated populations. *Econ. Bot.* 42, 461–477. doi: 10.1007/BF02862791

Wilson, H. D. (1988b). Quinoa biosystematics II: free living populations. *Econ. Bot.* 42, 478–494. doi: 10.1007/BF02862791

Wilson, H. D. (1990). Quinua and relatives (*Chenopodium* sect. *Chenopodium* subsect. Cellulata). *Econ. Bot.* 44, 92–110. doi: 10.1007/BF02860478

Zurita-Silva, A., Fuentes, F., Zamora, P., Jacobsen, S. E., and Schwember, A. (2014). Breeding quinoa (*Chenopodium quinoa* Willd.): potential and perspectives. *Mol. Breed.* 34, 13–30. doi: 10.1007/s11032-014-0023-5

Conflict of Interest Statement: The authors declare that the research was conducted in the absence of any commercial or financial relationships that could be construed as a potential conflict of interest.

The reviewer EL and handling Editor declared their shared affiliation, and the handling Editor states that the process nevertheless met the standards of a fair and objective review.

6

Variability in Seed Traits in a Collection of *Cannabis sativa* L. Genotypes

Edited by:
Antonio M. De Ron,
Spanish National Research Council - Misión Biológica de Galicia, Spain

Reviewed by:
Miroslav Griga,
AGRITEC Plant Research Ltd., Czech Republic
Albert Vandenberg,
University of Saskatchewan, Canada

***Correspondence:**
Remo Reggiani
reggiani@ibba.cnr.it

*Incoronata Galasso [1], Roberto Russo [1], Sergio Mapelli [1], Elena Ponzoni [1], Ida M. Brambilla [1], Giovanna Battelli [2] and Remo Reggiani [1]**

[1] Institute of Agricultural Biology and Biotechnology, CNR, Milano, Italy, [2] Institute of Science of Food Production, CNR, Milano, Italy

The seed of *Cannabis sativa* L. is an expanding source of proteins and oil for both humans and animals. In this study, the proximate composition of a collection of hemp cultivars and accessions of different geographical origins grown under the same conditions for 1 year was analyzed in order to identify potential accessions to improve hemp cultivars. Fatty acids, tocopherols, and antinutritional components, as well as concentrations of crude protein and oil were quantified. The seed oil concentrations varied between 285 and 360 g kg^{-1} dry seed (DS), while crude protein ranged between 316 and 356 g kg^{-1} dry matter (DM). The seed oil was mainly composed of unsaturated fatty acids and, as expected, the dominant fatty acids were linoleic and α-linolenic acid. A high variability among the cultivars and accessions was also detected for polyphenolic content which ranged from 5.88 to 10.63 g kg^{-1} DM, cv. Felina was the richest, whereas cv. Finola had the lowest polyphenolic content. Regarding antinutritional compounds in seed, a high variability was detected among all genotypes analyzed and phytic acid was particularly abundant (ranging between 43 and 75 g kg^{-1} DM). In conclusion, our results reveal noticeable differences among hemp seed genotypes for antinutritional components, oil and protein content. Collectively, this study suggests that the hemp seed is an interesting product in terms of protein, oil and antioxidant molecules but a reduction of phytic acid would be desirable for both humans and monogastric animals. The high variability detected among the different genotypes indicates that an improvement of hemp seed might be possible by conventional and/or molecular breeding.

Keywords: protein, antinutritional compounds, fatty acids, tocopherols, polyphenols, antioxidant activity, genotype diversity

INTRODUCTION

Cannabis sativa L. (hemp) is a wind-pollinated annual plant that originated in central Asia (Li, 1973, 1974; Mukherjee et al., 2008). Hemp, naturally, is a dioecious crop, but some monoecious cultivars have been obtained as a result of earlier breeding efforts. Hemp is an ancient crop that has been cultivated worldwide until the early twentieth century, after which its cultivation declined. Recently, interest in this multipurpose crop delivering fibers, shives, and seeds, has been renewed by an increasing demand not only for natural fibers but also for the high content and quality of seed protein and oil. Hemp seed contains 25–35% oil, 20–25% protein, 20–30% carbohydrates, 10–15%

insoluble fiber, vitamins, and minerals such as phosphorus, potassium, magnesium, sulfur, and calcium (Deferne and Pate, 1996; Callaway, 2004; House et al., 2010). The increasing demand for vegetable oils and proteins, along with current awareness about their nutritional and functional role in human diet, has made indispensable to characterize new plant sources. In this regard, hemp seed contains all the essential amino acids and fatty acids necessary to maintain healthy human life, and it might be a new good source of nutrients for both humans and livestock (Osburn, 1992; Deferne and Pate, 1996; Callaway, 2004; Russo and Reggiani, 2015a).

The principal value of hemp seed oil resides in its fatty acid composition. It contains the two dietary essential fatty acids: linoleic acid (LA; 18:2ω6) and the α-linolenic acid (ALA; 18:3ω3) in the ratio of 2.5–3:1, which has been claimed as ideal for human nutrition (Simopoulos, 2008). In addition, the biological metabolites of LA and ALA, the γ-linolenic acid (GLA; 18:3ω6) and stearidonic acid (SDA; 18:4ω3) are also present in hemp seed oil, which makes the nutritional value of hemp seed superior to other seed oils (Leizer et al., 2000). Both GLA and SDA act as precursors for the rapid synthesis of longer chain fatty acids which are involved in many important biological processes (Guil-Guerrero et al., 2010).

Furthermore, hemp seeds are also an excellent nutritional source of high quality proteins (Russo and Reggiani, 2015a), as they are easily digested, absorbed, and utilized (House et al., 2010). The two main proteins in hemp seeds are the globulin edestin and albumin (Callaway, 2004; Tang et al., 2006). Edestin, which accounts for about 60–80% of the total protein content (Odani and Odani, 1998; Tang et al., 2006), has been the subject of intensive investigation for a long time. Recently, an accurate characterization of edestin at biochemical and molecular level has been done by Kim and Lee (2011) and Docimo et al. (2014). This storage protein, which contains exceptionally high levels of arginine and glutamic acids (Russo and Reggiani, 2015a), is easily digested and rich in all essential amino acids (Callaway, 2004; House et al., 2010; Docimo et al., 2014).

However, the nutritional quality of plant proteins, as measured by their amino acid composition and digestibility, is influenced by numerous factors. The amino acid composition may be influenced by genotypic variability or agronomic conditions such as soil fertility and postharvest processing that alters the ratio of seed components (e.g., shelling). The digestibility of proteins may be affected by protein structure, presence of antinutritional compounds and high temperature processing (Sarwar, 1997). Recently, it has been demonstrated that hemp meal (a by-product obtained from mechanical or solvent extraction of the seed oil) contains some antinutritional compounds that need to be considered carefully in feeding applications (Russo and Reggiani, 2013). In hemp seed, the content of phytic acid (inositol hexaphosphate, IP_6) and trypsin inhibitors appeared fairly high compared to a number of other antinutritional compounds analyzed (Russo and Reggiani, 2015b). Phytic acid is the main organic form of phosphorus present in plant seeds. Its presence reduces protein digestibility and increases the excretion of endogenous nitrogen, amino acids and minerals, in particular bivalent cations (Cowieson et al., 2004). The trypsin inhibitors are present in many species of *Leguminosae*, *Brassicaceae*, and *Gramineae*, and reduce the biological activity of trypsin enzyme which is involved in the breakdown of many different proteins (Guillamón et al., 2008).

Therefore, in this work several hemp cultivars and accessions of different origins, grow in the same environment for 1 year, have been characterized for quantity and quality of seed oil, protein content, antinutritional compounds, polyphenols, tocopherols, and antioxidant activity. The objective of this study was the evaluation of these biochemical traits in order to identify potential accessions to improve hemp cultivars. To achieve this goal, we analyzed the seeds of 20 genotypes composed by the most used non-drug hemp cultivars that are allowed to be cultivated in Europe and a collection of hemp accessions obtained from genebanks.

MATERIALS AND METHODS

Plant Material

Seeds of hemp cultivars (cvs) and accessions were kindly provided by the genebank of IPK Gatersleben (Leibniz Institute of Plant Genetics and Crop Plant Research) (http://gbis.ipk-gatersleben.de/), the Centro di Ricerca per le Colture Industriali (CRA-CIN, Italy) and by seed companies (**Table 1**). All genotypes were sown at the beginning of spring (30 March 2014) in pots of about 1 m^2 with three replications. The pots were fertilized with 80 kg ha^{-1} N after the seedlings had emerged. The average air temperature was 19.4°C (30 March–31 October 2014). During summer the plants were irrigated when necessary. In order to avoid cross-pollination among the different cvs and accessions, all pots were covered with two layers of cloth (~ 200 mesh; Virexgomma, Somma Lombardo, Varese, Italy) during flowering time. Seeds were harvested at maturity from August to October. The harvesting date depended on the genotype. After harvesting, seeds were forced-air-dried (30°C) in a ventilated chamber until 8% seed moisture was reached. Thousand seed weight (TSW) was established on five replicates of 100 seeds each. Seeds were manually counted and weighed. During the flowering stage, a pool of inflorescences (flowers and leaves) were collected from each pot. The Δ^9-tetrahydrocannabinol (THC) concentration was measured in each genotype by the HPTLC densitometry method according to Fischedick et al. (2009).

Total Protein and Antinutritional Compound Analysis

Samples of whole seeds were ground in a mortar and defatted by extracting with hexane (1:10, w/v). Protein was extracted from defatted flours using a Plant Total Protein Extraction Kit (Sigma-Aldrich, St. Louis, USA). The Plant Total Protein Extraction Kit includes two reagents, a plant specific protease inhibitor cocktail and a chaotropic reagent with increased solubilizing power to extract more hydrophobic proteins. Protein content was determined via a Quantum Protein Kit (Euroclone, Milan, Italy), using Bovine Serum Albumin (BSA) as standard.

TABLE 1 | List of hemp accessions (from CAN19 to CAN58) and cultivars (from cv. Finola to cv. Kc Dora) analyzed in this study.

Hemp genotypes	Sex type	Origin	Flowering time	Supply source
C. sativa CAN19	D*	Italy	End July	IPK[a]
C. sativa CAN20	D	Korea	End July	IPK
C. sativa CAN24	D	Italy	End July	IPK
C. sativa CAN26	D	Turkey	End July	IPK
C. sativa CAN39	D	China	End July	IPK
C. sativa CAN40	D	Italy	End July	IPK
C. sativa CAN48	D	Italy	End July	IPK
C. sativa CAN51	D	Argentina	End July	IPK
C. sativa CAN58	D	Spain	End August	IPK
C. sativa cv. Finola	D	Finland	End June	Seed company[b]
C. sativa cv. Carmagnola	D	Italy	Beginning August	Seed company[c]
C. sativa cv. Carmagnola Selezionata (CS)	D	Italy	Beginning August	CRA-CIN[d]
C. sativa cv. Fibranova	D	Italy	Beginning August	CRA-CIN[d]
C. sativa cv. Fedora	M**	France	Middle July	Seed company[e]
C. sativa cv. Futura 75	M	France	Middle July	Seed company[e]
C. sativa cv. Felina 32	M	France	Middle July	Seed company[e]
C. sativa cv. Ferimon	M	France	Middle July	Seed company[e]
C. sativa cv. Codimono	M	Italy	Middle July	CRA-CIN[f]
C. sativa cv. Carmaleonte	M	Italy	Middle July	CRA-CIN[f]
C. sativa cv. Kc Dora	M	Hungary	End July	Seed company[g]

[a]*IPK, Leibniz Institute of Plant Genetics and Crop Plant Research, Gatersleben, Germany.*
[b]*Exchange with other scientists. Seed company: Finola Ltd, http://www.finola.fi/.*
[c]*Assocanapa srl, Carmagnola (Torino, Italy) http://www.assocanapa.org/.*
[d]*CRA-CIN, Centro di ricerca per le colture industriali, Rovigo (IT) from dr. G. Grassi.*
[e]*Cooperative Centrale des Producteurs de Semences de Chanvre (Le Mans, France).*
[f]*CRA-CIN, Centro di ricerca per le colture industriali (CRA-CIN), Bologna (IT) from dr. A. Di Candilo.*
[g]*Agromag Kft, Hungary, http://www.agromag.hu/en.*
D, Dioecious plant; M**, Monoecious plant.*

Phytic acid was extracted from defatted flour as described by De Boland et al. (1975) with minor modifications. Briefly, the phytic acid was extracted with 0.4 N HCl plus 0.7 M Na_2SO_4 and then precipitated by 15 mM $FeCl_3$ in 0.2 N HCl. The phosphorus content of the precipitate was colorimetrically determined, after acid digestion with sulfuric acid, according to Chen et al. (1956). Phytic acid content was determined by multiplying the phytate phosphorus content by a constant factor of 3.55 (Raboy and Dickinson, 1984).

Trypsin inhibitor (TI) was extracted from defatted flour with 2 mM glycine buffer pH 11 containing 2 mM NaCl, 10 mM urea, and 25 mM EDTA. Trypsin inhibitor activity was measured as described by Russo and Reggiani (2013). To 200 μL of extract were added 200 μL (40 $\mu g.mL^{-1}$) of trypsin and then pre-incubated at 37°C for 3 min. After this, 500 μL (0.4 $mg.mL^{-1}$) of N-benzoyl-DL-arginine-p-nitroanilide (pre-warmed to 37°C) were added to start the reaction. After incubation at 37°C for 10 min, 100 μL 30% (v/v) acetic acid was added to terminate the reaction and the mix centrifuged. Activity of trypsin was measured by the absorbance at 410 nm due to p-nitroaniline released. One unit of TI was defined as 0.01 decreases in absorbance at 410 nm under assay conditions compared with the control (without inhibitor).

Total Phenolic Content, Antioxidant Activity and Tocopherols

Phenols were extracted from defatted flours with 80% ethanol at 70°C (Russo and Reggiani, 2015c). The total phenolic (TP) content was determined using the Folin-Ciocalteu method according to Velioglu et al. (2006). One hundred microliters of the ethanolic extract or caffeic acid standard (50, 100, 200, 400 μg) was combined with 500 μL of Folin-Ciocalteu reagent and allowed to stand at room temperature for 5 min. Then, 400 μL of 60 $g.L^{-1}$ sodium carbonate solution were added to the mix and the tubes were heated at 45°C for 15 min. The absorbance was measured at 765 nm after sitting for 30 min in the dark. The results were expressed as caffeic acid equivalents per g of dry weight (g CAE kg^{-1} DW).

The antioxidant activity of the ethanolic extracts was determined by the Antioxidant Assay Kit (Sigma-Aldrich, Milan, Italy) according to the manufacturer's protocol. In the kit, trolox (a water-soluble vitamin E analog) was present as a standard for antioxidant control. Trolox has been broadly applied in assaying food samples (Re et al., 1999; Huang et al., 2005; Russo and Reggiani, 2015c). The results were expressed as mol trolox equivalents per kg of dry weight (mol TE kg^{-1} DW).

Tocopherols were analyzed in HPLC by direct injection of oil (Gimeno et al., 2000). The tocopherol isomers were separated using an HPLC Jasco Tritotar III and Jasco MD910 Diode Array Detector (DAD). An amount of 5–10 μl of pure oil was loaded into a Kinetex 2.6 μm C18 100A (Phenomenex, Castel Maggiore, Italy) column (100 × 4.6 mm) and eluted with 1.5 $ml.min^{-1}$ MeOH 95%. The DAD spectra gave information on the purity of tocopherols. Absorbance at 280 nm was elaborated by the Borwin software system to determine tocopherol amounts in comparison with standards (Sigma-Aldrich, St. Louis, USA).

Oil Content and Fatty Acid Composition

Lipid extraction was performed according to Hara and Radin (1978). The fatty acid composition was determined in triplicate by gas chromatography (GC) of the oils after transesterification according to Christie (1982) with the modifications described by Chouinard et al. (1999). Gas chromatography analyses were carried out on an Agilent 6890 GC system (Paolo alto, CA) equipped with autosampler, on column injector, and a FID detector. The separation was performed on a 100% dimethylpolysiloxane column (CP-Sil 88 for FAME, 100 m × 0.25 mm × 25 μm) adopting the following conditions. The carrier gas was H_2 at a constant pressure of 28 psi. Column temperature was set at 190°C for 22 min, and then increased to 240°C at a rate of 10°C min^{-1} and maintained for 13 min. Fatty acid methyl esters (FAMEs) were identified by comparison of their retention times with known standards (37-component FAME mix, Supelco 47885-U) and expressed as relative percentage of total fatty acids.

Statistical Analyses

All analyses were carried out in triplicate. Analysis of Variance (ANOVA) was applied to establish significant differences ($P \leq 0.01$) between hemp genotypes in the levels of seed traits using SPSS version 16.0 software. Mean separation was performed using Duncan's test and referring to $P \leq 0.05$ probability level. Pearson's correlations between seed traits were also calculated.

RESULTS

All hemp plants considered in this study were analyzed for THC content (**Figure 1**). They showed THC concentrations between 0.07 and 0.19% in their inflorescences, which are below the authorized limit of 0.20% on a DM basis set in Italy.

Protein and Antinutritional Compounds

Hemp seed meal was found to be a rich protein source with a mean protein content of 337 $g.kg^{-1}$ of DW (**Table 2**). ANOVA analysis revealed significant differences in protein content among cvs and accessions. The CAN20 and CAN40 accessions showed the highest protein concentration in the meal (356 and 354 $g.kg^{-1}$ DW, respectively). Pearson analysis demonstrated that the total protein content was highly correlated with TSW indicating that protein is an important component of seed weight (**Table 3**). A positive correlation is also evident among protein content and antioxidant activity (**Table 3**).

Phytate content was very high in all tested meals (mean 5.84%). The lowest phytate content was detected in Kc Dora 4.37% and CAM40 4.38%, whereas cv Felina meal showed almost twice that amount (7.55%; **Table 2**). Regarding the antinutritional

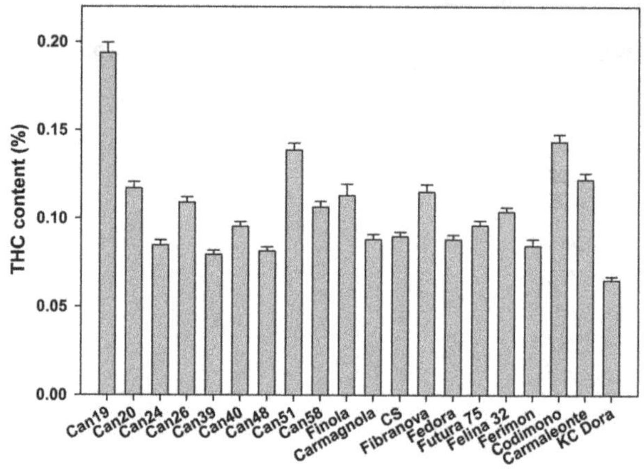

FIGURE 1 | THC content in inflorescences of hemp accessions (from CAN19 to CAN58) and cultivars (from cv. Finola to cv. Kc Dora).

TI, **Table 2** shows that its activity varied from 10.8 to 27.8 $unit.mg^{-1}$ of defatted flour. Among hemp genotypes, Fibranova exhibited the lowest TI activity (**Table 2**). ANOVA analysis also evidenced significant differences in TI activity between cvs and accessions (cvs are on average richer in TI). Pearson analysis indicated that TI activity is highly correlated to phytate content (**Table 3**).

Tocopherols, Total Phenolic Content, and Antioxidant Activity

All hemp seeds analyzed in this study contained primarily γ-tocopherol and lesser quantities of α- and δ-tocopherols (data not shown). **Table 2** reports the total content of tocopherols including all isomers identified. The average of tocopherol concentration for hemp genotypes was 0.88 $g.Kg^{-1}$ of oil. ANOVA analysis showed that the differences in tocopherols among all genotypes and groups (cvs vs. accessions) are significant at 0.01 level. A similar result was observed for TP content (**Table 2**). The highest TP content was detected in some monoecious cultivars (see those with "a" letters in parenthesis by the Duncan's range test). According to **Table 2**, TP varied from 5.88 to 10.63 mg CAE g^{-1} DW of defatted flour. Pearson analysis showed that TP content was positively correlated to antioxidant activity ($r = 0.775$, **Table 3**) and tocopherols ($r = 0.275$) at the 0.01 and 0.05 levels, respectively. These data support the notion that phenols and tocopherols are effective scavengers of free radicals. As can be seen in **Table 2**, antioxidant activity ranged from 0.92 to 1.70 mol TE kg^{-1} DW of defatted flour. The cv Felina exhibited the highest antioxidant activity (**Table 2**).

Oil Content and Composition

The oil content of the hemp seed genotypes ranged from 28.5 to 36.0% (**Table 2**) with an average of 31.9%. The highest oil content was measured in CAN58 (36%) and the lowest in Carmaleonte (28.5%). ANOVA revealed significant differences among genotypes for oil content at the 0.01 level.

TABLE 2 | Seed traits in hemp accessions (from CAM19 to CAN58) and cultivars (from Finola to Kc Dora).

	TSW[a]	Oil[b]	ALA(ω3)[c]	LA(ω6)[c]	Toc[d]	Protein[d]	Phytate[d]	TI[e]	TP[f]	Antiox[g]
CAN19	17.5 (efgh)	32.5 (defg)	15.9 (n)	60.2 (b)	0.76 (gh)	345 (cd)	53.0 (f)	18.4 (f)	7.34 (gh)	1.66 (abc)
CAN20	19.1 (cd)	29.5 (hij)	18.5 (g)	59.4 (de)	0.82 (efgh)	356 (a)	60.1 (de)	13.8 (k)	9.01 (bc)	1.67 (abc)
CAN24	13.9 (i)	32.7 (cdef)	28.6 (a)	46.6 (m)	0.85 (defg)	331 (ef)	60.3 (de)	17.1 (g)	6.93 (h)	1.22 (g)
CAN26	21.7 (b)	34.9 (ab)	19.3 (f)	57.1 (g)	0.85 (defg)	345 (c)	44.6 (h)	15.8 (hij)	8.65 (cd)	1.63 (bcd)
CAN39	9.4 (k)	30.4 (ghij)	15.8 (o)	55.2 (j)	0.60 (i)	320 (gh)	53.7 (f)	11.3 (l)	7.90 (ef)	1.44 (f)
CAN40	19.3 (cd)	34.8 (abc)	20.5 (c)	54.0 (l)	0.69 (hi)	354 (ab)	43.8 (h)	16.3 (ghi)	7.83 (fg)	1.46 (f)
CAN48	22.8 (a)	34.0 (abcd)	12.8 (q)	58.7 (f)	0.75 (gh)	339 (cde)	73.3 (a)	18.9 (ef)	8.39 (de)	1.59 (de)
CAN51	21.6 (b)	35.4 (ab)	20.0 (e)	56.1 (i)	0.79 (fgh)	336 (de)	65.5 (b)	20.2 (cd)	7.86 (fg)	1.46 (f)
CAN58	16.7 (h)	36.0 (a)	20.2 (d)	54.4 (k)	0.86 (defg)	346 (bc)	61.7 (cd)	16.2 (ghij)	9.03 (bc)	1.66 (abc)
Finola	11.6 (j)	32.7 (cdef)	21.1 (b)	59.6 (cd)	0.90 (defg)	317 (gh)	63.0 (bcd)	16.2 (ghij)	5.88 (i)	0.92 (h)
Carmagnola	21.4 (b)	31.0 (fghi)	18.4 (g)	56.4 (h)	1.10 (bc)	334 (ef)	52.7 (f)	19.4 (de)	9.03 (bc)	1.62 (bcd)
CS	19.3 (cd)	31.6 (efgh)	17.8 (i)	59.7 (cd)	1.10 (b)	316 (h)	58.5 (e)	21.1 (c)	9.22 (b)	1.60 (de)
Fibranova	18.3 (de)	30.7 (fghij)	15.7 (o)	58.6 (f)	1.35 (a)	325 (fg)	53.1 (f)	10.8 (l)	9.40 (b)	1.62 (bcd)
Fedora	17.8 (efg)	29.2 (ij)	16.3 (l)	59.7 (c)	0.93 (def)	339 (cde)	64.4 (bc)	21.0 (c)	7.51 (fg)	1.56 (e)
Futura 75	17.9 (ef)	29.8 (hij)	17.5 (j)	60.1 (b)	0.96 (cde)	337 (cde)	75.0 (a)	27.8 (a)	10.29 (a)	1.68 (ab)
Felina 32	16.8 (gh)	29.6 (hij)	16.0 (n)	60.2 (b)	0.98 (bcd)	343 (cd)	75.5 (a)	26.6 (b)	10.63 (a)	1.70 (a)
Ferimon	17.1 (fgh)	30.1 (hij)	16.2 (m)	62.0 (a)	0.95 (de)	344 (cd)	72.7 (a)	26.4 (b)	10.11 (a)	1.68 (ab)
Codimono	17.3 (efgh)	33.8 (bcde)	18.2 (h)	56.4 (hi)	0.97 (bcde)	336 (de)	48.1 (g)	15.4 (ij)	9.09 (bc)	1.67 (abc)
Carmaleonte	19.3 (cd)	28.5 (j)	13.1 (p)	59.1 (e)	0.61 (i)	345 (cd)	44.3 (h)	15.2 (j)	10.11 (a)	1.67 (abc)
Kc Dora	19.5 (c)	31.5 (fgh)	16.8 (k)	56.2 (hi)	0.88 (defg)	332 (ef)	43.7 (h)	16.9 (gh)	9.00 (bc)	1.66 (abc)
Mean ± SEM	17.9 ± 0.4	31.9 ± 0.3	17.9 ± 0.4	57.5 ± 0.4	0.88 ± 0.04	337 ± 2	58.4 ± 1.4	18.2 ± 0.6	8.66 ± 0.16	1.56 ± 0.02
P genotype	<0.01	<0.01	<0.01	<0.01	<0.01	<0.01	<0.01	<0.01	<0.01	<0.01
P group*	ns	<0.01	<0.05	<0.01	<0.01	<0.01	ns	<0.01	<0.01	ns

[a] *Data expressed as g.1000 seeds.*
[b] *Data expressed as % of seed weight.*
[c] *Data expressed as % of total fatty acids.*
[d] *Data expressed as $g.kg^{-1}$ of oil.*
[e] *Data expressed as $unit.mg^{-1}$.*
[f] *Data expressed as g CAE kg^{-1}.*
[g] *Data expressed as mol TE kg^{-1}.*
*Means with different letters in parentheses within the same row differ significantly by Duncan's range test ($P \leq 0.05$); *group cvs vs. group accessions. TSW, Thousand seed weight; Toc, tocopherols; TI, Trypsin inhibitor activity; TP, Total phenolic content; Antiox, Antioxidant activity.*

TABLE 3 | Pearson correlation coefficient (r) among seed traits in twenty hemp genotypes.

	TSW	Oil	ALA(ω3)	LA(ω6)	Toc	Protein	Phytate	TI	TP
Oil	0.192	1							
ALA(ω3)	−0.280*	0.386**	1						
LA(ω6)	0.205	−0.424**	−0.749**	1					
Toc	0.132	−0.106	0.034	0.189	1				
Protein	0.402**	0.074	−0.080	0.065	−0.314*	1			
Phytate	−0.090	−0.168	−0.058	0.315*	0.150	−0.035	1		
TI	0.189	−0.174	−0.089	0.386**	0.163	0.127	0.679**	1	
TP	0.346**	−0.367**	−0.486**	0.389**	0.272*	0.259*	0.140	0.379**	1
Antiox	0.540**	−0.198	−0.580**	0.372**	0.152	0.463**	−0.035	0.237	0.775**

***Correlation is significant at the 0.01 level; *Correlation is significant at the 0.05 level.*
TSW, Thousand seed weight; Toc, tocopherols; TI, Trypsin inhibitor activity; TP, Total phenolic content; Antiox, Antioxidant activity.

The fatty acid composition of hemp seed oils is shown in **Table 4**. The principal saturated fatty acid was the palmitic acid (PA; 16:0) in all genotypes ranging between 5.98 and 8.60%, followed by stearic acid (SA; 18:0), varying from 2.26 to 4.61%. The saturated fatty acid fraction represents on average 10.8% of the total fatty acids (**Table 5**). The major monounsaturated fatty acid was oleic acid (OA; 18:1*cis*9, average of 13.4%) (**Table 4**) and CAN39 seeds showed the highest OA (16.8%) content. With respect to the essential fatty acids, the seeds of the analyzed genotypes contained on average 55.7% of LA and 17.4% of

TABLE 4 | Oil composition in twenty hemp seed accessions (from CAN19 to CAN58) and cultivars (from cv. Finola to cv. Kc Dora).

	PA 16:0	PAL 16:1	SA 18:0	OA 18:1*cis*9	LA 18:2ω6	GLA 18:3ω6	ALA 18:3ω3	SDA 18:4ω3	ARA 20:0	GOA 20:1*cis*11	BA 22:0
CAN19	6.05 (ij)	0.12 (j)	2.60 (h)	13.6 (g)	56.7 (c)	3.48 (c)	14.9 (m)	0.99 (bc)	0.86 (abc)	0.34 (abcd)	0.34 (ab)
CAN20	5.98 (j)	0.12 (j)	3.03 (ef)	11.5 (m)	57.9 (a)	1.46 (i)	18.0 (f)	0.51 (f)	0.90 (ab)	0.34 (abcd)	0.32 (abc)
CAN24	8.60 (a)	0.15 (def)	3.03 (ef)	11.5 (m)	46.1 (j)	0.49 (p)	28.4 (a)	0.23 (jk)	0.85 (abc)	0.37 (ab)	0.34 (ab)
CAN26	7.30 (b)	0.16 (cde)	2.62 (h)	12.3 (k)	56.1 (d)	0.94 (k)	18.9 (e)	0.34 (h)	0.67 (cdef)	0.31 (bcde)	0.27 (bcde)
CAN39	7.09 (bc)	0.16 (cde)	3.69 (b)	16.8 (a)	54.5 (g)	0.65 (n)	15.7 (j)	0.16 (l)	0.77 (bcde)	0.29 (cdef)	0.24 (bcde)
CAN40	6.79 (cde)	0.13 (hij)	3.34 (d)	13.9 (f)	51.8 (i)	2.27 (f)	19.8 (c)	0.75 (de)	0.71 (bcdef)	0.28 (def)	0.27 (bcde)
CAN48	7.35 (b)	0.19 (b)	2.91 (g)	16.3 (b)	57.1 (b)	1.54 (h)	12.4 (n)	0.32 (hi)	1.00 (a)	0.39 (a)	0.42 (a)
CAN51	5.98 (j)	0.12 (j)	2.95 (fg)	13.5 (g)	55.4 (f)	0.74 (m)	19.7 (c)	0.27 (ij)	0.75 (bcde)	0.31 (bcde)	0.30 (bcd)
CAN58	6.97 (cd)	0.12 (j)	4.61 (a)	12.5 (j)	53.7 (h)	0.62 (n)	19.9 (b)	0.24 (j)	0.73 (bcdef)	0.28 (def)	0.25 (bcde)
Finola	5.99 (j)	0.14 (fghi)	2.26 (i)	9.2 (n)	55.0 (f)	4.54 (a)	19.6 (d)	1.54 (a)	1.02 (a)	0.34 (abcd)	0.35 (ab)
Carmagnola	7.31 (b)	0.20 (b)	3.49 (c)	13.2 (h)	55.3 (f)	1.19 (j)	17.9 (f)	0.49 (f)	0.57 (ef)	0.23 (f)	0.15 (e)
CS	6.16 (hij)	0.15 (def)	2.32 (i)	12.7 (i)	58.2 (a)	1.46 (i)	17.3 (g)	0.49 (f)	0.60 (ef)	0.31 (bcde)	0.27 (bcde)
Fibranova	6.94 (cde)	0.14 (fghi)	2.87 (g)	14.4 (e)	57.4 (b)	1.21 (j)	15.3 (k)	0.42 (g)	0.77 (bcde)	0.33 (abcd)	0.30 (bcd)
Fedora	6.38 (gh)	0.17 (c)	2.97 (fg)	13.2 (h)	56.5 (c)	3.23 (d)	15.3 (k)	0.98 (c)	0.73 (bcdef)	0.36 (abc)	0.19 (de)
Futura 75	6.34 (ghi)	0.24 (a)	3.03 (ef)	11.9 (l)	58.0 (a)	2.14 (g)	16.8 (h)	0.71 (e)	0.52 (f)	0.25 (ef)	0.15 (e)
Felina 32	6.03 (ij)	0.16 (cd)	2.57 (h)	13.9 (f)	57.9 (a)	2.28 (f)	15.2 (kl)	0.77 (d)	0.62 (def)	0.32 (bcde)	0.20 (cde)
Ferimon	6.44 (fgh)	0.16 (cd)	2.53 (h)	11.4 (m)	58.1 (a)	3.85 (b)	15.1 (l)	1.04 (b)	0.74 (bcde)	0.30 (bcde)	0.29 (bcd)
Codimono	6.72 (def)	0.13 (ghij)	2.94 (fg)	14.3 (e)	55.8 (e)	0.57 (o)	18.0 (f)	0.18 (kl)	0.77 (bcde)	0.31 (bcde)	0.33 (ab)
Carmaleonte	7.34 (b)	0.15 (de)	3.13 (e)	15.7 (c)	56.7 (c)	2.43 (e)	12.6 (n)	0.55 (f)	0.88 (abc)	0.22 (f)	0.35 (ab)
Kc Dora	6.64 (efg)	0.14 (efgh)	3.45 (c)	15.4 (d)	55.3 (f)	0.90 (l)	16.5 (i)	0.26 (j)	0.82 (abcd)	0.30 (bcde)	0.32 (abc)
Mean ± SEM	6.72 ± 0.15	0.15 ± 0.01	3.02 ± 0.12	13.4 ± 0.43	55.7 ± 0.64	1.80 ± 0.27	17.4 ± 0.79	0.56 ± 0.08	0.76 ± 0.03	0.31 ± 0.01	0.28 ± 0.02
P genotype	<0.01	<0.01	<0.01	<0.01	<0.01	<0.01	<0.01	<0.01	<0.01	<0.01	<0.01
P group*	ns	<0.01	<0.05	ns	<0.01	<0.01	<0.01	<0.01	ns	ns	ns

*Data are expressed as % of oil fraction; *group cvs vs. group accessions. PA, palmitic acid; PAL, palmitoleic acid; SA, stearic acid; OA, oleic acid; LA, linoleic acid; GLA, γ-linolenic acid; ALA, α-linolenic acid; SDA, stearidonic acid; ARA, arachidic acid; GOA, gondoic acid; BA, behenic acid.*

Means with different letters in parentheses within the same row differ significantly by Duncan's range test ($P \leq 0.05$).

TABLE 5 | Saturated fatty acid (SFA), monounsaturated fatty acid (MUFA), polyunsaturated fatty acid (PUFA), and LA:ALA (ω6:ω3) ratio in twenty hemp accessions (from CAN19 to CAN58) and cultivars (from cv. Finola to cv. Kc Dora).

	SFA	MUFA	PUFA	LA:ALA ratio
CAN19	9.8	14.1	76.1	3.78
CAN20	10.2	11.9	77.8	3.22
CAN24	12.8	12.0	75.2	1.63
CAN26	10.9	12.8	76.4	2.96
CAN39	11.8	14.2	71.0	3.49
CAN40	11.1	14.3	74.6	2.63
CAN48	11.7	16.9	71.4	4.60
CAN51	10.0	13.9	76.1	2.80
CAN58	12.6	12.9	74.6	2.69
Finola	9.6	9.7	80.7	2.82
Carmagnola	11.5	13.6	74.9	3.06
CS	9.4	13.2	77.5	3.36
Fibranova	10.9	14.8	74.3	3.72
Fedora	10.3	13.7	76.0	3.67
Futura 75	10.0	12.4	77.6	3.44
Felina 32	9.4	14.4	76.2	3.77
Ferimon	10.0	11.9	78.1	3.84
Codimono	10.8	14.7	74.5	3.11
Carmaleonte	11.7	16.1	72.2	4.52
Kc Dora	11.2	15.9	72.9	3.35
Mean ± SEM	10.8 ± 0.2	13.7 ± 0.4	75.4 ± 0.5	3.32 ± 0.15

Data are expressed as % of total fatty acids.

ALA, resulting in a ω6:ω3 ratio = 3.3:1 (**Table 5**). Significant differences among the genotypes analyzed were observed for LA (from 46.1 to 58.1%), the fatty acid present in hemp seed in the greatest proportion. Furthermore, noticeable differences in ALA concentration were observed, with the greatest amount measured in CAM24 (28.4%) and the lowest quantity in CAN48 and Carmaleonte (~12%).

The other fatty acids: palmitoleic (POA; 16:1), GLA, SDA, arachidic (ARA; 20:0), gondoic (GOA; 20:1*cis*11), and behenic acid (BA; 22:0) were found in small amounts. However, the percentage of GLA was the highest among the other minor unsaturated fatty acids and varied from 0.49 to 4.54%. The highest content of GLA was detected in Finola (4.54%) which also contained the highest quantity of SDA (1.54%), followed by Ferimon, CAN19 and Fedora with a content in GLA of 3.85, 3.48, 3.23%, respectively and of 1.04, 0.99, 0.98% of SDA, respectively (**Table 4**). ANOVA revealed significant variability among genotypes for all the fatty acids at the 0.01 level. Pearson correlation analysis showed that ALA is positively correlated with oil content and inversely correlated with LA, TP content and antioxidant activity (**Table 3**), whereas LA was positively correlated with antinutritional compounds (phytate and TI), TP content and antioxidant activity.

DISCUSSION

The possibility of using industrial hemp cvs with low levels of THC (<0.20% THC on a DM basis as established by the European Union and also by Italian law) is leading to a re-introduction of this plant into the Italian production systems. In the 2015 about 2,000 hectares (ha) have been cultivated with hemp in Italy and in all countries of European Union its cultivation has reached 25,224 ha (European Industrial Hemp Association, http://eiha.org/downloads/). The interest on hemp is due to the huge number of products that can be obtained from this plant (Small and Marcus, 2002), and recently it is also largely focused on seeds, which are rich in healthy nutritional fats and proteins for both humans and animals (Mustafà et al., 1999; Rodriguez-Leyva and Pierce, 2010; Girgih et al., 2014).

In this work we have compared the seed composition of several cvs that are allowed to be cultivated in Europe with a collection of hemp accessions obtained from the IPK genebank and grown for 1 year in the same environment. Although hemp tend to perform better in its area of development (Dempsey, 1975), in our trial, except CAM58 which richest the flowering stage (see **Table 1**) and, consequently, seed maturation very late and cv Finola which was able to produce a discrete quantity of seeds but the plant remained very short, all the remaining hemp genotypes showed a good environmental adaptation.

In general, a high variability ($P < 0.01$) for all biochemical traits among the genotypes analyzed was observed (**Table 2**). Some of them, such as TSW, total protein, oil and ALA content appeared higher in hemp seed accessions than in cvs. In fact, protein content was >31% in all meals analyzed but CAN20 and CAN40 had the highest protein content ~35% and the latter also had an oil content of up to 34.8%.

As reported by others (Kriese et al., 2004; Chen et al., 2010; Vonapartis et al., 2015), the most abundant fatty acids in our hemp seed collection proved to be LA (average 55.7%), ALA (17.4%), and OA (13.4%), which together represent 86.5% of total fatty acids (**Tables 4, 5**). However, a high variability for LA and ALA content was detected among all hemp genotypes (**Table 4**) and the most different was CAN24 which showed 46.1 and 28.4% of LA and ALA respectively, giving a ratio ω6:ω3 (LA:ALA) of 1.63:1, which has never been reported before. In fact, a significant number of previous studies on hemp oil composition (Oomah et al., 2002; Callaway, 2004; Vonapartis et al., 2015) as well as most of the hemp genotypes analyzed in this study showed a ω6:ω3 ratio of roughly 3:1, reported as optimal for human nutrition (Erasmus, 1993; Leizer et al., 2000).

Simopoulos (2008) reported a ω6:ω3 ratio of 15–16.7:1 in Western European and American food supplies. This very high ratio, in today's diets, promotes the pathogenesis of many diseases, including cancer, cardiovascular and inflammatory and autoimmune diseases, whereas increased levels of ω3 polyunsaturated fatty acids (PUFAs) exert suppressive effects (Simopoulos, 2008; Gómez Candela et al., 2011). Therefore, the oil of CAN24 might be used, particularly in Western diets characterized by high LA and low ALA fatty acid intake, which is considered detrimental to health (Simopoulos, 2008; Gómez Candela et al., 2011).

In addition to LA and ALA, hemp seed oil also contains their direct metabolites GLA and SDA, which serve as intermediaries in the formation of longer-chain fatty acids and vital hormone-like prostaglandins in the body (Guil-Guerrero et al., 2010). The results of this study confirm that hemp seed oil is a good source of PUFA, especially GLA and SDA, but the content detected among the hemp seed genotypes varies greatly, even more than that detected for LA and ALA. As expected, the cv Finola showed the highest content of GLA (4.54%) and SDA (1.54%; **Table 4**) which is in agreement with results from other authors (Callaway, 2004; Vonapartis et al., 2015). However, an interesting level of GLA and SDA was also detected in CAN19, Fedora and Ferimon showing a content of GLA >3.0% and of SDA ~1%.

Hemp seeds showed a substantial content of tocopherols and total phenols (**Table 2**). In particular, TP content was higher in hemp seed than in flax, which is a species extremely rich in total phenols (Russo and Reggiani, 2015c). The high antioxidant activity detected in almost all hemp seeds, due to tocopherols and TP content, can ensure high oxidative stability that would makes hemp oil suitable for food and industrial applications, as well as improving the nutritional quality of human diets.

However, despite the functional and nutraceutical properties of this crop, the seeds are not totally free of antinutritional compounds, such as phytic acid, condensed tannins, cyanogenic glycosides, saponin and TI (Russo and Reggiani, 2013; Pojić et al., 2014). The quite high content (mean 5.84%) of phytate in our hemp collection confirmed the findings of Russo and Reggiani (2013) and the variability between the different hemp genotypes was statistically significant (**Table 2**). In some genotypes, the phytate content was more than 7% (those with letter "a" by Duncan's range test). Therefore, the use of such meal appears rather limited for monogastric animals and in human nutrition since a high level of phytic acid will affect protein digestibility, organoleptic properties, and bioavailability of macro- and microelements (Guillamón et al., 2008). Furthermore the high phytate content found, particularly in French hemp seed cultivars, will greatly limit the use of this protein source in novel food or feed formulations.

In addition, TI is responsible for the reduced digestibility of seed proteins and, for this reason, it constitutes one of the main antinutritional factor in plant seeds (Peric et al., 2013). As shown in **Table 2**, considerable variability in TI activity among all hemp meals was evident. However, the levels of TI in hemp seeds were lower in comparison with those observed in some cereals and soybean (Sosulki et al., 1988; Tsukamoto et al., 1995). Pearson coefficient ($r = 0.679$) indicated a strong positive correlation between TI activity and phytate content (**Table 3**) and this is in agreement with previous results reported by Russo and Reggiani (2015b) who analyzed six hemp cultivars grown in two different experimental fields for two consecutive years.

In conclusion our analysis confirmed that hemp seed is an excellent product in terms of protein content, oil composition and antioxidant molecules for food and feed formulations, but the content of some antinutritional compounds, particularly phytate will influence negatively its potential nutritional value. Therefore, an improvement for this trait would be necessary. Based on our data, some hemp accessions present in our germplasm collection, appear to contain more interesting traits than the cultivated hemp cvs. In fact, hemp accessions appeared more interesting then hemp cvs with regard to protein and oil content. In particular the Italian CAN40 showed a high content of both, therefore, might be used to improve oil and protein content in hemp cvs through hybridization and selection. Furthermore, the Italian accession CAM40 might be also a good candidate to reduce the content of phytate. In general, we observed that, among the genotypes analyzed, all Italian dioic and monoic cvs showed less phytate content than the monoic French cvs. This trend, except for CAN48, is also observed in the other Italian accessions. Therefore, our analysis evidenced that, although we analyzed a small segment of hemp germplasm collection and for only 1 year, probably an improvement of the cultivated hemp cvs for some traits might be possible using the large genetic variability present in the germplasm of hemp.

AUTHOR CONTRIBUTIONS

This work was carried out in collaboration among authors. The authors of this manuscript worked together to design, conduct, analyze and interpret the findings of these experiments. All the authors read and approved the final manuscript.

ACKNOWLEDGMENTS

This study was partially supported by the Research Project FilAgro—Strategie innovative e sostenibili per la filiera agroalimentare—Accordo Quadro N. 18093/RCC del 5/8/2013.

REFERENCES

Callaway, J. C. (2004). Hemp seed as a nutritional resource: an overview. *Euphytica* 140, 65–72. doi: 10.1007/s10681-004-4811-6

Chen, T., He, J., Zhang, J., Zhang, H., Qian, P., Hao, J., et al. (2010). Analytical characterization of hemp seed (seed of *Cannabis sativa* L.) oil from eight regions in China. *J. Diet Suppl.* 7, 117–129. doi: 10.3109/19390211003781669

Chen, P. S., Toribara, T. Y., and Warner, H. (1956). Microdetermination of phosphorus. *Anal. Chem.* 28, 1756–1758. doi: 10.1021/ac60119a033

Christie, W. W. (1982). A simple procedure for rapid transmethylation of glycerolipids and cholesteryl esters. *J. Lipid Res.* 23, 1072–1075.

Chouinard, P. Y., Corneau, L., Barbano, D. M., Metzger, L. E., and Bauman, D. E. (1999). Conjugated linoleic acids alter milk fatty acid composition and inhibit milk fat secretion in dairy cows. *J. Nutr.* 129, 1579–1584.

Cowieson, A. J., Acamovic, T., and Bedford, M. R. (2004). The effects of phytase and phytic acid on the loss of endogenous amino acids and minerals from chickens. *Brit. Poultry Sci.* 45, 101–108. doi: 10.1080/00071660410001668923

De Boland, A. R., Garner, G. B., and O'Dell, B. L. (1975). Identification and properties of phytate in cereal grains and oilseed products. *J. Agric. Food Chem.* 23, 1186–1189. doi: 10.1021/jf60202a038

Deferne, J. L., and Pate, D. W. (1996). Hemp seed oil: a source of valuable essential fatty acids, *J. Int. Hemp Assn.* 3, 4–7.

Dempsey, J. M. (ed.). (1975). "Hemp," in *Fibre Crops* (Gainesville, FL: University of Florida Press), 46–89.

Docimo, T., Caruso, I., Ponzoni, E., Mattana, M., and Galasso, I. (2014). Molecular characterization of edestin gene family in *Cannabis sativa* L. *Plant Physiol. Bioch.* 84, 142–148. doi: 10.1016/j.plaphy.2014.09.011

Erasmus, U. (1993). *Fats that Heal, Fats that Kill.* Books Alive, 7436 Fraser Drive, V5J 5B9 (Burnaby, B.C).

Fischedick, J. T., Glas, R., Hazekamp, A., and Verpoorte, R. (2009). A qualitative and quantitative HPTLC densitometry method for the analysis of cannabinoids in *Cannabis sativa* L. *Phytochem. Analysis* 20, 421–426. doi: 10.1002/pca.1143

Gimeno, E., Calero, E., Castellote, A. I., Lamuela-Raventós, R. M., de la Torre, M. C., and López-Sabater, M. C. (2000). Simultaneous determination of α-tocopherol and β-carotene in olive oil by reversed-phase high-performance liquid chromatography. *J. Chromatogr. A* 881, 255–259. doi: 10.1016/S0021-9673(00)00272-7

Girgih, A. T., He, R., Malomo, S., Offengenden, M., Wu, J., and Aluko, R. E. (2014). Structural and functional characterization of hemp seed (*Cannabis sativa* L.) protein-derived antioxidant and antihypertensive peptides. *J. Funct. Foods* 6: 384–394. doi: 10.1016/j.jff.2013.11.005

Gómez Candela, C., Bermejo López, L. M., and Loria Kohen, V. (2011). Importance of a balanced omega 6/omega 3 ratio for the maintenance of health: nutritional recommendations. *Nutr. Hosp.* 26, 323–329. doi: 10.1590/S0212-16112011000200013

Guil-Guerrero, J. L., Rincon-Cervera, M. A., and Venegas-Venegas, E. (2010). Gamma-linolenic and stearidonic acids: purification and upgrading of C18-PUFA oils. *Eur. J. Lipid Sci. Technol.* 112, 1068–1081. doi: 10.1002/ejlt.200900294

Guillamón, E., Pedrosa, M. M., Burbano, C., Cuadrado, C., and Muzquiz, M. (2008). The trypsin inhibitors present in seed of different grain legume species and cultivar. *Food Chem.* 107, 68–74. doi: 10.1016/j.foodchem.2007.07.029

Hara, A., and Radin, N. S. (1978). Lipid extraction of tissues with a low-toxicity solvent. *Anal. Biochem.* 90, 420–426. doi: 10.1016/0003-2697(78)90046-5

House, J. D., Neufeld, J., and Leson, G. (2010). Evaluating the quality of protein from hemp seed (*Cannabis sativa* L.) products through the use of the protein digestibility corrected amino acid score method. *J. Agric. Food Chem.* 58, 11801–11807. doi: 10.1021/jf102636b

Huang, D., Ou, B., and Prior, R. L. (2005). The chemistry behind antioxidant capacity assays. *J. Agr. Food Chem.* 53, 1841–1856. doi: 10.1021/jf030723c

Kim, J. J., and Lee, M. J. (2011). Isolation and characterization of edestin from Cheungsam hempseed. *J. Appl. Biol. Chem.* 54, 84–88. doi: 10.3839/jabc.2011.015

Kriese, U., Schumann, E., Weber, W. E., Beyer, M., Bruhl, L., and Matthaus, B. (2004). Oil content, tocopherol composition and fatty acid patterns of the seeds of 51 *Cannabis sativa* L. genotypes. *Euphytica* 137, 339–351. doi: 10.1023/B:EUPH.0000040473.23941.76

Leizer, C., Ribnicky, D., Poulev, A., Dushenkov, S., and Raskin, I. (2000). The composition of hemp seed oil and its potential as an important source of nutrition. *J. Nutraceut. Funct. Med. Foods* 2, 35–53. doi: 10.1300/J133v02n04_04

Li, H. L. (1973). An archaeological and historical account of cannabis in China. *Econ. Botany* 28, 437–444.

Li, H. L. (1974). The origin and use of Cannabis in eastern Asia linguistic-cultural implications. *Econ. Botany* 28, 293–301.

Odani, S., and Odani, S. (1998). Isolation and primary structure of a methionine and cystine-rich seed protein of *Cannabis sativa* L. *Biosci. Biotechnol. Biochem.* 62, 650–654.

Oomah, B. D., Busson, M., Godfrey, D. V., and Drover, J. C. (2002). Characteristics of hemp (*Cannabis sativa* L.) seed oil. *Food Chem.* 76, 33–43. doi: 10.1016/S0308-8146(01)00245-X

Osburn, L. (1992). Hemp seed: the most nutritionally complete food source in the world. Part two: hemp seed oils and the flow of live force. *Hemp Line J.* 1, 12–13.

Mukherjee, A., Roy, S. C., De Bera, S., Jiang, H. E., Li, X., Li, C. S., et al. (2008). Results of molecular analysis of an archaeological hemp (*Cannabis sativa* L.) DNA sample from North West China. *Genet. Resour. Crop Evol.* 55, 481–485. doi: 10.1007/s10722-008-9343-9

Mustafà, A. F., McKinnon, J. J., and Christensen, D. A. (1999). The Nutritive Value of Hemp Meal for Ruminants. *Canadian. J. Anim. Sci.* 79, 91–95. doi: 10.4141/A98-031

Peric, V., Srebric, M., and Mladenovic-Drinic, S. (2013). Trypsin inhibitors in soybean. soybean: a dawn to the legume world: the future of soybean research is already here. *J. Int. Legume Soc.* 1, 23–24. Available online at: http://ils.nsseme.com/

Pojić, M., Mišan, A., Sakač, M., Dapčević Hadnadev, T., Šarić, B., Milovanović, I., et al. (2014). Characterization of byproducts originating from hemp oil processing. *J. Agric. Food Chem.* 62, 12436–12442. doi: 10.1021/jf5044426

Raboy, V., and Dickinson, D. B. (1984). Effect of phosphorus and zinc nutrition on soybean seed phytic acid and zinc. *Plant Physiol.* 75, 1094–1098.

Re, R., Pellegrini, N., Proteggente, A., Pannala, A., Yang, M., and Rice-Evans, C. (1999). Antioxidant activity applying an improved ABTS radical cation decolorization assay. *Free Radic. Biol. Med.* 26, 1231–1237. doi: 10.1016/S0891-5849(98)00315-3

Rodriguez-Leyva, D., and Pierce, N. G. (2010). The cardiac and haemostatic effects of dietary. hemp seed. *Nutrit. Metabol.* 7:32. doi: 10.1186/1743-7075-7-32

Russo, R., and Reggiani, R. (2013). Variability of antinutritional compounds in hemp seed meal of Italian and French varieties. *Plant* 1, 25–29. doi: 10.11648/j.plant.20130102

Russo, R., and Reggiani, R. (2015a). "Protein concentration and amino acid profile in hemp seed and flaxseed meal," in *Eucarpia International Symposium on Protein Crops*, ed A. De Ron (Pontevedra: Spanish Association for Legumes), 193–195.

Russo, R., and Reggiani, R. (2015b). Evaluation of protein concentration, amino acid profile and antinutritional compounds in hemp seed meal from dioecious and monoecious varieties. *Am. J. Plant Sci.* 6, 14–22. doi: 10.4236/ajps.2015.61003

Russo, R., and Reggiani, R. (2015c). Phenolics and antioxidant activity in flax varieties with different productive attitude. *Int. Food Res. J.* 22, 1736–1739. Available online at: http://www.ifrj.upm.edu.my

Sarwar, G. (1997). The protein digestibility-corrected amino acid score method overestimates quality of proteins containing antinutritional factors and of poorly digestible proteins supplemented with limiting amino acids in rats. *J. Nutr.* 127, 758–764.

Simopoulos, A. P. (2008). The importance of the omega-6/omega-3 fatty acid ratio in cardiovascular disease and other chronic diseases. *Exp. Biol. Med.* 233, 674–688. doi: 10.3181/0711-MR-311

Small, E., and Marcus, D. (2002). "Hemp: a new crop with new uses for North America," in *Trends in New Crops and New Uses*, eds J. Janick and A. Whipkey (Alexandria, VA: ASHS Press), 284–326.

Sosulki, F. W., Minja, L. A., and Christensen, D. A. (1988). Trypsin inhibitors and nutritive value in cereals. *Plant Foods and Hum. Nutr.* 38, 23–34. doi: 10.1007/BF01092307

Tang, C. H., Ten, Z., Wang, X. S., and Yang, X. Q. (2006). Physicochemical and functional properties of hemp (*Cannabis sativa* L.) protein isolate. *J. Agric. Food Chem.* 54, 8945–8950. doi: 10.1021/jf0619176

Tsukamoto, C., Shimada, S., Igita, K., Kudou, S., Kokubun, M., Okubo, K., et al. (1995). Factor affecting isoflavone content in soybean seeds: changes in isoflavones, saponins, and composition of fatty acids at different temperatures during seed development. *J. Agric. Food Chem.* 43, 1184–1192. doi: 10.1021/jf00053a012

Velioglu, Y. S., Mazza, G., Gao, L., and Oomah, B. D. (2006). Antioxidant activity and total phenolics in selected fruits, vegetables, and grain products. *J. Agr. Food Chem.* 46, 4113–4117. doi: 10.1021/jf9801973

Vonapartis, E., Aubin, M. P., Seguin, P., Mustafa, A. F., and Charron, J. B. (2015). Seed composition of ten industrial hemp cultivars approved for production in Canada. *J. Food Comp. Anal.* 39, 8–12. doi: 10.1016/j.jfca.2014.11.004

Conflict of Interest Statement: The authors declare that the research was conducted in the absence of any commercial or financial relationships that could be construed as a potential conflict of interest.

7

Bacterial Communities in the Rhizosphere of Amilaceous Maize (*Zea mays* L.) as Assessed by Pyrosequencing

Edited by:
Antonio M. De Ron,
Spanish National Research Council,
Spain

Reviewed by:
Birgit Mitter,
Austrian Institute of Technology,
Austria
Christel Baum,
University of Rostock, Germany
Diego Flores,
Colegio de Postgraduados, Mexico

***Correspondence:**
David Correa-Galeote
david.correa@eez.csic.es
Eulogio J. Bedmar
eulogio.bedmar@eez.csic.es

David Correa-Galeote[1], Eulogio J. Bedmar[1]*, Antonio J. Fernández-González[1], Manuel Fernández-López[1] and Gregorio J. Arone[2]*

[1] Department of Soil Microbiology and Symbiotic Systems, Estación Experimental del Zaidín, Agencia Estatal Consejo Superior de Investigaciones Científicas, Granada, Spain, [2] Department of Agricultural Sciences, National University of Huancavelica, Huancavelica, Peru

Maize (*Zea mays* L.) is the staple diet of the native peasants in the Quechua region of the Peruvian Andes who continue growing it in small plots called chacras following ancestral traditions. The abundance and structure of bacterial communities associated with the roots of amilaceous maize has not been studied in Andean chacras. Accordingly, the main objective of this study was to describe the rhizospheric bacterial diversity of amilaceous maize grown either in the presence or the absence of bur clover cultivated in soils from the Quechua maize belt. Three 16S rRNA gene libraries, one corresponding to sequences of bacteria from bulk soil of a chacra maintained under fallow conditions, the second from the rhizosphere of maize-cultivated soils, and the third prepared from rhizospheric soil of maize cultivated in intercropping with bur clover were examined using pyrosequencing tags spanning the V4 and V5 hypervariable regions of the gene. A total of 26031 sequences were found that grouped into 5955 distinct operational taxonomic units which distributed in 309 genera. The numbers of OTUs in the libraries from the maize-cultivated soils were significantly higher than those found in the libraries from bulk soil. One hundred ninety seven genera were found in the bulk soil library and 234 and 203 were in those from the maize and maize/bur clover-cultivated soils. Sixteen out of the 309 genera had a relative abundance higher than 0.5% and the were (in decreasing order of abundance) Gp4, Gp6, *Flavobacterium*, Subdivision3 genera incertae sedis of the Verrucomicrobia phylum, *Gemmatimonas*, *Dechloromonas*, *Ohtaekwangia*, *Rhodoferax*, *Gaiella*, *Opitutus*, Gp7, *Spartobacteria* genera incertae sedis, *Terrimonas*, Gp5, *Steroidobacter* and *Parcubacteria* genera incertae sedis. Genera Gp4 and Gp6 of the Acidobacteria, *Gemmatimonas* and *Rhodoferax* were the most abundant in bulk soil, whereas *Flavobacterium*, *Dechloromonas* and *Ohtaekwangia* were the main genera in the rhizosphere of maize intercropped with bur clover, and Gp4, Subdivision3 genera incertae sedis of phylum Verrucomicrobia, Gp6 and *Rhodoferax* were the main genera in the rhizosphere of maize plants. Taken together, our results suggest that bur clover produces specific changes in rhizospheric bacterial diversity of amilaceous maize plants.

Keywords: maize rhizospheric soil, bacterial community, pyrosequncing, 16S rRNA gene, Andean chacras

INTRODUCTION

Maize (*Zea mays* L.) reached the southern Andean highlands after its domestication in Mexico some 8700 years before the present (Piperno et al., 2009; van Heerwaarden et al., 2011; Grobman et al., 2012), and is since then the staple diet of the Quechua natives. Despite 500 years of Western colonization and national modernization efforts, native peasants continue growing amilaceous maize as did their ancestors, mostly without chemical fertilization and no irrigation. Most farmers (94.3%) do not use certified seeds because the fields are planted with their own community seeds (Instituto Nacional de Estadística e Informática [INEI], 2012), which include more than 1600 entries grouped in 55 races (Manrique, 1997).

It is well established that, in addition to mineral fertilizers, which are not usually employed in the Quechua agricultural practices, N inputs to agriculture can be obtained from symbiotic nitrogen fixation by rhizobial bacteria inside root nodules of leguminous plants (Sprent, 2009), and that rhizobia associated with feed/fodder legumes contribute a substantial proportion of this fixed nitrogen (Herridge et al., 2008). In the traditional Quechua farming, amilaceous maize seeds are used for sowing at each growing season and plants grow associated with innately emerged bur clover (*Medicago hispida*, Gaernt. syn. *M. polymorpha* L.), an annual shrub of the *Fabaceae* family within the tribe *Trifolieae* which, though native to the Mediterranean basin, is distributed worldwide. In the Peruvian Andes bur clover grows up to almost 4000 m above sea level where it is also widely used as a pasture and as a cover crop, greatly contributing to the fodder units required for feeding of the national cattle. The microsymbiont from nodules of bur clover has been isolated and identified as *Ensifer medicae*, and the rhizobial-*M. hispida* symbiotic association has been suggested to be the main N source for growth of maize plants in the Andean chacras (Arone et al., 2014).

The plant rhizosphere is a complex environment where soil microbial communities play a key role in ecosystems functions and are among the most complex, diverse and important assemblages in the biosphere. Accordingly, the study of plant-associated microorganisms is of great interest for their contribution to plant nutrition, hormonal control of plant growth, disease suppression, etc. (Bastida et al., 2009). Many studies on the composition and community structure of plant rhizosphere have been performed by using culture-dependent techniques and low-resolution molecular methods such as DGGE, TGGE, PLFA and SSCP profiles of microbial communities. Although isolation of culturable bacteria is appropriate for functional analysis, this approach often shows a rather limited diversity because a high percentage of naturally occurring bacteria remains in a non-culturable state (Oliver, 2000; Torsvik et al., 2002). Culture-independent molecular methods provide additional information on the diversity of bacterial communities by analyzing and comparing a very large number of samples. Among them, the bar-coded pyrosequencing technology introduced by 454 Life Science (Margulies et al., 2005; Rothberg and Leamon, 2008) describes microbial community composition in complex habitats such as the deep sea (Sogin et al., 2006; Huse et al., 2008; Polymenakou et al., 2015), human microbiome (Sundquist et al., 2007; Dethlefsen et al., 2008; Claesson et al., 2009; Hang et al., 2014), coastal microbial mats (Bolhuis and Stal, 2011; Tytgat et al., 2014) and dental implants (Kumar et al., 2012; Dabdoub et al., 2013), and soil (Uroz et al., 2010; Mao et al., 2011; Barberán et al., 2012; Sun et al., 2014; Sengupta and Dick, 2015).

Previous studies have analyzed bacterial taxa associated with maize. Some of these studies have been focused on the culturable fraction (Rai et al., 2007; Rijavec et al., 2007; Pereira et al., 2009) and others have assessed bacterial diversity independently of culture (Schmalenberger and Tebbe, 2003; Herschkovitz et al., 2005a,b; Sanguin et al., 2006a,b; Pereira et al., 2011; Ikeda et al., 2013). New generation sequencing has also been used to analyze the diversity and heritability of the maize rhizosphere microbiome under field conditions (Peiffer et al., 2013) and the importance of rare taxa for bacterial diversity in the rizosphere of Bt- and conventional maize varieties (Dohrmann et al., 2013). Because bacterial diversity in roots of amilaceous maize grown in Andean chacras has not been studied using high throughput sequencing, the main objective of this work was to describe the abundance and structure of bacterial communities associated with the roots of amilaceous maize by pyrosequencing. Given that amilaceous maize grows together with bur clover, we hypothesized that plant cultivation type could affect abundance and composition of maize rhizospheric bacteria. To test this hypothesis, we have analyzed changes on rhizospheric bacterial diversity in the absence and the presence of bur clover.

MATERIALS AND METHODS

Site Description and Soil Sampling

Chacras are small parcels of soil (200–10000 m^2) used by Quechua peasants to grow maize, quinoa, wheat, potatoes, and others cereals and vegetables. Soils were taken from three chacras within the same soil plot located near Allpas (12° 50′ 27″ S, 74° 34′ 14″ W, at 3537 m above sea level), a village in the province of Acobamba (Huancavelica, Peru). At sampling time, one chacra had been cultivated with amilaceous maize (M soil) for 3 years, other with maize and bur clover (MT soil) for 5 years, and the third was bulk soil from a chacra under fallow conditions (B soil) for at least 3 years. Seeds of *Z. mays* L. morphotype Qarway were planted by mid October 2012 and soil samples collected 120 days later, when the plants were at the grain filling stage. At the second hilling stage, 75–90 days after planting, bur clover looms up without previous sowing and has to be removed by hand to maintain the chacras free of the legume. Lateral roots (∼2 mm diameter, 2–3 cm long; 5–10 cm depth) were taken from maize plants grown at four different sites within each chacra and four replicates were sampled for each site. After cleaning of the soil attached to the roots, the remaining adhering rhizospheric soil was carefully removed and pooled together to obtain 2 g. Bulk soil samples (four sites, four replicates/site; 0–5 cm depth) were taken from the chacra under fallow conditions. Soil samples were kept at −20°C until further processing. The three chacras belong to the same soil plot, which has a sandy-loam texture (62.5% sand, 30.0% silt, 7.5% clay); the average pH was 6.3 and that of the soil organic matter was 28.7 g/soil kg.

Extraction of DNA from Soil

DNA was extracted from 250 mg of unfrozen soil as previously indicated (Correa-Galeote et al., 2013). Essentially, samples were homogenized in 1 ml of extraction buffer containing 100 mM Tris (pH 8.0), 100 mM EDTA, 100 mM NaCl, 1% (w/v) polyvinylpyrrolidone and 2% (w/v) sodium dodecyl sulfate using a 2-ml mini-bead-beater tube containing 0.5 g and 0.1 g of 106-μm- and 2-mm-diameter glass beads, respectively, for 60 s at 1600 rpm. Cell debris was eliminated by centrifugation (14000 rpm for 5 min at 4°C). Proteins were removed by treatment with 5 M sodium acetate. After treatment for 12 h with ice-cold isopropanol, nucleic acids were precipitated by centrifugation (14000 rpm for 30 min at 4°C), washed with 70% ice-cold ethanol, recentrifruged (14000 rpm for 15 min at 4°C) and air-dried for 30 min. Finally, DNA was purified using GeneClean columns (Qiagen). Quality and size of DNA were checked by electrophoresis on 1% agarose and quantified by spectrophotometry at 260 nm using a Nanodrop spectrophotometer (NanoDrop ND1000).

Amplification and Pyrosequencing of DNA from Maize Roots

Polymerase chain reaction (PCR) amplification of the hypervariable V4–V5 regions of the 16S rRNA gene was performed over each individual DNA extraction from soils using universal primers U519F and U926R (Baker et al., 2003) joined to a multiplex identifier sequence (Binladen et al., 2007; Parameswaran et al., 2007). For each sample, amplicons were generated in several replicate PCRs using mixtures (25 μl) that contained 25 pmol of each primer, 1.8 mM $MgCl_2$, 0.2 mM dNTPs, 1 × the corresponding *Taq* buffer, 1 U of *Taq* Master (5 Prime, USA) and 10 ng of the DNA template. The PCR program consisted of an initial denaturation step at 94°C for 4 min, 25 cycles of denaturation at 94°C for 15 s, primer annealing at 55°C for 45 s and extension at 72°C for 1 min, followed by a final step of heating at 72°C for 10 min. Amplicons of the same treatment were pooled to reduce per-PCR variability and purified using the ultracentrifugal filters Ultracel-100 K membranes (Amicon) according to the manufacturer's instructions. After quantification by Nanodrop ND1000 and visualization of the DNA by agarose electrophoresis, the samples were combined in equimolar amounts and pyrosequenced in a Roche Genome Sequencer FLX system using 454 Titanium chemistry at LifeSequencing S.L. (Valencia, Spain).

Taxonomic Assignment of Sequence Reads and Diversity Indexes

Raw sequences were processed through the Ribosomal Database Project (RDP) pyrosequencing pipeline[1] release 11 (Cole et al., 2014). Sequences were trimmed for primers, filtered and assigned to three libraries (BG, MG and MTG containing 16S rRNA gene sequences from B, M and MT soils, respectively) according to their tags. Sequences shorter than 150 base pair, with quality scores <20 or containing any unresolved nucleotides were removed from the dataset. Chimeras were identified using the Uchime tool from FunGene database (Edgar et al., 2011) and removed from the dataset. Sequences were aligned using the Infernal alignment tool in RDP (Nawrocki et al., 2009). Aligned sequences were clustered into operational taxonomic units (OTUs) defined at 97% similarity cutoff using Complete Linkage Clustering RDP tool and their relative abundances calculated. The number of sequences in each OTU was employed to calculate the Good's coverage index, which is considered a relative measure of how well the sequences obtained represent the entire populations (Hughes and Bohannan, 2004). The RDP Classifier, a Bayesian rRNA classifying algorithm (Wang et al., 2007), was used to assign phylogenetic groups based on sequence similarity. Matches with an RDP confidence estimate below 60% were designated as unclassified bacteria. Shannon (H′) and Simpson (S′) diversity indexes and Jaccard indexes (J_{class} and J_{abund}) were used to analyze the alpha- and beta-diversity, respectively (Chao et al., 2005).

[1]http://pyro.cme.msu.edu

Statistical Analyses

Relative abundances of the main genera were compared using the *G*-test (w/Yates') + Fischer's test in the STAMP software (Parks et al., 2014). Comparisons among the total number of sequences in the genomic libraries were done using the Newcombe–Wilson method (Newcombe, 1998; Parks and Beiko, 2010).

Accession Numbers

Pyrosequencing reads have been sent to GenBank to obtain their under accession numbers. They are available as KX479121-KX485313, KX559454-KX570596 and KX470813-KX479120 for the genomic libraries MTG, MG and BG, respectively.

RESULTS

Pyrosequencing and Sequence Analysis

A total of 38786 sequences were obtained from the three 16S rDNA samples sent to pyrosequencing, of which 26031 were retained after filtering and removing chimeras. The mean number of total retained sequences per library was 8677, ranging from 6286 to 11387. Average length of retained sequences was

TABLE 1 | Number of OTUs, values of Good's coverage index, and Shannon and Simpson biodiversity indexes in genomic libraries from bulk soil (BG) and rhizospheric soil of maize (MG) and maize intercropped with clover (MTG) grown in Andean chacras.

Diversity index	MTG	MG	BG
Number of OTUs	1846	3287	2318
Good's coverage	85.58	84.92	86.44
Shannon	6.55	7.21	7.06
Simpson	0.0056	0.0031	0.0016

TABLE 2 | Number of taxa and of 16S rRNA sequences in genomic libraries from bulk soil (BG) and rhizospheric soil of maize (MG) and maize intercropped with clover (MTG) grown in Andean chacras.

	Genomic Libraries					
	MTG		MG		BG	
	Number of taxa	Number of sequences (%)	Number of taxa	Number of sequences (%)	Number of taxa	Number of sequences (%)
Phylum	21	5711 (90.85)	23	9594 (84.26)	21	6898 (82.53)
Class	41	5066 (80.59)	44	8390 (73.68)	42	8358 (75.93)
Order	43	3933 (62.57)	51	5136 (45.10)	45	3391 (39.97)
Family	43	3607 (57.38)	51	4141 (36.37)	45	2620 (31.35)
Genus	203	3131 (49.81)	234	5650 (49.62)	197	4003 (47.89)
Unclassified sequences		575 (9.15)		1793 (15.75)		1460 (17.47)

The percentage of sequences in relation with the total number of sequences is shown in brackets.

374 ± 5 base pair (mean ± SD). All the sequences aligned correctly in the expected position of the 16S rDNA sequence of *Escherichia coli* and were grouped at 97% similarity in 5955 distinct OTUs, of which 1846, 3287 and 2318 were found in MTG, MG and BG libraries, respectively (**Table 1**), which also contained 575 (9.15%), 1793 (15.75%), and 1460 (17.47%) unclassified sequences, respectively (**Table 2; Figure 1**). The remaining valid sequences distributed into 21, 23, and 21 phyla, respectively (**Table 2**), of which phyla Proteobacteria, Bacteroidetes, Acidobacteria, Actinobacteria, Firmicutes, Verrucomicrobia, Gemmatimonadetes, Planctomycetes, Chloroflexi, Armatimonadetes, Parcubacteria, Nitrospirae, Candidatus Saccharibacteria, candidate division WPS-2, candidate division WPS-1, Latescibacteria, BRC1 and Microgenomates (in decreasing abundance) were found in all the three libraries (**Figure 1**). Phyla Spirochaetes and Chlamydiae were found in MTG and MG and so were Ignavibacteriae and Cyanobacteria in libraries MG and BG, respectively (**Figure 1**). Hydrogenedentes was detected only in MTG, Deinococcus–Thermus in MG and Elusimicrobia in BG (**Figure 1**). Numbers of classes, orders, families and genera are also shown in **Table 2**.

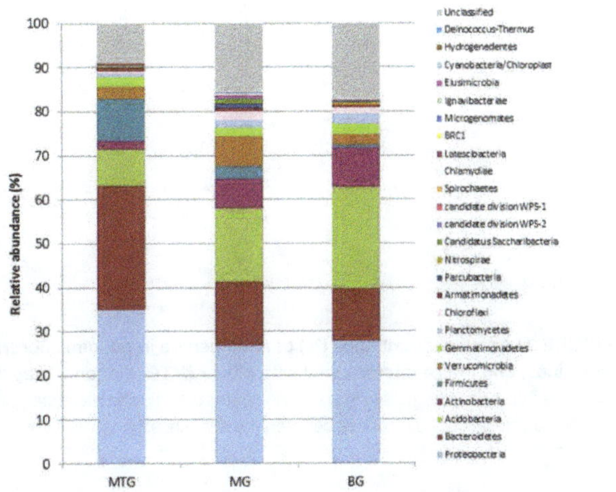

FIGURE 1 | Relative abundance of bacterial phyla in genomic libraries from bulk soil (BG) and rhizospheric soil of maize (MG) and maize intercropped with clover (MTG) grown in Andean chacras.

Coverage and Diversity Indexes

With values of the Good's coverage index higher than 80% at 90% confidence interval (**Table 1**), the Shannon index of 5.13 for OTUs in the MTG clone library was lower than those of 7.06 and 7.21 for the FG and MG libraries, respectively. The Simpson index, however, showed clear differences among the OTUs in each library, with values of 0.0016, 0.0031 and 0.0056 for the MTG, MG and BG clone libraries, respectively (**Table 1**).

Sequences in library BG distributed in 197 genera, a value lower than those of 203 and 234 found in libraries MTG and MG, respectively (**Table 2**). Of the total number of genera, the BG library shared 132 and 155 with the MTG and MG libraries, respectively, and libraries MTG and MG had 159 genera in common (**Table 3**). Values of the J_{class} index were similar between the BG and MG libraries (0.56) and the MTR and MG libraries (0.57), values that were higher than that (0.49) between the MTG and BG libraries (**Table 3**). The J_{abund} indexes for the multiple comparisons among genomic libraries were similar (**Table 3**).

TABLE 3 | Number of shared genera among genomic libraries, and Jaccard similarity indexes using presence/absence (J_{class}) and relative abundances (J_{abund}) of the genera in libraries from bulk soil (BG) and rhizospheric soil of maize (MG) and maize intercropped with clover (MTG) grown in Andean chacras.

	Number of shared genera	J_{class}	J_{abund}
MTG-BG	132	0.49	0.94
MG-BG	155	0.56	0.92
MTG-MG	159	0.57	0.98

Taxonomic Composition and Statistical Analyses

From the total 26031 sequences, only 12784 could be assigned to 309 different genera (Supplementary Table S1). Sixteen (16) out of the 309 genera had a relative abundance higher than 0.5%, and represented the 64.23% of the total number of sequences classified at the genus level. Altogether, these genera were (in

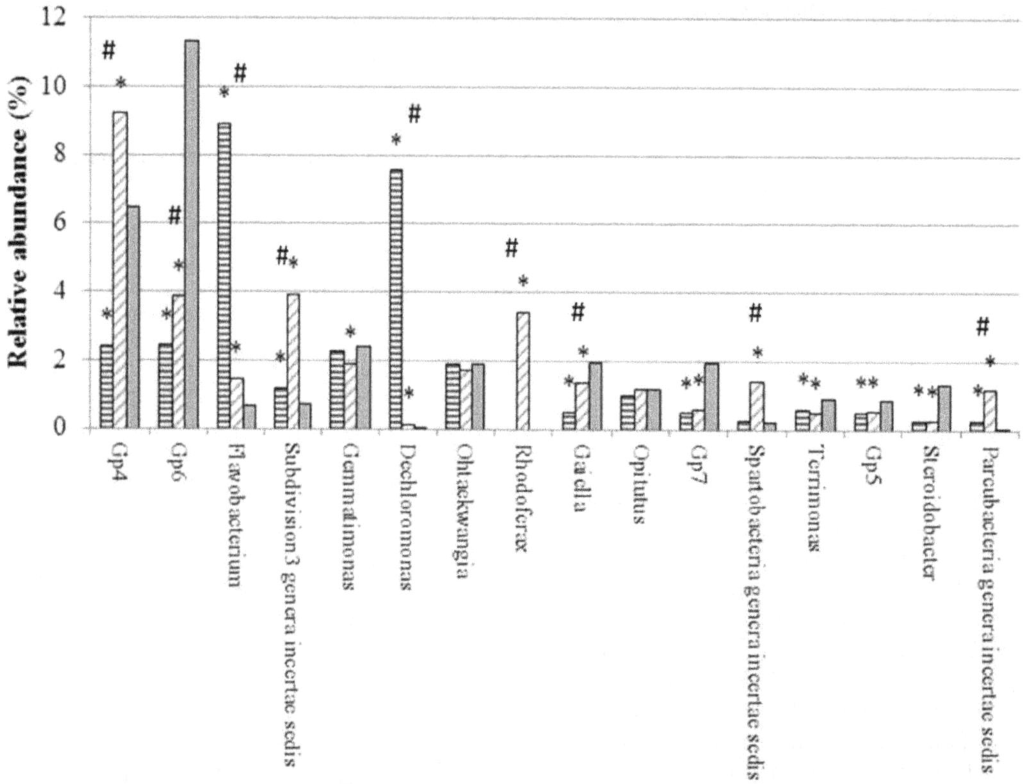

FIGURE 2 | Relative abundance (%) of main genera in genomic libraries from bulk soil (BG, solid gray) and rhizospheric soil of maize (MG, horizontal gray lines) and maize intercropped with clover (MTG, diagonal gray lines) grown in Andean chacras. Relative abundances were compared using the combined G + Fischer tests ($\alpha \leq 0.05$). *indicates statistically differences between the sequences in the MTG and MG libraries as compared to the BG library; # indicates statistically differences between the sequences in the MTG and MG libraries.

decreasing order of abundance) Gp4, Gp6, *Flavobacterium*, Subdivision3 genera incertae sedis of the Verrucomicrobia phylum, *Gemmatimonas*, *Dechloromonas*, *Ohtaekwangia*, *Rhodoferax*, *Gaiella*, *Opitutus*, Gp7, *Spartobacteria* genera incertae sedis, *Terrimonas*, Gp5, *Steroidobacter* and *Parcubacteria* genera incertae sedis (Supplementary Table S1) (**Figure 2**).

Relative abundances of genera *Flavobacterium*, Subdivision3 genera incertae sedis of the Verrucomicrobia phylum, *Dechloromonas* and *Parcubacteria* in each the MTG and MG libraries were statistically higher than those in the BG library. In contrast, the number of sequences of genera Gp6, *Gaiella*, Gp7, *Terrimonas*, Gp5 and *Steroidobacter* in the MTG and MG libraries were lower than those in the BG library.

Numbers of sequences of genera *Flavobacterium* and *Dechloromonas* in the MTG library were statistically higher than those in the MG library (**Figure 2**). On the other hand, the relative abundance of genera Gp4, Gp6, Subdivision3 genera incertae sedis of phylum Verrucomicrobia, *Rhodoferax*, *Gaiella* and *Spartobacteria* in MTG were significantly lower than those in the MG library (**Figure 2**). Differences in the relative abundance of sequences corresponding to genera *Gemmatimonas*, *Ohtaekwangia*, *Opitutus*, Gp7, *Terrimonas*, Gp5, *Steroidobacter* and *Parcubacteria* were not found between the MTG and MG libraries (**Figure 2**).

DISCUSSION

Using 454 next generation sequencing we assessed the composition and abundance of bacterial rhizospheric communities of amilaceous maize plants grown in Andean chacras following Quechua traditional practices. The total number of genera found in our genomic libraries were higher than those obtained in previous studies based on culture-dependent and culture-independent methods (Gomes et al., 2001; Schmalenberger and Tebbe, 2002; Baumgarte and Tebbe, 2005; Brusetti et al., 2005; Aira et al., 2010; Miethling-Graff et al., 2010; Pereira et al., 2011; García-Salamanca et al., 2013; Li et al., 2014), and similar to those found in the rhizosphere of different maize varieties after pyrosequencing (Dohrmann et al., 2013).

Altogether, a 12.71% of the total sequences found inside roots corresponded to unclassified bacteria, which indicates the presence of hitherto uncultured bacterial groups. A high percentage of unclassified sequences have also been reported in the rhizosphere of maize plants when pyrosequencing was used to study bacterial biodiversity (Dohrmann et al., 2013).

A variety of bacteria have been reported to be rhizospheric, among them mostly Proteobacteria, Firmicutes, Actinobacteria, Acidobacteria and Bacteroidetes (Sharma et al., 2005; DeAngelis et al., 2009; Peiffer et al., 2013; Uroz et al., 2013). In our study, regardless of soil provenance, members of phylum Proteobacteria were the most abundant (29.10%) followed by those of Bacteroidetes (17.03%), Acidobacteria (16.62%), Actinobacteria (6.40%), Verrucomicrobia (4.48%) and Firmicutes (3.55%) (**Figure 1**). Although the amount of bacterial phyla found in the three genomic libraries was similar (**Table 2**), the number of OTUs and values of the Shannon and Simpson indexes for the MTG library were lower than those for the BG library (**Table 1**), which suggests that intercropping reduced richness of the community. These results agree with those previously published that show that microbial diversity in maize rhizospheric soils was lower than in bulk soils (García-Salamanca et al., 2013; Peiffer et al., 2013). This reduction can be explained if one considers that maize plants intercropped with bur clover somehow select from the bulk soil those bacteria able to promote plant growth and development; e.g., due to the symbiosis with *E. medicae* (Arone et al., 2014) free-living N_2 bacteria would not be required for plant growth. Our results also show that diversity in the MG library was higher than that in the BG library (**Table 1**). It is possible that the absence of bur clover promts the plants to increase bacterial diversity around the roots to improve their nutrition. Alternatively, it is also possible that more time is required for the maize crop to become well established in soils, which, in turn, could alter the soil microbiota. On the other hand, clear dominance of specific rhizospheric bacteria was not found, and multiple comparisons analyses of the values of the β-diversity indexes indicated the three genomic libraries have a similar bacterial community core. The absence of specific phylogenetic groups between bulk and rhizospheric soil from an oak forest after pyrosequencing of the 16S rRNA gene has been reported (Uroz et al., 2010).

Plants influence the abundance and composition of the bacterial rhizospheric community by secreting a variety of compounds through their roots into the surrounding soil to feed and manipulate the microbes that live there (Johnston-Monje et al., 2016). If one considers the bulk soil of the chacras as the reservoir of microbial communities, whereas maize increased relative abundance of genera Gp4, *Flavobacterium*, Subdivision3 genera incertae sedis of the Verrucomicrobia phylum, *Dechloromonas*, *Rhodoferax*, *Spartobacteria* and *Parcubacteria* incertae sedis, the maize-bur clover consortium increased relative abundance of *Flavobacterium*, Subdivision3 genera incertae sedis of the Verrucomicrobia phylum, *Dechloromonas* and *Parcubacteria* incertae sedis.

It is well established that plant genotype plays a significant role in shaping plant-associated microbial communities and in determining the biological outcome of such associations (Smith and Goodman, 1999; Ding et al., 2013). In this sense, 16S rRNA-based pyrosequencing studies on microbial diversity from rhizosphere of different maize genotypes have shown that variation in microbial diversity could be attributed to host genetics (Bouffaud et al., 2012; Peiffer et al., 2013). Because seeds of amilaceous maize used for planting were the same and the environmental conditions (including soil type and physicochemical properties), were, also, the same for the three chacras, our results suggest that the plant could be a main factor controlling bacterial diversity in their rhizosphere, whether alone (maize) or in consortium with other plants (maize and bur clover). Since richness of bacteria colonizing maize roots was based on pyrosequencing, it is not known whether the detection of bacteria based on DNA signature alone represent active microbes that are interacting with the host plant.

AUTHOR CONTRIBUTIONS

Conceived and designed the experiments: DC-G, EJB, GJA. Performed the experiments: DC-G, GJA. Analyzed the data: DC-G, EJB. Contributed reagents/materials/analysis tools: AJF-G, MF-L. Improve the paper: AJF-G, MF-L, GJA. Wrote the paper: DC-G, EJB.

ACKNOWLEDGMENTS

This study was supported by ERDF-cofinanced grant PE2012-AGR1968 from Consejería de Economía, Innovación y Ciencia (Junta de Andalucía, Spain) and the CSIC-sponsored I-COOP Agrofood project 2014CD0013.

REFERENCES

Aira, M., Gómez-Brandón, M., Lazcano, C., Bååth, E., and Domínguez, J. (2010). Plant genotype strongly modifies the structure and growth of maize rhizosphere microbial communities. *Soil Biol. Biochem.* 42, 2276–2281. doi: 10.1016/j.soilbio.2010.08.029

Arone, G., Calderón, C., Moreno, S., and Bedmar, E. J. (2014). Identification of Ensifer strains isolated from root nodules of *Medicago hispida* grown in association with *Zea mays* in the Quechua region of the Peruvian Andes. *Biol. Fertil. Soils* 50, 185–190. doi: 10.1007/s00374-013-0825-3

Baker, G. C., Smith, J. J., and Cowan, D. A. (2003). Review and re-analysis of domain-specific 16S primers. *J. Microbiol. Methods* 55, 541–555. doi: 10.1016/j.mimet.2003.08.009

Barberán, A., Bates, S. T., Casamayor, E. O., and Fierer, N. (2012). Using network analysis to explore co-occurrence patterns in soil microbial communities. *ISME J.* 6, 343–351. doi: 10.1038/ismej.2011.119

Bastida, F., Moreno, J. L., Nicolás, C., Hernández, T., and García, C. (2009). Soil metaproteomics: a review of an emerging environmental science. Significance, methodology and perspectives. *Eur. J. Soil Sci.* 60, 845–859. doi: 10.1111/j.1574-6941.2011.01284.x

Baumgarte, S., and Tebbe, C. C. (2005). Field studies on the environmental fate of the Cry1Ab Bt-toxin produced by transgenic maize (MON810) and its effect on bacterial communities in the maize rhizosphere. *Mol. Ecol.* 14, 2539–2551. doi: 10.1111/j.1365-294X.2005.02592.x

Binladen, J., Gilbert, M. T., Bollback, J. P., Panitz, F., Bendixen, C., Nielsen, R., et al. (2007). The use of coded PCR primers enables high-throughput sequencing of multiple homolog amplification products by 454 parallel sequencing. *PLoS ONE* 2:e197. doi: 10.1371/journal.pone.0000197

Bolhuis, H., and Stal, L. J. (2011). Analysis of bacterial and archaeal diversity in coastal microbial mats using massive parallel 16S rRNA gene tag sequencing. *ISME J.* 5, 1701–1712. doi: 10.1038/ismej.2011.52

Bouffaud, M.-L., Kyselková, M., Gouesnard, B., Grundmann, G., Muller, D., and Moënne-Loccoz, Y. (2012). Is diversification history of maize influencing selection of soil bacteria by roots? *Mol. Ecol.* 21, 195–206. doi: 10.1111/j.1365-294X.2011.05359.x

Brusetti, L., Francia, P., Bertolini, C., Pagliuca, A., Borin, S., Sorlini, C., et al. (2005). Bacterial communities associated with the rhizosphere of transgenic Bt 176 maize (*Zea mays*) and its non transgenic counterpart. *Plant Soil* 226, 11–21. doi: 10.1007/s11104-005-5399-x

Chao, A., Chazdon, R. L., Colwell, R. K., and Shen, T. J. (2005). A new statistical approach for assessing similarity of species composition with incidence and abundance data. *Ecol. Lett.* 8, 148–159. doi: 10.1111/j.1461-0248.2004.00707.x

Claesson, M. J., O'Sullivan, O., Wang, Q., Nikkilä, J., Marchesi, J. R., Smidt, H., et al. (2009). Comparative analysis of pyrosequencing and a phylogenetic microarray for exploring microbial community structures in the human distal intestine. *PLoS ONE* 4:e6669. doi: 10.1371/journal.pone.0006669

Cole, J. R., Wang, Q., Fish, J. A., Chai, B., McGarrell, D. M., Sun, Y., et al. (2014). Ribosomal database project: data and tools for high throughput rRNA analysis. *Nucleic Acids Res.* 42, D633–D642. doi: 10.1093/nar/gkt1244

Correa-Galeote, D., Tortosa, G., and Bedmar, E. J. (2013). Determination of denitrification genes abundance in environmental samples. *Metagenomics* 2, 235702. doi: 10.4303/mg/235702

Dabdoub, S., Tsigarida, A. A., and Kumar, P. S. (2013). Patient-specific analysis of periodontal and peri-implant microbiomes. *J. Dent. Res.* 92, 168S–175S. doi: 10.1177/0022034513504950

DeAngelis, K. M., Brodie, E. L., DeSantis, T. Z., Andersen, G. L., Lindow, S. E., and Firestone, M. K. (2009). Selective progressive response of soil microbial community to wild oat roots. *ISME J.* 3, 168–178. doi: 10.1038/ismej.2008.103

Dethlefsen, L., Huse, S., Sogin, M. L., and Relman, D. A. (2008). The pervasive effects of an antibiotic on the human gut microbiota, as revealed by deep 16S rRNA sequencing. *PLoS Biol.* 6:e280. doi: 10.1371/journal.pbio.0060280

Ding, T., Palmer, M. W., and Melcher, U. (2013). Community terminal restriction fragment length polymorphisms reveal insights into the diversity and dynamics of leaf endophytic bacteria. *BMC Microbiol.* 13:1. doi: 10.1186/1471-2180-13-1

Dohrmann, A. B., Küting, M., Jünemann, S., Jaenicke, S., Schlüter, A., and Tebbe, C. C. (2013). Importance of rare taxa for bacterial diversity in the rhizosphere of Bt-and conventional maize varieties. *ISME J.* 7, 37–49. doi: 10.1038/ismej.2012.77

Edgar, R. C., Haas, B. J., Clemente, J. C., Quince, C., and Knight, R. (2011). UCHIME improves sensitivity and speed of chimera detection. *Bioinformatics* 27, 2194–2200. doi: 10.1093/bioinformatics/btr381

García-Salamanca, A., Molina-Henares, M. A., Dillewijn, P., Solano, J., Pizarro-Tobías, P., Roca, A., et al. (2013). Bacterial diversity in the rhizosphere of maize and the surrounding carbonate-rich bulk soil. *Microb. Biotechnol.* 6, 36–44. doi: 10.1111/j.1751-7915.2012.00358.x

Gomes, N. C. M., Heuer, H., Schönfeld, J., Costa, R., Mendonca-Hagler, L., and Smalla, K. (2001). Bacterial diversity of the rhizosphere of maize (*Zea mays*) grown in tropical soil studied by temperature gradient gel electrophoresis. *Plant Soil* 232, 167–180. doi: 10.1023/A:1010350406708

Grobman, A., Bonavia, D., Dillehay, T. D., Piperno, D. R., Iriarte, J., and Holst, I. (2012). Preceramic maize from Paredones and Huaca Prieta, Peru. *Proc. Natl. Acad. Sci. U.S.A.* 109, 1755–1759. doi: 10.1073/pnas.1120270109

Hang, J., Desai, V., Zavaljevski, N., Yang, Y., Lin, X., Satya, R. V., et al. (2014). 16S rRNA gene pyrosequencing of reference and clinical samples and investigation of the temperature stability of microbiome profiles. *Microbiome* 2, 31. doi: 10.1186/2049-2618-2-31

Herridge, D. F., Peoples, M. B., and Boddey, R. M. (2008). Global inputs of biological nitrogen fixation in agricultural systems. *Plant Soil* 311, 1–18. doi: 10.1007/s11104-008-9668-3

Herschkovitz, Y., Lerner, A., Davidov, Y., Okon, Y., and Jurkevitch, E. (2005a). *Azospirillum brasilense* does not affect population structure of specific rhizobacterial communities of inoculated maize (*Zea mays*). *Environ. Microbiol.* 7, 1847–1852. doi: 10.1111/j.1462-2920.2005.00926.x

Herschkovitz, Y., Lerner, A., Davidov, Y., Rothballer, M., Hartmann, A., Okon, Y., et al. (2005b). Inoculation with the plant-growth-promoting rhizobacterium *Azospirillum brasilense* causes little disturbance in the rhizosphere and rhizoplane of maize *(Zea mays)*. *Microb. Ecol.* 50, 277–288. doi: 10.1007/s00248-004-0148-x

Hughes, J. B., and Bohannan, B. J. M. (2004). "Application of ecological diversity statistics in microbial ecology," in *Molecular Microbial Ecology Manual, Section 7*, eds G. A. Kowalchuk, F. J. de Bruijn, I. M. Head, A. D. Akkermans, and J. D. van Elsas (Berlin: Springer), 3223–3246.

Huse, S. M., Dethlefsen, L., Huber, J. A., Welch, D. M., Relman, D. A., and Sogin, M. L. (2008). Exploring microbial diversity and taxonomy using SSU rRNA hypervariable tag sequencing. *PLoS Genet.* 4:e1000255. doi: 10.1371/journal.pgen.1000255

Ikeda, A. C., Bassani, L. L., Adamoski, D., Stringari, D., Cordeiro, V. K., Glienke, C., et al. (2013). Morphological and genetic characterization of endophytic bacteria isolated from roots of different maize genotypes. *Microb. Ecol.* 65, 154–160. doi: 10.1007/s00248-012-0104-0

Instituto Nacional de Estadística e Informática [INEI] (2012). *IV Censo Nacional Agropecuario*. Perú: Instituto Nacional de Estadística e Informática.

Johnston-Monje, D., Lundberg, D. S., Lazarovits, G., Reis, V. M., and Raizada, M. N. (2016). Bacterial populations in juvenile maize rhizospheres originate from both seed and soil. *Plant Soil* 1–19. doi: 10.1007/s11104-016-2826-0

Kumar, P. S., Mason, M. R., Brooker, M. R., and O'Brien, K. (2012). Pyrosequencing reveals unique microbial signatures associated with healthy and failing dental implants. *J. Clin. Periodontol.* 39, 425–433. doi: 10.1111/j.1600-051X.2012.01856.x

Li, X., Rui, J., Mao, Y., Yannarell, A., and Mackie, R. (2014). Dynamics of the bacterial community structure in the rhizosphere of a maize cultivar. *Soil Biol. Biochem.* 68, 392–401. doi: 10.1016/j.soilbio.2013.10.017

Manrique, A. (1997). *El maíz en el Perú*. 2ª Edn. Peru: CONCYTEC.

Mao, Y., Yannarell, A. C., and Mackie, R. I. (2011). Changes in N-transforming archaea and bacteria in soil during the establishment of bioenergy crops. *PLoS ONE* 6:e24750. doi: 10.1371/journal.pone.0024750

Margulies, M., Egholm, M., Altman, W. E., Attiya, S., Bader, J. S., Bemben, L. A., et al. (2005). Genome sequencing in microfabricated high-density picolitre reactors. *Nature* 15, 376–380.

Miethling-Graff, R., Dockhorn, S., and Tebbe, C. C. (2010). Release of the recombinant Cry3Bbl protein of Bt maize MON88017 into field soil and detection of effects on the diversity of rhizosphere bacteria. *Eur. J. Soil Biol.* 46, 41–48. doi: 10.1016/j.ejsobi.2009.10.003

Nawrocki, E. P., Kolbe, D. L., and Eddy, S. R. (2009). Infernal 1.0: inference of RNA alignments. *Bioinformatics* 25, 1335–1337. doi: 10.1093/bioinformatics/btp157

Newcombe, R. G. (1998). Two-sided confidence intervals for the single proportion: comparison of seven methods. *Stat. Med.* 17, 857–872. doi: 10.1002/(SICI)1097-0258(19980430)17:8<857::AID-SIM777>3.0.CO;2-E

Oliver, J. D. (2000). "The public health significance of viable but nonculturable bacteria," in *Nonculturable Microorganisms in the Environment*, eds R. R. Colwell and D. J. Grimes (Washigton, DC: ASM Press), 277–300.

Parameswaran, P., Jalili, R., Tao, L., Shokralla, S., Gharizadeh, B., Ronaghi, M., et al. (2007). A pyrosequencing-tailored nucleotide barcode design unveils opportunities for large-scale sample multiplexing. *Nucleic Acids Res.* 3, e130. doi: 10.1093/nar/gkm760

Parks, D. H., and Beiko, R. G. (2010). Identifying biologically relevant differences between metagenomic communities. *Bioinformatics* 26, 715–721. doi: 10.1093/bioinformatics/btq041

Parks, D. H., Tyson, G. W., Hugenholtz, P., and Beiko, R. G. (2014). STAMP: statistical analysis of taxonomic and functional profiles. *Bioinformatics* 30, 3123–3124. doi: 10.1093/bioinformatics/btu494

Peiffer, J. A., Spor, A., Koren, O., Jin, Z., Tringe, S. G., Dangl, J. L., et al. (2013). Diversity and heritability of the maize rhizosphere microbiome under field conditions. *Proc. Natl. Acad. Sci. U.S.A.* 110, 6548–6553. doi: 10.1073/pnas.1302837110

Pereira, C. J., Pereira, E. J. G., Cordeiro, E. M. G., Della Lucia, T. M. C., Tótola, M. R., and Guedes, R. N. C. (2009). Organophosphate resistance in the maize weevil *Sitophilus zeamais*: magnitude and behavior. *Crop Protect.* 28, 168–173. doi: 10.1016/j.cropro.2008.10.001

Pereira, P., Ibáñez, F., Rosenblueth, M., Etcheverry, M., and Martínez-Romero, E. (2011). Analysis of the bacterial diversity associated with the roots of maize *(Zea mays L.)* through culture-dependent and culture-independent methods. *ISRN Ecol.* 2011:938546. doi: 10.5402/2011/938546.

Piperno, D. R., Ranere, A. J., Holst, I., Iriarte, J., and Dickau, R. (2009). Starch grain and phytolith evidence for early ninth millennium BP maize from the Central Balsas River Valley, Mexico. *Proc. Natl. Acad. Sci. U.S.A.* 106, 5019–5024. doi: 10.1073/pnas.0812525106

Polymenakou, P. N., Christakis, C. A., Mandalakis, M., and Oulas, A. (2015). Pyrosequencing analysis of microbial communities reveals dominant cosmopolitan phylotypes in deep-sea sediments of the eastern Mediterranean Sea. *Res. Microbiol.* 166, 448–445. doi: 10.1016/j.resmic.2015.03.005

Rai, R., Dash, P. K., Prasanna, B. M., and Singh, A. (2007). Endophytic bacterial flora in the stem tissue of a tropical maize *(Zea mays L.)* genotype: isolation, identification and enumeration. *World J. Microbiol. Biotechnol.* 23, 853–858. doi: 10.1007/s11274-006-9309-z

Rijavec, T., Lapanje, A., Dermastia, M., and Rupnik, M. (2007). Isolation of bacterial endophytes from germinated maize kernels. *Can. J. Microbiol.* 53, 802–808. doi: 10.1139/W07-048

Rothberg, J. M., and Leamon, J. H. (2008). The development and impact of 454 sequencing. *Nat. Biotechnol.* 26, 1117–1124. doi: 10.1038/nbt 1485

Sanguin, H., Herrera, A., Oger-Desfeux, C., Dechesne, A., Simonet, P., Navarro, E., et al. (2006a). Development and validation of a prototype 16S rRNA-based taxonomic microarray for Alphaproteobacteria. *Environ. Microbiol.* 8, 289–307. doi: 10.1111/j.1462-2920.2005.00895.x

Sanguin, H., Remenant, B., Dechesne, A., Thioulouse, J., Vogel, T. M., Nesme, X., et al. (2006b). Potential of a 16S rRNA-based taxonomic microarray for analyzing the rhizosphere effects of maize on *Agrobacterium* spp. and bacterial communities. *Appl. Environ. Microbiol.* 72, 4302–4312. doi: 10.1128/AEM.02686-05

Schmalenberger, A., and Tebbe, C. C. (2002). Bacterial community composition in the rhizosphere of a transgenic, herbicide-resistant maize *(Zea mays)* and comparison to its non-transgenic cultivar Bosphore. *FEMS Microbiol. Ecol.* 40, 29–37. doi: 10.1111/j.1574-6941.2002.tb00933.x

Schmalenberger, A., and Tebbe, C. C. (2003). Bacterial diversity in maize rhizospheres: conclusions on the use of genetic profiles based on PCR-amplified partial small subunit rRNA genes in ecological studies. *Mol. Ecol.* 12, 251–262. doi: 10.1046/j.1365-294X.2003.01716.x

Sengupta, A., and Dick, W. A. (2015). Bacterial community diversity in soil under two tillage practices as determined by pyrosequencing. *Microb. Ecol.* 70, 853–859. doi: 10.1007/s00248-015-0609-4

Sharma, S., Aneja, M. K., Mayer, J., Munch, J. C., and Schloter, M. (2005). Characterization of bacterial community structure in rhizosphere soil of grain legumes. *Microb. Ecol.* 49, 407–415. doi: 10.1007/s00248-004-0041-7

Smith, K. P., and Goodman, R. M. (1999). Host variation for interactions with beneficial plant-associated microbes. *Annu. Rev. Phytopathol.* 37, 473–491. doi: 10.1146/annurev.phyto.37.1.473

Sogin, M. L., Morrison, H. G., Huber, J. A., Welch, D. M., Huse, S. M., Neal, P. R., et al. (2006). Microbial diversity in the deep sea and the underexplored "rare biosphere". *Proc. Natl. Acad. Sci. U.S.A.* 103, 12115–12120. doi: 10.1073/pnas.0605127103

Sprent, J. I. (2009). *Legume Nodulation: A Global Perspective.* Oxford: Wiley-Blackwell.

Sun, J., Zhang, Q., Zhou, J., and Wei, Q. (2014). Pyrosequencing technology reveals the impact of different manure doses on the bacterial community in apple rhizosphere soil. *Appl. Soil Ecol.* 78, 28–36. doi: 10.1016/j.apsoil.2014.02.004

Sundquist, A., Bigdeli, S., Jalili, R., Druzin, M. L., Waller, S., Pullen, K. M., et al. (2007). Bacterial flora-typing with targeted, chip-based pyrosequencing. *BMC Microbiol.* 7:108. doi: 10.1186/1471-2180-7-108

Torsvik, V., Øvreås, L., and Thingstad, T. F. (2002). Prokaryotic diversity-magnitude, dynamics, and controlling factors. *Science* 296, 1064–1066. doi: 10.1126/science.1071698

Tytgat, B., Verleyen, E., Obbels, D., Peeters, K., De Wever, A., D'hondt, S., et al. (2014). Bacterial diversity assessment in Antarctic terrestrial and aquatic microbial mats: a comparison between bidirectional pyrosequencing and cultivation. *PLoS ONE* 9:e97564. doi: 10.1371/journal.pone.0097564

Uroz, S., Buée, M., Murat, C., Frey-Klett, P., and Martin, F. (2010). Pyrosequencing reveals a contrasted bacterial diversity between oak rhizosphere and surrounding soil. *Environ. Microbiol. Rep.* 2, 281–288. doi: 10.1111/j.1758-2229.2009.00117.x

Uroz, S., Ioannidis, P., Lengelle, J., Cébron, A., Morin, E., Buée, M., et al. (2013). Functional assays and metagenomic analyses reveals differences between the microbial communities inhabiting the soil horizons of a Norway spruce plantation. *PLoS ONE* 8:e55929. doi: 10.1371/journal.pone.0055929

van Heerwaarden, J., Doebley, J., Briggs, W. H., Glaubitz, J. C., Goodman, M. M., Gonzalez, J. D. J. S., et al. (2011). Genetic signals of origin, spread, and introgression in a large sample of maize landraces. *Proc. Natl. Acad. Sci. U.S.A.* 108, 1088–1092. doi: 10.1073/pnas.1013011108

Wang, Q., Garrity, G. M., Tiedje, J. M., and Cole, J. R. (2007). Naive Bayesian classifier for rapid assignment of rRNA sequences into the new bacterial taxonomy. *Appl. Environ. Microb.* 73, 5261–5267. doi: 10.1128/AEM.00062-07

Conflict of Interest Statement: The authors declare that the research was conducted in the absence of any commercial or financial relationships that could be construed as a potential conflict of interest.

8

Cottonseed Protein, Oil, and Mineral Status in Near-Isogenic *Gossypium hirsutum* Cotton Lines Expressing Fuzzy/Linted and Fuzzless/Linted Seed Phenotypes under Field Conditions

Edited by:
Antonio M. De Ron,
National Spanish Research Council (CSIC), Spain

Reviewed by:
Josefina Hernández Nistal,
Universidad de Santiago de Compostela, Spain
Miroslav Griga,
AGRITEC Plant Research Ltd., Czech Republic

***Correspondence:**
Nacer Bellaloui,
Crop Genetics Research Unit, Research Plant Physiologist, United States Department of Agriculture, Agricultural Research Service, 141 Experiment Station Road, PO Box 345
Stoneville, MS 38776, USA
nacer.bellaloui@ars.usda.gov

Nacer Bellaloui**, Salliana R. Stetina and Rickie B. Turley

Crop Genetics Research Unit, Plant Physiology, United States Department of Agriculture, Agricultural Research Service, Stoneville, MS, USA

Cotton is an important crop in the world and is a major source of oil for human consumption and cotton meal for livestock. Cottonseed nutrition (seed composition: protein, oil, and minerals) determines the quality of seeds. Therefore, maintaining optimum levels of cottonseed nutrition is critical. Physiological and genetic mechanisms controlling the levels of these constituents in cottonseed are still largely unknown. Our previous research conducted under greenhouse conditions showed that seed and leaf nutrition differed between fuzzless and fuzzy seed isolines. Therefore, the objective of this research was to investigate the seed fuzz phenotype (trait) effects on seed protein, oil, N, C, S, and minerals in five sets of near-isogenic mutant cotton lines for seed fuzz in a 2-year experiment under field condition to evaluate the stability of the effect of the trait on seed nutrition. The isolines (genotypes) in each set differ for the seed fuzz trait (fuzzless/linted seed line, *N* lines, and fuzzy/linted seed line, *F* lines). Results showed that seed protein was higher in the fuzzy genotype in all sets, but seed oil was higher in fuzzless genotype in all sets. The concentrations of seed Ca and C were higher in all fuzzless genotypes, but N, S, B, Fe, and Zn were higher in most of the fuzzy genotypes. Generally, minerals were higher in leaves of *F* lines, suggesting the translocation of minerals from leaves to seeds was limited. The research demonstrated that fiber development could be involved in cottonseed composition. This may be due to the involvement of fiber development in carbon and nitrogen metabolism, and the mobility of nutrients from leaves (source) to seed (sink). This information is beneficial to breeders to consider fuzzless cottonseed for potential protein and oil use and select for higher oil or higher protein content, and to physiologists to further understand the mobility of minerals to increase the quality of cottonseed nutrition for food and feed.

Keywords: cottonseed protein, cottonseed oil, cottonseed composition, minerals, near-isogenic cotton, cotton mutants

Introduction

Cotton is a major crop in the world (Yu et al., 2012). Cotton fiber is a source of natural textile, and cottonseed is a source of oil for human consumption, cotton meal and minerals for livestock feed (Yu et al., 2012; He et al., 2013). Therefore, maintaining high quality fiber and cottonseed nutritional value is critical. There are four cultivated cotton species, *Gossypium hirsutum* (about 95% of the cultivated cotton), *G. barbadense*, *G. arboreum*, and *G. herbaceum* together (the last three represent about 5%) (Turley et al., 2007; Padmalatha et al., 2012; Bellaloui and Turley, 2013; Stetina et al., 2014). Cotton fibers are developed from the ovule epidermis and are single-celled seed trichomes, and about 30% of the seed epidermal cells differentiate into spinable fibers (Arpat et al., 2004; Wilkins and Arpat, 2005). The development of cotton fibers comprises four stages named fiber cell initiation, fiber cell elongation/primary cell wall (occurs up to 20 days post-anthesis), fiber cell synthesis/secondary cell wall (occurs between 5 and 15 days post-anthesis, and the secondary cell wall synthesis begins at about 20 days post-anthesis and reaches up to 45 days post-anthesis), and maturation (occurs from 45 to 50 days post-anthesis) where fibers dehydrate and produce mature cotton lint (Ji et al., 2003; Arpat et al., 2004; Wilkins and Arpat, 2005).

Upland cotton (*Gossypium hirsutum*) is the predominant species commercially cultivated in the United States and produces two types of fiber on cottonseed, lint and fuzz. Lint fiber is economically valuable, and it is longer and develops faster than the fuzz (Stewart, 1975; Seagull and Giavalis, 2004; Stetina et al., 2014). During ginning process, lint fiber is removed and fuzz fiber left behind (Bechere et al., 2009). There are two naturally occurring mutations in cotton allowing the development of lint fiber, but preventing development of fuzz fiber, and these are referred to as the fuzzless seed alleles and designated as genotypes $N_{1}\_$ (dominant fuzzless allele) or n_2n_2 (recessive fuzzless allele) (Turley and Kloth, 2002; Turley et al., 2007; Stetina et al., 2014).

It is reported that the fuzzless trait in cottonseed is a novel tool to use to understand the biology, genetics, and biochemical and metabolic processes (Turley et al., 2007; Padmalatha et al., 2012). In spite of the few reports available on the effects of fuzzless trait on the biology, genetics, and molecular biology (Padmalatha et al., 2012), very limited information is available on the effects of fuzzless trait on cottonseed composition (Yu et al., 2012; Bellaloui and Turley, 2013; He et al., 2013). For example, the effect of fuzzless trait was investigated on genetic products (Turley and Kloth, 2002; Gou et al., 2007; Turley et al., 2007; Zhao et al., 2009; Pang et al., 2010a,b; Liu et al., 2012), physiological (Ruan et al., 2003; Shi et al., 2006; Yang et al., 2006; Wang et al., 2010; Wang and Ruan, 2010; Zhang et al., 2011), and molecular (Machado et al., 2009; Guan et al., 2011; Walford et al., 2011). It was also reported that carbohydrate and energy metabolisms are involved in fiber development and carbon skeletons for the synthesis of cell wall polysaccharides and fatty acids (Gou et al., 2007; Yang et al., 2008; Pang et al., 2010a,b), and actin cytoskeleton to trigger the secondary cell wall synthesis (Li et al., 2002, 2005; Wang et al., 2010; Wang and Ruan, 2010).

TABLE 1 | Effect of source of variance (F- and P-values) on cottonseed protein concentration (g kg^{-1}), oil concentration (g kg^{-1}), N, C, S, P, K, Ca, Mg percentages (%), and Fe, B, Mn, and Zn concentration (mg kg^{-1}).

Effect	Protein		Oil		Minerals																					
					N		C		S		P		K		Ca		Mg		Fe		B		Mn		Zn	
	F	P	F	P	F	P	F	P	F	P	F	P	F	P	F	P	F	P	F	P	F	P	F	P	F	P
Year	15.8	***	1.4	NS	15.8	**	8.2	*	3.3	NS	42.8	***	0.4	NS	62.0	**	44.3	***	16.9	**	36.7	**	1.0	NS	2.3	NS
Line	20.4	***	13.4	***	20.4	***	17.3	***	5.0	***	5.9	***	4.3	***	62.6	***	8.9	**	16.1	***	24.5	***	5.1	***	59.3	***
Year × Line	2.3	**	3.6	***	2.3	**	3.9	***	1.4	NS	1.8	*	1.5	NS	1.6	NS	2.2	*	3.5	**	1.4	NS	0.4	NS	13.0	***

Cotton was grown in the field in 2012 and 2013 in Stoneville, MS, USA.
**Significance at $P \leq 0.05$; **significance at $P \leq 0.01$; ***significance at $P \leq 0.001$.*

TABLE 2 | Effect of source of variance (F- and P-values) on leaf macro-nutrients (%) and micro-nutrients (mg kg^{-1}).

Effect	Macro-nutrients												Micro-nutrients							
	N		S		P		K		Ca		Mg		Fe		B		Mn		Zn	
	F	P	F	P	F	P	F	P	F	P	F	P	F	P	F	P	F	P	F	P
Year	18.1	**	87.7	***	11.6	***	6.2	*	51.1	***	102.3	***	1.9	NS	85.0	***	1.6	NS	25.0	***
Line	15.5	***	32.5	***	43.5	NS	32.3	NS	12.7	***	53.2	***	55.1	***	21.3	***	4.1	***	10.8	***
Year × Line	2.6	**	3.7	***	1.7	NS	1.9	*	0.6	NS	5.4	***	3.9	***	4.4	***	2.4	**	4.1	***

Cotton was grown in the field in 2012 and 2013 in Stoneville, MS, USA.
**Significance at $P \leq 0.05$; **significance at $P \leq 0.01$; significance at ***$P \leq 0.001$.*

Cottonseed quality is also determined by its content of mineral and non-mineral nutrients such as N, C, S, K, Ca, Zn, and Fe because of their direct or indirect contribution to: protein synthesis (such as N, K, S); oil (such as C and N); carbohydrates (such as C, K, B); metabolite synthesis (Cu, Zn, Fe, Mg); integrity of cell membrane and cell wall structure (Ca and B); cell membrane, lipid synthesis, energy transfer and phosphorylation reactions, carbohydrate metabolism, and nutrient active uptake processes (P); and osmoregulation, stomatal closure, carbohydrate movement, and nutrient mobility (K). The physiological and biochemical roles of these nutrients in plant growth and development were previously reported (Mengel and Kirkby, 1982; Marschner, 2012). The benefit of minerals to human health was also previously reported. For example, an unbalanced diet of micronutrients such as Fe, Zn, leads to human malnutrition (Samman et al., 1998; Fletcher et al., 2004; Lu et al., 2008). Therefore, maintaining an optimum level of Fe, Zn, Cu, Mo, Mn (Zhang et al., 2004; Heinemann et al., 2005) in seeds of major crops that are used for protein and oil sources is critical. It was reported that nutrient uptake, translocation, redistribution, and accumulation are processes controlling the concentrations of minerals in seeds (Grusak and DellaPenna, 1999; White and Broadley, 2005), and most of the genetic basis of these process are not known (Ding et al., 2010).

To our knowledge this is the first comprehensive report to use near-isogenic cotton mutant genotypes for the fuzzless trait to evaluate its effects on cottonseed nutrition (protein, oil, N, S, C, and minerals) under field conditions. Our hypothesis was that because several metabolic processes including carbon metabolism and polysaccharide cell wall synthesis are involved in fiber development (Li et al., 2002, 2005; Gou et al., 2007; Yang et al., 2008; Wang et al., 2010; Wang and Ruan, 2010; Pang et al., 2010a,b), cottonseed nutrition will be influenced as protein and oil are results of carbon, nitrogen, and carbohydrate metabolism. Since we observed differences in leaf and seed nutrients between lines, we hypothesized that nutrients in lint fiber may be impacted as well. Therefore, we investigated the effect of fuzz trait on nutrient content in lint for 1 year only to obtain preliminary results for future research planning.

Materials and Methods

Germplasm Development

The development of near-isogenic lines was described in detail elsewhere (Stetina et al., 2014). The near-isogenic lines (NILs), one expressing the fuzzy/linted and one expressing fuzzless/linted phenotype were developed in five upland cotton backgrounds (Stetina et al., 2014). The following are five sets of *F* and *N* lines/genotypes: Sure-Grow (SG) 747 *F* vs. SG 747 *N*; MD51*ne F* vs. MD51*ne N*; STV 7A*gl F* vs. STV 7A*gl N*; DP 5690 *F* vs. DP 5690 *N*; DES 119 *F* vs. DES 119 *N*. Genotype SA 243 ("Ballard naked seed," PI 528610) was the fuzzless parent with the dominant fuzzless seed allele, N_1N_1 (Turley et al., 2007), and DP 444 BG/RR (Monsanto Company, St. Louis, MO, USA) was used as the commercial check.

TABLE 3 | Effect of fuzzless/linted (*N*) and fuzzy/linted (*F*) cottonseed phenotypes on cottonseed protein concentration (g kg^{-1}), oil concentration (g kg^{-1}), N, C, S, P, K, Ca, Mg percentages (%), and Fe, B, Mn, and Zn concentration (mg kg^{-1}).

Line/genotype	Protein	Oil	C	N	S	Ca	K	Mg	P	B	Fe	Mn	Zn
SA 243	226 ef	270 ef	51.3 d	3.27 g	0.26 e	0.13 f	0.94 e	0.34 g	0.56 g	10.0 f	44.3 e	13.4 c	29.9 e
DP 444 BG/RR	287a	282 cd	51.3 d	4.10 bc	0.34 c	0.14 e	1.03 bc	0.40 b	0.64 bc	12.8 a	52.8 b	13.3 cd	56.9 c
DES 119 *N*	240 d	293 b	52.8 b	3.48 ef	0.30 d	0.18 a	0.95 e	0.41b	0.65 b	11.7 b	49.0 c	14.0 b	33.9 e
DES 119 *F*	276 b	260 fg	50.8 e	4.32 a	0.40 a	0.12 g	1.05 b	0.43 a	0.72 a	12.8 a	62.4 a	14.6 a	81.7 a
DP 5690 *N*	217 g	323 a	53.3 a	3.25 g	0.27 e	0.15 c	0.91 f	0.35 f	0.55 g	9.3 g	39.9 g	12.2 f	25.9 f
DP 5690 *F*	252 c	275 fg	50.9 e	3.95 c	0.37 b	0.11 h	0.99 d	0.38 cd	0.59 f	11.6 b	48.7 cd	11.5 g	76.8 b
MD 51*ne N*	221 fg	274 fg	52.2 c	3.27 g	0.26 e	0.18 a	0.95 e	0.37 e	0.59 ef	10.7 d	44.6 e	13.6 c	30.8 e
MD 51*ne F*	274 b	257 gh	50.9 e	4.15 ab	0.35 bc	0.14 d	1.05 bc	0.39 c	0.62 cd	11.7 b	50.6 c	12.2 f	75.6 b
SG 747 *N*	229 e	288 bc	52.1 c	3.31 fg	0.26 e	0.17 b	0.94 e	0.37 de	0.61 de	11.0 c	42.2 f	12.9 de	31.2 e
SG 747 *F*	285 a	250 h	50.3 g	4.11 bc	0.31 d	0.12 fg	1.02 c	0.38 c	0.65 bc	12.9 a	50.1 c	12.7 e	50.7 d
STV 7A*gl N*	197 h	264 e	52.7 b	3.57 de	0.26 e	0.15 c	0.95 e	0.38 cd	0.59 ef	10.4 e	39.3 g	11.3 g	30.0 ef
STV 7A*gl F*	255 c	234 i	50.5 f	3.72 d	0.30 d	0.11 h	1.10 a	0.40 b	0.65 b	11.8 b	47.0 d	11.4 g	55.9 c

Cotton was grown in the field in 2012 and 2013 in Stoneville, MS, USA. Means within a column followed by the same letter are not significantly different at 5% as determined by Fisher's LSD test.

TABLE 4 | Effect of fuzzless/linted (*N*) and fuzzy/linted (*F*) cottonseed phenotypes on leaf Ca, K, Mg, P, N, and S percentages (%), and B, Fe Mn, and Zn concentration (mg kg^{-1}).

Line/genotype	Ca	K	Mg	P	N	S	B	Fe	Mn	Zn
SA 243	3.39 de	2.01 f	1.09 c	0.42 g	4.23 d	0.96 e	40.9 e	78.6 g	166.6 cd	37.4 d
DP 444 BG/RR	3.36 e	1.88 g	0.95 d	0.39 h	3.84 f	0.84 f	40.2 e	77.6 g	158.3 f	37.4 cd
DES 119 *N*	4.31 b	3.48 b	1.22 b	0.75 a	4.41 c	1.61 b	51.9 b	120.2 b	170.8 b	42.4 a
DES 119 *F*	3.09 f	2.38 e	0.78 e	0.58 d	3.40 g	0.96 e	40.2 e	90.1 de	172.8 ab	30.8 e
DP 5690 *N*	3.58 cd	2.78 c	1.00 d	0.65 c	4.04 c	1.70 a	48.0 c	91.3 d	173.3 a	39.0 b
DP 5690 *F*	3.68 c	1.70 h	0.67 f	0.65 c	3.80 c	1.31 d	41.3 e	87.2 e	167.8 c	37.4 d
MD 51*ne N*	4.70 a	2.84 c	1.68 a	0.70 b	4.90 b	1.54 b	55.1 a	141.0 a	162.3 e	42.3 a
MD 51*ne F*	3.66 c	1.55 i	0.98 d	0.51 f	3.80 f	0.97 e	43.3 d	117.2 b	162.0 e	31.6 e
SG 747 *N*	4.45 b	2.61 d	1.61 a	0.64 c	4.60 b	1.42 c	50.7 b	119.7 b	165.1 d	43.5 a
SG 747 *F*	3.38 e	1.60 hi	0.93 d	0.43 g	3.81 f	0.89 ef	40.3 e	82.0 f	167.9 c	36.7 d
STV 7A*gl N*	4.88 a	3.81 a	1.22 b	0.75 a	4.81 a	1.38 cd	50.3 b	139.4 a	162.5 e	37.2 d
STV 7A*gl F*	3.40 de	2.48 e	0.75 e	0.55 e	2.98 h	0.85 f	40.2 f	107.2 c	162.5 e	38.8 bc

Cotton was grown in the field in 2012 and 2013 in Stoneville, MS, USA.
Means within a column followed by the same letter are not significantly different at 5% as determined by Fisher's LSD test.

Field Management and Growth Conditions

A field experiment was conducted in 2012 and 2013 at Stoneville, MS, USA. Field management and growth conditions were previously detailed in Stetina et al. (2014). Briefly, cottonseeds were planted on four-row plots spaced 1.02 m apart and each plot was 9.14 m long with a 3.04 m alley between plots. Planting dates were 25 April 2012 and 13 May 2013. Insecticides and fungicides were applied to control pest and diseases. Field management was conducted according to the standard agronomic practices for cotton production in the Mississippi Delta region (http://msucares.com/crops/cotton/index.html). Cotton bolls were harvested from 10 adjacent plants at three (2012) and four (2013) intervals to prevent the loss of fuzzless cottonseed content due to rain and wind. The harvest intervals were on 28 August, 6 September, and 27 September in 2012, and on 5 September, 12 September, 17 September, and 27 September in 2013. A defoliant was used prior to the last harvest date each year. Cottonseed samples were processed at the USDA ARS Cotton Ginning Laboratory at Stoneville, MS, USA and saw-ginned on 20 November 2012 and 12 November 2013. Cottonseed were collected and acid-delinted for seed composition analyses. Soil samples were taken across the field by dividing the field into four main sections and taking 20–25 samples from each section at depth of 30.5 cm. Therefore, one composite sample (combined 20–25 samples) represented each section. Analysis of the four samples, representing the four sections, showed uniformity in main soil nutrients. Therefore, the soil nutrients shown in here are average of the four sections representing the entire field. For leaf sampling, the most recent fully expanded leaves were taken from each plot. Fourteen leaves were taken from the middle two rows (7 leaves were taken along each one middle row to ensure equal distribution of leaves along each row). Leaves were oven-dried at 65°C and ground into fine particle with Laboratory Mill 3600 (Perten, Springfield, IL). Seed and leaf C, N, K, Ca, Mn, and Zn were analyzed at the University of Georgia's Soil, Plant, and Water Laboratory, Athens, GA. Seed protein, oil, B, Fe, P, and leaf B, Fe, and P were analyzed as described below.

Boron Analysis

Boron concentrations in the most recently fully expanded leaves at boll stage and mature seeds were determined according to Lohse (1982) using the azomethine-H method, and samples were prepared according to John et al. (1975). Boron concentration was determined spectrophotometrically by reading the samples at 420 nm using a Beckman Coulter DU 800 spectrophotometer (Fullerton, CA). The concentration of B was measured after color development, and B concentration was expressed as mg B kg^{-1} dwt.

Iron Analysis

Iron concentrations in the most recently fully expanded leaves at boll stage and mature seeds were measured according to Bandemer and Schaible (1944) and Loeppert and Inskeep (1996). The concentration was determined by acid wet digestion (Analytical Methods Committee, 1959), extraction, and reaction of the reduced ferrous Fe using 10 ml of 0.02 M 1,10-phenanthroline, and the samples were prepared for measurement using a quinol solution of 1% (w/v) reagent. Concentrations of Fe in leaves and seeds were measured spectrophotometrically at 510 nm using a Beckman Coulter DU 800 spectrophotometer (Fullerton, CA).iron was expressed as mg Fe kg^{-1} dwt.

Phosphorus Analysis

Phosphorus concentrations in leaves and mature seeds were determined according to Cavell (1955) using the yellow phosphor-vanado-molybdate complex. Briefly, 2 g of dried ground samples were ashed and 10 ml of 6 M HCl was added. The samples, then were placed in a water bath to evaporate the solution to dryness. Then, a 2 ml of 36% v/v HCl was added and the samples were boiled. Ten milliliters of distilled water was added to the samples and the solution was brought to boil for a few seconds, and then diluted to 50 ml with distilled water, and then filtered. To measure P concentration, a reagent of 5 ml of 5 M HCl and 5 ml of ammonium molybdate–ammonium metavanadate was freshly prepared and added to the filtrate. The concentration of P was measured spectrophotometrically using a Beckman Coulter DU 800 spectrophotometer at 400 nm. The concentration of P in the sample was expressed as a percentage.

Analyses of N, S, C in Leaves and Seed

The most recently fully expanded leaves during boll stage and mature seeds were analyzed for minerals, N, S, and C concentrations by digesting 0.6 g of dried, ground seed in HNO_3 in a microwave digestion system. The concentrations of minerals in the samples were determined using inductively coupled plasma spectrometry (ICP) (Bellaloui and Turley, 2013; Bellaloui et al., 2014). For N, S, and C measurements, a 0.25 g ground-dried sample was combusted in an oxygen atmosphere at 1350°C to convert elemental N, S, and N, S, and C were measured by an elemental

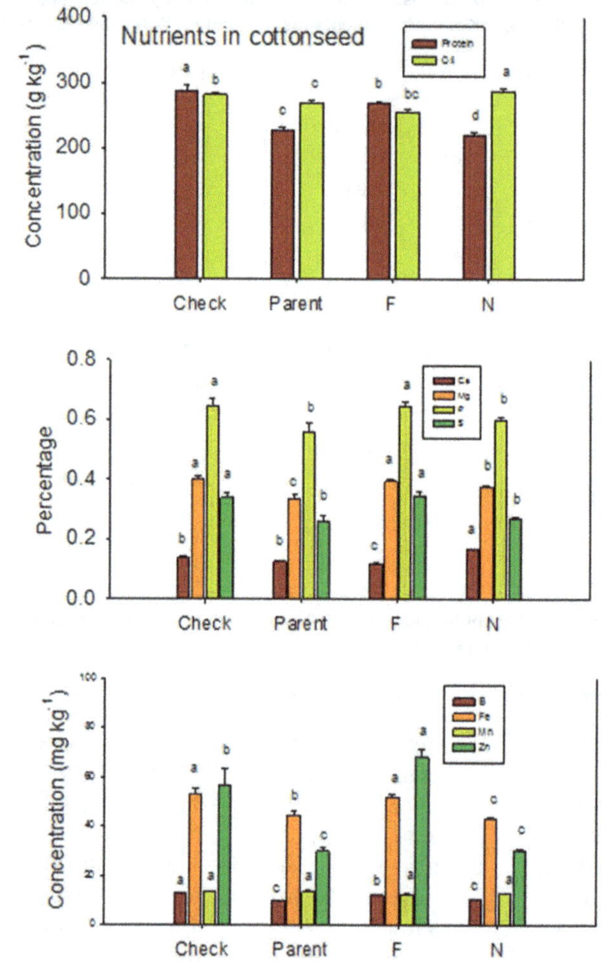

FIGURE 1 | Effects of cottonseed fuzz trait on cottonseed protein, oil, and nutrients in near-isogenic *Gossypium hirsutum* lines expressing fuzzy/linted (*F*) and fuzzless/linted (*N*) seed phenotypes, the fuzzless/linted parent SA 243 (parent), and the fuzzy/linted commercial cultivar DP 444 BG/RR (check). Cotton was grown in the field in 2012 and 2013 in Stoneville, MS, USA. Bars show means ± standard error of the mean (SE). Within each constituent, means with the same letter are not significantly different at 5% as determined by Fisher's LSD test.

analyzer using thermal conductivity cells (LECOCNS-2000 Elemental Analyzer, LECO Corporation, St. Joseph, MI) (Bellaloui et al., 2011, 2014).

Analyses of Minerals, N, S, and C in Soil

Mineral concentrations in soil, N, S, and C were analyzed at The University of Georgia's Soil, Plant, and Water Laboratory, Athens, GA. The concentrations of the minerals K and Mn were determined using a 5-g soil: 20 ml Mehlich-1 solution and analyzed using inductively coupled plasma (ICP) spectrometry (Bellaloui et al., 2009). Percentages of N, S, and C were determined in a 0.25-g sample of soil by combusting samples in an oxygen atmosphere at 1350°C, and converting elemental N, S, and C into N_2, SO_2, and CO_2 gasses, and N, S, and C were determined by elemental analyzer using thermal conductivity cells (LECOCNS-2000 Elemental Analyzer LECO Corporation, St. Joseph, MI,

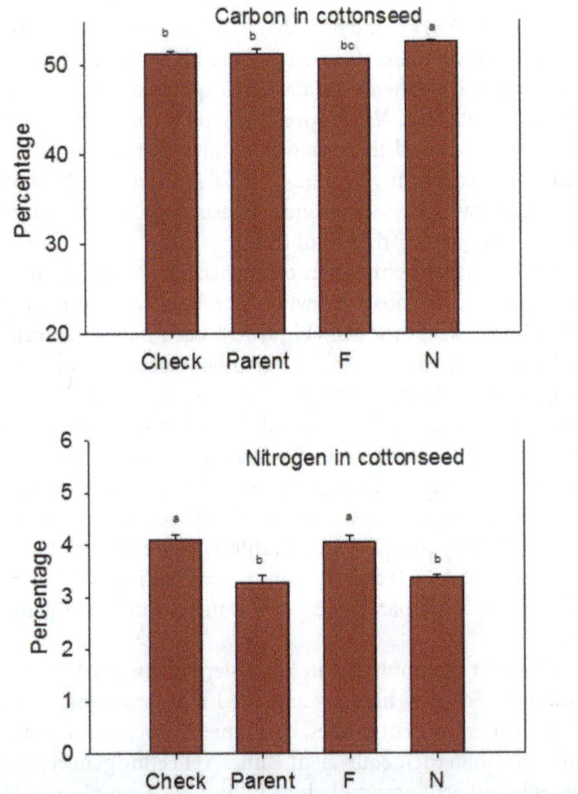

FIGURE 2 | Effects of cottonseed fuzz trait on cottonseed carbon (C) and nitrogen (N) in near-isogenic *Gossypium hirsutum* lines expressing fuzzy/linted (*F*) and fuzzless/linted (*N*) seed phenotypes, the fuzzless/linted parent SA 243 (parent), and the fuzzy/linted commercial cultivar DP 444 BG/RR (check). Cotton was grown in the field in 2012 and 2013 in Stoneville, MS, USA. Bars show means ± standard error of the mean (SE). Within each constituent, means with the same letter are not significantly different at 5% as determined by Fisher's LSD test.

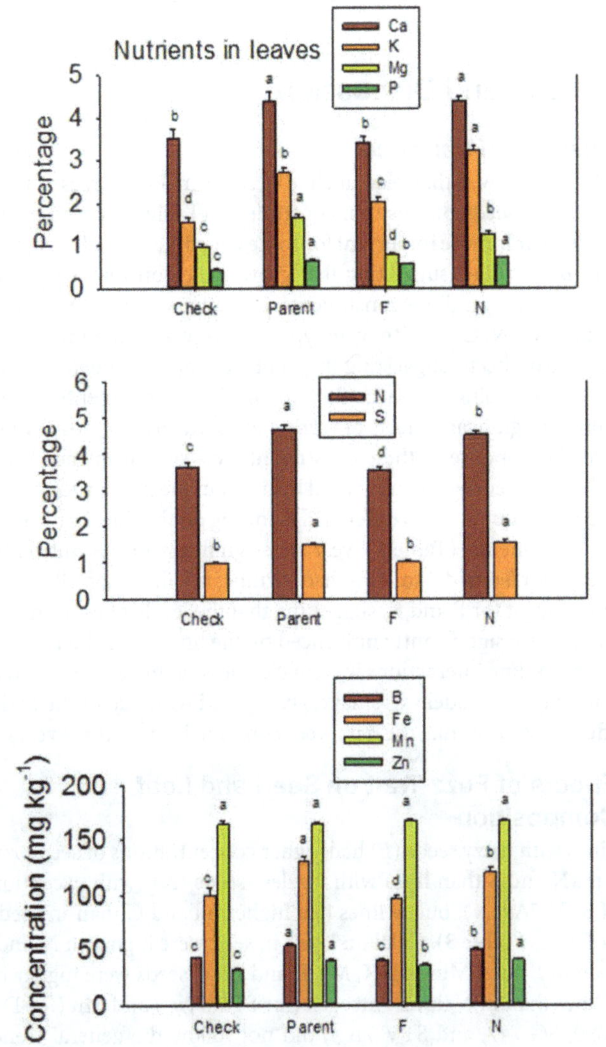

FIGURE 3 | Effects of cottonseed fuzz trait on cotton leaf nutrients in near-isogenic *Gossypium hirsutum* lines expressing fuzzy/linted (*F*) and fuzzless/linted (*N*) seed phenotypes, the fuzzless/linted parent SA 243 (parent), and the fuzzy/linted commercial cultivar DP 444 BG/RR (check). Cotton was grown in the field in 2012 and 2013 in Stoneville, MS, USA. Bars show means ± standard error of the mean (SE). Within each constituent, means with the same letter are not significantly different at 5% as determined by Fisher's LSD test.

USA) (Bellaloui et al., 2009). Concentrations of N, S, and C were expressed as percentages.

Cottonseed Protein and Oil Analysis

Mature cottonseeds were collected from each plot and analyzed for protein and oil. Briefly, approximately 25 g of seed was ground using a Laboratory Mill 3600 (Perten, Springfield, IL). Protein and oil in cottonseed were analyzed by near infrared reflectance according to Wilcox and Shibles (2001) and Bellaloui and Turley (2013) using a diode array feed analyzer AD 7200 (Perten, Springfield, IL). Calibrations were developed using Perten's Thermo Galactic Grams PLS IQ software, and the calibration equation was established according to AOAC methods (Association of Official Analytical Chemists (AOAC), 1990a,b). Cottonseed protein and oil were expressed on a seed dry matter basis (Wilcox and Shibles, 2001; Boydak et al., 2002; Bellaloui and Turley, 2013).

Experimental Design and Data Analysis

This experiment was a part of a large experiment that was designed in a split-plot. The main plot was line (genotype), and plots were arranged in a randomized complete block design with 4 replications. The subplot was ginning method (Stetina et al., 2014). In the current study, we were interested in the conventional saw ginning method only. Therefore, the only factors under the study were line, year, and their interactions. Analysis of variance was conducted using Proc Mixed model in SAS (Statistical Analysis System; SAS Institute, Inc., Cary, NC, 2002-2012). Year (Y), Line, and Y × Line interactions were modeled as fixed effects, and replicates and their interactions were considered random effects. Means were separated by Fisher's protected least significant difference test at the 5% level of significance using SAS (2002–2012).

Correlation was performed using PROC CORR in SAS (2002–2012).

Results and Discussion

Analysis of Variance

ANOVA showed that year and line were the major sources of variability for seed composition constituents (**Table 1**). Based on *F*-value, year is more important for some constituents such as P, Mg, and B than line, suggesting that growing season may have also an influence on these constituents. For others constituents such as protein, N, C, and Zn, genotypic effects were more important than year effects, suggesting the significant effects of genetic factors in controlling the accumulation of these constituents. There were no significant effects of year on oil, S, K, and Mn, suggesting that the response of these constituents was similar in each year. All constituents were influenced by line, but the magnitude of the accumulation in seeds differed, depending on the line. For leaf N, S, and minerals (**Table 2**), year was significant for all nutrients, except for Fe and Mn. Line had significant effects for all nutrients, except for P and K, suggesting that the levels of nutrients in leaves were significantly influenced by the line. Since the *F*-value of year × line interactions was smaller (less significant: its contribution to the model was smaller) compared with that of the main effect of year or line, the data were combined across the 2 years.

Effects of Fuzz Trait on Seed and Leaf Composition

Lines with fuzzy seeds (*F*) had higher concentrations of seed protein, N and S than lines with fuzzless seeds (*N*) (with exception of STV 7A*gl N*), but *N* lines had higher oil and C than in seeds of *F* lines (**Table 3**). Calcium levels in seed were higher in *N* lines than in *F* lines. Minerals K, Mg, P, and Zn in seeds were higher in *F* lines than in *N* lines. Nitrogen (for STV 7A*gl*) and Mn (for DP 5690, SG 747, and STV 7A*gl*) did not follow the general trend of other nutrients of higher nutrient accumulation in *F* lines. Instead N did not differ between *N* and *F* lines in STV 7A*gl*, and Mn was higher in the *N* line for DP 5690, but did not differ between *N* and *F* lines in SG 747 and STV 7A*gl*. Except for Mn in DES 119, MD 51*ne*, and STV 7A*gl*, and Ca, P, and N for DP 5690, N and S and all minerals in leaves were higher in *N* lines than in *F* lines (**Table 4**). When the data were expressed across all lines for the fuzz trait (*F* vs. *N*) (**Figures 1, 2**), oil, C, and Ca, were higher in cottonseed in *N* lines than in *F* lines, but protein and other minerals, except Mn, were higher in *F* lines than in *N* lines, supporting data expressed on line set basis. Fuzzless and fuzzy lines had similar concentrations of Mn.

The higher accumulation of nutrients of *N* lines than *F* lines in leaves was also observed when all *F* lines were compared with all *N* lines, except for Mn (**Figure 3**). Each line set accumulated different levels of protein, oil, or other nutrients, and the level ranged from about 197 g kg $^{-1}$ to 287 g kg^{-1} for protein, and 234 g kg^{-1} to 293 g kg^{-1} for oil. Similar wider ranges of other nutrients were observed between *F* and *N* lines. Cotton leaves accumulated higher concentrations of nutrients in *N* lines than in *F* lines, except for Ca and P in DP 5690, and Mn in DES 119, MD 51*ne*, and STV 7A*gl* (**Table 4**). Generally, both the fuzzless parent (SA 243) and the commercial cultivar DP 444 BG/RR accumulated comparable levels of nutrients to the isogenic lines sets.

Nutrient accumulation in lint differed between *N* and *F* lines (**Table 5**). Some *N* lines accumulated more nutrients in the lint than their equivalent *F* lines, and other *N* lines accumulated less nutrients than their equivalent *F* lines, reflecting genotypic differences in nutrient accumulations in lint. For example, Ca, K, S, B, Fe, and Na contents were higher in *N* lines than in *F* lines, except in MD 51*ne* and SG 747 for K, DP 5690 and DES 119 for S, SG 747 and DP 5690 for B, DP 5690 for Fe, and MD 51*ne* for Na. All lines had at least one nutrient where *F* line was higher than *N* line, with the exception of MD 51*ne* where the *N* line was always higher or equal to the *F* line. Generally, a similar trend was also found when data were expressed on fuzz trait basis (all *F* lines vs. all *N* lines) (**Figure 4**).

The higher protein and lower oil concentrations in cottonseed may suggest there is a potential commercial use for fuzzless seed as a source for food (oil) and feed (cottonseed meal). Our results

TABLE 5 | Effect of fuzzless/linted (*N*) and fuzzy/linted (*F*) cottonseed phenotypes on cotton lint nutrients Ca, K, Mg, P, C, N, S percentages (%), and Fe, B, Mn, Na, and Zn concentration (mg kg^{-1}).

Line	Ca	K	Mg	P	C	N	S	B	Fe	Mn	Na	Zn
SA 243	0.07 f	0.48 a	0.06 de	0.03 bc	44.3 a	0.74 g	0.030 e	3.1 de	7.1 ef	3.2 e	40.5 c	3.8 ef
DP 444 BG/RR	0.09 e	0.51 b	0.07 bc	0.03 bc	44.0 bc	0.97 bcd	0.033 de	3.0 e	6.4 f	3.7 ed	31.6 ef	5.3 ab
DES 119 *N*	0.15 c	0.53 a	0.07 bc	0.02 d	44.1 bc	0.88 def	0.038 bc	4.1 bc	9.6 c	5.0 bc	52.4 a	4.1 de
DES 119 *F*	0.06 f	0.46 e	0.06 c	0.03 a	43.8 de	1.05 ab	0.035 cd	2.7 e	7.0 ef	4.1 d	41.2 bc	5.5 a
DP 5690 *N*	0.11 d	0.50 b	0.07 b	0.02 cd	43.9 cde	0.79 fg	0.030 e	3.3 de	6.8 ef	4.6 c	53.0 a	3.3 f
DP 5690 *F*	0.07 ef	0.46 e	0.07 b	0.02 d	44.1 bc	0.96 bcde	0.030 e	2.9 e	8.5 d	4.8 c	45.9 b	4.1 de
MD 51*neN*	0.19 a	0.47 de	0.06 c	0.03 b	44.0 cde	0.84 f	0.053 a	4.5 b	11.1 b	6.6 a	33.2 ef	5.2 ab
MD 51*ne F*	0.09 e	0.47 cde	0.05 e	0.02 cd	44.1 bc	0.87 ef	0.038 bc	3.0 e	6.7 ef	4.7 c	34.1 de	4.5 cd
SG 747 *N*	0.11 d	0.43 f	0.06 c	0.02 d	44.0 cde	0.95 cde	0.038 bc	3.0 e	9.1 cd	3.7 ed	39.2 cd	5.2 ab
SG 747 *F*	0.09 e	0.43 f	0.06 d	0.03 b	44.0 cb	0.99 abc	0.033 de	3.6 cd	7.4 e	4.1 d	26.3 fg	4.3 cde
STV 7Agl *N*	0.17 b	0.54 a	0.08 a	0.02 cd	43.8 e	1.09 a	0.040 b	5.9 a	14.0 a	5.4 b	33.7 ef	4.5 cd
STV 7Agl *F*	0.07 ef	0.48 cde	0.07 bc	0.03 b	44.1 bc	0.79 fg	0.030 e	2.8 e	6.7 ef	3.4 e	23.6 g	4.7 bc

Cotton was grown in the field in 2013 only in Stoneville, MS, USA.
Means within a column followed by the same letter are not significantly different at 5% as determined by Fisher's LSD test.

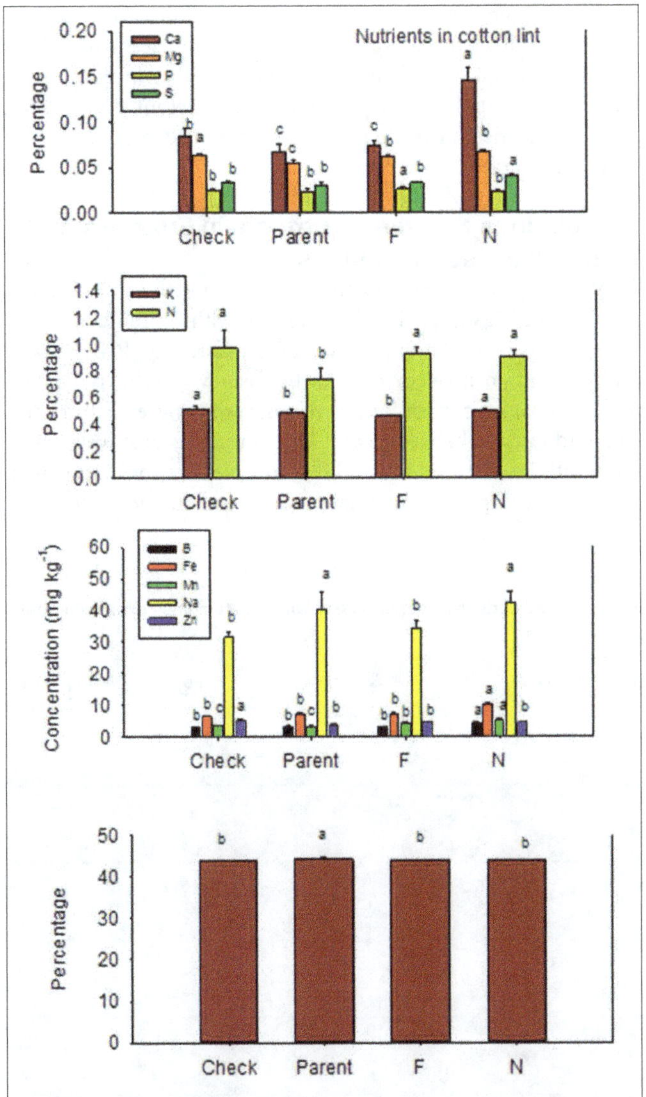

FIGURE 4 | Effects of cottonseed fuzz trait on cotton lint nutrients in near-isogenic *Gossypium hirsutum* lines expressing fuzzy/linted (*F*) and fuzzless/linted (*N*) seed phenotypes, the fuzzless/linted parent SA 243 (parent), and the fuzzy/linted commercial cultivar DP 444 BG/RR (check). Cotton was grown in the field in 2013 only in Stoneville, MS, USA. Bars show means ± standard error of the mean (SE). Within each constituent, means with the same letter are not significantly different at 5% as determined by Fisher's LSD test.

are in agreement with previous reports (Bechere et al., 2009). They developed homozygous naked-tufted M_8 mutant lines and evaluated the developed lines for lint yield, fiber quality, seed oil content, ginning efficiency, and yarn spinning. They found that lint yield in the naked-tufted seed mutants was lower when compared with their original fuzzy parents. Also, they found that seed oil content in naked-tufted seed mutants was higher than fuzzy parents. Bellaloui and Turley (2013), working on similar material with the same trait, but under greenhouse conditions, and found, generally, that fuzzless cotton seeds accumulated higher oil and lower protein compared with their equivalent fuzzy seeds, showing an inverse relationship between protein and oil. The inverse relationship between protein and oil was previously reported in other species such as corn (Kebede et al., 2013), soybean (Burton, 1985; Ray et al., 2006; Bellaloui et al., 2013a,b), and cottonseed (Pettigrew and Dowd, 2011). Pettigrew and Dowd (2011) investigated the effect of irrigation and planting dates on seed oil and protein in six conventional cotton (*Gossypium hirsutum*) cultivars (DP 445BR, DP 555BR, FM 800BR, FM 960BR, ST 4892BR, and ST 5599BR), planted at late April and late May, and under irrigated and non-irrigated (dryland) field conditions in Stoneville, MS, from 2005 to 2008. They found irrigation increased the concentrations of seed oil, but decreased seed protein. However, under dryland conditions, protein was higher than oil at late April and late May planting. Their results showed opposite trend between protein and oil, agreeing with the current results.

The higher accumulation of some nutrients in seeds of *F* lines than in *N* lines, and the higher accumulation of some other nutrients in *N* lines than in *F* lines was also previously reported (Bellaloui and Turley, 2013), and this could be due to genotypic differences. The consistent higher accumulation of nutrients in leaves of *N* lines than in *F* lines across line sets suggests limited translocation of nutrients from leaves to seed, and this may explain the lower accumulation of minerals in seeds of some fuzzless isolines. For example, B showed higher accumulation in leaves and lower accumulation in seed of *N* lines than in the *F* lines. It is known that cotton has higher B requirement for growth, development, and seed quality. Although B is mobile in sugar-alcohol containing species such as rice, pears, almond, and celery, it looks like B in cotton has limited mobility (Brown and Shelp, 1997; Brown et al., 2002; Dordas, 2006) from leaves to seed. Although high concentrations of B occurred in leaves (ranged from 40 to 55 mg B kg^{-1}), cottonseed contained limited concentrations of B (<20 mg B kg^{-1}). Soil used in our experiment had adequate B concentration (ranged from 1.5 to 2.0 mg kg^{-1}) to support the crop. The limited mobility of B in cotton may be one reason of why cotton sometimes responds to foliar B application. Dordas (2006) investigated the foliar application of B on cotton and found there was an increase in cotton lint, and concluded that it is possible that the critical levels of B in cotton have been assessed by visual symptoms and not by yield of comparative field studies. It must be noted also that some other studies did not find a positive response to foliar B applications (Heitholt, 1994).

The other mineral that showed an interesting pattern was Ca. The accumulation of Ca in leaves, seeds, and lint was higher in *N* lines than in *F* lines, and this could be explained by the fact that Ca taken up was not used for further physiological and biochemical processing involved in lint fiber development and structure such as the synthesis of cell wall polysaccharides and fatty acids (Gou et al., 2007; Yang et al., 2008; Pang et al., 2010a,b) and the secondary cell wall synthesis (Li et al., 2002, 2005; Wang et al., 2010; Wang and Ruan, 2010). Also, the down-regulation of calcium and phytohormone mediated signals observed at fiber initiation stage in the fuzzless mutants (Padmalatha et al., 2012) could explain the higher levels of Ca in leaves, seeds, and lint due to lower requirements of Ca by fuzzless seed lines. The significant role of Ca in cotton was previously reported and involved in pollen germination (Brewbaker and Kwack, 1963), pollen tube growth (Zhang et al., 1997), stimulation of fertilization (Faure et al., 1994; Tian and Russell, 1997), and egg activation (Digonnet et al., 1997), and this is because Ca is required for vesicle fusion at the tip of the elongating tube (Pierson et al., 1996). It was also found that Ca enhances the antioxidant enzyme activity and protects the plant under oxidative stress conditions through reactive oxygen species (ROS) scavenging (Jiang and Huang, 2001). The apparent explanation that seed Ca in *N* lines was higher than in *F* lines could be due to higher requirements of Ca by *N* lines for physiological and structural functions, may be, for seed protection from severe environmental conditions such as drought and high heat as seeds in *N* lines are covered with little lint compared with *F* lines. If this is the case, then germination rate between *N* and *F* lines should be different as seed of *N* lines may be more hard seeded because of the higher Ca accumulation. The higher Ca content in leaves and lint may support the higher Ca requirement by *N* lines compared with F lines. Because the current results showed higher Ca in leaves and seeds in *N* lines than in *F* lines, it would be worthwhile to further investigate the effects of Ca supply on Ca partitioning in different plant tissues, including seeds and relate that to the rate of ROS enzymes and germination rates.

Correlations Between Nutrients in Fuzzless and Fuzzy Near-Isogenic Lines

In *N* lines, protein was positively correlated with Ca, B, Fe, Mn, and Zn, and oil was positively correlated with C (**Table 6**). There were positive correlations between Ca and K, Mg, P, B, Fe, Mn, and Zn, and positive correlations between Mg and P, B, Fe, Mn, and Zn. Positive correlations were observed between P and B, Fe, Mn, and Zn, and between S and Mn. Boron was positively correlated with Fe, Mn, and Zn, and Fe had a positive correlation with Mn and Zn. A positive correlation between Mn and Zn was also

TABLE 6 | Pearson correlation coefficient (*R*- and *P*-values) between nutrients in cottonseed of near-isogenic *Gossypium hirsutum* lines expressing the fuzzless/linted seed phenotype across 2012 and 2013.

		Protein	Oil	Ca	K	Mg	P	C	N	S	B	Fe	Mn	Zn
Protein	*R*	1												
	P													
Oil	*R*	NS	1											
	P													
Ca	*R*	0.63	NS	1										
	P	***												
K	*R*	NS	NS	0.36	1									
	P			*										
Mg	*R*	NS	NS	0.52	0.54	1								
	P			***	***									
P	*R*	NS	NS	0.58	0.61	0.92	1							
	P			***	***	***								
C	*R*	NS	0.41	NS	NS	NS	NS	1						
	P		**		NS		NS							
N	*R*	NS	NS	NS	NS	NS	NS	NS	1					
	P						NS							
S	*R*	NS	NS	NS	NS	NS	NS	NS	NS	1				
	P						NS							
B	*R*	0.49	NS	0.70	NS	0.77	0.73	NS	NS	NS	1			
	P	**		***		***	***							
Fe	*R*	0.51	NS	0.65	0.40	0.75	0.71	NS	NS	NS	0.74	1		
	P	***		***	*	***	***				***			
Mn	*R*	0.40	NS	0.49	NS	0.38	0.35	NS	NS	0.30	0.51	0.45	1	
	P	*		**		*	*			*	***	**		
Zn	*R*	0.31	NS	0.65	NS	0.59	0.50	NS	NS	NS	0.75	0.62	0.41	1
	P	*		***		***	**				***	***	**	

N = 40; $^{*}p \leq 0.5$; $^{**}p \leq 0.01$; $^{***}p \leq 0.001$.

TABLE 7 | Table 6 Pearson correlation coefficient (*R*- and *P*-values) between nutrients in cottonseed of near-isogenic *Gossypium hirsutum* lines expressing the fuzzy/linted seed phenotype across 2012 and 2013.

		Protein	Oil	Ca	K	Mg	P	C	N	S	B	Fe	Mn	Zn
Protein		1												
Oil	*R*	NS	1											
	P													
Ca	*R*	0.41		1										
	P	**												
K	*R*	NS	NS	NS	1									
	P													
Mg	*R*	NS	NS	NS	NS	1								
	P													
P	*R*	NS	NS	NS	NS	0.95	1							
	P					***								
C	*R*	NS	NS	NS	NS	NS	NS	1						
	P													
N	*R*	NS	NS	NS	NS	NS	NS	NS	1					
	P													
S	*R*	NS	NS	NS	NS	NS	NS	NS	NS	1				
	P													
B	*R*	0.50	NS	0.41	NS	0.46	0.44	NS	NS	NS	1			
	P	***		*		**	**							
Fe	*R*	NS	NS	NS	NS	0.62	0.58	NS	NS	NS	0.54	1		
	P					***	***				***			
Mn	*R*	NS	NS	0.36	NS	0.54	0.51	NS	NS	NS	0.61	0.67	1	
	P			*		***	***				***	***		
Zn	*R*	NS	0.37	NS	NS	NS	NS	NS	NS	NS	NS	0.31		1
	P		*									*		

N = 40; $^{*}p \le 0.5$; $^{**}p \le 0.01$; $^{***}p \le 0.001$.

observed. In *F* lines, positive correlations between protein and B, and between protein and Ca were observed (**Table 7**). Oil was positively correlated with Zn, and Ca had a positive correlation with B and Mn. Both Mg and P had positive correlation with B, Fe, and Mn. Boron had positive correlations with Fe and Mn, and Fe had positive correlations with Mn and Zn.

Correlation between nutrients in *N* lines and *F* lines showed significant positive correlation between some nutrients in both *N* and *F* lines, but some nutrients did not correlate (**Tables 6, 7**). For example, in *N* lines protein was positively correlated with Ca, B, Fe, Mn, and Zn, suggesting the indirect involvement of these minerals with protein synthesis. Also, the positive correlations between Ca and oil, and between C and oil support the observation of similar trends of these nutrients in *N* lines, and the involvement of carbon in oil metabolism. Nutrients such as P, Ca, and B; Ca, B, and Mn; Mg, B, Fe, and Mn; P, B, and Mn; Fe, Mn, and Zn showed consistency of positive correlation in both *N* and *F* lines. Other nutrients were not consistent between *N* and *F* lines, and this is due to genotypic effects. The positive and negative correlation between cation and anion nutrients were previously reported, although this correlation depends on growth conditions, genotype, and nutrient supply (Mengel and Kirkby, 1982; Marschner, 2012).

Nutrient uptake, translocation, redistribution, and accumulation are processes controlling the accumulation of minerals in seeds (Grusak and DellaPenna, 1999; White and Broadley, 2005), and most of the genetic basis of these process are still not known (Ding et al., 2010). Previous research on fuzzless cottonseed showed that fiber development involves physiological, biochemical, and molecular processes (Turley et al., 2007;

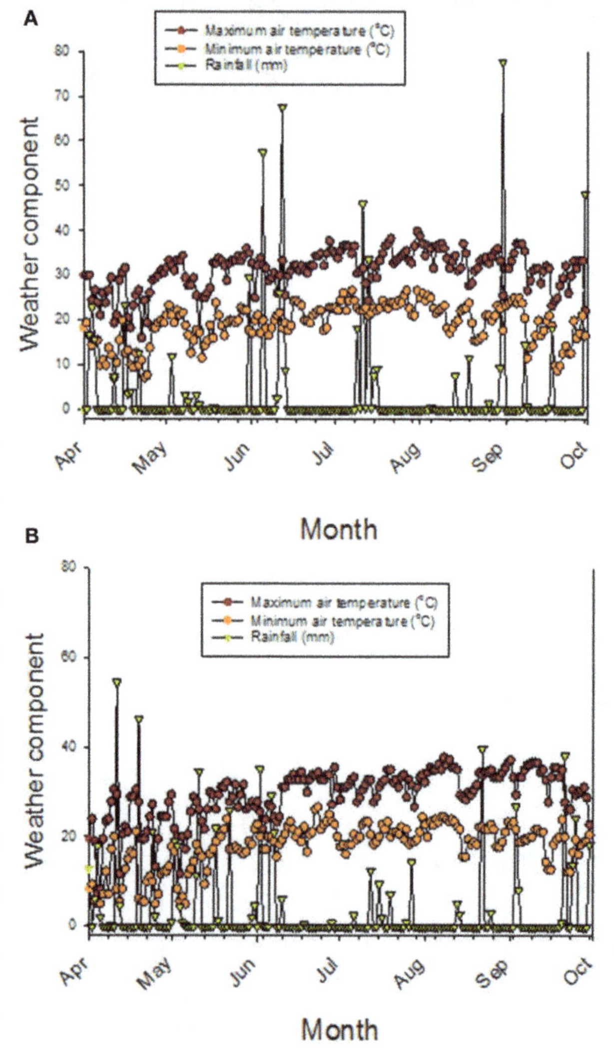

FIGURE 5 | Maximum and minimum air temperatures, and precipitation (rainfall) in 2012 (A) and in 2013 (B). Weather data obtained from MSUCares, Stoneville (2015).

Padmalatha et al., 2012). For example, it was found that the fiber development involved processes related to genetics (Turley and Kloth, 2002; Gou et al., 2007; Turley et al., 2007; Zhao et al., 2009; Liu et al., 2012), phytohormones (Yang et al., 2006; Zhang et al., 2011), transcription factors such as MYB25 (Machado et al., 2009), turgor mechanism (Ruan et al., 2003; Wang et al., 2010; Wang and Ruan, 2010), soluble sugars, potassium, and organic acids, ion-transporters (H+-ATPases and K+-transporter) (Wang et al., 2010; Wang and Ruan, 2010); upregulation of potassium and sugar transporters to maintain the turgor pressure (Ruan et al., 2004), carbohydrate and energy metabolism and carbon skeletons for the synthesis of cell wall polysaccharides and fatty acids (Gou et al., 2007; Yang et al., 2008; Pang et al., 2010a,b), and the secondary cell wall synthesis (Li et al., 2002, 2005; Wang et al., 2010; Wang and Ruan, 2010). Because these processes were involved in fiber development, then it is expected to have differences in phenotypic traits such as protein, oil, and minerals between fuzzy and fuzzless cottonseed. This is because of the involvement of carbon skeleton and energy metabolism (Gou et al., 2007; Yang et al., 2008; Pang et al., 2010a,b), K (Ruan et al., 2004; Wang et al., 2010), and Ca (Padmalatha et al., 2012) in fiber development. The only difference between each line set investigated here is cottonseed fuzz trait, so differences in cottonseed composition constituents observed in our study could be explained by the trait as each isoline set has the same genetic background (Bellaloui and Turley, 2013). It must be noted here that information explaining nutrient dynamics in fuzzy and fuzzless cottonseed is almost non-existent (Bellaloui and Turley, 2013), and most of research on fuzzless cottonseed has been on cell biology, genetic, and molecular biology (Padmalatha et al., 2012; Bellaloui and Turley, 2013). However, it is possible that cottonseed differences could be explained by stored energy metabolism differences such as sugars, hormones signaling, and lower lint yields (Turley et al., 2007; Stetina et al., 2014) in fuzzless mutants. These major down regulations, mentioned above, and lower lint yield may have led to a nutrient imbalance, resulting in differences in nutrient uptake, transport, mobility from leaves (source) to seed (sink), and nutrient accumulation in leaves and cottonseed between *N* and *F* isogenic lines during boll development. The weather data (**Figures 5A,B**) showed that during May-August (reproduction stage period for cotton), the maximum and minimum temperatures in 2012 were higher than in 2013. For example, maximum temperatures in 2012 in May, June, July, August, and September were, respectively, 30.9, 31.7, 33.9, and 33.8°C vs. 26.2, 30.4, 31.5, and 33.8°C in 2013. Rainfall (mm) was higher in April and May in 2013 than in 2012, but in June, July, August, generally the rainfall was higher in 2012 than in 2013. Bearing in mind the rainfall in 1 day can reach upto 77 mm (for example in 2012 in August) or upto 40 mm (for example, in August in 2013). It is clear that the different patterns in temperature and rainfall between years (**Figures 5A,B**) will impact soil moisture, and consequently nutrient uptake, transport, and accumulation of nutrients in seeds. The effect of rainfall and temperatures in each year could be minimum since the experiment was irrigated and all genotypes were exposed to the same growing conditions. However, we still do not know the response of *F* and *N* genotypes or genotypes in each isoline set (fuzzy or fuzzless) to soil moisture resulted from rainfall or higher temperatures and their effects on nutrients uptake and transport. Therefore, we cannot exclude the effect of weather conditions on seed nutrients accumulation in *F* and *N* genotypes or on genotypes in the two sets of isolines. Further research is needed to determine the response of *F* and *N* genotypes to growing conditions such as drought and heat under field conditions.

It appears that the higher accumulation of nutrients of most *N* lines in their leaves and lower accumulation of nutrients (N, S, and minerals) in seeds of some of the *F* lines could be due to limited mobility of the nutrients from leaves to seed and lower nutrients demand by seeds of these *N* line, resulting in higher accumulation of nutrients in leaves. Some *N* lines did not follow this pattern, and this could be due to genotypic differences and fuzz trait. The real mechanism of how the fuzz development is involved in differential nutrients between *N* and *F* lines are still not understood (Bellaloui and Turley, 2013), and further research on the effects of nutrient supply on nutrient mobility is needed. Also, research is underway, and more biochemical and cellular methods including gene expression profiling based on RNA sequencing will be used to identify the gene responsible for this trait.

Conclusions

The current research demonstrated that seed protein was higher in fuzzy genotype in all near-isogenic sets, but seed oil was higher in fuzzless genotype in all near-isogenic sets, suggesting possible commercial use for fuzzless seed as source of both protein or oil as the level of protein and oil was competitive with the commercial check. The higher carbon in all fuzzless cottonseed may indicate that higher energy-storage occurred in fuzzless cottonseed. The higher accumulation of most of nutrients in leaves of fuzzless lines and lower accumulation of these nutrients in cottonseed suggested limited mobility and translocation of these nutrients in fuzzless lines compared with fuzzy lines. The research demonstrated that fuzz fiber development altered cottonseed composition, and this may be due to the involvement of fuzz fiber development in carbon and nitrogen metabolism, and the mobility of nutrients from leaves (source) to seed (sink). This information is beneficial to breeders for selection for higher oil or higher protein content, and to physiologist to further understand the mechanisms of mineral movement within plants. Understanding the mechanisms of nutrient mobility from leaves to seed would help for selecting for cottonseed with higher nutritional quality for human consumption and animal feed.

Acknowledgments

Technical support for this project provided by S. Mosley, K. Jordan, and R. Manning is appreciated. Funding for this research was provided by USDA, Agricultural Research Service projects 6066-21220-012-00D, 6066-21000-051-00D, and 6066-22000-074-00D. Mention of trade names or commercial products in this publication is solely for the purpose of providing specific information and does not imply recommendation or endorsement by the Unites States Department of Agriculture (USDA). USDA is an equal opportunity provider and employer.

References

Analytical Methods Committee (1959). *Analysts.* Her majesty's stationery office, London. 84, 214.

Arpat, A. B., Waugh, M., Sullivan, J. P., Gonzales, M., Frisch, D., Main, D., et al. (2004). Functional genomics of cell elongation in developing cotton fibers. *Plant Mol. Biol.* 54, 911–929. doi: 10.1007/s11103-004-0392-y

Association of Official Analytical Chemists (AOAC), (1990a). "Method 988.05," in *Official Methods of Analysis, 15th Edn.*, ed K. Helrich (Arlington, VA: AOAC), 70.

Association of Official Analytical Chemists (AOAC), (1990b). "Method 920.39," in *Official Methods of Analysis, 15th Edn.*, ed K. Helrich (Arlington, VA: AOAC), 79.

Bandemer, S. L., and Schaible, P. J. (1944). Determination of iron. A study of the o-phenanthroline method. *Ind. Eng. Chem. Anal. Ed.* 16, 317–319. doi: 10.1021/i560129a013

Bechere, E., Auld, D. L., and Hequet, E. (2009). Development of 'naked-tufted' seed coat mutants for potential use in cotton production. *Euphytica* 167, 333–339. doi: 10.1007/s10681-009-9890-y

Bellaloui, N., Gillen, A. M., Mengistu, A., Kebede, H., Fisher, D. K., Smith, J. R., et al. (2013a). Responses of nitrogen metabolism and seed nutrition to drought stress in soybean genotypes differing in slow-wilting phenotype. *Front. Plant Sci.* 4:498. doi: 10.3389/fpls.2013.00498

Bellaloui, N., Hanks, J. E., Fisher, D. K. and Mengistub, A. (2009). Soybean seed composition is influenced by within-field variability in soil nutrients. *Plant Manag. Netw.* 8. doi: 10.1094/CM-2009-1203-01-RS

Bellaloui, N., Hu, Y., Mengistu, A., Kassem, M. A., and Abel, C. A. (2013b). Effects of foliar boron application on seed composition, cell wall boron, and seed δ15N and δ13C isotopes in water-stressed soybean plants. *Front. Plant Sci.* 4:270. doi: 10.3389/fpls.2013.00270

Bellaloui, N., Mengistu, A., Walker, R. R., and Young, L. D. (2014). Soybean seed composition affected by seeding rates and row spacing in the Midsouth USA. *Crop Sci.* 54, 1782–1795. doi: 10.2135/cropsci2013.07.0463

Bellaloui, N., Smith, J. R., Gillen, A. M., and Ray, J. D. (2011). Effects of maturity, genotypic background, and temperature on seed mineral composition in near-isogenic soybean lines in the early soybean production system. *Crop Sci.* 51, 1161–1171. doi: 10.2135/cropsci2010.04.0187

Bellaloui, N., and Turley, R. B. (2013). Effects of fuzzless cottonseed phenotype on cottonseed nutrient composition in near isogenic cotton (*Gossypium hirsutum* L.) mutant lines under well-watered and water stress conditions. *Front. Plant Sci.* 4:516. doi: 10.3389/fpls.2013.00516

Boydak, E., Alpaslan, M., Hayta, M., Gercek, S., and Simsek, M. (2002). Seed composition of soybeans grown in the Harran region of Turkey as affected by row spacing and irrigation. *J. Agric. Food Chem.* 50, 4718–4720. doi: 10.1021/jf0255331

Brewbaker, J. L., and Kwack, B. H. (1963). The essential role of calcium ion in pollen germination and pollen tube growth. *Am. J. Bot.* 50, 859–865. doi: 10.2307/2439772

Brown, P. H., Bellaloui, N., Wimmer, M. A., Bassil, E. S., Ruiz, J., Hu, H., et al. (2002). Boron in plant biology. *Plant Biol.* 4, 205–223. doi: 10.1055/s-2002-25740

Brown, P. H., and Shelp, B. J. (1997). Boron mobility in plants. *Plant Soil* 193, 85–101. doi: 10.1023/A:1004211925160

Burton, J. W. (1985). "Breeding soybean for improved protein quantity and quality," in *World Soybean Research Conference III: Proceedings*, ed R. Shibles (Ames; Boulder: Westview Press), August 12–17. 361–367.

Cavell, A. J. (1955). The colorimetric determination of phosphorus in plant materials. *J. Sci. Food Agric.* 6, 479–480. doi: 10.1002/jsfa.2740060814

Digonnet, C., Aldon, D., Leduc, N., Dumas, C., and Rougier, M. (1997). First evidence of a calcium transient in flowering plants at fertilization. *Dev. Camb. Engl.* 124, 2867–2874.

Ding, G., Yang, M., Hu, Y., Liao, Y., Shi, L., and Xu, F. (2010). Quantitative trait loci affecting seed mineral concentrations in *Brassica napus* grown with contrasting phosphorus supplies. *Ann. Bot.* 105, 1221–1234. doi: 10.1093/aob/mcq050

Dordas, C. (2006). Foliar boron application affects lint and seed yield and improves seed quality of cotton grown on calcareous soils. *Nutr. Cycl. Agroecosyst.* 76, 19–28. doi: 10.1007/s10705-006-9037-7

Faure, J. E., Digonnet, C., and Dumas, C. (1994). An *in vitro* system for adhesion and fusion of maize gametes. *Science* 263, 1598–1600. doi: 10.1126/science.263.5153.1598

Fletcher, R. J., Bell, I. P., and Lambert, J. P. (2004). Public health aspects of food fortification: a question of balance. *Proc. Nutr. Soc.* 63, 605–614. doi: 10.1079/PNS2004391

Gou, J. Y., Wang, L. J., Chen, S. P., Hu, W. L., and Chen, X. Y. (2007). Gene expression and metabolite profiles of cotton fiber during cell elongation and secondary cell wall synthesis. *Cell Res.* 17, 422–434. doi: 10.1038/sj.cr.7310150

Grusak, M. A., and DellaPenna, D. (1999). Improving the nutrient composition of plants to enhance human nutrition and health. *Ann. Rev. Plant Physiol. Plant Mol. Biol.* 50, 133–161. doi: 10.1146/annurev.arplant.50.1.133

Guan, X., Lee, J. J., Pang, M., Shi, X., Stelly, D. M., and Chen, Z. J. (2011). Activation of Arabidopsis seed hair development by cotton fiber-related genes. *PLoS ONE* 6:e21301. doi: 10.1371/journal.pone.0021301

He, Z., Shankle, M., Zhang, H., Way, T. R., Tewolde, H., and Uchimiya, M. (2013). Mineral composition of cottonseed is affected by fertilization management practices. *Agron. J.* 105, 341–350. doi: 10.2134/agronj2012.0351

Heinemann, R. J. B., Fagundes, P. L., Pinto, E. A., Penteado, M. V. C., and Lanfer-Marquez, U. M. (2005). Comparative study of nutrient composition of commercial brown, parboiled and milled rice from Brazil. *J. Food Comp. Anal.* 18, 287–296. doi: 10.1016/j.jfca.2004.07.005

Heitholt, J. J. (1994). Supplemental boron, boll retention percentage, ovary carbohydrates and lint yield in modern cotton genotypes. *Agron. J.* 86, 492–497. doi: 10.2134/agronj1994.00021962008600030007x

Ji, S. J., Lu, Y. C., Feng, J. X., Wei, G., Li, J., Shi, Y. H., et al. (2003). Isolation and analyses of genes preferentially expressed during early cotton fiber development by subtractive PCR and cDNA array. *Nucleic Acids Res.* 31, 2534–2543. doi: 10.1093/nar/gkg358

Jiang, Y., and Huang, B. (2001). Effects of calcium on antioxidant activities and water relations associated with heat tolerance in two cool-season grasses. *J. Exp. Bot.* 52, 341–349. doi: 10.1093/jexbot/52.355.341

John, M. K., Chuah, H. H., and Neufeld, J. H. (1975). Application of improved azomethine-H method to the determination of boron in soils and plants. *Anal. Lett.* 8:559–568. doi: 10.1080/00032717508058240

Kebede, H., Abbas, H. K., Fisher, D. K., and Bellaloui, N. (2013). Relationship between aflatoxin contamination and physiological responses of corn plants under drought and heat stress. *Toxins* 4, 1385–1403. doi: 10.3390/toxins4111385

Li, X. B., Cai, L., Cheng, N. H., and Liu, J. W. (2002). Molecular characterization of the cotton GhTUB1 gene that is preferentially expressed in fiber. *Plant Physiol.* 130, 666–674. doi: 10.1104/pp.005538

Li, X. B., Fan, X. P., Wang, X. L., Cai, L., and Yang, W. C. (2005). The cotton ACTIN1 gene is functionally expressed in fibers and participates in fiber elongation. *Plant Cell* 17, 859–875. doi: 10.1105/tpc.104.029629

Liu, K., Han, M., Zhang, C., Yao, L., Sun, J., and Zhang, T. (2012). Comparative proteomic analysis reveals the mechanisms governing cotton fiber differentiation and initiation. *J. Proteomics* 75, 845–856. doi: 10.1016/j.jprot.2011.09.025

Loeppert, R. L., and Inskeep, W. P. (1996). "Colorimetric determination of ferrous iron and ferric iron by the 1,10-phenanthroline method," in *Methods of Soil Analysis: Part 3, Chemical Methods*, ed J. M. Bigham (Madison, WI: SSSA), 659–661.

Lohse, G. (1982). Microanalytical azomethine-H method for boron determination in plant tissue. *Commun. Soil Sci. Plant Anal.* 13, 127–134. doi: 10.1080/00103628209367251

Lu, K., Li, L., Zheng, X., Zhang, Z., Mou, T., and Hu, Z. (2008). Quantitative trait loci controlling Cu, Ca, Zn, Mn and Fe content in rice grains. *J. Genet.* 87, 305–310. doi: 10.1007/s12041-008-0049-8

Machado, A., Wu, Y., Yang, Y., Llewellyn, D. J., and Dennis, E. S. (2009). The MYB transcription factor GhMYB25 regulates early fiber and trichome development. *Plant J.* 59, 52–62. doi: 10.1111/j.1365-313X.2009.03847.x

Marschner, P. (2012). *Marschner's Mineral Nutrition of Higher Plants. 3rd Edn.* San Diego, CA: Academic Press.

Mengel, K., and Kirkby, E. A. (1982). *Principles of Plant Nutrition, 3rd Edn.* Worblaufen; Bern: International Potash Institute.

MSUCares, Stoneville, M. S. (2015). (http://ext.msstate.edu/anr/drec/weather.cgi). Verified on February 10, 2015.

Padmalatha, K. V., Patil, D. P., Kumar, K., Dhandapani, G., Kanakachari, M., Phanindra, M. L. V., et al. (2012). Functional genomics of fuzzless-lintless mutant of *Gossypium hirsutum* L. cv. MCU5 reveal key genes and pathways involved in cotton fiber initiation and elongation. *BMC Genomics* 13:624–638. doi: 10.1186/1471-2164-13-624

Pang, C. Y., Wang, H., Pang, Y., Xu, C., Jiao, Y., Qin, Y. M. X., et al. (2010a). Comparative proteomics indicates that biosynthesis of pectic precursors is important for cotton fiber and Arabidopsis root hair elongation. *Mol. Cell. Proteomics* 9, 2019–2033. doi: 10.1074/mcp.M110.000349

Pang, C. Y., Wang, H., Song, W. Q., and Zhu, Y. X. (2010b). The cotton ATP synthase δ1 subunit is required to maintain a higher ATP/ADP ratio that facilitates rapid fiber cell elongation. *Plant Biol.* 12, 903–909. doi: 10.1111/j.1438-8677.2009.00313.x

Pettigrew, W. T., and Dowd, M. K. (2011). Varying planting dates or irrigation regimes alters cottonseed composition. *Crop Sci.* 51, 2155–2164. doi: 10.2135/cropsci2011.02.0085

Pierson, E. S., Miller, D. D., Callaham, D. A., van Aken, J., Hackett, G., and Hepler, P. K. (1996). Tip-localized calcium entry fluctuates during pollen tube growth. *Dev. Biol.* 174, 160–173. doi: 10.1006/dbio.1996.0060

Ray, J. D., Fritschi, F. B., and Heatherly, L. G. (2006). Large application of fertilizer N at planting affects seed protein and oil concentrations in the early soybean production system. *Field Crops Res.* 99, 67–74. doi: 10.1016/j.fcr.2006.03.006

Ruan, Y. L., Llewellyn, D. J., and Furbank, R. T. (2003). Suppression of sucrose synthase gene expression represses cotton fiber cell initiation, elongation, and seed development. *Plant Cell* 15, 952–964. doi: 10.1105/tpc.010108

Ruan, Y. L., Xu, S. M., White, R., and Furbank, R. T. (2004). Genotypic and developmental evidence for the role of plasmodesmatal regulation in cotton fiber elongation mediated by callose turnover. *Plant Physiol.* 136, 4104–4113. doi: 10.1104/pp.104.051540

Samman, S., Naghii, M. R., Lyons Wall, P. M., and Verus, A. P. (1998). The nutritional and metabolic effects of boron in humans and animals. *Biol. Trace Elem. Res.* 66, 227–235. doi: 10.1007/BF02783140

SAS (2002-2012). *Statistical Analysis System.* (Cary, NC: SAS Institute Inc).

Seagull, R. W., and Giavalis, S. (2004). Pre- and post- anthesis application of exogenous hormones alters fiber production in *Gossypium hirsutum* L. cultivar Maxxa GTO. *J. Cotton Sci.* 8, 105–111. Available online at: https://www.cotton.org/journal/2004-08/2/upload/jcs08-105.pdf

Shi, Y. H., Zhu, S. W., Mao, X. Z., Feng, J. X., Qin, Y. M., Zhang, L., et al. (2006). Transcriptome profiling, molecular biological, and physiological studies reveal a major role for ethylene in cotton fiber cell elongation. *Plant Cell.* 18, 651–664. doi: 10.1105/tpc.105.040303

Stetina, S. R., Turley, R. B., Bellaloui, N., and Boykin, J. C. (2014). Yield and fiber quality of five pairs of near-isogenic cotton (*Gossypium hirsutum* L.) lines expressing fuzzless/linted and fuzzy/linted seed phenotypes. *J. Crop Improv.* 28, 680–699. doi: 10.1080/15427528.2014.931901

Stewart, J. M. (1975). Fiber initiation on the cotton ovule (*Gossypium hirsutum*). *Am. J. Bot.* 62, 723–730. doi: 10.2307/2442061

Tian, H. Q., and Russell, S. D. (1997). Micromanipulation of male and female gametes of *Nicotiana tabacum*: II. Preliminary attempts for *in vitro* fertilization and egg cell culture. *Plant Cell Rep.* 16, 657–661. doi: 10.1007/BF01275510

Turley, R. B., and Kloth, R. H. (2002). Identification of a third fuzzless seed locus in upland cotton (*Gossypium hirsutum* L.). *J. Hered.* 93, 359–364. doi: 10.1093/jhered/93.5.359

Turley, R. B., Vaughn, K. C., and Scheffler, J. A. (2007). Lint development and properties of fifteen fuzzless seed lines of Upland cotton (*Gossypium hirsutum* L.). *Euphytica* 156, 57–65. doi: 10.1007/s10681-006-9351-9

Walford, S. A., Wu, Y., Llewellyn, D. J., and Dennis, E. S. (2011). GhMYB25-like: a key factor in early cotton fiber development. *Plant J.* 65, 785–797. doi: 10.1111/j.1365-313X.2010.04464.x

Wang, J., Wang, H. Y., Zhao, P. M., Han, L. B., Jiao, G. L., Zheng, Y. Y., et al. (2010). Overexpression of a profilin (GhPFN2) promotes the progression of developmental phases in cotton fibers. *Plant Cell Physiol.* 51, 1276–1290. doi: 10.1093/pcp/pcq086

Wang, L., and Ruan, Y. L. (2010). Unraveling mechanisms of cell expansion linking solute transport, metabolism, plasmodesmtal gating and cell wall dynamics. *Plant Signal. Behav.* 5, 1561–1564. doi: 10.4161/psb.5.12.13568

White, P. J., and Broadley, M. R. (2005). Biofortifying crops with essential mineral elements. *Trends Plant Sci.* 10, 586–593. doi: 10.1016/j.tplants.2005.10.001

Wilcox, J. R., and Shibles, R. M. (2001). Interrelationships among seed quality attributes in soybean. *Crop Sci.* 41, 11–14. doi: 10.2135/cropsci2001.41111x

Wilkins, T. A., and Arpat, A. B. (2005). The cotton fiber transcriptome. *Physiol. Plant.* 124, 295–300. doi: 10.1111/j.1399-3054.2005.00514.x

Yang, S. S., Cheung, F., Lee, J. J., Ha, M., Wei, N. E., Sze, S. H., et al. (2006). Accumulation of genome-specific transcripts, transcription factors and phytohormonal regulators during early stages of fiber cell development in allotetraploid cotton. *Plant J.* 47, 761–775. doi: 10.1111/j.1365-313X.2006.02829.x

Yang, Y. W., Bian, S. M., Yao, Y., and Liu, J. Y. (2008). Comparative proteomic analysis provides new insights into the fiber elongating process in cotton. *J. Proteome Res.* 7, 4623–4637. doi: 10.1021/pr800550q

Yu, J., Yu, S., Fan, S., Song, M., Zhai, H., Li, X., et al. (2012). Mapping quantitative trait loci for cottonseed oil, protein and gossypol content in a *Gossypium hirsutum* × *Gossypium barbadense* backcross inbred line population. *Euphytica* 187, 191–201. doi: 10.1007/s10681-012-0630-3

Zhang, J. S., Yang, H. Y., Zhu, L., and Tong, H. (1997). Ultracytochemical localization of calcium in the pollen tube track of cotton gynoecium. *Acta Bot. Sin.* 39, 121–125.

Zhang, M. W., Guo, B. J., and Peng, Z. M. (2004). Genetic effects on Fe, Zn, Mn and P contents in indica black pericarp rice and their genetic correlations with grain characteristics. *Euphytica* 135, 315–323. doi: 10.1023/B:EUPH.0000013340.98344.60

Zhang, M., Zheng, X., Song, S., Zeng, Q., Hou, L., Li, D., et al. (2011). Spatiotemporal manipulation of auxin biosynthesis in cotton ovule epidermal cells enhances fiber yield and quality. *Nat. Biotechnol.* 29, 453–458. doi: 10.1038/nbt.1843

Zhao, P. M., Wang, L. L., Han, L. B., Wang, J., Yao, Y., Wang, H. Y., et al. (2009). Proteomic identification of differentially expressed proteins in the ligon lintless mutant of upland cotton (*Gossypium hirsutum* L.). *J. Proteome Res.* 9, 1076–1087. doi: 10.1021/pr900975t

Conflict of Interest Statement: The authors declare that the research was conducted in the absence of any commercial or financial relationships that could be construed as a potential conflict of interest.

9

Elevated Genetic Diversity in an $F_{2:6}$ Population of Quinoa (*Chenopodium quinoa*) Developed through an Inter-ecotype Cross

Edited by:
Antonio M. De Ron,
Misión Biológica de Galicia-CSIC,
Spain

Reviewed by:
Pedro Revilla,
Misión Biológica de Galicia-CSIC,
Spain
Svetla Sofkova-Bobcheva,
Maritsa Vegetable Crops Research
Institute, Bulgaria

***Correspondence:**
Eric N. Jellen
jellen@byu.edu

Ouafae Benlhabib[1], Noura Boujartani[1], Peter J. Maughan[2], Sven E. Jacobsen[3] and Eric N. Jellen[2]*

[1] Departement de Production, Protection et Biotechnologies Vegetales, Agronomic and Veterinary Institute Hassan II, Rabat, Morocco, [2] Department of Plant and Wildlife Sciences, Brigham Young University, Provo, UT, USA, [3] Department of Plant and Environmental Sciences, Faculty of Life Sciences, University of Copenhagen, Copenhagen, Denmark

Quinoa (*Chenopodium quinoa*) is a seed crop of the Andean highlands and Araucanian coastal regions of South America that has recently expanded in use and production beyond its native range. This is largely due to its superb nutritional value, consisting of protein that is rich in essential amino acids along with vitamins and minerals. Quinoa also presents a remarkable degree of tolerance to saline conditions, drought, and frost. The present study involved 72 $F_{2:6}$ recombinant-inbred lines and parents developed through hybridization between highland (0654) and coastal (NL-6) germplasm groups. The purpose was to characterize the quinoa germplasm developed, to assess the discriminating potential of 21 agro-morpho-phenological traits, and to evaluate the extent of genetic variability recovered through selfing. A vast amount of genetic variation was detected among the 72 lines evaluated for quantitative and qualitative traits. Impressive transgressive segregation was measured for seed yield (22.42 g/plant), while plant height and maturity had higher heritabilities (73 and 89%, respectively). Other notable characters segregating in the population included panicle and stem color, panicle form, and resistance to downy mildew. In the Principal Component analysis, the first axis explained 74% of the total variation and was correlated to plant height, panicle size, stem diameter, biomass, mildew reaction, maturation, and seed yield; those traits are relevant discriminatory characters. Yield correlated positively with panicle length and biomass. Unweighted Pair Group Method with Arithmetic Mean-based cluster analysis identified three groups: one consisting of late, mildew-resistant, high-yielding lines; one having semi-late lines with intermediate yield and mildew susceptibility; and a third cluster consisting of early to semi-late accessions with low yield and mildew susceptibility. This study highlighted the extended diversity regenerated among the 72 accessions and helped to identify potentially adapted quinoa genotypes for production in the Moroccan coastal environment.

Keywords: quinoa, *Chenopodium*, downy mildew resistance, breeding, genetic diversity, *Peronospora variabilis*

INTRODUCTION

Quinoa (*Chenopodium quinoa*) is an ancient Andean seed crop of exceptional nutritional, and in particular protein, quality (Koziol, 1992). Its cultivation has expanded within the past decade beyond its traditional range to more than 70 countries (Food Agriculture Organization of United Nations [FAO], 2012), although the Andean nations of Peru and Bolivia are still primary quinoa producers. Due to quinoa's exceptional nutritional value, its capacity for adaptation to diverse agro-ecological conditions and its high commercial value, quinoa is expected to play an essential role in the UN-FAO strategy to sustainably feed the world's growing population (Jacobsen et al., 2013).

Biologically, *C. quinoa* belongs to an allotetraploid ($2n = 4x = 36$) complex found throughout the New World and whose root is the North American weed pitseed goosefoot (*C. berlandieri*). At least two other independent domestication events are presumed to have given rise to Mesoamerican vegetable and seed domesticates (*C. berlandieri* ssp. *nuttaliae*; Wilson and Heiser, 1979) as well as an extinct seed crop of the ancient cultures of eastern North America, *C. berlandieri* ssp. *jonesianum* (Smith and Yarnell, 2009; Kistler and Shapiro, 2011). After dispersing to the southeastern plains of South America, the weedy taxon of that region, avian goosefoot (*C. hircinum*), is presumed to have been brought into cultivation in the region of Lake Titicaca as *C. quinoa* and, at a later time, dispersed long-range to the Araucanian region on the Pacific slope in what is now Chile (Wilson, 1990; Bruno and Whitehead, 2003). Molecular genetic analyses have confirmed that the lowland Chilean coastal material is highly diverse (Fuentes et al., 2009), yet represents a branch that is separate from the even more variable quinoa germplasm of the High Andes (Christensen et al., 2007). The coastal Chilean germplasm includes genotypes that are highly resistant to heat and are day length insensitive - critical characters for quinoa breeders and agronomists seeking to expand quinoa's production into lowland subtropical and warm-season temperate environments around the world (Jacobsen et al., 2003).

The current work is part of a quinoa introduction and selection program initiated in 2000 in Morocco and involving Brigham Young University, the Institut Agronomique et Veterinaire Hassan II, and Copenhagen University of Denmark. Germplasm adapted to Moroccan production environments was to be identified in this long-term program through traditional breeding methodologies based upon the initial creation of highly diverse quinoa populations composed of the following: (1) in the first phase, introduction of internationally available cultivated quinoa varieties, landraces, and breeding lines; followed by, (2) introduction of breeding populations combining adaptive cultivated characteristics of lowland Chilean-origin × agronomically superior Andean Highland types; and then, (3) introduction of early-generation populations derived from disease-, pest-, and heat-resistant wild lowland × cultivated parent crosses. The primary objectives of this study, falling under phase 2 of the aforementioned project, were to (1) describe agro-morpho-phenological traits an $F_{2:6}$ population of recombinant-inbred lines (RIL) developed from an NL-6 (female, lowland) × 0654 (male, highland) cross; (2) estimate the extent of genetic diversity present by the F_6 generation in this population; and (3) estimate the discriminating potential and heritability of several of the evaluated traits.

MATERIALS AND METHODS

Plant Material

The quinoa population in this study was developed at BYU, Provo (USA), through a cross between the Dutch variety NL-6 (female) and Peruvian highland line 0654 (male) initially made by Alejandro Bonifacio, and was previously described in Maughan et al. (2012) as 'Pop39'. The NL-6 parent, which is derived from lowland Chilean germplasm, has yellow panicles, short stature, is early maturing, and is sufficiently heat-tolerant to allow for its cultivation at low elevations. Line 0654 is a highland Peruvian valley ecotype with red-purple panicles, late maturity, tall stature, and is heat susceptible. The population included in the current field evaluation consisted of 70 $F_{2:6}$ lines plus the two parents.

Field Characterization and Statistical Analyses

Characterization of Pop39 was performed on an experimental plot at the IAV Hassan II in Rabat, Morocco, near the laboratory to facilitate daily measurements of growth and phenology. The trial was sown indoors on January 29, transplanted to sandy loam soil in the field February 26–27, maintained through manual weeding, irrigated regularly as needed, and treated twice against aphids. No fungal control against mildew was applied. To protect against avian seed predation, an anti-sparrow net was used to cover the plots. The quinoa population was planted in two completely randomized blocks, with about 20 plants and five measurements per accession per block. Elementary plots were two rows of one-meter long and spaced 35 cm apart.

Sixteen quantitative and seven qualitative traits related to plant morphology, phenology and agronomic performance were measured. Quantitative measurements included seed diameter; plant height (at 60 days = PH60, 75 days = PH75, 90 days = PH90, and at maturity = PHM); panicle or inflorescence length (IL) and width (IW); sensitivity to downy mildew (SM); number of days to maturity (DM), and phenological stages and yield (GY) components, all of which were measured during the growing season and/or at harvest. Additional biomass data logged included main stem diameter (SD); root length (RL); aboveground fresh (FWA) and dry (DWA) weights; root fresh (FWR) and dry (DWR) weights; and overall biomass (BM). Qualitative traits used in the evaluation includes color of the seeds (ranged into a 1–10 scale by using classes of Photoshop color numerical codes), leaves (green/dark green) and stem (green/purple), and the panicle color (yellow/pink/orange/red/purple) and shape (globular versus amaranthiform).

The evaluation of downy mildew (*Peronospora variabilis*) resistance (Choi et al., 2010) was performed in the laboratory on young leaves according to the inoculation method of Mhada et al. (2015). Pathogen development was evaluated every 48 h on three leaves per accession using a 0–5 notation scale where score 0 corresponds to no lesion, 1 to small and disperse lesions with less than 1 mm diameter and no sporulation on the lower side of the leaf, 2 to clearly individualized lesions increasing in number and size with a diameter between 0.5 and 1 cm without showing any sporulation on the lower side of the leaf, 3 to brown lesions, covering less than 50% of the leaf surface with a beginning of the sporulation at the lower side, 4 to lesions of larger size, covering more than 50% of the leaf area and 5 to lesions covering more than 91% of the leaf area, with a high sporulation rating on the lower and upper area.

Collected data were subjected to descriptive analyses, analysis of variance (ANCOVA) with two factors, the accessions as a fixed factor and the blocks as a random factor, and means comparisons using the Fisher test with XLSTAT software (XLSTAT-Pro 7.5). PAST software (v. 2.16, Hammer et al., 2001) was used to perform multivariate analyses to subtract the principal components which account for much of the variance, to compute the correlations matrix and to set up the dendrogram that gathers the $F_{2:6}$ accessions into clusters using the Unweighted Pair Group Method with Arithmetic Mean (UPGMA) procedure (Sokal and Michener, 1958).

RESULTS

Qualitative Traits

The 72 studied lines represented a wide range of morphological variability that was reflected in seed and panicle coloration, plant and panicle size and shape, seed diameter, grain yield, maturity, and resistance to downy mildew (**Figure 1**). The population exhibited segregation for seed color, with 46% of the RIL's having white seeds and 22% having red seeds – the two main coloration patterns. The remaining lines were yellow (15%), white-speckled red (3%), yellow speckled with red (7%), and light-yellow or orange (3% each).

Forty-six percent of the RIL's in the population had green stems and light green leaves (NL-6 parental type), 43% had red stems and dark green leaves (0654-type), and the rest of the lines (8%) were heterogeneous, having green stems streaked with red and light green leaves or fully red stems and light green leaves.

With respect to panicle variation, 35% of the RIL's had yellow panicles like P1, 25% were purple like P2, 19% had red panicles, and 3% had pink panicles. Several RIL's exhibited panicle color segregation or heterogeneity, including yellow/red, yellow/purple, orange/ red, yellow/purple, and orange/purple. In addition to panicle color, the amaranth panicle shape (amaranthiform) was present in 60% of the RIL's, with the remaining 39% exhibiting globular (glomerulate) panicles.

Branching pattern of the main stem was another discriminating trait. Sixty-nine percent of the RIL's did not

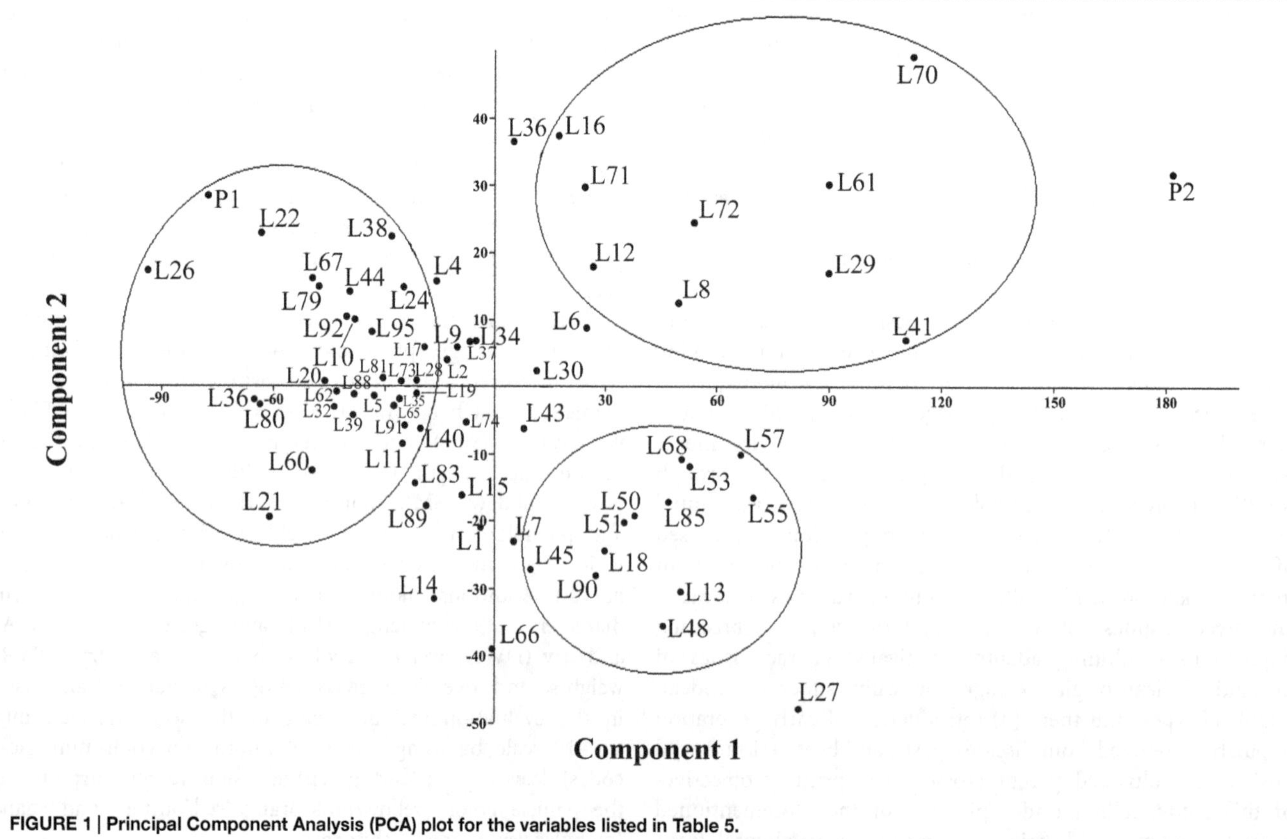

FIGURE 1 | Principal Component Analysis (PCA) plot for the variables listed in Table 5.

exhibit branching. Since branching is a wild-type trait and neither parent normally has a branched stem, the presence of significant stem branching in ~30% of the lines indicates the two parents may have complementary genes suppressing stem branching.

Quantitative Traits

All 16 quantitative traits showed very highly significant differences among the RIL's for practically all the studied characters, providing strong evidence for the important diversity generated within the Pop39 and highlighting the relevant discriminating power of the variables used. Seed diameter, for example, varied from 1.71 to 2.11 mm and there were highly significant differences among lines, with 28% of the RIL's having larger seeds than the 0654 parent.

Sorting RIL's by plant height at maturity differentiated the population into five classes or groups, producing a distribution skewed slightly toward shorter plant height. The three intermediate classes - classes 2, 3, and 4 -encompassed more than 80% of the lines (**Table 1**). At maturity, the NL-6 parent was 63.90 cm tall and thus fell into in class 1; the 0654 parent at 214.88 cm height was at the upper end of class 5. Consequently, transgressive segregants for plant height were not detected in the population.

Descriptive and variance analyses of biomass and its components highlighted the important diversity present among the $F_{2:6}$ lines. The main stem diameter varied between 0.54 and 1.89 cm (**Table 2**). The NL-6 parent had a diameter of 0.71 cm and for parent 0654 it was 1.15 cm. The main root length of the whole population averaged 10.15 cm and varied among lines between 8.50 and 12.80 cm. The panicle length also presented a wide variation among lines; it was three times greater in line L70 (120.70 cm) compared to L26 (36.90 cm). The panicle width also varied, ranging from 3.61 to 17.50 cm in lines L26 and L90, respectively (**Table 2**).

Lines could be classified into five biomass groups, with the NL-6 parent falling into class 2 (13.16–20.10 g) and 0654 grouping with class 5 (50.52–79.39 g). The aboveground and root biomasses had values for the 0654 parent of 196.85 and 12.94 g, respectively (**Table 2**). The 0654 parent also presented the highest stem dry weight (70.79 g). Total plant dry biomass varied between 6.19 and 79.39 g for lines L26 and 0654, respectively. Standard deviations were variable among characters, being very high for the fresh weight of the aboveground plant and the length of the panicle. In contrast, standard deviations were minimal for stem diameter, plant size, and root dry weight (**Table 2**).

Five seed yield classes were identified, with >80% of the lines falling into the three lowest classes and yielding less than 12 g. The 0654 parent grouped in class 1 and yielded only 2.20 g, while NL-6 yielded 12.18 g (class 4). Five of the RIL's in class 5 produced 17–23 g of seeds per plant, indicating the potential for positive transgressive segregation for yield in Pop39. The grain yield measured per individual ranged between 0.13 and 22.42 g; the average yield of the whole F_6 population was of 8.70 g/plant (**Table 1**). Thus, variation for seed production was highly significant in the population.

TABLE 1 | Main traits class's interval, average, and lines' percentage in quinoa $F_{2:6}$ RIL Pop39.

Traits	Class	Interval	Average	Number of lines	Lines' %
PHM (cm)	1	61–90	76.24	6	8.33
	2	90–115	105.13	23	31.94
	3	115–140	124.94	21	29.17
	4	140–170	159.31	15	20.83
	5	170–215	193.06	7	9.72
GY (g/pl)	1	0.13–4.50	2.54	15	20.83
	2	4.5–8.0	6.64	21	29.17
	3	8.0–12.0	9.62	22	30.56
	4	12.0–17.0	14.19	9	12.50
	5	17.0–23.0	20.48	5	6.94
DM (days)	1	124–131	127.15	10	13.89
	2	131–140	134.47	15	20.83
	3	140–153	147.2	24	33.33
	4	153–70	162.01	18	25.00
	5	170–192	187.31	5	6.94
RPV (scores)	1	0.5–1.2	1.0	8	11.11
	2	1.4–1.7	1.5	12	16.67
	3	1.8–2.1	2.0	13	18.06
	4	2.2–2.3	2.3	15	20.83
	5	2.4–2.5	2.5	7	9.72
	6	2.6–3.0	2.7	17	23.61

PHM, Plant height at maturity; GY, grain yield; DM, days to maturity; YPV, resistance to peronospora variabilis. RPV scores: 1 = very resistance, 2 = resistance, 3 = slightly resistance, 4 = slightly susceptible, 5 = susceptible, 6 = very susceptible.

TABLE 2 | Biomass and its components in $F_{2:6}$ quinoa Pop39.

Character	Mean	Max	Highest line	Min	Lowest line	SD
SD (cm)	0.83	1.89	L53	0.54	L26	0.42
RL (cm)	10.15	12.80	L8	8.50	L39	1.85
IL (cm)	66.85	120.70	L70	36.90	L26	26.18
IW (cm)	9.35	17.50	L90	3.61	L26	5.13
FWA (g)	51.52	196.85	0654	10.17	L21	43.21
FWR (g)	4.60	12.94	0654	0.83	L21	4.02
DWA (g)	18.14	70.79	0654	2.68	L26	16.39
DWR (g)	2.27	7.47	L90	0.47	L26	2.01
BM (g/plt)	29.11	79.39	0654	6.19	L26	21.91

SD, main stem diameter; RL, root length; IL, inflorescence length; IW, inflorescence width; FWA, above ground fresh weight; FWR, root fresh weight; DWA, above ground dry weight; DWR, root dry weight; BM, biomass. SD, standard deviation.

The 70 F_6 quinoa lines grouped into five precocity classes when sorted by their number of days to maturity (**Table 1**), with a population DM average of 148 days and a standard deviation of 16.78 days. About 65% of the $F_{2:6}$ lines required less than 5 months to mature (<150 days). The remaining 35% of lines required >150 days to mature and grouped in classes 4 and 5. The ANOVA for DM, as with the other characters, had a highly significant *p*-value. The NL-6 parent was earlier than any of the RIL's, having reached maturity after only 124 days (**Table 1**). The 0654 parent, L48, and L90 were the latest-maturing genotypes at 192 days.

TABLE 3 | ANOVA for resistance to mildew (*Peronospora variabilis*) infection in Pop39.

	Sum of squares	Mean squares	Mean	*SD*	F Fisher	*Pr > F*
2 days	30.388	0.428	0.488	0.554	1.732	0.003
4 days	85.860	1.209	1.484	0.901	1.969	<0.0001
6 days	133.950	1.887	2.223	1.040	2.772	<0.0001
8 days	144.923	2.041	2.758	1.097	2.595	<0.0001
10 days	181.966	2.563	3.428	1.095	4.908	<0.0001
12 days	191.798	3.144	4.173	1.143	7.947	<0.0001

Downy Mildew Resistance

Analysis of variance for resistance to mildew showed significant differences among lines from the second day of the inoculation (**Table 3**). *Perenospora variabilis* symptoms did not show up right after the inoculation on all the RIL's – indicating that initial reaction to the pathogen depends on the genotype. The 70 F6 lines were classified in six groups (**Table 1**), with the most resistant ones in class 1 (L55, L7, L27, L41, and L89) and the most susceptible in class 6 (L37, L12, L18, L92, L44, and L95). The NL-6 parent was slightly susceptible to mildew while 0654 was among the most resistant. Fifty percent of the accessions (36/72) were classified between the two parents, indicating that transgressive segregation for resistance was substantial in Pop39. The resistant 0654 parent did not show any sporulation at day 12 from the inoculation; this behavior was also the case for other lines of the three first classes. Spores appeared on the 10th day on NL-6, which was also the case in lines falling into the other three susceptible classes. The most highly susceptible lines developed their first symptoms well before 10 days after inoculation.

Collected data showed that the pathogen infected every single RIL; however, the most susceptible genotypes rapidly developed symptoms and allowed for sporulation, whereas resistant lines delayed the progression and spread of disease, preventing spore growth and dissemination.

Heritability

Broad-sense heritability and variances for morphological characters are presented in **Table 4**. Phenotypic variances were greatest for PH (1154.7) and FWA (1066.97). The highest value of genotypic variance was also noted for PH (838.53); SD presented the lowest genotypic variance (0.017). Highest heritability values were observed for DM (89%) and PH (73%). Heritabilities were intermediate for DWA (55%), IL (46%), and BM (46%). Root length had the lowest heritability at 10%. Grain yield heritability was measured at 42% in this population and environment.

Principal Component Analysis

Multivariate analysis was undertaken to highlight genotype groups with similar traits and significant correlations among characters. The principal components analysis (PCA) was computed to assess the contributions of individual variables to the global variance. Fifteen quantitative traits were used and data analysis showed that the four first components explained 96% of the variance. These 15 traits' contributions are presented in **Table 5** for each of the first four axes. The first principal component (PC1) explained 74% of the total variance. The variables positively correlated to PC1 were PH (0.92); IL and IW (0.85 and 0.64, respectively); SD (0.69); BM (0.94); and DM (0.65). Resistance to mildew was negatively correlated to PC1 (−0.46). Grain yield was correlated to PC2 (0.6), to which DM was negatively correlated (−0.47).

According to the correlation matrix in **Table 6**, the most significant relationships among traits were GY with IL (0.52) and BM (0.52); DM with PHM (0.71), DWA (0.72), DWR (0.64), and BM (0.54). Susceptibility to mildew (SM) was negatively correlated to PHM (0.54) and DM (0.47).

Figure 1 presents the PCA plot of Pop39 $F_{2:6}$ RIL's for PC1 and PC2. On the positive side of axis 1 are grouped 15 entries – 0654 (P2), L70, L41, L61, L29, L27, L55, L57, L72, L53, L68, L13, L8, L85, L48 – that shared the characteristics of tall plant height, long and wide panicles, large stalk diameter, large biomass, and relatively long growth cycle. On the negative side of axis 1, RIL's L60, L67, L21, L22, L80, L31, NL-6 (P1), and L26 presented lower values of the aforementioned characters. Lines on the positive side of axis 2 (L70, L16, L36, and NL-6) are characterized by higher grain yield and earlier maturity; on the negative side of axis 2 are RIL's that matured later and had lower seed yield, notably lines L13, L14, L48, L27, and L66.

Dendrogram

A dendrogram was computed based on UPGMA using PAST v. 2.16 (Hammer et al., 2001). The F2:6 quinoa RIL Pop39 separated into three clusters (**Figure 2**). Group A included short-stature lines (61–140 cm) with small panicles (36.9–69.4 cm) and low biomass (6.19–29.56 242 g). With the exception of lines L9 and L89, they were also early maturing and susceptible to *P. variabilis*. Cluster A also included the NL-6 parent (P1).

Cluster B included six lines (0654 or P2, L29, L41, L72, L61, L70), all of tall size (140–215 cm) and having both long (90.2–120.7 cm) and wide (12.63–16.14 cm) panicles. They were also of high biomass (50.52–7.39 g) and high grain yield (12–23 g). Parent 0654 was an exception because it yielded only 2.20 g/pl. However, 0654 and line L41 were very resistant to mildew, L70 and L29 had intermediate levels of resistance, while L61 was very susceptible.

Cluster C (**Figure 2**) consisted of 24 lines of somewhat large size (115–170 cm), long panicles (57.8–120.7 cm), high biomass (20.77–48.02 g), and late maturity (140–192 d). These RIL's were mostly resistant to downy mildew, except for highly susceptible lines L30, L12, and L18.

DISCUSSION

The presented cross operates between two contrasting lines, NL-6 adapted to European lowlands and of short size (63.90 cm), and 0654 from the Peruvian valleys and taller (214.88 cm). There is a large genetic distance between $F_{2:6}$ accessions, and much variability within the population. Several authors reported that the valley ecotype is generally taller than highland ecotype

TABLE 4 | Variance and broad-sense heritability components of 12 traits in quinoa $F_{2:6}$ RIL Pop39.

Character	Phenotypic variance	Genotypic variance	Environmental variance	Heritability
Plant height at maturity (cm)	1154.70	838.53	316.17	0.73
Inflorescence length (cm)	588.55	271.00	317.55	0.46
Inflorescence width (cm)	19.67	7.99	11.68	0.40
Seed diameter (cm)	0.047	0.017	0.03	0.36
Root length (cm)	208.30	0.30	208.00	0.10
Above ground fresh weight (g)	1066.97	423.33	643.64	0.39
Above ground dry weight (g)	174.46	96.68	77.78	0.55
Root fresh weight (g)	7.36	2.81	4.55	0.38
Root dryweight (g)	2.28	0.96	1.32	0.42
Biomass (g/pl)	335.23	155.25	179.98	0.46
Grain yield (g/pl)	40.38	17.08	23.30	0.42
Days to maturity (days)	262.76	235.37	27.39	0.89

TABLE 5 | Variables contribution to the four principal axes in quinoa $F_{2:6}$ RIL Pop39.

Character	Axis 1	Axis 2	Axis 3	Axis 4
Plant high at day 60	0.0650	−0.0968	0.6565	0.3918
Plant high at day 75	0.1629	−0.0308	0.8131	0.3110
Plant high at day 90	0.4522	−0.1115	0.8343	0.2193
Plant high at day maturity	0.9203	−0.3676	0.0480	0.0026
Inflorescence length (cm)	0.8503	0.1523	0.2153	−0.4385
Inflorescence width (cm)	0.6432	0.2206	0.0569	−0.1345
Seed diameter (cm)	0.6985	0.1068	0.1202	0.0360
Root length (cm)	0.3010	0.1763	0.1327	−0.0476
Above ground fresh weight (g)	0.9093	0.3786	−0.1027	0.1235
Above ground dry weight (g)	0.9591	0.0183	−0.1688	0.0238
Root fresh weight (g)	0.7962	0.2265	−0.0886	0.0139
Root dry weight (g)	0.8589	0.0792	−0.0944	−0.0632
Biomass (g/pl)	0.9433	0.2053	−0.0096	−0.0542
Grain yield (g/pl)	0.3200	0.6045	0.4417	−0.2231
Days to maturity (days)	0.6549	−0.4757	−0.4734	0.0613
Mildew reaction	−0.4647	0.2946	0.0707	−0.0330
Variance	74%	11%	9%	3%
Sum of the variance	74%	85%	93%	96%

(Gandarillas, 1979; Cáceres, 1993; Mujica and Jacobsen, 1999; Carmen, 2008).

The $F_{2:6}$ quinoa RIL's presented a large genetic variability, which is translated to qualitative traits of stem, leaf, inflorescence and seed color diversity, and also by inflorescence shape and size variation. Seed color varies from black, brown, red, pink, yellow, orange to white. The inflorescence presents a very wide range of colors from white, yellow, pink, red to dark red, brown, and purple. Within the $F_{2:6}$ accessions, the leaf color was light green in 47% RIL's or dark green in 45%; the stems were red (43%) or green (46%) and the inflorescence were mainly yellow (35%) or red (25%), but also orange, pink, purple.

Several authors have investigated the genetic control of a number of qualitative characters. A single gene with several alleles controls leaf color, red color is dominant over purple, dominating over green. The red color is dominant at leaves axils (Tapia et al., 1979; Izquierdo Fernández et al., 2001; Carmen, 2008). Concerning the panicle shape, a 3:1 segregation of glomerulate upon amaranth forms was reported (Carmen et al., 2008). Male sterility identified in quinoa is nucleocytoplasmic, with three fertile plants for one sterile (Izquierdo Fernández et al., 2001; Carmen et al., 2008).

There is a wide range of seed colors being white, yellow, pink, dark red, brown, and purple (Cusack, 1984; Risi and Galwey, 1989). The genetic determination of this trait requires studying several cross descendants' between homozygote lines with contrasted seed colors. The tetraploid status of the species has to be taken into account while analyzing the offspring.

The UPGMA clustering and PCA showed that the highland-phenotype, relatively unadapted lines in Rabat's coastal climate formed a distinct group far apart from the other clusters. Highland-phenotype accessions– those most like the Peruvian 0654 parent – are of tall stature and have a more indeterminate growth pattern when cultivated in the mild and humid coastal environment. Moreover, since 0654 is susceptible to severe decreases in seed yield when exposed to elevated temperatures during flowering (personal observations), low-yielding lines may have inherited this characteristic from 0654.

Maughan et al. (2012) reported that single-nucleotide polymorphism (SNP) markers identified in this same population (Pop39) displayed 95% normal (non-distorted) segregation, with most of the skewed markers favoring the NL-6 alleles. They further speculated that this pattern of segregation distortion favoring SNP alleles from the NL-6 parent might be due to heat-induced sterility coming from the 0654 parent, even though the population was advanced to homozygosity each generation under relatively mild (25°C) temperatures in the greenhouse. Consequently, it would be interesting in a future study to address the question of whether or not these NL-6 skewed regions carry alleles for heat tolerance.

This highland RIL's are, however, highly resistant to downy mildew; they develop small lesions and have a long latent period when inoculated with oospores of *P. variabilis* under controlled conditions. The pathogen affected most genotypes at different intensities. 0654 parent was ranked among the most resistant $F_{2:6}$ RIL's group, probably carrying the resistance

TABLE 6 | Correlation matrix among variables measured in $F_{2:6}$ RIL quinoa Pop39.

0	PHM	IL	FWA	DWA	FWR	DWR	BM	GY	DM
PHM									
IL	0,72421								
FWA	0,69635	0,75655							
DWA	0,85625	0,76911	0,88701						
FWR	0,62341	0,69904	0,8136	0,79851					
DWR	0,72522	0,76126	0,7972	0,88355	0,93488				
BM	0,78207	0,84317	0,9138	0,95209	0,81857	0,88611			
GY	0,10011	0,52115	0,43149	0,23385	0,3125	0,29107	0,51739		
DM	0,71335	0,38958	0,46564	0,71753	0,51038	0,6402	0,53786	−0,31201	
SM	−0,53602	−0,31489	−0,32164	−0,45151	−0,40435	−0,43732	−0,38074	0,06723	−0,468

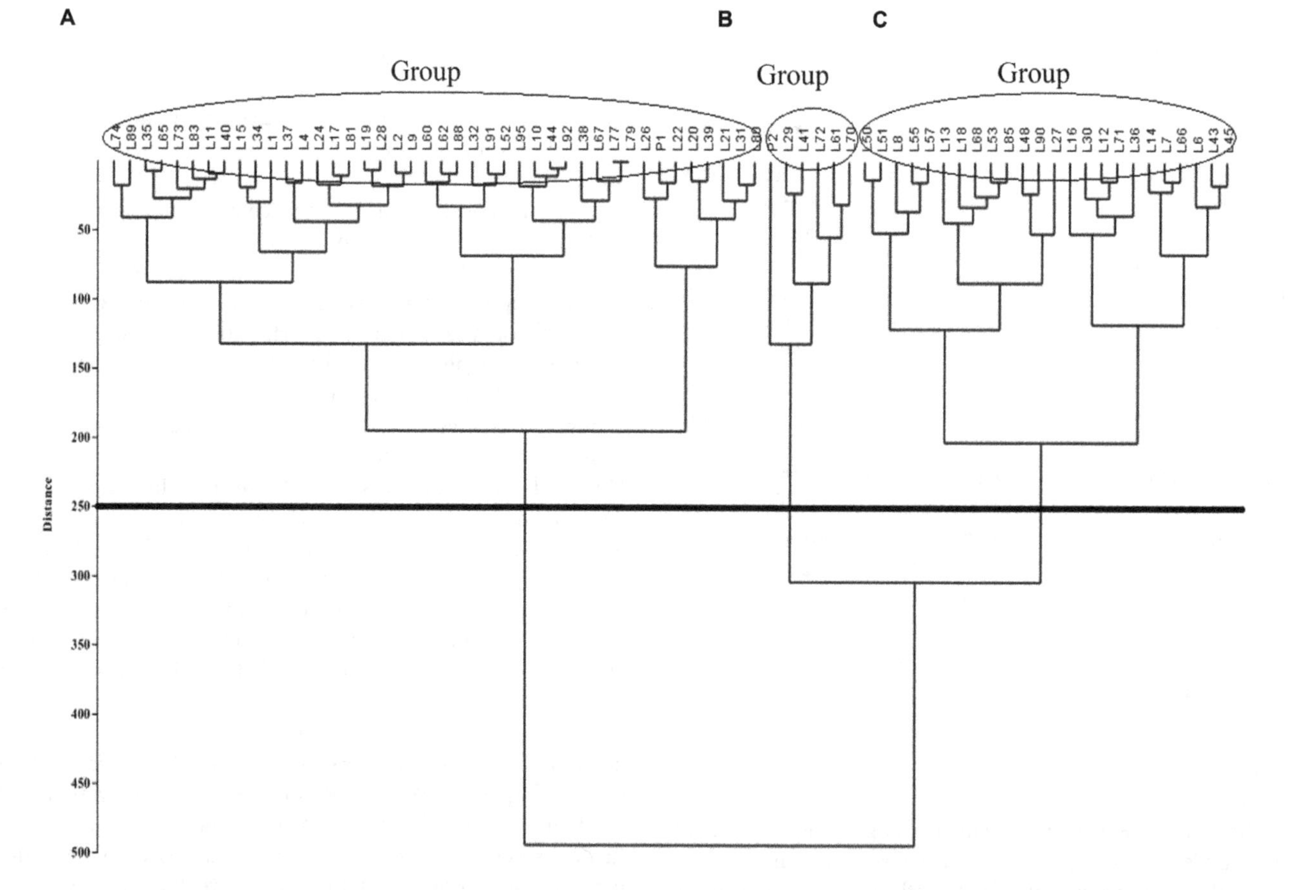

FIGURE 2 | Unweighted Pair Group Method with Arithmetic Mean (UPGMA)-based dendrogram showing three clusters (A–C) of RIL's in quinoa Pop39.

factor. This assumption is in agreement with Kitz (2008) from the same population. The highest resistance was that of RIL L55, which confirms our earlier greenhouse observations (unpublished).

According to Alfano and Collmer (1997), Campbell et al. (1999), and Eulgem et al. (2004), the valley ecotype is usually more resistant to downy mildew than early and short size lowland ecotype, corresponding to group A of the dendrogram. These results are in concordance with Bonifacio and Saravia (1999).

The data analyses confirmed the significant diversity of the quantitative traits. Seed yield was positively correlated to panicle length (0.52) and total biomass (0.52). Bhargava et al. (2007) state good correlation between grain yield and plant biomass while Carmen (2008) reports positive correlations with plant size, panicle width, total dry matter and crop earliness.

Mujica (1988) puts up two genetic indexes that are based on high inheritability traits that are highly correlated to grain yield; those are central glomeruli diameter (Index I1) and stem and panicle diameters (Index I2). The efficiency of the two indexes reached 144.48 and 148.32%, respectively; they represent relevant parameters for germplasm evaluation and selection studies (Mujica, 1988).

Bertero et al. (2004) reported striking disparities for quinoa character correlations between cold/highland and lowland/tropical environments. For example, they reported grain yield versus biomass correlations of 0.92 and 0.48 in cold/highland versus warm/lowland test sites, respectively. Although Rabat is on the coastal plain and at mid-latitude, its proximity to the ocean moderates the temperature; nonetheless, the observed correlation of yield versus biomass of 0.52 in this study follows the heat-stress pattern detected by Bertero et al. (2004). Biomass was still a close second behind (IL, 0.521) for magnitude of correlation with grain yield in this study.

Bertero et al. (2004) also reported a disparity for quinoa grain yield-maturity correlations, having observed values of 0.87 and 0.37 in cold/highland and warm/lowland environments, respectively. In the present study, we found that grain yield and days to maturity were negatively correlated at −0.31. This is most likely due to later-maturing lines encountering debilitating late spring-summer heat stress, especially for lines carrying alleles from the heat-susceptible parent 0654.

Bertero et al. (1999) reported strong effects on seed size and maturation from high temperatures and post-anthesis exposure to long days in Altiplano (variety 'Kanckolla') and Valley (variety 'Blanca de Junin') quinoa genotypes. Since our parental cultivar '0654' is of similar intermediate-late maturity and environmental adaptation as 'Blanca de Junin', it is not surprising that plant height, inflorescence size, biomass, seed diameter, and maturation all contributed heavily to Axis 1 of the PCA (**Table 5**). With a February sowing date at Rabat (~34° N latitude), lines having greater allelic contributions for daylength sensitivity from '0654' would be expected to have delayed maturation, higher biomass, etc., whereas daylength-neutral alleles from the coastal genetic background of 'NL-6' would have contributed to markedly earlier maturation and relatively reduced biomass parameters under the same lengthening-day conditions - and without transgressive segregation for maturation from 'NL-6' contributing to a longer growth cycle.

Clearly, salient breeding objectives in the mild coastal climate of Rabat should include higher seed yield, shorter growing cycle and downy mildew resistance - though given the correlation between more lengthy growth cycle and higher mildew resistance, and in light of Rabat's lack of severe summer heat or winter freeze constraints, mildew resistance probably should not be sacrificed for earlier maturation. This is based on the presumption that the humid climate in Rabat is conducive to mildew spore exposure and growth year-round. Nevertheless, evaluations of larger numbers of segregating progeny from lowland × highland quinoa populations like Pop39 should facilitate uncoupling of traits like heat tolerance/early maturity + mildew susceptibility.

The extensive morphological variation present in just 70 F2:6 lines of the NL-6 × 0654 RIL population, combined with the pattern of relatively moderate linkage distortion detected in this population using molecular markers (Maughan et al., 2012), bodes well for the continued use of lowland coastal × Andean highland quinoa breeding populations. The detection of positive transgressive segregants for seed size and yield, in particular, combined with an observed heritability for yield of approximately 40%, bodes well for future selection to develop higher-yielding quinoa lines for the Moroccan coastal environment.

AUTHOR CONTRIBUTIONS

OB and NB designed the study. PM and EJ prepared the population. NB and OB collected the data. NB, OB, and SJ analyzed the data. OB, SJ, PM, and EJ obtained funding to support this project. NB, OB, and EJ took primary responsibility for writing the manuscript.

FUNDING

Funding for advancement of the population to the F2:6 generation was provided by Brigham Young University and the Holmes Family Foundation through grants in support of undergraduate research.

ACKNOWLEDGMENTS

The SWUP-Med project of the European Union provided principal funding for this study. The original cross for this population was made by Alejandro Bonifacio and we are grateful for his contribution of the F_1 seed.

REFERENCES

Alfano, J. R., and Collmer, A. (1997). The type III (Hrp) secretion pathway of plant pathogenic bacteria: trafficking harpins, avr proteins, and death. *J. Bacteriol.* 179, 5655–5662.

Bertero, H. D., de la Vega, A. J., Correa, G., Jacobsen, S. E., and Mujica, A. (2004). Genotype and genotype-by-environment interaction effects for grain yield and grain size of quinoa (*Chenopodium quinoa* Willd.) as revealed by pattern analysis of international multi- environment trials. *Field Crops Res.* 89, 299–318.

Bertero, H. D., King, R. W., and Hall, A. J. (1999). Photoperiod-sensitive development phases in quinoa (*Chenopodium quinoa* Willd.). *Field Crops Res.* 60, 231–243. doi: 10.1016/S0378-4290(98)00128-2

Bhargava, A., Shukla, S., and Ohri, D. (2007). Genetic variability and inter-relationship among various morphological and quality traits in quinoa (*Chenopodium quinoa* Willd.). *Field Crops Res.* 101, 104–116. doi: 10.1016/j.fcr.2006.10.001

Bonifacio, A., and Saravia, R. (1999). "Evaluación de la resistencia al mildiu en quinua," in *Proceedings of the Tercer Taller de Preduza en Resistencia Duradera en Cultivos Altos en la Zona Andina*, Cochabamba, 49–59.

Bruno, M., and Whitehead, W. T. (2003). Chenopodium cultivation and formative period agriculture at Chiripa, Bolivia. *Lat. Am. Antiq.* 14, 339–355. doi: 10.2307/3557564

Cáceres, E. (1993). *Cultivos Andinos*. Bolivia: Instituto Nacional de Alimentacioìn y Nutrición, 126.

Campbell, N. A., Reece, J. B., and Mitchell, L. G. (1999). *Biology*, 5th Edn. Menlo Park, CA: Benjamin/Cummings, 679–680.

Carmen, D. C. (2008). *Diversité Génétique et Réponse aux Contraintes du Climat: Une Étude de cas à Partir de la Biologie des Populations de Quinoa (Chenopodium quinoa Willd.) de Bolivie*. Doctoral thesis, Faculty of Agricultural Sciences in Gembloux, Gembloux.

Carmen, D. C., Mahy, G., and Winkel, T. (2008). La quinoa en Bolivie: une culture ancestrale devenue culture de rente "bio-équitable". *Biotechnol. Agron. Soc. Environ.* 12, 421–435.

Choi, Y. J., Danielsen, S., Lubeck, M., Hong, S. B., Delhey, R., and Shin, H. D. (2010). Morphological and molecular characterization of the causal agent of downy mildew on quinoa (*Chenopodium quinoa*). *Mycopathologia* 169, 403–412. doi: 10.1007/s11046-010-9272-y

Christensen, S. A., Pratt, D. B., Pratt, C., Nelson, P. T., Stevens, M. R., Jellen, E. N., et al. (2007). Assessment of genetic diversity in the USDA and CIP-FAO international nursery collections of quinoa (*Chenopodium quinoa* Willd.) using microsatellite markers. *Plant Genet. Resour.* 5, 82–95. doi: 10.1017/S1479262107672293

Cusack, D. (1984). Quinoa: grain of the Incas. *Ecologist* 14, 21–31.

Eulgem, T., Weigman, V. J., Chang, H., McDowell, J. M., Holub, E. B., Glazebrook, J., et al. (2004). Gene resistance signatures from three genetically separable resistance gene signaling pathways for downy mildew resistance. *Plant Physiol.* 135, 1129–1144. doi: 10.1104/pp.104.040444

Food Agriculture Organization of United Nations [FAO] (2012). *International Year of Quinoa IYQ-2013*. Available at: http://www.rlc.fao.org/en/about-fao/iyq-2012/

Fuentes, F., Martinez, E. A., Hinrichsen, P. V., Jellen, E. N., and Maughan, P. J. (2009). Assessment of genetic diversity patterns in Chilean quinoa (*Chenopodium quinoa* Willd.) germplasm using multiplex fluorescent microsatellite markers. *Conserv. Genet.* 10, 369–377. doi: 10.1007/s10592-008-9604-3

Gandarillas, H. (1979). "La quinua (*Chenopodium quinoa* Willd.), Genética y origen," in *La Quinua y la Kañiwa Cultivos Andinos*, eds M. E. Tapia, H. Gandarillas, S. Alandia, A. Cardozo, and A. Mujica, (Bogota: CIID-IICA), 45–64.

Hammer, O., Harper, D. A. T., and Ryan, P. D. (2001). PAST: paleontological statistics software package for education and data analysis. *Palaeontol. Electronica* 4:9.

Izquierdo Fernández, J. I., Mujica, A., Jacobsen, S. E., Marathée, J. P., and Morón, C. (2001). *Quinua (Chenopodium quinoa wild.) Ancestral cultivo andino, alimento del presente y futuro. Cultivos andinos. Versión 1.0. (CD-Rom)*. Santiago: Food and Agriculture Organization.

Jacobsen, S. E., Mujica, A., and Jensen, C. R. (2003). The resistance of quinoa (*Chenopodium quinoa* Willd.) to adverse abiotic factors. *Food Rev. Int.* 19, 99–109. doi: 10.1081/FRI-120018872

Jacobsen, S. E., Sorensen, M., Pedersen, S. M., and Weiner, J. (2013). Feeding the world: genetically modified crops versus agricultural biodiversity. *Agron. Sustain. Dev.* 33, 651–662. doi: 10.1007/s13593-013-0138-9

Kistler, L., and Shapiro, B. (2011). Ancient DNA confirms a local origin of domesticated chenopod in eastern North America. *J. Archaeol. Sci.* 38, 3549–3554. doi: 10.1016/j.jas.2011.08.023

Kitz, L. (2008). *Evaluation of Downy Mildew (Peronospora farinosa F. Sp. CHENOPOD II) Resistance Among Quinoa Genotypes and Investigation of P. farinosa Growth Using Scanning Electron Microscopy*. Ph.D. thesis, Brighan Young University, Provo, UT.

Koziol, M. J. (1992). Chemical composition and nutritional value of quinoa (*Chenopodium quinoa* Willd.). *J. Food Composit. Anal.* 5, 35–68. doi: 10.1016/0889-1575(92)90006-6

Maughan, P. J., Smith, S. M., Rojas-Beltran, J. A., Elzinga, D., Raney, J. A., Jellen, E. N., et al. (2012). Single nucleotide polymorphism identification, characterization, and linkage mapping in quinoa. *Plant Genome* 5, 114–125. doi: 10.3835/plantgenome2012.06.0011

Mhada, M., Ezzahiri, B., and Benlhabib, O. (2015). Assessment of downy mildew resistance (Peronospora farinosa) in a quinoa (*Chenopodium quinoa* Willd.) germplasm. *Int. J. Biol. Med. Res.* 6, 4748–4752.

Mujica, A. (1988). *Genetic Parameters and Selection Indexes for Quinoa (Chenopodium quinoa Wild.)*. Doctoral thesis, Chapingo Autonomous University, Texcoco, 7.

Mujica, A., and Jacobsen, S. E. (1999). "Resistencia de la quinua a la sequía y otros factores abióticos adversos y su mejoramiento," in *I Curso Internacional Sobre Fisiología de la Resistencia a Sequía en Quinua (Chenopodium quinoa Willd.)* (Lima: CIP-DANIDA), 25–38.

Risi, I., and Galwey, N. M. (1989). "Chenopodium grains of the Andes: a crop for temperate latitudes," in *New Crops For Food And Industry*, eds G. E. Wickens, N. Haq, and P. Day (London: Chapman and Hall), 222–234.

Smith, B. D., and Yarnell, R. A. (2009). Initial formation of an indigenous crop complex in eastern North America at 3800 B.P. *Proc. Natl. Acad. Sci. U.S.A.* 106, 6561–6566. doi: 10.1073/pnas.0901846106

Sokal, R., and Michener, C. (1958). A statistical method for evaluating systematic relationships. *Univ. Kansas Sci. Bull.* 38, 1409–1438.

Tapia, M., Gandarillas, H., Alandia, S., Cardozo, A., Mujica, A., Ortiz, R., et al. (1979). *La Quinua y la Kamwa. Cultivos andinos*. Bogota: IICA, CIID, 228.

Wilson, H. D. (1990). Quinua and relatives (*Chenopodium* sect. *Chenopodium* subsect. *Cellulata*). *Econ. Bot.* 44, 92–110. doi: 10.1007/BF02860478

Wilson, H. D., and Heiser, C. B. (1979). The origin and evolutionary relationships of 'Huauzontle' (*Chenopodium nuttaliae* Safford), domesticated chenopod of Mexico. *Am. J. Bot.* 66, 198–206. doi: 10.2307/2442525

Conflict of Interest Statement: The authors declare that the research was conducted in the absence of any commercial or financial relationships that could be construed as a potential conflict of interest.

10

Where do Protein Bodies of Cereal Seeds Come From?

Edited by:
Antonio M. De Ron,
Spanish National Research Council,
Spain

Reviewed by:
Hao Peng,
Washington State University, USA
Carmen Cuadrado,
Instituto Nacional de Investigación y
Tecnología Agraria y Alimentaria,
Spain

***Correspondence:**
Alessandro Vitale
vitale@ibba.cnr.it

*Emanuela Pedrazzini, Davide Mainieri, Claudia A. Marrano and Alessandro Vitale**

Istituto di Biologia e Biotecnologia Agraria, Consiglio Nazionale delle Ricerche, Milano, Italy

Protein bodies of cereal seeds consist of ordered, largely insoluble heteropolymers formed by prolamin storage proteins within the endoplasmic reticulum (ER) of developing endosperm cells. Often these structures are permanently unable to traffic along the secretory pathway, thus representing a unique example for the use of the ER as a protein storage compartment. In recent years, marked progress has been made in understanding what is needed to make a protein body and in formulating hypotheses on how protein body formation might have evolved as an efficient mechanism to store large amounts of protein during seed development, as opposed to the much more common system of seed storage protein accumulation in vacuoles. The major key evolutionary events that have generated prolamins appear to have been insertions or deletions that have disrupted the conformation of the eight-cysteine motif, a protein folding motif common to many proteins with different functions and locations along the secretory pathway, and, alternatively, the fusion between the eight-cysteine motif and domains containing additional cysteine residues.

Keywords: cereals, evolution of subcellular compartments, prolamins, protein bodies, protein sorting, seeds

INTRODUCTION

Almost 50% of the global food protein supply comes from cereal seeds (faostat.fao.org). Most of these proteins accumulate in protein bodies (PB), large polymers formed in the endoplasmic reticulum (ER). First discovered in the endosperm cells of maize developing seeds (Duvick, 1961), PBs are the most striking example for an alternative, perhaps surprising, use of the ER.

The ER is the port of entry of the secretory pathway, which leads to the cell surface or the inner hydrolytic compartments using the Golgi apparatus and various endosomes as intermediate sorting stations. The ER takes care of promoting efficient, correct structural maturation of proteins that enter the secretory pathway and selects for disposal those with incurable structural defects (Anelli and Sitia, 2008). In this respect, it is therefore mainly a compartment of transit. Its use to temporarily store proteins is very uncommon and almost exclusive to plants (Herman and Schmidt, 2004). More strikingly, the permanent ER residence of proteins that do not play a role in typical ER functions seems a feature unique to the major storage proteins of *Poaceae* seeds, the most studied being those of cereals (Shewry and Halford, 2002).

ER-LOCATED PROTEIN BODIES EVOLVED WITH GRASSES

The *Poaceae* family (commonly termed grasses) originated less than 80 million years ago (Gaut, 2002), a long time after the most common classes of seed storage proteins, the 2S albumins and the 7S/11S globulins universally present in seeds and even in fern spores, had appeared

(Rödin and Rask, 1990; Shutov et al., 1998). 2S albumins and 7S/11S globulins accumulate in protein storage vacuoles (PSVs, Vitale and Hinz, 2005), which are therefore the first subcellular compartment that evolved to store seed proteins. PSV proteins start their life in the ER and reach their destination mainly by Golgi- and endosome-mediated traffic, even if direct traffic from the ER to PSVs also occurs (Vitale and Hinz, 2005). PSVs have higher luminal pH and lower hydrolytic activity than vegetative, lytic vacuoles and they are ready to receive hydrolytic enzymes during germination, when they are transformed into lytic vacuoles for the rapid degradation of storage proteins.

Prolamins evolved with grasses, where they are in most cases the major, or almost exclusive, storage proteins (Shewry and Halford, 2002). Unlike the soluble 2S albumins and 7S/11S globulins, most prolamins rapidly form very large polymers that become insoluble to various extents because of inter-chain disulfide bonds or hydrophobic interactions. These electron dense, round-shaped structures with diameters between 0.5 and 2 μm were originally termed protein granules (Duvick, 1961; Khoo and Wolf, 1970) and then renamed PB (Graham et al., 1962). PB often detach from the ER cisternae, but retain ribosomes on their cytosolic face and do not fuse with other subcellular structures (Burr and Burr, 1976; Yamagata and Tanaka, 1986). Rice, maize, and sorghum prolamins exclusively form PB, whereas other prolamins, such as those of wheat, can either form PB or traffic through the Golgi complex to PSVs, or undergo autophagy (Rubin et al., 1992). The destiny depends on the individual prolamin, but also on the seed developmental stage: in wheat, the ability to traffic is more prominent at early stages and the formation of PB in the ER lumen favored at later stages (Tosi et al., 2009).

A WIDE SUPERFAMILY OF SMALL CYS-RICH PROTEINS

Common features of many, but not all, prolamins (**Table 1**) are a repeated domain rich in Pro or Gln residues (hence the name) and a domain derived from the eight-cysteine motif (8CM, José-Estanyol et al., 2004), characterized by eight Cys residues arranged in a specific order, forming four intra-chain disulfide bridges necessary to maintain a scaffold of alpha-helical segments connected by variable loops. The 8CM motif is common to a wide group of proteins termed the prolamin superfamily, with different functions and localizations along the secretory pathway. Besides many prolamins, the superfamily includes cereal α-globulins, trypsin inhibitors and α-amylase inhibitors, as well as hybrid proline-rich proteins (HyPRPs), non-specific lipid transfer proteins (nsLTPs) and 2S albumins, the latter three classes being widespread in plants. HyPRPs are cell wall proteins. nsLTPs are also secreted, often remaining anchored to the plasma membrane via a GPI anchor. 2S albumins, cereal α-globulins, trypsin inhibitors and α-amylase inhibitors are seed proteins accumulated in PSV. In prolamins and in the PSV-located members of the group three conserved regions, termed A, B, and C, can be identified within the 8CM motif, which in this case is therefore also termed ABC domain (Kreis et al., 1985).

Based on the presence or not of the 8CM/ABC domain and additional Cys residues available for inter-chain disulfide bonds, prolamins have been divided into high molecular weight (HMW), cysteine rich (S-rich), cysteine poor (S-poor), and α-prolamins, the latter being the only group that does not seem related to 8CM proteins (Shewry and Halford, 2002). Based on amino acid sequence similarity and phylogenetic analysis, prolamins have also been divided into three groups named I, II, and III (Xu and Messing, 2009). The two criteria indeed largely overlap: α-prolamins are in group I, S-rich and S-poor in group II and HMW in group III. Group II members are present in all cereals. Group III is present in *Pooideae* (wheat, barley, *Brachypodium*) but not in *Panicoideae* (maize, sorghum, millet) and rice, whereas group I is only present in *Panicoideae* and perhaps *Oryzeae*.

Group III is considered the most ancient, followed by II and then I. Sequence similarity indicates that an ancestral group III prolamin may have originated by duplication of the α-globulin gene before the divergence of the different *Poaceae* subfamilies (Xu and Messing, 2009; Gu et al., 2010). Sequence alignments of orthologous regions of chromosomes suggest that the absence of group III prolamins in non-*Pooideae* families is due to gene loss after duplication (Xu and Messing, 2009).

TABLE 1 | Summary of the main characteristics of the rice, maize, and wheat prolamins of group III and II (Xu and Messing, 2009) analyzed in this review.

Plant	Prolamin group	Prolamin	Repeated domain	Extra cysteine residues	8CM	Requiring reducing agents to be solubilized	PB biogenesis in heterologous systems
Rice	II	10 kD oryzeins	NO	YES	Altered by deletions	n.d.	n.d.
	II	16 kD oryzeins	NO	YES	Altered by deletions	n.d.	n.d.
	II	13 kD oryzeins	NO	NO	Altered by deletions	YES	YES (13a polypeptide)
Maize	II	β-zein	NO	NO	Altered by deletions	YES	YES
	II	27 kD γ-zein	Medium	YES	Intact	YES	YES
Wheat	II	γ-gliadin	Medium	NO	Intact	NO	Dubious
	III	HMW glutenins	Long	NO	Altered by insertion of the long repetitive domain and deletions	YES	YES

n.d.: not determined.

THE EVOLUTIONARY SHIFT FROM PSV TO THE ER

Phylogenetic studies thus indicate that early prolamins may have evolved by insertions/mutations in a gene encoding a PSV protein, causing a tendency to shift accumulation from PSV to the ER itself. Retention in the port of entry of the secretory pathway is clearly more parsimonious than traffic to vacuoles and can therefore increase cellular fitness in tissues that spend considerable energy to accumulate storage material, but it raises two new requirements: (i) minimizing deleterious effects on ER functions, (ii) finding a new strategy for the degradation of storage proteins during germination. The rapid formation of large polymers seems to have satisfied the first requirement. For the second, a key event has probably been the change in the tissue accumulation from cotyledons to the endosperm: the latter undergoes programmed cell death at the end of maturation. The loss of membrane integrity allows access of hydrolytic enzymes synthesized and secreted by the surrounding aleurone during germination. Roles in the efficient formation of PBs are played by protein machineries specific of the endosperm (Vitale and Boston, 2008; Holding, 2014) and by elements in the coding mRNAs that target the transcripts to specific ER regions (Tian and Okita, 2014). However, a number of prolamins can efficiently form PBs also when expressed individually in vegetative tissues of transgenic plants (**Table 1**; Geli et al., 1994; Shani et al., 1994; Bagga et al., 1995; Bellucci et al., 2000; Saito et al., 2009), indicating that PB formation exploits the general folding machinery of the ER present in all tissues and that the key features to form a PB reside in prolamins themselves, as we discuss in the next two paragraphs.

RICE AND WHEAT

Rice prolamins, encoded by about twenty genes, are small polypeptides (10, 13, or 16 kD) assigned to group II prolamins (Xu and Messing, 2009) and lacking any Pro- or Gln- rich repeated domain, indicating that the repetitions found in many other prolamins are not strictly necessary to form PBs (Onda and Kawagoe, 2011; Saito et al., 2012). A rice α-globulin-GFP fusion that is correctly sorted to PSV instead accumulates into ER-located PBs when the formation of the disulfide bond between Cys79 in the B region and Cys171 in the C region is inhibited by mutagenesis (Kawagoe et al., 2005). The hypervariable loop between regions B and C is much shorter in rice prolamins than in α-globulin, whereas in wheat HMW prolamins it is extensively elongated by the insertion of the very large repetitive domain rich in Pro and Gln (Kawagoe et al., 2005; Onda and Kawagoe, 2011). The authors suggested that these changes in length inhibit the formation of a critical intra-chain disulfide bond of the 8CM motif, making two Cys residues available for inter-chain disulfide bonds that allow the formation of PBs (Kawagoe et al., 2005). These insertions/deletions in α-globulin may thus have been the oldest events leading to group III formation in *Pooideae* and group II formation in *Pooideae* and rice (**Table 1**). It should also be underlined that only two 13 kD rice prolamins have the complete set of Cys of the 8CM motif and no other Cys residue, the other rice polypeptides having from 0 to 13 Cys residues in total (Onda and Kawagoe, 2011). Moreover, the two HMW prolamins of wheat (x-type and y-type HMW glutenin) have four to seven Cys residues, depending on the alleles. When expressed individually in transgenic tobacco these wheat prolamins form large polymers that are soluble only when reduced (Shani et al., 1994). Therefore, the loss of key Cys residues of the 8CM motif or the acquisition of new Cys residues can also be crucial for PB formation. Furthermore, at least once the initial rice PB core is formed, the rice prolamins without Cys residues can join it through interactions that are most probably of hydrophobic nature (Onda and Kawagoe, 2011).

Rice 13a prolamin has four Cys residues and is located in mid regions of the mature PB. When expressed in yeast as a GFP fusion, it forms insoluble PBs, whereas rice α-globulin similarly tagged with GFP is delivered to the yeast vacuole (Masumura et al., 2015). Deletions studies made on GFP fusions indicated the 13a portions corresponding to the B or C, but not A, regions form structures that resemble PBs by fluorescence microscopy, consistently with the view that the Cys residues in the B and C domains are important for PB formation. Unlike the full-length protein, these deletion mutants are soluble also in the absence of reducing agent, suggesting that there is not a direct causal relationship between insolubility and PB assembly. The polymerization state was not investigated, but the authors concluded that hydrophobic interactions may be important for the assembly of 13a prolamin into PBs (Masumura et al., 2015).

MAIZE

Maize prolamins (zeins) are grouped into four classes (Holding, 2014). α-zeins, polypeptides between 19 and 22 kD encoded by about twenty genes, are the most abundant. Three genes encode the γ-zeins (16kD, 27 kD, which is another very abundant zein, and 50 kD), whereas β-zein (15 kD) and δ-zein (10 kD, 18 kD) are the products of one and two genes, respectively. During seed development, γ- and β- zeins, which belong to group II prolamins, are synthesized first. The 27 kD γ-zein and β-zein form PBs when expressed individually in transgenic vegetative tissues (**Table 1**; Geli et al., 1994; Bagga et al., 1995).

Treatments of isolated maize PBs with reducing agents solubilize the 27 and 50 kD γ-zeins but not the other zeins (Vitale et al., 1982). The 27 kD γ-zein is perhaps the best studied prolamin and plays a fundamental role in the correct formation of maize PBs. A natural duplication of its locus generates Quality Protein Maize (Wu et al., 2010; Liu et al., 2016), whereas a deletion encompassing the 27 and 50 kD γ-zein genes reduces PB number to 12% and alters PB morphology (Yuan et al., 2014). The hypothesis that the more recent group I prolamins exploited the PB forming ability of the more ancient ones is also supported by the finding that co-expression of 27 kD γ-zein enhances the accumulation of a 19 kD α-zein polypeptide expressed in transgenic tobacco (Coleman et al., 1996). 27 kD γ-zein is composed of an N-terminal domain characterized by eight repeats of the amphipathic sequence ValHisLeuProProPro and

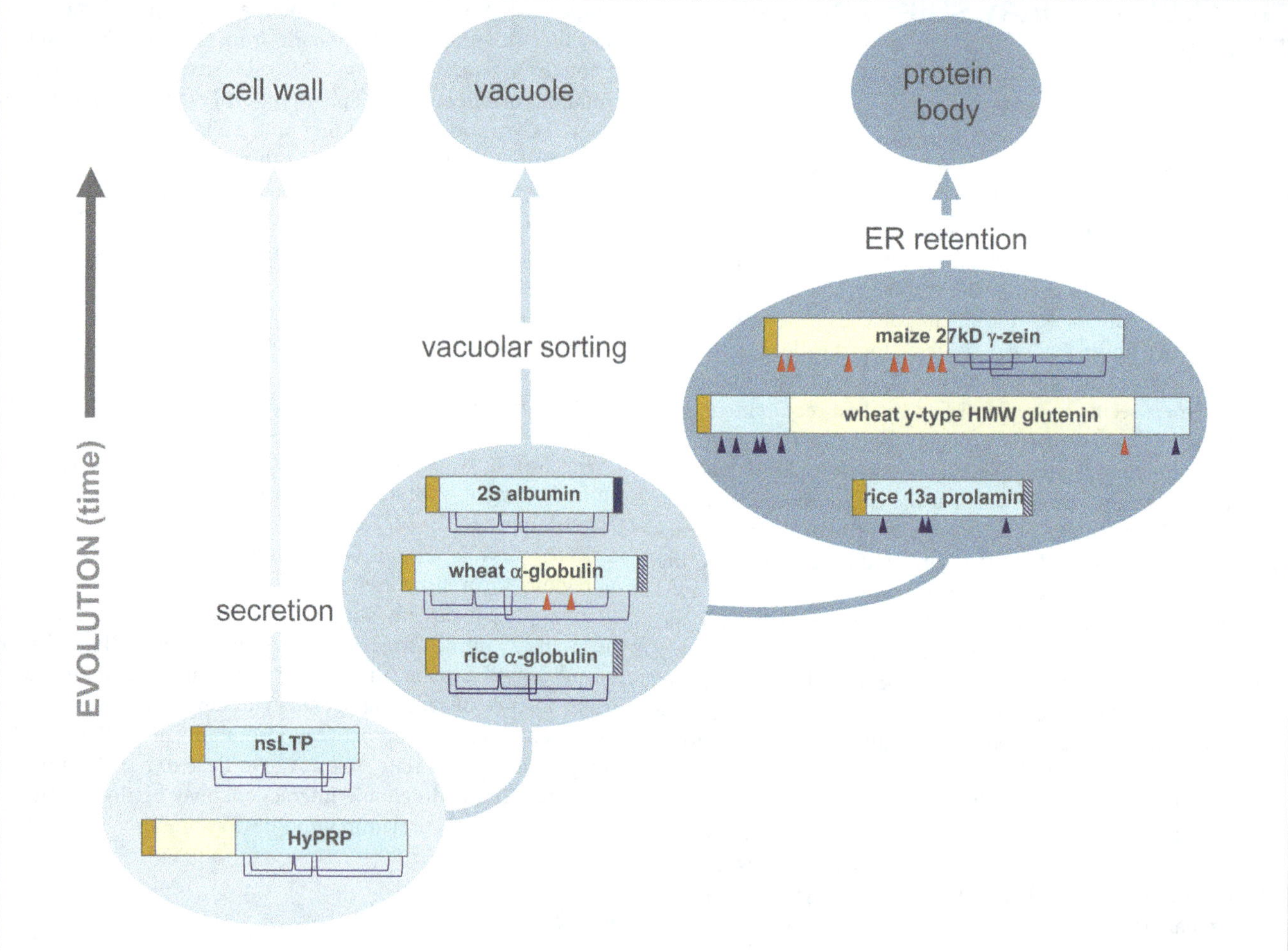

FIGURE 1 | A model of the evolutionary events at the origin of protein bodies. The model describes the origin of prolamins that are self-sufficient for PB formation and probably have a dominant effect on those that evolved later. Proteins with 8CM motifs are secreted or sorted to the vacuole, depending on the presence or not of vacuolar sorting signals. Insertions that do not alter the disulphide bonds of the 8CM motif do not affect protein destiny. Extensive polymerization that leads to PB formation can result from insertions that inhibit the correct formation of the four intra-chain bonds, deletions that involve the loss of critical Cys residues, or additions of domains containing new Cys residues. The repetitive domains may also mediate interactions with the ER membrane. Vacuolar sorting signals may be present or absent in PB proteins, but PB formation has anyway a dominant effect on intracellular traffic. Phylogenetic analysis and the results of protein engineering and expression of individual genes indicate that the PB forming prolamins evolved from superfamily members that are sorted to the vacuole, not from secreted members. The schematic protein structures are based on the following GenBank accessions: EU968356.1 (maize nsLTP), EU964401.1 (maize HyPRP), CAA40015.1 (2S albumin), ABG68034.1 (wheat α-globulin), BAA09308.1 (rice α-globulin), 27 kD zein, ABG68035.1 (wheat y-type HMW glutenin), and BAA36697.1 (rice 13a prolamin). Brown: signal peptide; light blue: conserved 8CM motif, or portions of it; yellow: Pro-rich or Glu-rich domain; blue: vacuolar sorting sequence; striped blue: putative vacuolar sorting sequence; blue lines: intra-chain disulphide bonds of the 8CM motif; blue triangles: Cys residues originating from the 8CM motif; red triangle: Cys residues that do not originate from the 8CM motif.

seven Cys residues, followed by a C-terminal 8CM/ABC domain. The N-terminal domain promotes PB formation when fused to other proteins that are otherwise available for intracellular traffic (Mainieri et al., 2004; Llop-Tous et al., 2010), although this dominant effect is not universal (de Virgilio et al., 2008; Ceresoli et al., 2016). Deletion of the N-terminal domain from γ-zein causes secretion of the remaining ABC domain, and the reciprocal deletion causes ER retention of the N-terminal domain, although in this case the typical round-shaped PBs are not formed (Geli et al., 1994). Zeolin, a fusion of the N-terminal domain with entire sequence of the 7S storage protein phaseolin (which does not contain Cys residues), forms PBs that, like those formed by the wild type γ-zein, are insoluble unless reduced, and it traffics along the secretory pathway when cells are treated *in vivo* with reducing agents (Mainieri et al., 2004; Pompa and Vitale, 2006). The importance of inter-chain disulfide bonds was confirmed by mutagenesis of the full-length γ-zein: when the seven Cys residues of the N-terminal domain were mutated to Ser, γ-zein was efficiently secreted (Mainieri et al., 2014). Progressive mutagenesis of these Cys residues gives rise to intermediate phenotypes, both in the full-length protein and a fusion between the N-terminal-domain and GFP: PBs become progressively smaller and ER retention is negatively affected (Llop-Tous et al., 2010; Mainieri et al., 2014). Deletion of the amphipathic repeats indicates that these are also important for PB formation (Llop-Tous et al., 2010). Altogether, these data indicate that the addition of an amphipathic sequence rich in Cys residues to an 8CM sequence has been crucial for this prolamin.

If six out of seven Cys residues of the N-terminal domain are mutated, the 27 kD γ-zein is in large part secreted, but a detectable proportion is sorted to the vacuole via a pathway that is sensitive to brefeldin A and wortmannin, two inhibitors of Golgi-mediated, vacuolar sorting (Mainieri et al., 2014). This indicates that limited multimerization avoids secretion and favors PSV sorting, whereas extensive polymerization promotes PB biogenesis. Protein sorting to PSVs requires sorting motifs recognized by specific receptors, but key recognition events could occur in the ER lumen and transient polymerization could contribute to sorting efficiency (Robinson et al., 2005; Robinson and Pimpl, 2014). Remarkably, the natural vacuolar sorting signal of phaseolin can be in part replaced by the insertion of a novel Cys residue that leads to polymerization of trimers (Pompa et al., 2010), underlining the evolutionary relationships between traffic to the vacuole and PB formation.

A WORKING MODEL FOR THE ORIGIN OF PROTEIN BODIES

The use of the ER to store proteins has evolved only recently and constitutes an interesting example of a simpler solution for protein accumulation (accumulation in the ER itself) evolving from more complex ones (sorting to vacuoles or secretion). **Figure 1** presents its possible evolutionary paths. All the evidence obtained studying prolamins that are self-sufficient to form PBs indicates that these structures have emerged by both exploiting and remodeling the 8CM scaffold. New availability of Cys residues involved in inter-chain disulfide bonds has been critical in this process and has resulted from either of two probably independent events: extensive insertions in the hypervariable region between the B and C domains characterize group III prolamins, whereas the addition of a new Cys-containing domain to an 8CM/ABC domain that can be virtually intact or mutated to variable extents characterizes many group II prolamins. The changes that have generated prolamins must thus have occurred more than once in evolution. Their success in being tolerated, without causing misfolding that would have led to degradation by ER quality control may be due to the rather simple folds of the small A, B, and C domains. The full explanation, however, must be more complex, since for example 7S globulins have a three-dimensional structure that does not tolerate great alterations (Hoffman et al., 1988; Pedrazzini et al., 1997; Lasserre et al., 2015), but zeolin is a very stable PB-forming protein fusion (Mainieri et al., 2014) and, conversely, a proportion of at least one rice 13 kD prolamin is detected in a ubiquitinated form in rice endosperm in which proteasome activity was inhibited, suggesting that normally this prolamin fails in part to fold properly and is degraded by ER quality control (Ohta and Taikawa, 2015).

It should be underlined that the formation of very large polymers is not *per se* sufficient to avoid traffic from the ER, at least in animal cells, a typical example being the secretion of procollagen fibrils (Bonfanti et al., 1998; Saito and Katada, 2015). Assembled PBs are heteropolymeric ordered structures in which certain polypeptides, such as γ- and β-zeins and rice 13–16 kD prolamins, are preferentially located at the PB surface (Lending and Larkins, 1989; Saito et al., 2012). Persistent interactions with chaperones and direct interactions with the luminal face of the ER membrane could thus play a major role in ER retention. Indeed, purified gliadins or a synthetic version of the Pro-rich repeat of 27 kD γ-zein directly interact *in vitro* with systems that mimic the inner face of the ER membrane (Kogan et al., 2004; Banc et al., 2009). It was suggested that also in gliadins the interactions could be mediated by the repeated region (Banc et al., 2009), even if the gliadin repeats are hydrophilic, unlike the amphipathic repeat of γ-zein, and the ability of γ-gliadin to form PB in the absence of other wheat prolamins is dubious (Napier et al., 1997; Tosi et al., 2009). Cell wall HyPRP proteins also have a hydrophilic, proline-rich repetitive N-terminal domain and an 8CM C-terminal domain (Dvorakova et al., 2007), but they are clearly unable to form PB. More detailed analysis and *in vivo* testing of the prolamin repetitive domains is therefore needed to determine their role in PB formation.

AUTHOR CONTRIBUTIONS

EP and AV wrote the manuscript. DM and CM contributed to the critical selection of the works discussed in the manuscript.

FUNDING

Supported by "Filagro" project of CNR-Regione Lombardia.

ACKNOWLEDGMENT

We thank Aldo Ceriotti, Massimilano Lauria, Angelo Viotti for the useful suggestions.

REFERENCES

Anelli, T., and Sitia, R. (2008). Protein quality control in the early secretory pathway. *EMBO J.* 27, 315–327. doi: 10.1038/sj.emboj.7601974

Bagga, S., Hank Adams, H., John, D., Kemp, J. D., and Sengupta-Gopalan, C. (1995). Accumulation of 15-kilodalton zein in novel protein bodies in transgenic tobacco. *Plant Physiol.* 107, 13–23.

Banc, A., Desbat, B., Renard, D., Popineau, Y., Mangavel, C., and Navailles, L. (2009). Exploring the interactions of gliadins with model membranes: effect of confined geometry and interfaces. *Biopolymers* 91, 610–622. doi: 10.1002/bip.21188

Bellucci, M., Alpini, A., Paolocci, F., Cong, L., and Arcioni, S. (2000). Accumulation of maize γ-zein and γ-zein:KDEL to high levels in tobacco leaves and differential increase of BiP synthesis in transformants. *Theor. Appl. Genet.* 101, 796–804. doi: 10.1007/s001220051546

Bonfanti, L., Mironov, A. A. Jr., Martinez-Menarguez, J. A., Martella, O., Fusella, A., Baldassarre, M., et al. (1998). Procollagen traverses the Golgi stack without leaving the lumen of cisternae: evidence for cisternal maturation. *Cell* 95, 993–1003. doi: 10.1016/S0092-8674(00)81723-7

Burr, B., and Burr, F. A. (1976). Zein synthesis in maize endosperm by polyribosomes attached to protein bodies. *Proc. Natl. Acad. Sci. U.S.A.* 73, 515–519. doi: 10.1073/pnas.73.2.515

Ceresoli, V., Mainieri, D., Del Fabbro, M., Weinstein, R., and Pedrazzini, E. (2016). A fusion between domains of the human bone morphogenetic protein-2 and maize 27 kD γ-Zein accumulates to high levels in the endoplasmic reticulum without forming protein bodies in transgenic tobacco. *Front. Plant Sci.* 7:358. doi: 10.3389/fpls.2016.00358

Coleman, C. E., Herman, E. M., Takasaki, K., and Larkins, B. A. (1996). The maize γ-zein sequesters α-zein and stabilizes its accumulation in protein bodies of transgenic tobacco endosperm. *Plant Cell* 8, 2335–2345. doi: 10.1105/tpc.8.12.2335

de Virgilio, M., De Marchis, F., Bellucci, M., Mainieri, D., Rossi, M., Benvenuto, E., et al. (2008). The human immunodeficiency virus antigen Nef forms protein bodies in leaves of transgenic tobacco when fused to zeolin. *J. Exp. Bot.* 59, 2815–2829. doi: 10.1093/jxb/ern143

Duvick, D. N. (1961). Protein granules of maize endosperm cells. *Cereal Chem.* 38, 374–385.

Dvorakova, L., Cvrckova, F., and Fischer, L. (2007). Analysis of the hybrid proline-rich protein families from seven plant species suggests rapid diversification of their sequences and expression patterns. *BMC Genomics* 8:412. doi: 10.1186/1471-2164-8-412

Gaut, B. S. (2002). Evolutionary dynamics of grass genomes. *New Phytol.* 154, 15–28. doi: 10.1046/j.1469-8137.2002.00352.x

Geli, M. I., Torrent, M., and Ludevid, D. (1994). Two structural domains mediate two sequential events in γ-zein targeting: protein endoplasmic reticulum retention and protein body formation. *Plant Cell* 6, 1911–1922. doi: 10.1105/tpc.6.12.1911

Graham, J. S. D., Jennings, A. C., Morton, R. K., Palk, B. A., and Raison, J. K. (1962). Protein bodies and protein synthesis in developing wheat endosperm. *Nature* 196, 967–969. doi: 10.1038/196967a0

Gu, Y. Q., Wanjugi, H., Coleman-Derr, D., Kong, X., and Anderson, O. A. (2010). Conserved globulin gene across eight grass genomes identify fundamental units of the loci encoding seed storage proteins. *Funct. Integr. Genomics* 10, 111–122. doi: 10.1007/s10142-009-0135-x

Herman, E. M., and Schmidt, M. A. (2004). Endoplasmic reticulum to vacuole trafficking of endoplasmic reticulum bodies provides an alternate pathway for protein transfer to the vacuole. *Plant Physiol.* 136, 3440–3446. doi: 10.1104/pp.104.051722

Hoffman, L. M., Donaldson, D. D., and Herman, E. M. (1988). A modified storage protein is synthesized, processed, and degraded in the seeds of transgenic plants. *Plant Mol. Biol.* 11, 717–729. doi: 10.1007/BF00019513

Holding, D. R. (2014). Recent advances in the study of prolamin storage protein organization and function. *Front. Plant Sci.* 5:276. doi: 10.3389/fpls.2014.00276

José-Estanyol, M., Gomis-Rüth, F., and Puigdomènech, P. (2004). The eight-cysteine motif, a versatile structure in plant proteins. *Plant Physiol. Biochem.* 42, 355–365. doi: 10.1016/j.plaphy.2004.03.009

Kawagoe, Y., Suzuki, K., Tasaki, M., Yasuda, H., Akagi, K., Katoh, E., et al. (2005). The critical role of disulfide bond formation in protein sorting in the endosperm of rice. *Plant Cell* 17, 1141–1153. doi: 10.1105/tpc.105.030668

Khoo, U., and Wolf, M. J. (1970). Origin and development of protein granules in maize endosperm. *Am. J. Bot.* 57, 1042–1050. doi: 10.2307/2441269

Kogan, M. J., Lopez, O., Cocera, M., Lopez-Iglesias, C., De La Maza, A., and Giralt, E. (2004). Exploring the interaction of the surfactant N-terminal domain of gamma-Zein with soybean phosphatidylcholine liposomes. *Biopolymers* 73, 258–268. doi: 10.1002/bip.10578

Kreis, M., Forde, B. G., Rahman, S., Miflin, B. J., and Shewry, P. R. (1985). Molecular evolution of the seed storage proteins of barley, rye and wheat. *J. Mol. Biol.* 183, 499–502. doi: 10.1016/0022-2836(85)90017-8

Lasserre, E., Ko, T., Dyer, J., and Murai, N. (2015). Modified bean seed protein phaseolin did not accumulate stably in transgenic tobacco seeds after methionine enhancement mutations. *Am. J. Plant Sci.* 6, 640–650. doi: 10.4236/ajps.2015.65069

Lending, C. R., and Larkins, B. A. (1989). Changes in the zein composition of protein bodies during maize endosperm development. *Plant Cell* 1, 1011–1023. doi: 10.2307/3869002

Liu, H., Shi, J., Sun, C., Gong, H., Fan, X., Qiu, F., et al. (2016). Gene duplication confers enhanced expression of 27-kDa γ-zein for endosperm modification in quality protein maize. *Proc. Natl. Acad. Sci. U.S.A.* 113, 4964–4969. doi: 10.1073/pnas.1601352113

Llop-Tous, I., Madurga, S., Giralt, E., Marzabal, P., Torrent, M., and Ludevid, M. D. (2010). Relevant elements of a maize γ-zein domain involved in protein body biogenesis. *J. Biol. Chem.* 285, 35633–35644. doi: 10.1074/jbc.M110.116285

Mainieri, D., Morandini, F., Maîtrejean, M., Saccani, A., Pedrazzini, E., and Vitale, A. (2014). Protein body formation in the endoplasmic reticulum as an evolution of storage protein sorting to vacuoles: insights from maize γ-zein. *Front. Plant Sci.* 5:331. doi: 10.3389/fpls.2014.00331

Mainieri, D., Rossi, M., Archinti, M., Bellucci, M., De Marchis, F., Vavassori, S., et al. (2004). Zeolin: a new recombinant storage protein constructed using maize γ-zein and bean phaseolin. *Plant Physiol.* 136, 3447–3456. doi: 10.1104/pp.104.046409

Masumura, T., Shigemitsu, T., Morita, S., and Satoh, S. (2015). Identification of the region of rice 13 kDa prolamin essential for the formation of ER-derived protein bodies using a heterologous expression system. *Biosci. Biotechnol. Biochem.* 79, 566–573. doi: 10.1080/09168451.2014.991684

Napier, J. A., Richard, G., Turner, M. F. P., and Shewry, P. R. (1997). Trafficking of wheat gluten proteins in transgenic tobacco plants: γ-gliadin does not contain an endoplasmic reticulum retention signal. *Planta* 203, 488–494. doi: 10.1007/s004250050218

Ohta, M., and Taikawa, F. (2015). OsHrd3 is necessary for maintaining the quality of endoplasmic reticulum-derived protein bodies in rice endosperm. *J. Exp. Bot.* 66, 4585–4593. doi: 10.1093/jxb/erv229

Onda, Y., and Kawagoe, Y. (2011). Oxidative protein folding: selective pressure for prolamin evolution in rice. *Plant Signal. Behav.* 6, 1966–1972. doi: 10.4161/psb.6.12.17967

Pedrazzini, E., Giovinazzo, G., Bielli, A., de Virgilio, M., Frigerio, L., Pesca, M., et al. (1997). Protein quality control along the route to the plant vacuole. *Plant Cell* 9, 1869–1880. doi: 10.2307/3870531

Pompa, A., De Marchis, F., Vitale, A., Arcioni, S., and Bellucci, M. (2010). An engineered C-terminal disulfide bond can partially replace the phaseolin vacuolar sorting signal. *Plant J.* 61, 782–791. doi: 10.1111/j.1365-313X.2009.04113.x

Pompa, A., and Vitale, A. (2006). Retention of a bean phaseolin/maize γ-zein fusion in the endoplasmic reticulum depends on disulfide bond formation. *Plant Cell* 18, 2608–2621. doi: 10.1105/tpc.106.042226

Robinson, D. G., Oliviusson, P., and Hinz, G. (2005). Protein sorting to the storage vacuoles of plants: a critical appraisal. *Traffic* 6, 615–625. doi: 10.1111/j.1600-0854.2005.00303.x

Robinson, D. G., and Pimpl, P. (2014). Receptor-mediated transport of vacuolar proteins: a critical analysis and a new model. *Protoplasma* 251, 247–264. doi: 10.1007/s00709-013-0542-7

Rödin, J., and Rask, L. (1990). Characterization of matteuccin, the 2.2S storage protein of the ostrich fern. Evolutionary relationship to angiosperm seed storage proteins. *FEBS J.* 192, 101–107. doi: 10.1111/j.1432-1033.1990.tb19201.x

Rubin, R., Levanony, H., and Galili, G. (1992). Evidence for the presence of 2 different types of protein bodies in wheat endosperm. *Plant Physiol.* 99, 718–724. doi: 10.1104/pp.99.2.718

Saito, K., and Katada, T. (2015). Mechanisms for exporting large-sized cargoes from the endoplasmic reticulum. *Cell. Mol. Life Sci.* 72, 3709–3720. doi: 10.1007/s00018-015-1952-9

Saito, Y., Kishida, K., Takata, K., Takahashi, H., Shimada, T., Tanaka, K., et al. (2009). A green fluorescent protein fused to rice prolamin forms protein body-like structures in transgenic rice. *J. Exp. Bot.* 60, 615–627. doi: 10.1093/jxb/ern311

Saito, Y., Shigemitsu, T., Yamasaki, R., Sasou, A., Goto, F., Kishida, K., et al. (2012). Formation mechanism of the internal structure of type I protein bodies in rice endosperm: relationship between the localization of prolamin species and the expression of individual genes. *Plant J.* 70, 1043–1055. doi: 10.1111/j.1365-313X.2012.04947.x

Shani, N., Rosenberg, N., Kasarda, D. D., and Galili, G. (1994). Mechanisms of assembly of wheat high molecular weight glutenins inferred from expression of wild-type and mutant subunits in transgenic tobacco. *J. Biol. Chem.* 269, 8924–8930.

Shewry, P. R., and Halford, N. G. (2002). Cereal seed storage proteins: structures, properties and role in grain utilization. *J. Exp. Bot.* 53, 947–958. doi: 10.1093/jexbot/53.370.947

Shutov, A. D., Braun, H., Chesnokov, Y. V., and Bäumlein, H. (1998). A gene encoding a vicilin-like protein is specifically expressed in fern spores. *Eur. J. Biochem.* 252, 79–89. doi: 10.1046/j.1432-1327.1998.2520079.x

Tian, L., and Okita, T. W. (2014). mRNA-based protein targeting to the endoplasmic reticulum and chloroplasts in plant cells. *Curr. Opin. Plant Biol.* 22, 77–85. doi: 10.1016/j.pbi.2014.09.007

Tosi, P., Parker, M., Gritsch, C. S., Carzaniga, R., Martin, B., and Shewry, P. R. (2009). Trafficking of storage proteins in developing grain of wheat. *J. Exp. Bot.* 60, 979–991. doi: 10.1093/jxb/ern346

Vitale, A., and Boston, R. S. (2008). Endoplasmic reticulum quality control and the unfolded protein response: insights from plants. *Traffic* 9, 1581–1588. doi: 10.1111/j.1600-0854.2008.00780.x

Vitale, A., and Hinz, G. (2005). Sorting of proteins to storage vacuoles: how many mechanisms? *Trends Plant Sci.* 10, 316–323. doi: 10.1016/j.tplants.2005.05.001

Vitale, A., Smaniotto, E., Longhi, R., and Galante, E. (1982). Reduced soluble proteins associated with maize endosperm protein bodies. *J. Exp. Bot.* 33, 439–448. doi: 10.1093/jxb/33.3.439

Wu, Y., Holding, D. R., and Messing, J. (2010). Gamma-zein is essential for endosperm modification in quality protein maize. *Proc. Natl. Acad. Sci. U.S.A.* 107, 12810–12815. doi: 10.1073/pnas.1004721107

Xu, J.-H., and Messing, J. (2009). Amplification of prolamin storage protein genes in different subfamilies of the Poaceae. *Theor. Appl. Genet.* 119, 1397–1412. doi: 10.1007/s00122-009-1143-x

Yamagata, H., and Tanaka, K. (1986). The site of synthesis and accumulation of rice storage proteins. *Plant Cell Physiol.* 27, 135–145.

Yuan, L., Dou, Y., Kianian, S. F., Zhang, C., and Holding, D. R. (2014). Deletion mutagenesis identifies a haploinsufficient role for γ-zein in opaque2 endosperm modification. *Plant Physiol.* 164, 119–130. doi: 10.1104/pp.113.230961

Conflict of Interest Statement: The authors declare that the research was conducted in the absence of any commercial or financial relationships that could be construed as a potential conflict of interest.

11

Exploitation of Common Bean Flours with Low Antinutrient Content for Making Nutritionally Enhanced Biscuits

Edited by:
Soren K. Rasmussen,
University of Copenhagen, Denmark

Reviewed by:
Michael A. Grusak,
USDA-ARS Children's Nutrition Research Center, USA
Tom Warkentin,
University of Saskatchewan, Canada

***Correspondence:**
Francesca Sparvoli
sparvoli@ibba.cnr.it

Francesca Sparvoli[1], Monica Laureati[2], Roberto Pilu[3], Ella Pagliarini[2], Ivan Toschi[3], Gianluca Giuberti[4], Paola Fortunati[4], Maria G. Daminati[1], Eleonora Cominelli[1] and Roberto Bollini[1]*

[1] CNR, Institute of Agricultural Biology and Biotechnology, Milan, Italy, [2] Department of Food, Environmental and Nutritional Sciences, University of Milan, Milan, Italy, [3] Department of Agricultural and Environmental Sciences - Production, Landscape, Agroenergy, University of Milan, Milan, Italy, [4] Alimentari e Ambientali, Facoltà di Scienze Agrarie, Istituto di Scienze degli Alimenti e della Nutrizione, Università Cattolica del Sacro Cuore, Piacenza, Italy

Consumption of legumes is associated with a number of physiological and health benefits. Legume proteins complement very well those of cereals and are often used to produce gluten-free products. However, legume seeds often contain antinutritional compounds, such as phytate, galactooligosaccharides, phenolic compounds, lectins, enzyme inhibitors, whose presence could affect their nutritional value. Screening natural and induced biodiversity for useful traits, followed by breeding, is a way to remove undesirable components. We used the common bean cv. Lady Joy and the *lpa1* mutant line, having different seed composition for absence/presence of lectins,α-amylase inhibitor, (α-AI) and phytic acid, to verify the advantage of their use to make biscuits with improved nutritional properties. We showed that use of unprocessed flour from normal beans (Taylor's Horticulture and Billò) must be avoided, since lectin activity is still present after baking, and demonstrated the advantage of using the cv. Lady Joy, lacking active lectins and having active α-AI. To assess the contribution of bean flour to biscuit quality traits, different formulations of composite flours (B12, B14, B22, B24, B29) were used in combinations with wheat (B14), maize (gluten-free B22 and B29), or with both (B12 and B24). These biscuits were nutritionally better than the control, having a better amino acid score, higher fiber amount, lower predicted glycemic index (pGI) and starch content. Replacement of cv. Lady Joy bean flour with that of *lpa1*, having a 90% reduction of phytic acid and devoid of α-AI, contributed to about a 50% reduction of phytic acid content. We also showed that baking did not fully inactivate α-AI, further contributing to lowering the pGI of the biscuits. Finally, data from a blind taste test using consumers indicated that the B14 biscuit was accepted by consumers and comparable in terms of liking to the control biscuit, although the acceptability of these products decreased with the increase of bean content. The B22 gluten-free biscuits, although received liking scores that were just above the middle point of the hedonic scale, might represent a good compromise between health benefits (absence of gluten and lower pGI), expectations of celiac consumers and likeness.

Keywords: amino acid score, α-amylase inhibitor, biscuits, consumers test, lectins, nutritional enhancement, phytic acid, predicted glycemic index

INTRODUCTION

Legumes are well-recognized functional foods and their use as ingredients for food formulations is getting increasing attention (Vaz Patto et al., 2015). A number of studies has provided indications that consumption of legumes is associated with several physiological and health benefits, such as prevention of cardiovascular disease, obesity, diabetes mellitus, and cancer. These health benefits are attributed to high content in legume seeds of important soluble and insoluble fiber, to their slowly digestible starch properties as well as to prebiotic oligosaccharides, phenolic compounds and some proteins such as the lupin γ-conglutinin, and the soybean 7S globulin α' chain (Duranti, 2006; Sparvoli et al., 2015). Some of these components regulate glycaemia and gastrointestinal function, whereas others provide antioxidant properties (Cardador-Martinez et al., 2002). In particular, due to their poor digestibility related to the inherent physical and structural properties of starch, legumes possess lower glycemic index (GI) when compared to cereal grains, a characteristic providing benefits for people with diabetes and/or cardiovascular disease (Hoover and Zhou, 2003; Sandhu and Lim, 2008). Moreover, being gluten-free, legumes could be used as ingredients to biofortify preparations for the celiac population.

Common bean (*Phaseolus vulgaris* L.) has significant cultural and historical importance as a staple food and is essential to human diets in many parts of the world. Bean seeds are a good source of energy, complex carbohydrates (dietary fibers, starch, and oligosaccharides), proteins, important minerals and vitamins (such as iron, zinc, B-vitamins, folates) as well as antioxidants and polyphenols required for human health. From a nutritional standpoint, bean seeds are higher in proteins than cereal grains (18–24 vs. 8–15%) and the amino acid profile of seed (storage) proteins well complements that of cereals, which are normally rich in sulfur amino acids and poor in lysine, tryptophan and threonine. However, seeds of common bean also contain a number of bioactive and/or antinutritional compounds, such as lectins, digestive enzyme inhibitors, phytate, galactooligosaccharides, phenolic compounds, whose presence could affect seed nutritional value (Sparvoli et al., 2015).

In bean seeds, major lectins are the erythroagglutinating and leucoagglutinating phytohemagglutinins (PHA-E and PHA-L, respectively). These PHAs, together with the evolutionary related α-amylase inhibitor (α-AI), belong to the APA gene family of storage proteins and are inherited as a single Mendelian locus (Lioi et al., 2003). Bean α-AI, also known as phaseolamin, is able to inhibit mammalian α-amylases, thus inhibiting starch digestion by blocking access to a basic active site of the α-amylase enzyme (Santimone et al., 2004). As α-AI prevents the digestion of complex carbohydrates, it is widely used as basic active ingredient of commercial starch blocker preparations for the control of body weight (Barrett and Udani, 2011). However, α-AI has also been shown to be effective in reducing post-prandial plasma levels of glucose, insulin, C-peptide and gastric inhibitory polypeptide in healthy subjects as well as in individuals affected by diabetes mellitus (Layer et al., 1985, 1986).

Phytic acid (*myo*-inositol-1,2,3,4,5,6-hexa*kis*phosphate) is highly negatively charged at physiological pH and easily precipitates in the form of phytate salts, binding important mineral cations such as iron, zinc, potassium, calcium, and magnesium. Monogastric animals, including humans, lack phytases in their digestive tract and fail to process the phytates present in seeds, thus phytic acid is poorly digested and decreases the nutritional value of the seeds (and derived foods) by limiting phosphorus and mineral bioavailability (Schlemmer et al., 2009).

The growing body of research on legumes health benefits has stimulated interest in increasing consumption of grain legumes and thus to expand their use in food products. In particular, biscuits are the largest category of snack items among baked food worldwide and they are one of the best ways to reach all segments of the population, due to their low manufacturing cost, convenience and long shelf life. For these reasons, biscuits represent a good way to propose composite flours (Chavan and Kadam, 1993).

Common beans have been regarded as useful ingredients for the production of snacks or baked products, like biscuits, bread or pasta, especially for the development of gluten-free products (Szafranski et al., 2005; Anton et al., 2009; Siddiq et al., 2013; Manonmani et al., 2014; Giuberti et al., 2015, 2016). Incorporation of common bean flour causes a positive impact on the levels of proteins, dietary fibers, resistant starch and predicted glycemic index, pGI (Anton et al., 2009; Giuberti et al., 2015, 2016). However, their use is limited by the presence of antinutritional factors. Consumption of raw or inadequately cooked beans is well known to cause poisoning, which clinical symptoms are extreme nausea, vomiting, diarrhea, severe acute gastroenteritis and intestinal malabsorption (Noah et al., 1980; Rodhouse et al., 1990, Petry et al., personal communication). The toxicity has been ascribed to the presence of active lectins (Kumar et al., 2013), that accumulate in substantial amounts in bean seeds. Thus, the use of bean flour needs typical processing involving cooking and grinding, or extrusion and grinding, both of which introduce a heating step that would inactivate enzyme inhibitors and lectins (Elkowicz and Sosulski, 1982; Alonso et al., 2000). Inactivation and/or removal of undesirable components are essential for improving bean overall nutritional quality and acceptability, aiming to favor their potentiality as human food and/or food ingredient.

Screening of natural and induced biodiversity for useful traits, followed by breeding, is a way for removal of undesirable components. Using this approach we identified genotypes showing differences in the content of PHA-E, PHA-L, α-AI and phytic acid, and we used them as parents in breeding programs to produce lines with different sets of storage proteins (Confalonieri et al., 1992; Bollini et al., 1999; Campion et al., 2009; **Figure 1**).

Our purpose was to verify the advantage of using these genetic materials to make composite flours for the preparation of biscuits with improved nutritional properties. We based our work on the cv. Lady Joy and the *lpa1* mutant line. The cv. Lady Joy has an active α-AI and contains only low amounts of the so-called "pinto lectin," an almost inactive PHA whose presence in common beans has been reported with a frequency lower than 10% (Pusztai et al., 1979, 1981; Voelker et al., 1986), consequently it has not erythroagglutinating activity (**Figure 1C**). The *lpa1* line carries a mutation in the gene coding for the phytic acid transporter

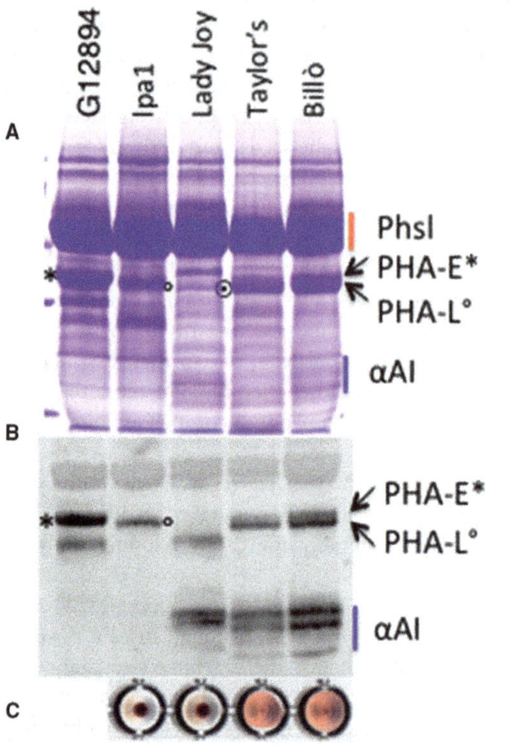

FIGURE 1 | SDS-PAGE analysis (A), immunoblot profiles (B), and hemagglutination test (C) of the seed extracts from the different genotypes used to prepare the bean containing biscuits. G12894, wild accession containing PHA-E. The major storage proteins are indicated. Red bar, Phsl, phaseolin polypeptides; blu bar, α-AI; asterisk, PHA-E; circle, PHA-L; dotted circle, pinto lectin.

(*Pvmrp1*) and shows a reduction of seed phytic acid of about 90% (Panzeri et al., 2011). Furthermore, this line is devoid of erythroagglutinating and α-AI activities (lacks PHA-E and α-AI), accumulating only low amounts of PHA-L (**Figure 1B**, compare *lpa1* with G12984, a wild bean accession only containing PHA-E, and the control cv. Taylor's Horticulture). We also used two wild type genotypes: the Italian landrace Billò, typically cultivated in Piedmont region (Piergiovanni and Lioi, 2010) and the cv. Taylor's Horticulture, that is the recurrent parent of the cv. Lady Joy.

Here, we present data demonstrating the advantage of using the cv. Lady Joy, and the *lpa1* line for the nutritional improvement of the biscuits. Having very low lectin activity or phytic acid content in their seeds, these bean flours do not need specific thermal treatments of the flour to inactivate PHA.

The resulting biscuits turned out to be nutritionally better than the control biscuit, having a better amino acid score, higher amount of fibers, lower starch content, and lower pGI. We also show that baking did not fully inactivate α-AI, further contributing to lowering the pGI of the biscuits. Finally, we provide data from a consumer blind taste test using consumers showing that the best biscuit formulation (B14) had a good acceptability score, not significantly different from control biscuit.

MATERIALS AND METHODS

Plant Materials

Common bean whole flours were obtained by finely grinding the seeds with a Waring Commercial Blender (http://www.waringcommercialproducts.com/catalog.php?pcID=84_95&products_id=296). Bean flours were used as such, as no processing involving cooking and grinding, or extrusion and grinding was applied prior their use. Seeds were obtained from genotypes differing for their PHA composition and/or presence or absence of phytic acid: (i) control genotypes: cv. Taylor's Horticultural (Asgrow) and the Italian landrace Billò, a typical landrace of Cuneo, Piedmont region, both containing PHA-E, PHA-L, and α-AI; (ii) cv. Lady Joy, lacking active PHA and containing α-AI (Confalonieri et al., 1992); (iii) *lpa1* mutant line, with 90% seed phytic acid reduction and lacking PHA-E and α-AI (Campion et al., 2009). Commercial unprocessed flours of soft wheat ("00" type, http://www.molinochiavazza.it/page.php?pagina=7) and maize ("fioretto" type, http://www.molinorossetto.com/it/farine-di-mais/1862-farina-fioretto-gialla.html) were purchased from a local supplier.

Preparation of Biscuits

Biscuits were prepared by PRIMO PAN bakery (Battifollo, CN, Italy), based on the recipe of a typical Italian biscuit, from Cuneo region, called "Pasta di meliga" (http://www.primopan.com/fogliemais.html). This typical biscuit is made with composite flour containing 67 and 33% of soft wheat and maize flours, respectively (**Table 1**, control). The ingredients for biscuits preparation included: 540 g of composite flour, 300 g raw cane sugar, 300 g butter, 2 medium eggs (ca. 110 g), 2 g common salt, 14 g baking powder, natural vanilla flavor (**Table 1**). Biscuits were baked at 180°C for 12 min. In addition to the control, five different types of biscuit samples were prepared (B12–B29) replacing maize and/or wheat with different amounts of common bean flours as described in **Table 1**.

Hemagglutination Test

Hemagglutinating activity in the extracts was estimated by a serial dilution procedure using human type A erythrocytes suspension (0.5% in phosphate buffered saline buffer pH 7.0, PBS) adding 50 μl of sample to the same volume of human erythrocytes. Agglutination was visually determined after 4 h incubation at room temperature. For each sample, serial dilutions in PBS, ranging from 1:10 to 1:1280, of seed and biscuit extracts were assayed. Results were recorded by monitoring how many 1:1 dilutions of the sample were necessary to prevent agglutination of red blood cells. Hemagglutinating unit (HAU) is defined as the reciprocal of the highest dilution still showing by eye agglutination of red cells, according to Trugo and von Baer (1998).

Protein Extraction, SDS-PAGE and Immunoblot Analyses

Total seed proteins extraction from dry bean seeds and biscuits, using 10 volumes of 20 mM borate pH 9, and

TABLE 1 | Main ingredients composition of biscuit samples.

Sample	Flours composition						Total flours (g)				Dough weight (g)
	g			% of dough weight				Other ingredients (g)			
	Wheat	Maize	Bean	Wheat	Maize	Bean		Eggs	Butter	Sugar	
Control	360	180	–	29	14	–	540	110	300	300	1250
B12	240	150	150	19	12	12	540	110	300	300	1250
B14	360	–	180	29	–	14	540	110	300	300	1250
B22	–	270	270	–	22	22	540	110	300	300	1250
B24	120	120	300	10	10	24	540	110	300	300	1250
B29	–	180	360	–	14	29	540	110	300	300	1250

separation by electrophoresis on an 15% sodium dodecylsulfate-polyacrylamide gel (SDS-PAGE) were as described by Bollini and Chrispeels (1978). Gels were stained with Coomassie Brilliant Blue R-250 or blotted on a supported nitrocellulose membrane (Hybond-C, GE-Healthcare). Immunoblot analysis was performed according to Burnette (1981) using rabbit antibodies against *P. vulgaris* recombinant α-AI at 1:1000 dilution. These antibodies allow the detection of both PHA and α-AI (Ceriotti et al., 1989). Peroxidase-linked anti-rabbit IgG was used as the secondary antibody.

Assay of α-Amylase Inhibitor Activity

Inhibitory activity against human salivary α-amylase (EC 3.2.1.1; Type IX-A) was measured by the increase of iodine staining after the action of the α-amylase enzyme on soluble starch as described in Altabella and Chrispeels (1990). Briefly, different volumes (from 10 to 100 μl) of biscuit flour or seed extracts diluted 50 or 200 fold in 20 mM borate buffer pH 9, respectively, were preincubated with a fixed amount of α-amylase (0.15 U) for 30 min at room temperature in a final volume reaction of 300 μl. Then, 200 μl of a 0.15% solution of potato starch was added and after 5 min at room temperature the reaction was stopped by adding 1 ml of Iodine reagent (Varner and Mense, 1972) and absorbance was measured at 620 nm. Results were expressed as units of α-amylase inhibited per mg of flour, where one unit of inhibitor activity is the amount which will bring about 50% inhibition of the α-amylase in 30 min under the above conditions according to Marshall and Lauda (1975).

Determination of Phytic Acid Phosphate

Phytic acid phosphate (PAP) fraction was determined by a ferric precipitation method (Pilu et al., 2003) and expressed as μg/mg of sample. The amount of corresponding phytate level was then obtained by multiplying these values for the conversion factor 3.5484.

Bromatological Analysis and *In Vitro* Starch Digestion of Experimental Biscuits

The chemical composition of biscuit samples was assayed according to AOAC standard methods (AOAC, 2000). Samples were milled and analyzed for dry matter (DM), ash, crude protein (CP), and ether extract (EE); starch content was measured enzymatically. Neutral detergent fiber (NDF), acid detergent fiber (ADF), and acid detergent lignin (ADL) were determined using the filter bag system (ANKOM Technology, New York, USA). In particular, NDF was assessed according to the procedure of Mertens (Mertens et al., 2002), while ADF and ADL were determined according to Vansoest et al. (1991). Gross energy was measured using an adiabatic calorimeter (IKA 4000, Staufen, Germany). Total amino acid composition was evaluated on defatted flours according to Commission Regulation (EC) No 152/2009 of 27 January 2009 laying down the methods of sampling and analysis for the official control of feed.

The multi-enzymatic protocol detailed by Giuberti et al. (2015) was used to evaluate the *in vitro* starch digestion of biscuits over time. The method, based on a 2-step enzymatic digestion incorporating a gastric (pH = 2) and a pancreatic (pH = 5.2) phase, has been shown to predict GI values with good accuracy for several cereal- and legume-based foods. Briefly, milled samples (800 mg) were carefully inserted into 50 ml tubes containing glass balls and pre-treated with a 0.05 M HCl solution (5 ml) containing pepsin (5 mg/ml; Sigma P-7000, Sigma Aldrich Co., Milan, Italy) for 30 min at 37°C under agitation. After the simulation of the gastric digestion, the pH of the solution was adjusted to 5.2 by adding 20 ml of 0.1 M acetate buffer. Then, 5 ml of an enzyme mixture (amylase activity of about 7000 U/ml; Giuberti et al., 2015) given by pancreatin (Merck 7130; Merck KGaA, Darmstadt, Germany), amyloglucosidase (Sigma A-7095; Sigma Aldrich Co., Milan, Italy) and invertase (Sigma I-4504; Sigma Aldrich Co., Milan, Italy) enzymes was added. From each tube, aliquots were carefully taken at 0 (prior to the enzyme addition simulating the pancreatic phase) and at 30, 60, 120, and 180 min after the enzyme addition. To each collected aliquot, absolute ethanol was added in order to stop the enzyme hydrolysis and the amount of released glucose was determined colorimetrically (glucose oxidase kit GODPOD 4058; Giesse Diagnostic snc, Rome, Italy). Commercial fresh white bread (starch content of 72.3% DM) was used as reference and a blank was also included to correct for the glucose in the amyloglucosidase solution. The percentage of digested starch at each time interval was calculated using a factor of 0.9 to convert mono to polysaccharides. For each treatment, samples were analyzed in duplicate. After the enzyme digestion, a hydrolysis index (HI) was derived from the ratio between the area under the hydrolysis curve (AUHC) of each biscuit samples and the corresponding AUHC of the reference fresh white bread as a percentage over the same period. From the obtained HI, a predicted GI value was then calculated with the formula pGI = 8.198 + 0.862 × HI (Granfeldt, 1994).

Consumer Test

The acceptability of the six different types of biscuits was assessed involving 102 subjects (55 females and 47 males, age: mean = 24.0; sd = 5.3). Consumers were recruited according to their liking for and regular consumption of biscuits. These subjects were asked to taste one biscuit of each sample, and to express their liking according to a 100-mm linear hedonic scale anchored at the extremes with "dislike extremely" (left of the scale, score = 0) and "like extremely" (right of the scale, score = 100). In order to balance the effects of serving order and carry-over, samples presentation order was systematically varied over participants (Macfie et al., 1989). Samples were served at room temperature (about 20°C) in plastic plates coded with 3-digit numbers and evaluated in individual booths in white light conditions. Participants were asked not to smoke, eat or drink anything, except water, for 1 h before the tasting sessions.

The study complied with the principles established by the Declaration of Helsinki and the protocol was approved by the Ethics Committee of the University of Milan. Written informed consent was obtained from each subject before the liking assessment was performed.

RESULTS

Evaluation of Erythroagglutinating Activity in Bean Based Biscuits

Usually, bean seeds accumulate substantial amounts of PHA (both L and E types) and α-AI. Since consumption of raw or inadequately cooked beans may cause poisoning, to verify the role of baking on PHA activity we used different type of bean flours with or without PHAs. We compared different bean genotypes: (i) cv. Lady Joy flour, devoid of active lectin, (ii) its parental line, cv. Taylor's Horticulture, containing both PHA-E and PHA-L, (iii) another commercial landrace, Billò, also containing both PHA-E and PHA-L, (iv) the *lpa1* genotype, containing only PHA-L (not erythroagglutinating).

Biscuits were prepared according to the recipe of a typical Italian biscuit, from Cuneo region, called "Pasta di meliga," in which we replaced maize with bean (B14 formulation, **Figure 2**, **Table 1**).

FIGURE 2 | Appearance of bean biscuits and bean seeds of the cv. Lady Joy and *lpa1* line.

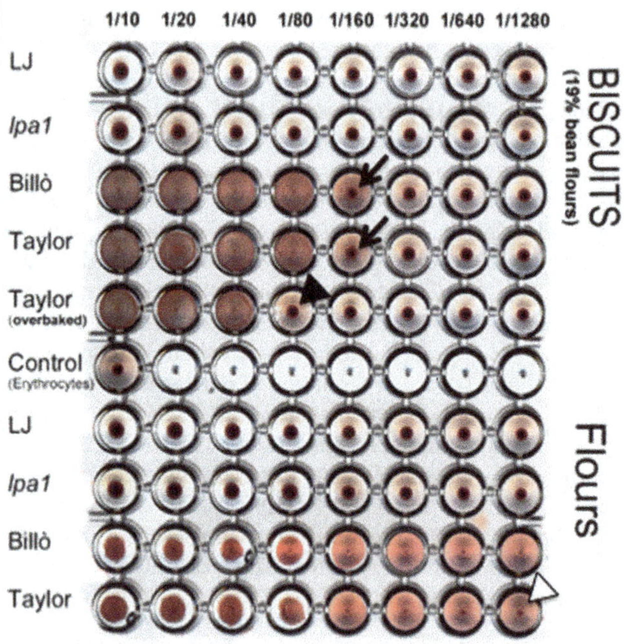

FIGURE 3 | Hemagglutination test of extracts from defatted flours of sample biscuits (B14 type) and corresponding bean seed flours of cv. Lady Joy (LJ), *lpa1* mutant line, Billò landrace and cv. Taylor's Horticulture. Control shows erythrocytes incubated with an equivalent voulme of PBS buffer, instead of the seed extract. Arrows and arrowheads indicate the agglutination limit. Samples dilutions are indicated.

The presence of active PHA was assessed using the erythroagglutination test (**Figure 3**). Results show that after baking, extracts of biscuits containing PHA (cv. Taylor's Horticulture and Billò) are still able to agglutinate erythrocytes after a serial dilution between 1:80 and 1:160 (**Figure 3**, black arrows). Three minutes of overbaking (baking from 12 to 15 min) decreased the erythroagglutinating activity to 1:40 dilution (**Figure 3**, black arrowhead), confirming that proper heat treatment is needed for inactivation. As expected, no agglutination was detected on extracts of biscuits containing cv. Lady Joy or *lpa1* flours.

To quantify the residual PHA activity we compared the agglutination results with those made on corresponding amounts of unprocessed bean flours. We calculated that defatted extracts of B14 biscuits contain about 19% of bean flour. Since at least three more 1:1 dilutions are needed to abolish agglutination in equivalent cv. Taylor's Horticulture and Billò extracts (**Figure 3**, white arrowhead), we estimated that the residual PHA activity in these biscuits was in the range of 5–10%.

Bean Biscuits Formulation

To assess the contribution of bean flour to biscuits quality traits, different formulations of composite flours were used in which wheat and/or maize flours were replaced by bean flour in the recipe of the biscuit "Pasta di meliga" (**Table 1**). To avoid any risk of poisoning due to residual active PHA after baking, we used the flour of cv. Lady Joy, which contains low amounts of the non-toxic "pinto lectin" and has an active α-AI. Since celiac disease has been associated with high incidences of diabetes mellitus (Lamacchia et al., 2014) and there is considerable interest in lowering the GI of gluten free foods for the maintenance of a good glycemic control, two types of biscuits, in which bean and maize flours completely replaced that of wheat (B22 and B29, **Table 1**), were also prepared.

Composite flours contained different percentages of bean flour: 26.7, 32.1, 50.0, 53.6, and 64.3% (B12, B14, B22, B24, B29, respectively), in combinations with wheat (B14), maize (B22 and B29), or with both wheat and maize flours (B12 and B24). Other ingredients included eggs, butter and sugar cane (**Table 1**). Biscuits were baked for 12 min at 180°C.

In order to assess the effect of baking on bean proteins, biscuit protein extracts were analyzed by SDS-PAGE (**Figure 4**). Results showed that in all the formulations the protein profile was similar to that of cv. Lad Joy seed extract, with phaseolin being the most abundant protein, as in common bean flours (**Figure 4**, compare LJ with B12, B14, B22, B24, B29). This result indicates that in the biscuits bean proteins have undergone poor hydrolysis.

Analysis of α-Amylase Inhibitor Activity and Phytic Acid Content in Biscuits

To verify the role of baking also on α-AI, its activity was analyzed in defatted extracts of B12, B14, B22, B24, B29 and control ("Pasta di meliga") biscuits as well as in B14 biscuits made with flours of the cv. Taylor's Horticultural (Lady Joy recurrent parent) and the landrace Billò.

No relevant activity was present in control biscuits (not containing bean flour), while consistent α-AI activity was measured in extracts of all biscuit types containing bean flours (**Table 2**). We observed that the level of α-AI activity did not correlate with the amount of bean flour in the biscuits. In fact, the highest α-AI activity was recorded in the B22 biscuit, which is not the one with the highest bean content. We then compared α-AI activities of the different biscuits with those present in the corresponding raw bean flours. Data showed that samples B12, B24, and B29 retained a α-AI activity between 49.4 and 62.5%, while higher activities were detected in samples B14 and B22, having 100 and 75.9% of the expected α-AI activities, respectively (**Table 2**).

We assayed α-AI activity also on extracts of B14 biscuits made with cv. Taylor's Horticulture and Billò flours. These biscuits retained about 34% of the starting α-AI activity, while overbaking reduced α-AI activity to about 10%. As a control, we measured the α-AI activity in the cv. Lady Joy B14 sample and confirmed that 100% of α-AI activity was present (**Table 2**).

Phytic acid is a well-known antinutrient and is very stable to heat treatment. The use of flours almost devoid of phytic acid could therefore significantly improve the nutritional quality of baked products. To evaluate this point we quantified phytic acid in extracts of B14 biscuits made with flour of the *lpa1* mutant, which is almost devoid of phytic acid, or with those of cv. Lady Joy, cv. Taylor's Horticulture, having normal phytic acid content (**Table 3**).

Results showed that biscuits made using *lpa1* flour contained less phytic acid (1.91 mg/g) than biscuits made with flours of the other two genotypes (4.63 and 3.78 mg/g). The reduction is 58.7% compared to phytic acid content of biscuits made with cv. Lady Joy, and 49.5% if compared to biscuits made with cv. Taylor's Horticulture (**Table 3**).

Proximate Composition, *In Vitro* Predicted Glycemic Index Classification and Amino Acid Composition

The different biscuits were also analyzed for their proximate composition, pGI, and total amino acid content (**Tables 4, 5**, respectively). Compared to control biscuits, the addition of bean flour increased protein content, from a minimum of 14% (B12) up to 45.7% (B29), as well as ashes, neutral detergent fiber (NDF), acid detergent fiber (ADF) and, to a lower extent, lignin and crude fiber. The addition of bean flour to the recipe also affected starch content, which decreased from a minimum of 25.8 (B12) to 37.0% (B29) (**Table 4**). Lastly, pGI values (calculated using commercial white wheat bread as reference; pGI = 94.4) decreased with the addition of bean flour in the composite, ranging from 88.9 to 61.9 for control and B29 biscuits, respectively.

As we found that B14 biscuits retain almost 100% of the activity of the α-AI present in bean flour, we compared pGI values of B14 biscuits made with different bean flours. In order to estimate the contribution of the active α-AI to the pGI, we used as a reference the pGI of biscuits made with the *lpa1* mutant, which is devoid of α-AI (**Table 2**). Our results showed that B14 biscuits containing active α-AI have lower pGI values than B14 biscuits made with flour of the *lpa1* mutant, thus indicating that active α-AI slightly contributes to the decrease of the pGI value (**Table 4**).

The presence of bean flour in the composite flours impacted also the amino acid composition of the biscuits: the highest effect was on the content of the essential amino acids Lys, Trp, and Arg, which compared to control biscuits raised one to two fold for Lys (118–275%), 39–134% for Trp, and 84–108% for Arg. The average (B12–B29) increase for all the remaining amino acids was ranging from about 30% (Leu) up to 53% (Phe). The only slight decrease was observed for Glu, whose content decreased between 1.5 and 17.3% (B12, B22, B24, and B29) with the exception of B14 (16.7% increase), suggesting that wheat significantly contributes more than maize to the content of this amino acid (compare B14 with B29). The overall content of sulfur amino acids (Met and Cys) was not very much affected by bean flour addition and their average content increased between 7.6 and 13.3% (B12–B29) with Met contributing more than Cys (average 30% more and 6% minus, respectively; see **Table A1**).

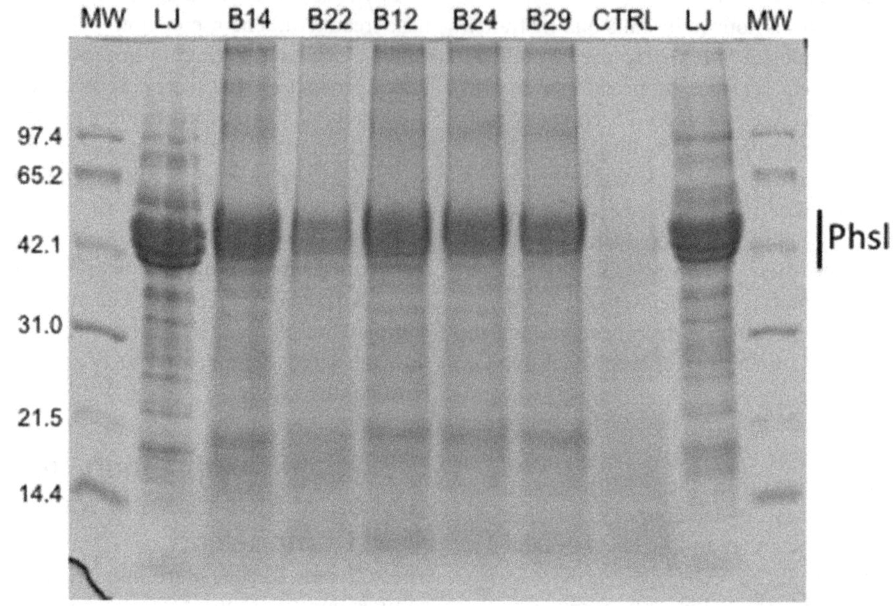

FIGURE 4 | SDS-PAGE analysis of extracts from defatted flours of sample biscuits (B14-B29, CTRL, control biscuit) and cv. Lady Joy seed extract (LJ). MW molecular weight markers. Bar indicates phaseolin polypeptides (Phsl).

TABLE 2 | α-amylase inhibitory activity in extracts of defatted biscuit flours and bean flours of cv. Lady Joy (LJ), cv. Tayolor's Horticultural (TH), Billò (B), the *lpa1* mutant line, and over baked (OB).

	Sample	% bean flour in defatted biscuit	U α-AI/100 mg bean flour	Expected U α-AI/100 mg defatted biscuit flour*	Measured U α-AI/100 mg defatted biscuit flour	% of expected U α-AI
Flour	cv. Lady Joy	–	1048.42 ± 94.5			
Biscuits	Control	0		0.00	37.56 ± 41.28	0.00
	B12	16		167.75	94.53 ± 20.44	56.35
	B14	19		199.20	203.98 ± 8.060	102.40
	B22	28		293.56	222.73 ± 40.77	75.87
	B24	32		335.50	209.61 ± 18.67	62.48
	B29	38		398.40	196.64 ± 37.90	49.36
Flours	cv. Lady Joy		1132.47 ± 161.7			
	cv. Taylor's H.		1226.04 ± 164.0			
	Billò		1181.99 ± 76.9			
	lpa1		0.00			
Biscuits	B14 (LJ)	19		215.17	234.96 ± 23.50	109.20
	B14 (TH)	19		232.95	160.68 ± 60.06	68.98
	B14-OB (TH)	19		232.95	48.65 ± 2.4	20.88
	B14 (B)	19		224.58	169.16 ± 48.87	75.32
	B14 (*lpa1*)	19		0.00	0.00	0.00

*Units of inhibited α-amylase are per 100 mg of flours (U α-AI/100 mg). *Values are the product of α-AI activity in bean flour x the percent of bean flour in defatted biscuit. B12-B29: see* ***Table 1****.*

Consumer Test

The introduction of a new ingredient, the bean flour, to a conventional food, such as biscuits, could result in a product that might not fully meet the acceptability of the consumers. To verify the liking of the biscuits containing increasing amounts of cv. Lady Joy bean flour, their acceptability was tested in comparison with the traditional "Pasta di meliga" (control), in a blind taste test on a consumer sample of 102 subjects. Hedonic data were subjected to ANOVA considering samples ($n = 6$) as source of variation and liking scores as dependent variable.

TABLE 3 | Phytic acid (PA) content in biscuit samples (defatted flours) and bean flours of the *lpa1* mutant line, cv. Lady Joy and cv. Tayolor's Horticultural.

	Sample	PA mg/g flour
Biscuits	B14 (*lpa1*)	1.91 ± 0.03
	B14 (LJ)	4.63 ± 0.81
	B14 (Taylor's H.)	3.78 ± 0.65
Flours	*lpa1*	2.38 ± 0.05
	Lady Joy cv.	9.32 ± 0.23
	Taylor's H. cv.	10.53 ± 1.01

Mean hedonic values for the six biscuits (one control and five containing different amounts of bean flour) are reported in **Table 6**. ANOVA results showed a significant effect of the factor samples [$F_{(5,\ 505)} = 11.52$, $p < 0.001$] on liking ratings. LSD *post-hoc* comparison indicated that the B14 biscuits received an acceptability score that was statistically comparable to the control biscuit, and the acceptability of B12 biscuits, although statistically different from the control, was not statistically different from that of B14 biscuits, indicating that these two types were well liked by consumers. However, liking scores decreased with the increase of the bean flour content as the B22 (gluten-free) and B24 formulations were significantly less liked and received liking scores that were just above the middle point of the hedonic scale, although they were not significantly different from the B12 sample. The B29 biscuit, gluten-free but with the highest addition of bean flour, was the least liked sample.

DISCUSSION

The increasing awareness of the link between diet and health is the driving force to develop novel foods to meet consumer's needs, including the identification, development and improvement of new ingredients for food formulations.

In this work we assessed the advantage of using flours of nutritionally enhanced common bean genotypes for the production of bean based biscuits, with the aim to contribute to the diversification of healthier and more nutritional diet.

Importance of Seed Nutrients Genetic Modulation for Biscuits Quality and Properties

A number of studies reported improved nutritive value of high protein biscuits made using composite flours, including different amounts of flour from soybean, field pea, lentil, chickpea, lupin, and common bean (Shrestha and Noomhorm, 2002; Rababah et al., 2006; Zucco et al., 2011). In general, the consumption of common bean has not gone beyond the traditional processing and uses, due to the antinutritional factors present in the seeds. Extrusion, thermal treatments, such as roasting, and soaking at high temperature, followed by cooking and dehydration, are often used to eliminate lectin and enzyme inhibitors activities (Hoojjat and Zabik, 1984; Costa et al., 2006; Boye et al., 2010; Kelkar et al., 2012; Siddiq et al., 2013). However, these

TABLE 4 | Proximate composition and *in vitro* predicted glycemic index of biscuit samples (g/100 g) [1 biscuit is about 11.6 ± 0.4 g; Lady Joy cv. (LJ), *lpa1* mutant line (*lpa1*), Tayolor's Horticultural cv. (TH), Billò (B)].

Sample	Dry matter	Ashes	Crude proteins	Ether extract	NDF[a]	ADF[b]	Lignin	Crude fiber	Total sugars	Starch	GE (kJ/g)[c]	pGI[d]
Control	97.1 ± 0.03	1.21 ± 0.03	7.07 ± 0.00	22.4 ± 0.26	1.40 ± 0.06	0.96 ± 0.07	0.77 ± 0.11	1.80 ± 0.13	31.2 ± 0.06	41.9 ± 0.96	22.4 ± 0.230	89.0 ± 0.35
B12	94.6 ± 0.04	1.87 ± 0.00	8.07 ± 0.23	24.4 ± 0.15	3.44 ± 0.17	2.07 ± 0.01	0.80 ± 0.00	1.70 ± 0.05	30.3 ± 0.14	31.1 ± 0.07	22.0 ± 0.005	80.6 ± 0.99
B14	95.7 ± 0.03	2.09 ± 0.00	9.24 ± 0.21	24.5 ± 0.49	2.23 ± 0.08	2.05 ± 0.03	0.77 ± 0.01	2.06 ± 0.26	30.9 ± 0.61	29.8 ± 0.02	22.6 ± 0.001	74.2 ± 1.34
B22	94.8 ± 0.00	2.21 ± 0.02	8.79 ± 0.02	24.5 ± 0.50	4.25 ± 0.68	3.04 ± 0.06	0.89 ± 0.18	1.90 ± 0.12	31.9 ± 0.76	30.5 ± 1.17	22.5 ± 0.090	70.7 ± 0.57
B24	95.3 ± 0.02	2.03 ± 0.03	8.94 ± 0.04	26.3 ± 0.10	3.45 ± 0.16	2.67 ± 0.03	0.92 ± 0.04	1.61 ± 0.23	31.1 ± 0.88	28.5 ± 0.31	22.6 ± 0.070	66.2 ± 0.92
B29	95.0 ± 0.12	2.32 ± 0.01	10.3 ± 0.26	23.3 ± 0.14	3.68 ± 0.17	2.98 ± 0.05	0.87 ± 0.22	2.28 ± 0.15	32.1 ± 0.77	26.4 ± 0.20	22.2 ± 0.180	61.9 ± 1.41
B14 (LJ)	92.9 ± 0.17	1.74 ± 0.14	11.2 ± 0.09	21.6 ± 0.09	nd	nd	nd	nd	30.2 ± 0.20	34.3 ± 0.58	nd	66.0 ± 0.71
B14 (*lpa1*)	94.4 ± 0.06	1.87 ± 0.14	12.0 ± 0.08	19.5 ± 0.58	nd	nd	nd	nd	29.5 ± 1.10	32.2 ± 0.53	nd	70.3 ± 0.21
B14 (TH)	94.5 ± 0.06	2.05 ± 0.14	11.3 ± 0.05	20.9 ± 0.33	nd	nd	nd	nd	30.3 ± 0.69	29.9 ± 0.53	nd	64.6 ± 1.70
B14 (B)	91.6 ± 0.08	1.74 ± 0.14	13.3 ± 0.02	21.6 ± 0.49	nd	nd	nd	nd	29.9 ± 0.78	29.6 ± 0.54	nd	64.5 ± 2.19

B12–B29: see ***Table 1****.*

Values are mean ± standard deviation of three independent determinations (dry weight basis), except for total sugars, the in vitro glycemic index classification, which were run in duplicate.

[a]*NDF, Neutral Detergent Fiber (cellulose, emicellulose, lignin).*

[b]*ADF, Acid Detergent Fiber (cellulose, lignin).*

[c] *GE, Gross Energy.*

[d]*pGI, predicted Glycemic Index, calculated using commercial white wheat bread as reference (pGI = 94.4).*

TABLE 5 | Total amino acid composition of biscuit samples (g/100 g defatted flour) and bean flours of cv. Lady Joy (LJ), cv. Taylor's Horticultural (TH), and *lpa1* mutant line.

Sample	Thr	Lys	Trp	Cys	Met	Phe	Val	His	Ile	Asp	Glu	Pro	Gly	Ala	Leu	Tyr	Arg	NH_4	Hexan extract % DM	Proteins % DM
Control	0.337	0.144	0.041	0.175	0.141	0.379	0.496	0.229	0.323	0.572	2.354	0.940	0.299	0.395	0.722	0.300	0.309	0.268	21.4	8.75
B12	0.371	0.314	0.057	0.170	0.170	0.496	0.625	0.220	0.404	0.745	2.318	0.942	0.357	0.455	0.809	0.352	0.569	0.302	23.6	10.05
B14	0.461	0.485	0.096	0.165	0.173	0.590	0.713	0.325	0.509	0.809	2.748	1.277	0.446	0.478	0.876	0.378	0.568	0.401	22.8	12.03
B22	0.472	0.454	0.058	0.156	0.189	0.575	0.544	0.353	0.475	1.020	1.947	1.312	0.405	0.580	1.005	0.405	0.557	0.238	23.4	11.28
B24	0.473	0.491	0.080	0.168	0.185	0.591	0.757	0.352	0.488	1.025	2.251	1.430	0.438	0.551	0.939	0.401	0.599	0.316	22.5	11.91
B29	0.532	0.540	0.084	0.164	0.194	0.656	0.724	0.370	0.531	1.141	2.239	1.392	0.469	0.614	1.063	0.438	0.643	0.251	24.0	12.85
LJ	1.005	1.602	0.223	0.261	0.326	1.327	1.272	0.798	1.086	2.442	3.799	3.973	0.932	1.037	1.792	0.703	1.463	0.313	–	24.86
TH	1.186	1.741	0.243	0.252	0.307	1.526	1.416	0.819	1.241	2.985	4.300	5.037	1.043	1.120	2.024	0.722	1.793	0.374	–	28.20
lpa1	1.120	1.861	0.228	0.282	0.334	1.521	1.377	0.875	1.269	3.010	4.638	5.334	1.073	1.104	2.130	0.748	1.851	0.396	–	29.11

TABLE 6 | Results of the consumer test ($n = 102$) for evaluating the acceptability of the different bean biscuits compared to the control biscuit "Pasta di meliga."

Sample	Mean ± SEM	LSD *post-hoc* groups
Control	69.8 ± 2.3	d
B12	60.7 ± 2.5	bc
B14	64.6 ± 2.5	cd
B22	54.5 ± 2.5	b
B24	55.1 ± 2.7	b
B29	45.5 ± 2.7	a

Identical letters indicate there is no statistically significant difference between the samples according to LSD post-hoc test. Consumers liking was expressed according to a 100-mm linear hedonic scale anchored at the extremes with "dislike extremely" (left of the scale, score = 0) and "like extremely" (right of the scale, score = 100).

types of processing are a source of additional costs for the producers and may lead to undesirable functional and sensory properties of the common bean flours (Batista et al., 2010). As a consequence, common bean is not a very common legume for flour fortification, despite its good protein and micronutrient content.

Here, we recommend the use of the cv. Lady Joy, containing an almost inactive and low abundant PHA (Pusztai et al., 1981; Confalonieri et al., 1992), to avoid any possible risk of lectin poisoning due to frequent and abundant consumption of these bean biscuits. In fact, we provide evidence that biscuits made with bean flours containing active PHA (Taylor's H. and Billò) still retain residual lectin activity (about 5–10% of the lectin in the original flours), while no agglutination is detected in biscuits made using flour of the cv. Lady Joy (**Figure 4**).

Another advantage of avoiding preventive thermal treatments of the bean flours (possible if using the cv. Lady Joy) was the finding that substantial levels of α-AI activity were retained in the bean biscuits, with the B14 (LJ) type showing the highest levels (**Table 2**). It is known that α-AI and, in general, storage proteins belonging to the *P. vulgaris* lectin family, are stable proteins resistant to thermal degradation (Sparvoli et al., 1997). Nevertheless, the finding that cv. Lady Joy biscuits B14 retained such a high α-AI activity was unexpected. A possible explanation could be that dry heating, as it occurs during baking, induces changes in the physicochemical properties of the dough and consequently of its components, causing enhancement of the α-AI activity as has been shown for microwave treatment of beans (Oomah et al., 2014). The presence of active α-AI was confirmed in cv. Taylor's Horticulture and Billò B14 biscuits, although to a lesser extent. This might be due to the fact that these two genotypes contain a different α-AI isoform that might differently respond to heat treatments (Bollini et al., 1999).

Biofortified Bean Biscuits: Improved Protein Content and Amino Acid Score, and Reduced Phytic Acid Content

All biscuit formulas with composite bean flours showed a better nutritional composition, compared to the control, i.e., higher levels of protein and fiber, better amino acid score, and lower starch content along with a reduction in the pGI values. A similar behavior should be expected in food preparation containing different percentage of legume flour, as already reported for rice spaghetti, semolina spaghetti and teff tagliatelle prepared with increasing levels of bean flour (Gallegos-Infante et al., 2010; Giuberti et al., 2015, 2016). In addition, present findings are consistent with data obtained by other authors (Rababah et al., 2006; Zucco et al., 2011). For example, fortification with 12% of broad bean flour resulted in a protein increase from 16.6 to 20.2% (Rababah et al., 2006). Our results also indicate that the presence of maize in composite flours decreased biscuit protein value, as B14 biscuits, not containing maize, have higher protein content than B22–B29 samples (**Tables 4, 5**). Indeed, this is in accordance to the fact that wheat grains have average protein content higher than that of maize, traditionally an oil crop.

The nutritional enhancement of bean-containing biscuits was evident also for the amino acid composition, as the addition of bean flour increased the content of limiting essential amino acids (Lys, Thr, Trp) without lowering the content of sulfur amino acids. Furthermore, the replacement of cv. Lady Joy bean flour with that of *lpa1* contributed to about a 50% reduction of phytic acid content in B14 biscuit formulations. Nutritional studies performed on volunteers indicated that iron absorption from *lpa1* beans is significantly higher than from their parents with normal phytic acid levels (Petry et al., 2013, 2015). These results indicate that the B14 formulation containing *lpa1* flour is suited to produce biofortified biscuits with an enhanced and more balanced protein content and with increased bioavailability of essential minerals.

One third of the global population is estimated to suffer from hidden hunger. Micronutrient deficiencies vary in severity and occur in every country, but they disproportionally affect women, children, and people living in developing countries (Von Grebmer et al., 2014). Thus, malnutrition and hidden hunger represent a big nutritional challenge. The bean biscuits we presented here could be further implemented and used as a biofortified food snack, especially for children, using the *lpa1* bean flour in the B29 formulation, together with a flour from a biofortified maize *lpa1* mutant (Cerino Badone et al., 2012; Landoni et al., 2013). In this case, we expect a novel food product having a very high mineral bioavailability and higher and balanced protein content (**Tables 2, 3, A1**).

Biofortified Bean Biscuits: Predicted Glycemic Index and Consumer's Test

The classification through the pGI can be useful to predict the likely *in vivo* glycemic response of food of interest and a reference material. Accordingly, foods can be categorized into low (<55), medium (55–69), and high GI (>70) (Foster-Powell et al., 2002). With respect to the control, the pGI values of experimental biscuits decreased with the addition of bean flour, as already reported by Giuberti et al. (2015) where the presence of 20 and 40% of bean flour in gluten-free rice pasta reduced the pGI from 61, in the control, to 37 and 31, respectively. With respect to cereal flours, starch from bean is characterized by higher level of amylose and by larger size of starch granules, both starch structural properties that can contribute to impede overall starch degradability (Sandhu and Lim, 2008). Moreover, comparative studies reported that cereal starches are characterized by the A-type power diffraction pattern, which has an open structure rendering cereal starches highly digestible (Biliaderis, 1991). On the contrary, legume starches tend to exhibit the C-type crystallinity, which is more resistant to digestion due to a close-packed structure (Biliaderis, 1991). Lastly, the increasing amounts of fiber and protein found in bean-enriched biscuits, along with the possible formation of amylose-lipid complexes during cooking, could have contributed to further reduce the accessibility of amylase to hydrolyze the starch (Zhang et al., 2015).

The active α-AI present in all bean biscuits (but the *lpa1*) represents an additional valuable feature, further contributing to the reduction of pGI (**Table 5**). Spadafranca et al. (2013) in a study performed on 12 volunteers to evaluate the effects of a standardized and purified *P. vulgaris* extract (PVE), containing 110.000 U of α-AI per 100 mg tablet, showed lower increments in glucose, insulin and C-peptide 3 h after meal consumption associated to 1 PVE tablet. We calculate that five B14 biscuits are equivalent to 1 PVE tablet, as they contain a comparable activity of α-AI.

On the basis of the above results, favorable implications for human health with a particular regard to the control of glycemia, especially in diabetic individuals and in the celiac population, can be hypothesized. For instance, for the celiac population, the enrichment of gluten-free baked products with fiber component is beneficial, since a general low intake of this food component has been reported (Lamacchia et al., 2014). In addition, although there is an on-going debate, the reduction of the dietary glycemic response following the consumption of lower GI food products can favorably influence several physiological parameters implicated as markers for conditions including overweight and obesity, diabetes mellitus and risk of coronary heart disease (Livesey et al., 2008; Brand-Miller et al., 2009). This aspect is also of particular interest concerning the celiac population, because several gluten-free foods exhibit higher GI than their gluten containing counterparts, due to the lack of the gluten matrix (Pellegrini and Agostoni, 2015).

In this work we considered two different gluten-free biscuit formulations, B22 and B29, and showed that they had lower pGI compared to control biscuit, confirming an additive contribution of bean flour. The acceptability of these products decreased with the increase of bean content, however the B22 gluten-free biscuits might represent a good compromise between health benefits (absence of gluten and lower pGI), expectations of celiac consumers and likeness. It should be pointed out that the consumers' test was a blind one and was not directed to celiac subjects. Appearance, flavor and texture are reported to be the most critical properties for consumer's acceptance of gluten-free products (Pagliarini et al., 2010). It is well known that gluten-free food products have a lower acceptability compared to those containing wheat (Pagliarini et al., 2010; Laureati et al., 2012). Moreover, it has been reported that differences in liking between celiac and non-celiac consumers may arise when judging gluten-free products (Laureati et al., 2012), with celiac consumers providing higher scores than non-celiac subjects to products of lower quality. In future studies, it is advisable to involve regular consumers of gluten-free products, since they are probably more sensitive to the differences in terms of liking to verify if a better score could be gained for B22 and B29 gluten-free biscuits.

In this context, future perspectives of this study should address the identification of the properties of appearance, smell, taste and texture that characterize bean-flour enriched biscuit formulations in order to define the sensory attributes that drive consumer's preference. Actually, this piece of research is intended as a pilot study carried out in order to give insights on the potentiality of Lady Joy bean flour as gluten-free ingredient, a fundamental topic, which truly needs to be investigated systematically.

CONCLUSIONS

Food technology is important for the development of novel foods and food ingredients. Here we show the importance of the genetic and breeding approach to contribute and possibly facilitate the improvement of the primary product. In this work we show the feasibility of making nutritionally improved biscuits by the addition of bean flour of selected genotypes. We took advantage of a PHA inactive genotype (Lady Joy), which allows the direct use of flour without any thermal treatment. This resulted in biscuits containing fair amount of active α-AI, which contributes to lowering the pGI. We also proved that the use of the *lpa1* line could significantly contribute to the reduction of total phytates in the biscuits, a trait that is expected to increase minerals bioavailability. We are now working to combine the properties of cv. Lady Joy (absence of active PHA and very active α-AI) with the low phytate content of the *lpa1* mutant to produce a new common bean line that could be useful to address the health issue of diabetic individuals with micronutrient (iron, zinc, etc) deficiencies and that if combined with maize flours in the B22 formulation could meet the requirement of celiac disease affected patients, that are more exposed to the risk of developing type 1 diabetes. Finally these products could be further diversified and improved by using nutritionally enhanced maize flours from *lpa1* mutants, anthocyanins enriched lines and "high quality protein" maize lines (Azevedo and Arruda, 2010; Landoni et al., 2013; Holding, 2014; Petroni et al., 2014).

AUTHOR CONTRIBUTIONS

FS conceptualized, designed, and supervised the project. MD, ML, RP, GG, PF carried out the experiments. FS, RB provided the bean genotypes and wrote the manuscript. FS, EC, ML, RP, EP, GG, RB regularly discussed the experiments, analyzed the results, provided useful suggestions during the project and critically revised the manuscript. All authors read and approved the final manuscript.

FUNDING

This work was supported by the Programme FILAGRO "Strategie innovative e sostenibili per la filiera agroalimentare," as part of the activities defined within the Accordo Quadro Consiglio Nazionale delle Ricerche and Regione Lombardia.

ACKNOWLEDGMENTS

We thank Dr. Alessandra Benati for her support in providing contacts with "PrimoPan" bakery.

REFERENCES

Alonso, R., Aguirre, A., and Marzo, F. (2000). Effects of extrusion and traditional processing methods on antinutrients and *in vitro* digestibility of protein and starch in faba and kidney beans. *Food Chem.* 68, 159–165. doi: 10.1016/S0308-8146(99)00169-7

Altabella, T., and Chrispeels, M. (1990). Tobacco plants transformed with the bean alpha-AI-gene express an inhibitor of insect alpha-amylase in their seeds. *Plant Physiol.* 93, 805–810. doi: 10.1104/pp.93.2.805

Anton, A. A., Fulcher, R. G., and Arntfield, S. D. (2009). Physical and nutritional impact of fortification of corn starch-based extruded snacks with common bean (*Phaseolus vulgaris* L.) flour: effects of bean addition and extrusion cooking. *Food Chem.* 113, 989–996. doi: 10.1016/j.foodchem.2008.08.050

AOAC (2000). *Official Method of Analysis, 7th Edn.* Washington, DC: Association of Official Analytical Chemists.

Azevedo, R. A., and Arruda, P. (2010). High-lysine maize: the key discoveries that have made it possible. *Amino Acids* 9, 979–989. doi: 10.1007/s00726-010-0576-5

Barrett, M. L., and Udani, J. K. (2011). A proprietary alpha-amylase inhibitor from white bean (*Phaseolus vulgaris*): a review of clinical studies on weight loss and glycemic control. *Nutr. J.* 10:24. doi: 10.1186/1475-2891-10-24

Batista, K., Prudencio, S., and Fernandes, K. (2010). Changes in the functional properties and antinutritional factors of extruded hard-to-cook common beans (*Phaseolus vulgaris*, L.). *J. Food Sci.* 75, C286–C290. doi: 10.1111/j.1750-3841.2010.01557.x

Biliaderis, C. G. (1991). The structure and interactions of starch with food constituents. *Can. J Physiol. Pharmacol.* 69, 60–78. doi: 10.1139/y91-011

Bollini, R., Carnovale, E., and Campion, B. (1999). Removal of anti nutritional factors from bean (*Phaseolus vulgaris* L.) seeds. *Biotech. Agr. Soc. Environ.* 3, 217–219.

Bollini, R., and Chrispeels, M. (1978). Characterization and subcellular-localization of vicilin and phytohemagglutinin, 2 major reserve proteins of *Phaseolus vulgaris* L. *Planta* 142, 291–298. doi: 10.1007/BF00385080

Boye, J., Zare, F., and Pletch, A. (2010). Pulse proteins: processing, characterization, functional properties and applications in food and feed. *Food Res. Int.* 43, 414–431. doi: 10.1016/j.foodres.2009.09.003

Brand-Miller, J., McMillan-Price, J., Steinbeck, K., and Caterson, I. (2009). Dietary glycemic index: health implications. *J. Am. Coll. Nutr.* 28, 446–449. doi: 10.1080/07315724.2009.10718110

Burnette, W. (1981). Western blotting - electrophoretic transfer of proteins from sodium dodecyl sulfate-polyacrylamide gels to unmodified nitrocellulose and radiographic detection with antibody and radioiodinated protein-A. *Anal Biochem.* 112, 195–203. doi: 10.1016/0003-2697(81)90281-5

Campion, B., Sparvoli, F., Doria, E., Tagliabue, G., Galasso, I., Fileppi, M., et al. (2009). Isolation and characterisation of an *lpa* (low phytic acid) mutant in common bean (*Phaseolus vulgaris* L.). *Theor. Appl. Genet.* 118, 1211–1221. doi: 10.1007/s00122-009-0975-8

Cardador-Martinez, A., Loarca-Pina, G., and Oomah, B. (2002). Antioxidant activity in common beans (*Phaseolus vulgaris* L.). *J. Agr. Food Chem.* 50, 6975–6980. doi: 10.1021/jf020296n

Cerino Badone, F., Amelotti, M., Cassani, E., and Pilu, R. (2012). Study of low Phytic Acid 1-7 (*lpa1*-7), a new ZmMRP4 Mutation in Maize. *J. Hered.* 103, 598–605. doi: 10.1093/jhered/ess014

Ceriotti, A., Vitale, A., and Bollini, R. (1989). Lectin-like proteins accumulate as fragmentation products in bean seed protein bodies. *FEBS Lett.* 250, 157–160. doi: 10.1016/0014-5793(89)80710-0

Chavan, J., and Kadam, S. (1993). Nutritional enrichment of bakery products by supplementation with nonwheat flours. *Crit. Rev. Food Sci. Nutr.* 33, 189–226. doi: 10.1080/10408399309527620

Confalonieri, M., Bollini, R., Berardo, N., Vitale, A., and Allavena, A. (1992). Influence of phytohemagglutinin on the agronomic performance of beans (*Phaseolus vulgaris* L.). *Plant Breed.* 109, 329–334. doi: 10.1111/j.1439-0523.1992.tb00192.x

Costa, G., Queiroz-Monici, K., Reis, S., and De Oliveira, A. (2006). Chemical composition, dietary fibre and resistant starch contents of raw and cooked pea, common bean, chickpea and lentil legumes. *Food Chem.* 94, 327–330. doi: 10.1016/j.foodchem.2004.11.020

Duranti, M. (2006). Grain legume proteins and nutraceutical properties. *Fitoterapia* 77, 67–82. doi: 10.1016/j.fitote.2005.11.008

Elkowicz, K., and Sosulski, F. W. (1982). Antinutritive factors in eleven legumes and their air-classified protein and starch fractions. *J. Food Sci.* 47, 1301–1304. doi: 10.1111/j.1365-2621.1982.tb07673.x

Foster-Powell, K., Holt, S. H. A., and Brand-Miller, J. C. (2002). International table of glycemic index and glycemic load values: 2002. *Amer. J. Clin. Nutr.* 76, 5–56.

Gallegos-Infante, J. S., Bello-Perez, L. A., Rocha-Guzman, N. E., Gonzales-Laredo, R. F., and Avial-Ontiveros, M. (2010). Effect of the addition of common bean (*Phaseolus vulgaris* L.) flour on the *in vitro* digestibility of starch and undigestible carbohydrates in spaghetti. *J. Food Sci.* 75, 151–156. doi: 10.1111/j.1750-3841.2010.01621.x

Giuberti, G., Gallo, A., Cerioli, C., Fortunati, P., and Masoero, F. (2015). Cooking quality and starch digestibility of gluten free pasta using new bean flour. *Food Chem.* 175, 43–49. doi: 10.1016/j.foodchem.2014.11.127

Giuberti, G., Gallo, A., Fiorentini, L., Fortunati, P., and Masoero, F. (2016). *In vitro* starch digestibility and quality attributes of gluten free 'tagliatelle' prepared with teff flour and increasing levels of a new developed bean cultivar. *Starch* 68, 374–378. doi: 10.1002/star.201500007

Granfeldt, Y. (1994). *Foods Factors Affecting Metabolic Responses to Cereal Products.* Ph.D. thesis, University of Lund.

Holding, D. R. (2014). Recent advances in the study of prolamin storage protein organization and function. *Front Plant Sci.* 20:276. doi: 10.3389/fpls.2014.00276

Hoojjat, P., and Zabik, M. (1984). Sugar-snap cookies prepared with wheat-navy bean-sesame seed flour blends. *Cereal Chem.* 61, 41–44.

Hoover, R., and Zhou, Y. (2003). *In vitro* and *in vivo* hydrolysis of legume starches by alpha-amylase and resistant starch formation in legumes - a review. *Carbohydr Polym.* 54, 401–417. doi: 10.1016/S0144-8617(03)00180-2

Kelkar, S., Siddiq, M., Harte, J., Dolan, K., Nyombaire, G., and Suniaga, H. (2012). Use of low-temperature extrusion for reducing phytohemagglutinin activity (PHA) and oligosaccharides in beans (*Phaseolus vulgaris* L.) cv. Navy and Pinto. *Food Chem.* 133, 1636–1639. doi: 10.1016/j.foodchem.2012.02.044

Kumar, S., Verma, A., Das, M., Jain, S., and Dwivedi, P. (2013). Clinical complications of kidney bean (*Phaseolus vulgaris* L.) consumption. *Nutrition* 29, 821–827. doi: 10.1016/j.nut.2012.11.010

Lamacchia, C., Camarca, A., Picascia, S., Di Luccia, A., and Ginafrani, C. (2014). Cereal-based gluten-free food: how to reconcile nutritional and technological properties of wheat proteins with safety for celiac disease patients. *Nutrients* 6, 575–590. doi: 10.3390/nu6020575

Landoni, M., Badone, F., Haman, N., Schiraldi, A., Fessas, D., Cesari, V., et al. (2013). Low phytic acid 1 mutation in maize modifies density, starch properties, cations, and fiber contents in the seed. *J. Agr. Food Chem.* 61, 4622–4630. doi: 10.1021/jf400259h

Laureati, M., Giussani, B., and Pagliarini, E. (2012). Sensory and hedonic perception of gluten-free bread: comparison between celiac and non-celiac subjects. *Food Res. Int.* 46, 326–333. doi: 10.1016/j.foodres.2011.12.020

Layer, P., Carlson, G., and Dimagno, E. (1985). Partially purified white bean amylase inhibitor reduces starch digestion *in vitro* and inactivates intraduodenal amylase in humans. *Gastroenterology* 88, 1895–1902. doi: 10.1016/0016-5085(85)90016-2

Layer, P., Zinsmeister, A., and Dimagno, E. (1986). Effects of decreasing intraluminal amylase activity on starch digestion and postprandial gastrointestinal function in humans. *Gastroenterology* 91, 41–48. doi: 10.1016/0016-5085(86)90436-1

Lioi, L., Sparvoli, F., Galasso, I., Lanave, C., and Bollini, R. (2003). Lectin-related resistance factors against bruchids evolved through a number of duplication events. *Theor. Appl. Genet.* 107, 814–822. doi: 10.1007/s00122-003-1343-8

Livesey, G., Taylor, R., Hulshof, T., and Howlett, J. (2008). Glycemic response and health – a systematic review and meta-analysis: relation between dietary glycemic properties and health outcomes. *Am. J. Clin. Nutr.* 87, 258–226.

Macfie, H. J. H., Bratchell, N., Greenhoff, K., and Vallis, L. V. (1989). Designs to balance the effect of order of presentation and first-order carry-over effects in hall tests. *J. Sens. Stud.* 4, 129–148. doi: 10.1111/j.1745-459X.1989.tb00463.x

Manonmani, D., Bhol, S., and Bosco, S. J. D. (2014). Effect of red kidney bean (*Phaseolus vulgaris* L.) flour on bread quality. *Open Access Libr. J.* 1, 1–6. doi: 10.4236/oalib.1100366

Marshall, J. J., and Lauda, C. M. (1975). Purification and properties of phaseolamin, an inhibitor of α-amylase, from the kidney bean, *Phaseolus vulgaris*. *J. Biol. Chem.* 250, 8030–8037.

Mertens, D., Allen, M., Carmany, J., Clegg, J., Davidowicz, A., Drouches, M., et al. (2002). Gravimetric determination of amylase-treated neutral detergent fiber in feeds with refluxing in beakers or crucibles: collaborative study. *J. AOAC Int.* 85, 1217–1240.

Noah, N., Bender, A., Reaidi, G., and Gilbert, R. (1980). Food poisoning from raw red kidney beans. *Br. Med. J.* 281, 236–237.

Oomah, B., Kotzeva, L., Allen, M., and Bassinello, P. (2014). Microwave and micronization treatments affect dehulling characteristics and bioactive contents of dry beans (*Phaseolus vulgaris* L.). *J. Sci. Food Agr.* 94, 1349–1358. doi: 10.1002/jsfa.6418

Pagliarini, E., Laureati, M., and Lavelli, V. (2010). Sensory evaluation of gluten-free breads assessed by a trained panel of celiac assessors. *Eur. Food Res. Technol.* 231, 37–46. doi: 10.1007/s00217-010-1249-z

Panzeri, D., Cassani, E., Doria, E., Tagliabue, G., Forti, L., Campion, B., et al. (2011). A defective ABC transporter of the MRP family, responsible for the bean *lpa1* mutation, affects the regulation of the phytic acid pathway, reduces seed *myo*-inositol and alters ABA sensitivity. *New Phytol.* 191, 70–83. doi: 10.1111/j.1469-8137.2011.03666.x

Pellegrini, N., and Agostoni, C. (2015). Nutritional aspects of gluten-free products. *J. Sci. Food Agric.* 95, 2380–2385. doi: 10.1002/jsfa.7101

Petroni, K., Pilu, R., and Tonelli, C. (2014). Anthocyanins in corn: a wealth of genes for human health. *Planta* 240, 901–911. doi: 10.1007/s00425-014-2131-1

Petry, N., Boy, E., Wirth, J. P., and Hurrell, R. F. (2015). Review: the potential of the common bean (*Phaseolus vulgaris*) as a vehicle for iron biofortification. *Nutrients* 11, 1144–1173. doi: 10.3390/nu7021144

Petry, N., Egli, I., Campion, B., Nielsen, E., and Hurrell, R. (2013). Genetic reduction of phytate in common bean (*Phaseolus vulgaris* L.) seeds increases iron absorption in young women. *J. Nutr.* 143, 1219–1224. doi: 10.3945/jn.113.175067

Piergiovanni, A. R., and Lioi, L. (2010). Italian common bean landraces: history, genetic diversity and seed quality. *Diversity* 2, 837–862. doi: 10.3390/d2060837

Pilu, R., Panzeri, D., Gavazzi, G., Rasmussen, S., Consonni, G., and Nielsen, E. (2003). Phenotypic, genetic and molecular characterization of a maize low phytic acid mutant (*lpa241*). *Theor. Appl. Genet.* 107, 980–987. doi: 10.1007/s00122-003-1316-y

Pusztai, A., Clarke, E., King, T., and Stewart, J. (1979). Nutritional-evaluation of kidney beans (*Phaseolus vulgaris*) - chemical composition, lectin content and nutritional-value of selected cultivars. *J. Sci. Food Agr.* 30, 843–848. doi: 10.1002/jsfa.2740300902

Pusztai, A., Grant, G., and Stewart, J. (1981). A new type of *Phaseolus vulgaris* (cv Pinto-III) seed lectin - isolation and characterization. *Biochim. Biophys. Acta* 671, 146–154. doi: 10.1016/0005-2795(81)90128-8

Rababah, T., Al-Mahasneh, M., and Ereifej, K. (2006). Effect of chickpea, broad bean, or isolated soy protein additions on the physicochemical and sensory properties of biscuits. *J. Food Sci.* 71, S438–S442. doi: 10.1111/j.1750-3841.2006.00077.x

Rodhouse, J., Haugh, C., Roberts, D., and Gilbert, R. (1990). Red kidney bean poisoning in the UK - an analysis of 50 suspected incidents between

1976 and 1989. *Epidemiol. Infect.* 105, 485–491. doi: 10.1017/S0950268800 04810X

Sandhu, K., and Lim, S. (2008). Digestibility of legume starches as influenced by their physical and structural properties. *Carbohyd. Polym.* 71, 245–252. doi: 10.1016/j.carbpol.2007.05.036

Santimone, M., Koukiekolo, R., Moreau, Y., Le Berre, V., Rouge, P., Marchis-Mouren, G., et al. (2004). Porcine pancreatic alpha-amylase inhibition by the kidney bean (*Phaseolus vulgaris*) inhibitor (alpha-AI1) and structural changes in the alpha-amylase inhibitor complex. *Biochim. Biophys. Acta* 1696, 181–190. doi: 10.1016/j.bbapap.2003.11.001

Schlemmer, U., Frølich, W., Prieto, R. M., and Grases, F. (2009). Phytate in foods and significance for humans: food sources, intake, processing, bioavailability, protective role and analysis. *Mol. Nutr. Food Res.* 53(Suppl. 2), S330–S375. doi: 10.1002/mnfr.200900099

Shrestha, A., and Noomhorm, A. (2002). Comparison of physico-chemical properties of biscuits supplemented with soy and kinema flours. *Int. J. Food Sci. Technol.* 37, 361–368. doi: 10.1046/j.1365-2621.2002. 00574.x

Siddiq, M., Kelkar, S., Harte, J., Dolan, K., and Nyombaire, G. (2013). Functional properties of flour from low-temperature extruded navy and pinto beans (*Phaseolus vulgaris* L.). *Lwt-Food Sci. Technol.* 50, 215–219. doi: 10.1016/j.lwt.2012.05.024

Spadafranca, A., Rinelli, S., Riva, A., Morazzoni, P., Magni, P., Bertoli, S., et al. (2013). *Phaseolus vulgaris* extract affects glycometabolic and appetite control in healthy human subjects. *Br. J. Nutr.* 109, 1789–1795. doi: 10.1017/S0007114512003741

Sparvoli, F., Bollini, R., and Cominelli, E. (2015). "Nutritional value," in *Grain Legumes,* ed A. De Ron (New York, NY: Springer Science+Business Media), 291–326.

Sparvoli, F., Daminati, M., Cantoni, R., and Bollini, R. (1997). "Possible role(s) of glycosylation on the biological properties of lectins and related proteins," in *Biomedical and Therapeutical Applications of Dietary Lectins. COST Action 98, 7th Workshop, Vol. 6,* eds S. Bardzoc and A. Pusztai (Lake Balaton: European Commission), 43–49.

Szafranski, M., Whittington, J. A., and Bessinger, C. (2005). Pureed cannellini beans can be substituted for shortening in brownies. *J. Am. Diet. Assoc.* 105, 1295–1298. doi: 10.1016/j.jada.2005.05.002

Trugo, L., and von Baer, D. (1998). *Analytical Methods for the Analysis of Antinutritional Factors in Legume Seeds.* Wageningen: Wageningen Press.

Vansoest, P., Robertson, J., and Lewis, B. (1991). Methods for dietary fiber, neutral detergent fiber, and nonstarch polysaccharides in relation to animal nutrition. *J. Dairy Sci.* 74, 3583–3597. doi: 10.3168/jds.S0022-0302(91)78551-2

Varner, J. E., and Mense, R. M. (1972). Characteristics of the process of enzyme release from secretory plant cells. *Plant Physiol.* 49, 187–189. doi: 10.1104/pp.49.2.187

Vaz Patto, M. C., Amarowicz, R., Aryee, A. N. A., Boye, J. I., Chung, H. J., Martín-Cabrejas, M. A., et al. (2015). Achievements and challenges in improving the nutritional quality of food legumes. *Crit. Rev. Plant Sci.* 34, 105–143. doi: 10.1080/07352689.2014.897907

Voelker, T., Staswick, P., and Chrispeels, M. (1986). Molecular analysis of 2 phytohemagglutinin genes and their expression in *Phaseolus vulgaris* cv Pinto, a lectin-deficient cultivar of the bean. *EMBO J.* 5, 3075–3082.

Von Grebmer, K., Saltzman, A., Birol, E., Wiesmann, D., Prasai, N., Yin, S., et al. (2014). *2014 Global hunger index: The Challenge of Hidden Hunger.* Bonn; Washington, DC; Dublin: Welthungerhilfe; International Food Policy Research Institute; Concern Worldwide.

Zhang, B., Dhital, S., and Gidley, M. J. (2015). Densely packed matrices as rate determining features in starch hydrolysis. *Trends Food Sci. Technol.* 43, 18–31. doi: 10.1016/j.tifs.2015.01.004

Zucco, F., Borsuk, Y., and Arntfield, S. (2011). Physical and nutritional evaluation of wheat cookies supplemented with pulse flours of different particle sizes. *Lwt Food Sci. Technol.* 44, 2070–2076. doi: 10.1016/j.lwt.2011.06.007

Conflict of Interest Statement: The authors declare that the research was conducted in the absence of any commercial or financial relationships that could be construed as a potential conflict of interest.

APPENDIX

TABLE A1 | Percent amino acid variations in biscuit samples containing different amounts of cv. Lady Joy bean flour compared to control biscuits.

Sample	Thr	Lys	Trp	Cys	Met	Met+Cys	Phe	Val	His	Ile	Asp	Glu	Pro	Gly	Ala	Leu	Tyr	Arg	Proteins
B12	10.09	118.06	39.02	−2.86	20.57	7.59	30.87	26.01	−3.93	25.08	30.24	−1.53	0.21	19.40	15.19	12.05	17.33	84.14	14.86
B14	36.80	236.81	134.15	−5.71	22.70	6.96	55.67	43.75	41.92	57.59	41.43	16.74	35.85	49.16	21.01	21.33	26.00	83.82	37.49
B22	40.06	215.28	41.46	−10.86	34.04	9.18	51.72	9.68	54.15	47.06	78.32	−17.29	39.57	35.45	46.84	39.20	35.00	80.26	28.91
B24	40.36	240.97	95.12	−4.00	31.21	11.71	55.94	52.62	53.71	51.08	79.20	−4.38	52.13	46.49	39.49	30.06	33.67	93.85	36.11
B29	57.86	275.00	104.88	−6.29	37.59	13.29	73.09	45.97	61.57	64.40	99.48	−4.89	48.09	56.86	55.44	47.23	46.00	108.09	46.86
Average (B12–B29)	37.03	217.22	82.93	−5.94	29.22	9.75	53.46	35.60	41.48	49.04	65.73	−2.27	35.17	41.47	35.59	29.97	31.60	90.03	32.85

12

Genetic Diversity of Cultivated Lentil (*Lens culinaris* Medik.) and its Relation to the World's Agro-ecological Zones

Edited by:
Antonio M. De Ron,
Spanish National Research Council,
Spain

Reviewed by:
Matthew Nicholas Nelson,
University of Western Australia,
Australia
Sripada M. Udupa,
ICARDA, Morocco

***Correspondence:**
Kirstin E. Bett
k.bett@usask.ca

Hamid Khazaei[1], *Carolyn T. Caron*[1], *Michael Fedoruk*[1], *Marwan Diapari*[1,2], *Albert Vandenberg*[1], *Clarice J. Coyne*[3], *Rebecca McGee*[3] *and Kirstin E. Bett*[1*]

[1] Department of Plant Sciences, University of Saskatchewan, Saskatoon, SK, Canada, [2] London Research and Development Centre, Agriculture and Agri-Food Canada, London, ON, Canada, [3] USDA-ARS, Pullman, WA, USA

Assessment of genetic diversity and population structure of germplasm collections plays a critical role in supporting conservation and crop genetic enhancement strategies. We used a cultivated lentil (*Lens culinaris* Medik.) collection consisting of 352 accessions originating from 54 diverse countries to estimate genetic diversity and genetic structure using 1194 polymorphic single nucleotide polymorphism (SNP) markers which span the lentil genome. Using principal coordinate analysis, population structure analysis and UPGMA cluster analysis, the accessions were categorized into three major groups that prominently reflected geographical origin (world's agro-ecological zones). The three clusters complemented the origins, pedigrees, and breeding histories of the germplasm. The three groups were (a) South Asia (sub-tropical savannah), (b) Mediterranean, and (c) northern temperate. Based on the results from this study, it is also clear that breeding programs still have considerable genetic diversity to mine within the cultivated lentil, as surveyed South Asian and Canadian germplasm revealed narrow genetic diversity.

Keywords: lentil, genetic diversity, population structure, germplasm, SNP markers, agro-ecological zones

INTRODUCTION

Cultivated lentil (*Lens culinaris* Medikus ssp. *culinaris*) is the third most important cool-season grain legume in the world after chickpea (*Cicer arietinum* L.) and pea (*Pisum sativum* L.) (FAO, 2015). Legumes are important components in farming systems, providing environmental and ecological benefits through crop rotation, especially by contributing to soil fertility and rhizosphere diversity through biological N_2 fixation. Global annual lentil production was around 5 million metric tons (Tg) from nearly 4.3 million ha (Mha) in 2013. Canada was the largest producer, contributing 38% of the world's production, followed by India, Turkey, and Australia (FAO, 2015). Lentil was one of the first domesticated grain legumes, originating from the Near East center of origin (Zohary, 1999). Lentil subsequently spread to central Asia and the Mediterranean Basin (Cubero, 1981; Lev-Yadun et al., 2000). It is a relatively new crop in North America, first introduced into northwest USA in the 1930s and into the northern temperate prairies of North America in the late 1960s (Muehlbauer et al., 1995). Globally today, lentil is grown in three major distinct agro-ecological zones: Mediterranean, sub-tropical savannah, and northern temperate (Tullu et al., 2011). These zones each exhibit different day lengths and temperatures, which limits the exchange of germplasm between agro-ecological adaptation zones.

Success in crop breeding is a function of heritability, genetic diversity, and selection. Natural agro-biodiversity stored in genebanks can be used to expand the diversity in crops. These collections are a vital source for discovering useful genes/alleles, which serve as a cornerstone for any pre-breeding program. There are currently 58,405 *Lens* accessions held in various genebanks worldwide (FAO, 2010). International Center for Agricultural Research in the Dry Areas (ICARDA) hosts the largest collection (19%) followed by the National Bureau of Plant Genetic Resources, India (17%) and the Australian temperate field crops collection (9%). Currently, the most accessible and accessed lentil collection is held by the USDA-ARS (United States Department of Agriculture—Agricultural Research Service; https://npgsweb.ars-grin.gov/).

Assessments of genetic diversity and relationships among preserved germplasm have important implications both for facilitating reliable documentation of genetic resources and for identifying material with possible utility for specific breeding purposes, particularly in cultivated lentil and other species with a narrow genetic base. Lentil is an autogamous diploid species with seven chromosome pairs and a relatively large genome of ~4 Gbp in the haploid complement (Arumuganathan and Earle, 1991). Considerable genetic diversity has been reported in *Lens* genetic resources for agro-morphological and phenological characteristics (e.g., Erskine and Choudhary, 1986; Erskine et al., 1989; Lazaro et al., 2001; Zaccardelli et al., 2012; Cristóbal et al., 2014). Molecular markers, as the more reliable and powerful of genetic tools, have been deployed to characterize lentil genetic resources. Various molecular marker techniques and types have been used for this purpose. These include restriction fragment length polymorphisms (RFLPs, Havey and Muehlbauer, 1989), amplified fragment length polymorphisms (AFLPs, Sharma et al., 1996; Alghamdi et al., 2014), random amplified polymorphic DNAs (RAPDs, Abo-Elwafa et al., 1995; Ford et al., 1997; Ferguson et al., 1998; Sonnante and Pignone, 2001), and inter simple sequence repeats (ISSRs, Scippa et al., 2008; Toklu et al., 2009; El-Nahas et al., 2011). Simple sequence repeats (SSRs) have the most widely used DNA markers for assessing genetic diversity in lentil (see Liu et al., 2008; Babayeva et al., 2009; Kaur et al., 2011; Zaccardelli et al., 2012; Dikshit et al., 2015; Idrissi et al., 2015).

In recent years, genome-wide nucleotide-level surveys from different individuals within or across species have received increasing emphasis (Yang et al., 2015). Development of gene-based single nucleotide polymorphisms (SNP) markers is effective for detecting genetic diversity in plant species (Frascaroli et al., 2013; Semagn et al., 2014). SNPs are the most abundant type of polymorphism in all genomes, which allows high-throughput genotyping that is low cost, locus specific, and co-dominant with simple documentation. So far, only a limited number of SNP markers have been used to study the genetic diversity in lentil (Lombardi et al., 2014; Basheer-Salimia et al., 2015). The population structure of global lentil accessions has not been extensively characterized based on their agro-ecological adaptation zones. The main aims of this study were to assess the population structure and genetic variation of a group of 352 lentil germplasm accessions of Canadian breeding lines (northern temperate adaptation) and *ex situ* germplasm collections of a diverse origin using a relatively large number of SNP markers spanning the genome.

MATERIALS AND METHODS

Plant Material and DNA Extraction

A total of 352 lentil accessions originating from 54 countries were collected from various sources including breeding lines obtained from the Crop Development Centre (CDC) collection in Saskatoon, Canada, from ICARDA, and the USDA-ARS. The accession numbers and origins of the selected accessions are given in Supplementary Table 1. Accessions were assigned to different major agro-ecological zones: Mediterranean, sub-tropical savannah (particularly northeast India, Nepal's lowland, and western Bangladesh), and Northern temperate according to global agro-ecological zones v3.0 (IIASA/FAO, 2012). In the Mediterranean adaptation zone, sowing occurs after the autumn equinox following a hot dry summer, prior to the winter solstice. Similarly, in the sub-tropical savannah (South Asia) planting occurs after the autumn equinox to take advantage of declining day lengths and temperatures during the juvenile phase and increases during the reproductive phase. Temperate adaptation zones require planting to occur after the spring equinox following a cold winter, prior to the summer solstice. Germplasm originating from Iran and Turkey were not assigned to specific agro-ecological zones due to different agro-ecological climates within the country borders for lentil production areas and a lack of specific collection location that would facilitate this classification. Furthermore, ICARDA breeding lines and USDA lines designated W6 # (hereafter referred to as USA breeding lines) were not assigned to an agro-ecological zones, since they were not specifically attributed a particular zone.

Canadian lines, along with ICARDA germplasm, were grown in Saskatoon, Canada and leaf tissues from at least five different plants were collected for genomic DNA extraction using a modified CTAB extraction method (Doyle and Doyle, 1990). DNA samples from leaf samples of greenhouse grown plants were provided by the USDA-ARS for the USA germplasm (PI and W6 lines; see Simon and Hannan, 1995).

SNP Discovery and Genotyping

The Lc1536 GoldenGate high-throughput assay (Illumina, San Diego, CA) described by Sharpe et al. (2013) was used to genotype the 352 lentil accessions. The SNP genotyping was performed on an Illumina BeadStation 500G (Illumina, San Diego, CA) at the National Research Council (NRC), in Saskatoon, Canada with the protocol supported by Illumina (Fan et al., 2003). A robust set of 1440 of the SNP markers was used for further analyses.

Statistical Analysis

The SNP marker data were analyzed using PowerMarker v. 3.25 (Liu and Muse, 2005) to calculate minor allele frequency (MAF), heterozygosity, gene diversity, and polymorphic information content (PIC).

Population Structure and Genetic Diversity

The program STRUCTURE v. 2.3.4 (Pritchard et al., 2000) was used to calculate the most probable number of sub-populations (K) in the panel. Five independent runs were conducted for each K ranging from 1 to 10 with both a burn-in time and Markov Chain Monte Carlo (MCMC) replication number of 500,000 using an admixture model. Selection of the best K-value was based on the procedure presented in Evanno et al. (2005) by submitting the results for each K to the STRUCTURE HARVESTER website, which returned the $L(K)$ and ΔK-value (Earl and vonHoldt, 2012), as well as a basic understanding of the nature of the germplasm. The results from STRUCTURE were presented at the country level, using the "rworldmap" package (South, 2011) in R (R Core Development Team, 2015).

Genetic structure of the lentil population was analyzed by performing principal coordinate analysis (PCoA) using GenAlEx v. 6.5 (Peakall and Smouse, 2012) based on standardized covariance of genetic distance for co-dominant markers. GenAlEx v. 6.5 was also used to calculate Analysis of Molecular Variance (AMOVA) among and within assigned groups. The genetic distance between genotypes and countries were computed using Nei's standard genetic distance (Nei, 1973) with PowerMarker software. A dendrogram was constructed from Nei's distance matrix using UPGMA and the resulting tree was visualized using iTOL v. 3.0 (Letunic and Bork, 2011).

The sequences of all the markers used in this study are described in Sharpe et al. (2013). All data are also available in KnowPulse (http://knowpulse.usask.ca) and Supplementary Table 2.

RESULTS

SNP Markers Information

Of 1400 SNP markers, 5.46% were monomorphic among the accessions, 2.14% generated ambiguous products, and 7.21% were rejected on the grounds of excess missing data points (>30% missing data per marker). This filtering resulted in 1194 high quality polymorphic SNP markers for use in the clustering analyses. Around 40% of these SNP markers were previously mapped and shown to be evenly distributed throughout the genome (Sharpe et al., 2013). The overall PIC value was 0.3092 $\pm$ 0.0789. Measurements of the average observed heterozygosity over all loci and gene diversity were 0.0375 $\pm$ 0.0755 and 0.3932 $\pm$ 0.1160, respectively (Supplementary Table 3).

Population Structure and Genetic Relationship

We ran STRUCTURE for a range of K (number of fixed subgroups or cluster) from 1 to 10 on the entire set of accessions. The estimated log probability of the data [Ln$P(K)$] for each K plateaued at $K = 3$. The maximum ΔK-value was also reached at $K = 3$ (Supplementary Figure 1) suggesting three distinct groups. Geographical distribution of the 352 lentil accessions along with their projected population structure are shown in **Figure 1**. These three clusters closely reflected the origins, pedigrees and breeding history of germplasm used in this study. Accessions collected from southern Asia and the Middle East were assigned to the same gene pool. Lentil accessions originating mainly from the Mediterranean, northeast Africa (along the Nile valley from Egypt to Ethiopia), and South America were assigned to the second group, whereas the third group consisted mostly of genotypes from northern latitudes (Canada and Russian). Results from principal coordinates analysis (**Figure 2**) were consistent with those of STRUCTURE and UPGMA cluster analysis by revealing three clusters (Supplementary Figure 2). The AMOVA based on PhiPT-values revealed that genetic variation mainly occurred within groups (86%), while the variation between the groups was 14% (Supplementary Table 4).

Based on the 1194 polymorphic SNP markers, three pairs of accessions were genetically indistinguishable despite having different accession numbers: PI 163589 and PI 320945, PI 431675 and PI 431731, and PI 297284 and PI 297285.

Based on Nei's genetic distance matrix, a close relationship exists between Indian material and germplasm from Nepal, Pakistan, and Afghanistan (0.0575, 0.0202, and 0.1310, respectively). Generally, South Asian and Middle Eastern (i.e., Iran and Turkey) germplasm grouped together, however, some of the Iranian and Turkish germplasm skewed to those from Mediterranean and northern climates. Within the second group, germplasm from Chile exhibited the closest relationship to germplasm from Morocco (0.0828). Canadian cultivars and breeding lines, representative of group 3, were closely related to much of the "W6" numbered lines which used in USDA lentil breeding program as parents in recombinant inbred lines development (USA breeding materials; 0.0968) and Russian (0.0961) germplasm (Supplementary Table 5). Most of the Canadian breeding lines were clustered together with the exception of CDC Plato, CDC Imigreen and CDC Cherie. Conversely, USA breeding lines and ICARDA breeding lines were relatively well distributed among the three groups (**Figure 2**).

DISCUSSION

The deep population structure of cultivated lentil and its importance in explaining genetic diversity underscores the value of using global lentil genetic resources to broaden the genetic base in breeding programs and to improve our knowledge of adaptation in this species. Given the sample size (individuals and country of origins) and relatively high marker density, this study identified three major clusters of germplasm reflecting the geographical origins, pedigrees, and breeding history of accessions. We categorized these clusters as (a) subtropical, South Asian, (b) Mediterranean, including Southern Europe and North Africa, and (c) northern temperate climates. These three groups reflect the main climatic regions in which lentil is widely grown as described by Tullu et al. (2011).

Lentil accessions from South Asia (primarily Nepal, India, and Pakistan) have a narrow genetic base and are genetically more isolated relative to other origins included in our study. This is

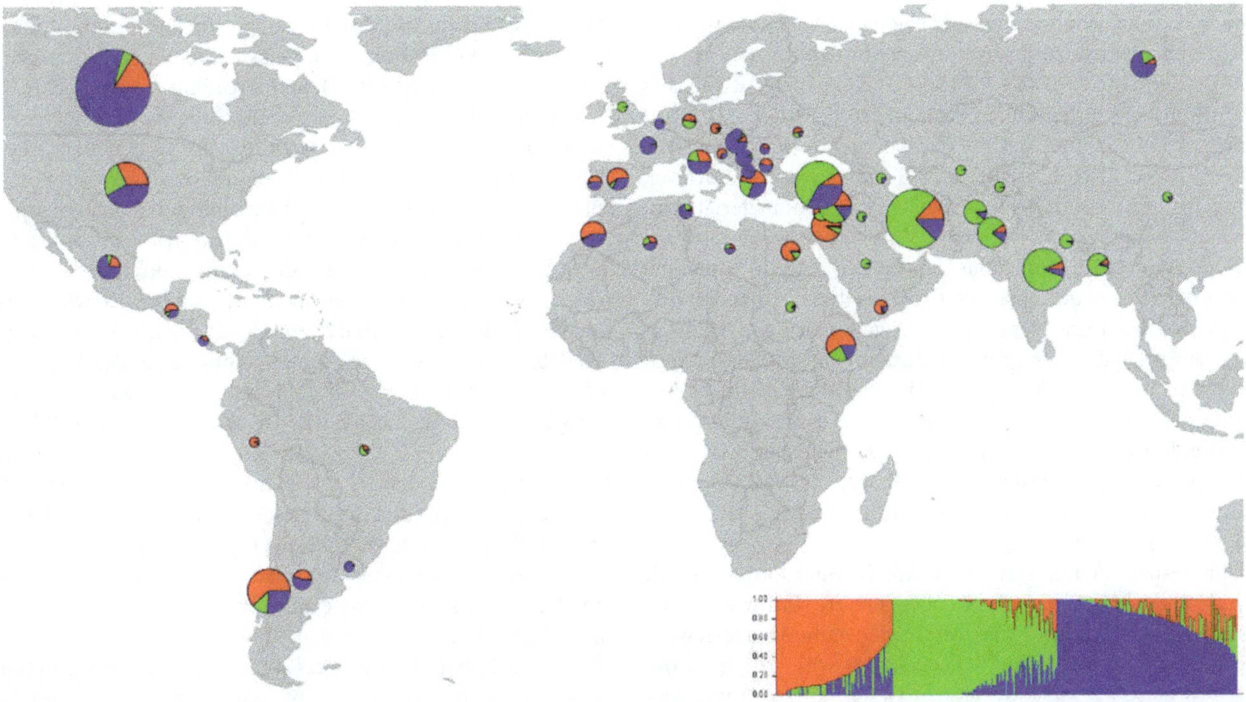

FIGURE 1 | Map of the world showing the country-specific distribution of 352 lentil accessions based on population structure ($K = 3$). The size and color of pie chart is corresponding to sample size and the percentage of samples in each group, respectively.

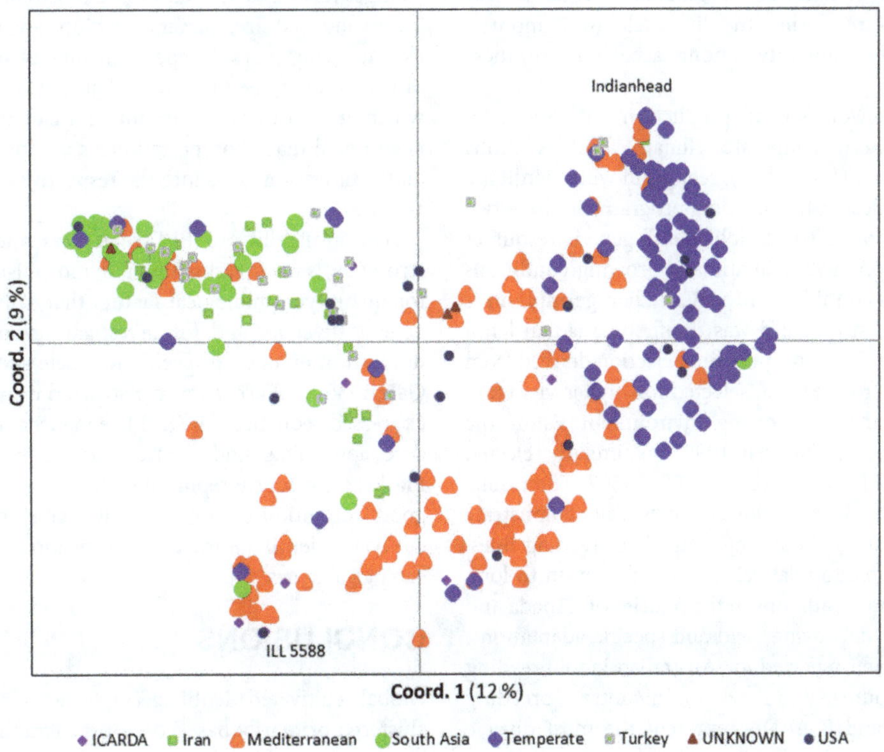

FIGURE 2 | Principal coordinate analysis (PCoA) of the 352 lentil accessions using 1194 polymorphic SNP markers. Each symbol/color combination represents different country of origin. The most two distinct accessions, Indianhead and ILL 5588, are labeled.

likely a result of their specific phenological adaptation to the sub-tropical savannah environment and a potential genetic bottleneck during a time when lentil was introduced to South Asia around 2000 BC (Erskine et al., 1989). Most previous diversity studies of lentil diversity using molecular markers to date revealed two distinct groups: South Asian and all other origins (Ferguson et al., 1998; Alo et al., 2011; Kumar et al., 2014). In our study, however, germplasm from South Asia grouped with parts of material from Iran and Turkey, likely a stopping point on their move eastward on trade routes extending from the center of origin. A manual inspection of genetic distances between these regions shows that germplasm from Nepal, India, Afghanistan, and Pakistan are closely related, while relatively distant from Middle Eastern germplasm (Iran and Turkey). This implies there might be two distinct sub-groups within this particular cluster (South Asian and Middle Eastern). Germplasm from Afghanistan was found to be related to germplasm from South Asia as was previously reported using another set of molecular markers (e.g., Ferguson et al., 1998; Kumar et al., 2014).

Our results demonstrate that the Mediterranean, North African, and Chilean germplasm collections are similar with only a few deviations, following the classification of the Mediterranean agro-ecological zone. This is consistent with results from Ferguson et al. (1998) and Lombardi et al. (2014). Lentil was domesticated in the Eastern Mediterranean around 7–8 BC, after which it spread to Europe. Lentil was introduced to South America by the Spanish via Chile (after 1500 AD). Lentil is mainly grown as a winter crop in these regions, under conditions of declining day length followed by gradual increase in day length and temperature during the life cycle (6–7 month). This may explain the similarity among accessions of these regions.

Nearly 50% of the world's lentil production now originates from Canada (northern temperate climate) and Australia (Mediterranean climate) (FAO, 2015), regions in which lentil is a relatively new crop. Clearly, the breeding programs in these two regions have drawn from distinct global lentil genetic resources that originate in regions with climatic and growing conditions that match their local conditions. Most Canadian germplasm is related to Laird and Eston. Laird was the first Canadian lentil cultivar, released in 1978. It was a pure line selection derived from PI 343028, originally from Russia, selected for higher yield and larger seeds in the Canadian growing environment. Eston, the second cultivar released in Canada in 1980, was similarly selected for adaptation from a Turkish accession, PI 179307. The results from PCoA and genetic distance cluster analyses all demonstrate a narrower genetic variability among Canadian breeding lines. This may be attributable to relatively recent adaptation to long day northern temperate conditions in the prairies of Canada and selection pressure for improving yield and specific adaptations. A similar trend has been reported for Australian lentil breeding lines and cultivars (Lombardi et al., 2014). In contrast, breeding lines from the USA and ICARDA represent the most diverse material in this study and elsewhere (Alghamdi et al., 2014). ICARDA breeding strategies are more internationally-focused, covering a wide variety of regions and adaptive traits as part of CGIAR's policy. For example, ILL 7502 and ILL 7537 were bred for Western Asian climates, while ILL 8008 was targeted for South Asian climates (Shiv Kumar, personal communication).

Lentil domestication already has led to ~40% loss in genetic diversity (Alo et al., 2011) and breeding in specific regions has narrowed this even more. Breeding new genotypes for traits of interest requires sampling broad genetic diversity. The various statistic methods we employed here support the presence of considerable genetic variability in global germplasm that is not being accessed in some regions. The results from population structure and PCoA in this study show, to some extent, a separation by origin of the accessions with closely related pedigrees typically group together. Phenotyping of available genetic diversity has demonstrated the importance of incorporating exotic germplasm into breeding programs focusing on biotic and abiotic stresses. For example, ILL 5588 (also known as PI 592998), an ICARDA accession originally collected from Jordan, is a known source of Ascochyta blight resistance in lentil (Erskine et al., 1996). It is clearly distinct from most of the temperate germplasm (**Figure 2**) but it has been used in the pedigrees of some Canadian lines, including CDC Plato and CDC Cherie. This may explain why they did not cluster with the other Canadian lines.

The grouping of some accessions outside of their geographic origin may be the result of outcrossing, migration, and adaptation during cultivation of the crop by local farmers. For example, Moroccan germplasm expresses slightly less variation compared to those from Turkey due to narrower environmental conditions (Idrissi et al., 2015). The major agro-morphological changes related to adaptation are mostly improvements to yield, increasing seed size, tolerance to biotic and abiotic stresses as well as improving market-dependent quality traits. Another source of division may be the growth habit of spring and winter types, which are most adapted to different climatic regions. It has also been noted that photoperiod plays a critical role to characterize lentil cultivation areas into the respective climatic regions (Tullu et al., 2011).

The availability of EST sequences and SNP discovery are strong tools for investigating polymorphism in different species, for quantifying biological factors that influence the patterns of genetic diversity and for investigating bottlenecks due to the domestication of crop species. An allele-specific Illumina Golden Gate 1536-SNP array was constructed using SNPs discovered in expressed sequence tag (EST) sequences from nine *L. culinaris* accessions. This study has confirmed that the sub-set of SNP markers previously reported by Sharpe et al. (2013) can provide good resolution at low cost for genetic characterization of cultivated lentil germplasm in relation to the world's agro-ecological zones.

CONCLUSIONS

Global cultivated lentil germplasm selected for this study clustered primarily based on eco-geographical origin into three basic groups: subtropical savannah, Mediterranean, and northern temperate. The narrow genetic base of some groups of germplasm (e.g., Canadian and South Asian) raises concern over the loss/penalty in yield due to biotic and abiotic stresses, particularly with the threat to global food security from climate change. This highlights the importance of harnessing the potential of lentil wild species in breeding programs by introgression of favorable genes from other regions. Based on the results from this study, it is also clear that breeding programs still have a lot of genetic diversity to mine within the cultivated species.

AUTHOR CONTRIBUTIONS

KB and AV designed the research; HK, and CTC analyzed data; MD and MF contributed to data generation; CJC and RM contributed germplasm and reagents; HK, CTC, KB, and AV contributed to writing of the manuscript.

ACKNOWLEDGMENTS

This research was supported by Natural Sciences and Engineering Research Council of Canada (NSERC) and the Saskatchewan Pulse Growers. The authors would like to acknowledge Rob Stonehouse and Lacey-Anne Sanderson for their kind technical support during the experiments.

REFERENCES

Abo-Elwafa, A., Muraik, K., and Shimada, T. (1995). Intra- and inter-specific variation in *Lens* species revealed by RAPD markers. *Theor. Appl. Genet.* 90, 335–340. doi: 10.1007/BF00221974

Alghamdi, S. S., Khan, A. M., Ammar, M. H., El-Harty, E. H., Migdadi, H. M., Abd El-Khalik, S. M., et al. (2014). Phenological, nutritional and molecular diversity assessment among 35 introduced lentil (*Lens culinaris* Medik.) genotypes grown in Saudi Arabia. *Int. J. Mol. Sci.* 15, 277–295. doi: 10.3390/ijms15010277

Alo, F., Furman, B. J., Akhunov, E., Dvorak, J., and Gepts, P. (2011). Leveraging genomic resources of model species for the assessment of diversity and phylogeny in wild and domesticated lentil. *J. Hered.* 102, 315–329. doi: 10.1093/jhered/esr015

Arumuganathan, K., and Earle, E. D. (1991). Nuclear DNA content of some important plant species. *Plant Mol. Biol.* 9, 208–218. doi: 10.1007/BF02672069

Babayeva, S., Akparov, Z., Abbasov, M., Mammadov, A., Zaifizadeh, M., and Street, K. (2009). Diversity analysis of Central Asia and Caucasian lentil (*Lens culinaris* Medik.) germplasm using SSR fingerprinting. *Genet. Resour. Crop. Evol.* 56, 293–298. doi: 10.1007/s10722-009-9414-6

Basheer-Salimia, R., Camilli, B., Scacchi, S., Noli, E., and Awad, M. (2015). Assessment of genetic diversity in lentils (*Lens culinaris* Medik.) based on SNPs. *Genet. Mol. Res.* 14, 5870–5878. doi: 10.4238/2015.June.1.4

Cristóbal, M. D., Pando, V., and Herrero, B. (2014). Morphological characterization of lentil (*Lens culinaris* Medik.) landraces from Castilla Y León, Spain. *Pak. J. Bot.* 46, 1373–1380.

Cubero, J. I. (1981). "Origin, taxonomy and domestication," in *Lentils*, eds C. Webb and G. Hawtin (London: Commonwealth Agricultural Bureaux), 15–38.

Dikshit, H. K., Singh, A., Singh, D., Aski, M. S., Prakash, P., Jain, N., et al. (2015). Genetic diversity in Lens species revealed by EST and genomic simple sequence repeat analysis. *PLoS ONE* 10:0138101. doi: 10.1371/journal.pone.0138101

Doyle, J. J., and Doyle, J. L. (1990). Isolation of plant DNA from fresh tissue. *Focus* 12, 13–15.

Earl, D. A., and vonHoldt, B. M. (2012). STRUCTURE HARVESTER: a website and program for visualizing STRUCTURE output and implementing the Evanno method. *Conserv. Genet. Resour.* 4, 359–361. doi: 10.1007/s12686-011-9548-7

El-Nahas, A., El-Shazly, H., Ahmed, S., and Omran, A. (2011). Molecular and biochemical markers in some lentil (*Lens culinaris* Medik.) genotypes. *Ann. Agric. Sci.* 56, 105–112. doi: 10.1016/j.aoas.2011.11.001

Erskine, W., Adham, Y., and Holly, L. (1989). Geographic distribution of variation in quantitative traits in a world lentil collection. *Euphytica* 43, 97–103. doi: 10.1007/BF00037901

Erskine, W., Bayaa, B., and Saxena, M. C. (1996). Registration of ILL 5588 lentil germplasm resistant to vascular wilt and ascochyta blight. *Crop Sci.* 36, 1080. doi: 10.2135/cropsci1996.0011183x0036000400055x

Erskine, W., and Choudhary, M. A. (1986). Variation between and within lentil landraces from Yemen republic. *Euphytica* 35, 695–700. doi: 10.1007/BF00028577

Evanno, G., Regnaut, S., and Goudet, J. (2005). Detecting the number of clusters of individuals using the software structure: a simulation study. *Mol. Ecol.* 14, 2611–2620. doi: 10.1111/j.1365-294X.2005.02553.x

Fan, J. B., Oliphant, A., Shen, R., Kermani, B. G., Garcia, F., Gunderson, K. L., et al. (2003). Highly parallel SNP genotyping. *Cold Spring Harb. Symp. Quant. Biol.* 68, 69–78. doi: 10.1101/sqb.2003.68.69

FAO (2010). *The Second Report on the State of the World's Plant Genetic Resources for Food and Agriculture*. Rome.

FAO (2015). *FAOSTAT. Food and Agriculture Organization of the United Nations*. Rome. Available online at: http://faostat.fao.org

Ferguson, M. E., Robertson, L. D., Ford-Lloyd, B. V., Newbury, H. J., and Maxted, N. (1998). Contrasting genetic variation amongst lentil landraces from different geographical origins. *Euphytica* 102, 265–273. doi: 10.1023/A:1018331432580

Ford, R., Pang, E. C. K., and Taylor, P. W. J. (1997). Diversity analysis and species identification in lens using PCR generated markers. *Euphytica* 96, 247–255. doi: 10.1023/A:1003097600701

Frascaroli, E., Schrag, T. A., and Melchinger, A. E. (2013). Genetic diversity analysis of elite European maize (*Zea mays* L.) inbred lines using AFLP, SSR, and SNP markers reveals ascertainment bias for a subset of SNPs. *Theor. Appl. Genet.* 126, 133–141. doi: 10.1007/s00122-012-1968-6

Havey, M. J., and Muehlbauer, F. J. (1989). Variability for restriction fragment lengths and phylogenies in lentil. *Theor. Appl. Genet.* 77, 839–843. doi: 10.1007/BF00268336

Idrissi, O., Udupa, S. M., Houasli, C., De Keyser, E., Van Damme, P., and De Riek, J. (2015). Genetic diversity analysis of Moroccan lentil (*Lens culinaris* Medik.) landraces using simple sequence repeat and amplified fragment length polymorphisms reveals functional adaptation towards agro-environmental origins. *Plant Breed* 134, 322–332. doi: 10.1111/pbr.12261

IIASA/FAO (2012). *Global Agro-ecological Zones (GAEZv3.0)*. Rome; Laxenburg: IIASA: FAO.

Kaur, S., Cogan, N., Pembleton, L., Shinozuka, M., Savin, K., et al. (2011). Transcriptome sequencing of lentil based on second-generation technology permits large-scale unigene assembly and SSR marker discovery. *BMC Genomics* 12:265. doi: 10.1186/1471-2164-12-265

Kumar, S., Hamweih, A., Manickavelu, A., Kumar, J., Sharma, T. R., and Baum, M. (2014). "Advances in lentil genomics," in *Legumes in Omics Era*, eds S. Gupta, N. Nadarajan, and D. S. Gupta (New York, NY: Springer), 111–130.

Lazaro, A., Ruiz, M., De la Rosa, L., and Martin, I. (2001). Relationships between agro/morphological characters and climatic parameters in Spanish landraces of lentil (*Lens culinaris* Medik.). *Genet. Resour. Crop Evol.* 48, 239–249. doi: 10.1023/A:1011234126154

Letunic, I., and Bork, P. (2011). Interactive tree of life v2: online annotation and display of phylogenetic trees made easy. *Nucleic Acids Res.* 39, W475–W478. doi: 10.1093/nar/gkr201

Lev-Yadun, S., Gopher, A., and Abbo, S. (2000). The cradle of agriculture. *Science* 288, 1602–1603. doi: 10.1126/science.288.5471.1602

Liu, J., Guan, J. P., Xu, D. X., Zhang, X. Y., Gu, J., and Zong, X. X. (2008). Genetic diversity and population structure in lentil (*Lens culinaris* Medik.)

germplasm detected by SSR markers. *Acta Agron. Sin.* 34, 1901–1909. doi: 10.1016/S1875-2780(09)60015-8

Liu, K., and Muse, S. V. (2005). PowerMarker: integrated analysis environment for genetic marker data. *Bioinformatics* 21, 2128–2129 doi: 10.1093/bioinformatics/bti282

Lombardi, M., Materne, M., Cogan, N. O., Rodda, M., Daetwyler, H. D., Slater, A. T., et al. (2014). Assessment of genetic variation within a global collection of lentil (*Lens culinaris* Medik) cultivars and landraces using SNP markers. *BMC Genet.* 15:150. doi: 10.1186/s12863-014-0150-3

Muehlbauer, F. J., Kaiser, W. J., Clement, S. L., and Summerfield, R. J. (1995). Production and breeding of lentil. *Adv. Agron.* 54, 283–332. doi: 10.1016/S0065-2113(08)60902-5

Nei, M. (1973). Analysis of gene diversity in subdivided populations. *Proc. Natl. Acad. Sci. U.S.A.* 70, 3321–3323. doi: 10.1073/pnas.70.12.3321

Peakall, R., and Smouse, P. E. (2012). GenAlEx 6.5: genetic analysis in Excel. Population genetic software for teaching and research-an update. *Bioinformatics* 28, 2537–2539. doi: 10.1093/bioinformatics/bts460

Pritchard, J. K., Stephens, M., and Donnelly, P. (2000). Inference of population structure using multilocus genotype data. *Genetics,* 155, 945–959.

R Core Development Team (2015). *R: A Language and Environment for Statistical Computing.* Vienna: R Foundation for Statistical Computing. Available online at: http://www.R-project.org

Scippa, G. S., Trupiano, D., Rocco, M., Viscosi, V., Di Michele, M., D'Andrea, A., et al. (2008). An integrated approach to the characterization of two autochthonous lentil (*Lens culinaris*) landraces of Molise (south-central Italy). *Heredity* 101, 136–144. doi: 10.1038/hdy.2008.39

Semagn, K., Babu, R., Hearne, S., and Olsen, M. (2014). Single nucleotide polymorphism genotyping using Kompetitive Allele Specific PCR (KASP): overview of the technology and its application in crop improvement. *Mol. Breed.* 33, 1–14. doi: 10.1007/s11032-013-9917-x

Sharma, S. K., Knox, M. R., and Ellis, T. H. N. (1996). AFLP analysis of the diversity and phylogeny of *Lens* and its comparison with RAPD analysis. *Theor. Appl. Genet.* 93, 751–758. doi: 10.1007/BF00224072

Sharpe, A., Ramsay, L., Sanderson, L.-A., Fedoruk, M. J., Clarke, W. E., Li, R., et al. (2013). Ancient orphan crop joins modern era: gene-based SNP discovery and mapping in lentil. *BMC Genom* 14:192. doi: 10.1186/1471-2164-14-192

Simon, C. J., and Hannan, R. M. (1995). Development and use of core subsets of cool-season food legume germplasm collections. *HortScience* 30, 907.

Sonnante, G., and Pignone, D. (2001). Assessment of genetic variation in a collection of lentil using molecular tools. *Euphytica* 120, 301–307. doi: 10.1023/A:1017568824786

South, A. (2011). rworldmap: a new R package for mapping global data. *R J.* 3, 35–43.

Toklu, F., Karaköy, T., Hakle, I., Bicer, T., Brandolini, A., Kilian, B., et al. (2009). Genetic variation among lentil (*Lens culinaris* Medik.) landraces from Southeast Turkey. *Plant Breed.* 128, 178–186. doi: 10.1111/j.1439-0523.2008.01548.x

Tullu, A., Diederichsen, A., Suvorova, G., and Vandenberg, A. (2011). Genetic and genomic resources of lentil: status, use and prospects. *Plant Genet. Resour.* 9, 19–29. doi: 10.1017/S1479262110000353

Yang, H., Li, C., Lam, H. M., Clements, J., Yan, G., and Zhao, S. (2015). Sequencing consolidates molecular markers with plant breeding practice. *Theor. Appl. Genet.* 128, 779–795. doi: 10.1007/s00122-015-2499-8

Zaccardelli, M., Lupo, F., Piergiovanni, A. R., Laghetti, G., Sonnante, G., Daminati, M. G., et al. (2012). Characterization of Italian lentil (*Lens culinaris* Medik.) germplasm by agronomic traits, biochemical and molecular markers. *Genet. Resour. Crop Evol.* 59, 727–738. doi: 10.1007/s10722-011-9714-5

Zohary, D. (1999). Monophyletic vs. polyphyletic origin of the crops on which agriculture was founded in the Near East. *Genet. Resour. Crop Evol.* 46, 133–142. doi: 10.1023/A:1008692912820

Conflict of Interest Statement: The authors declare that the research was conducted in the absence of any commercial or financial relationships that could be construed as a potential conflict of interest.

13

Additive Interactions of Unrelated Viruses in Mixed Infections of Cowpea (*Vigna unguiculata* L. Walp)

Edited by:
Antonio M. De Ron,
Spanish National Research Council,
Misión Biológica de Galicia - CSIC,
Spain

Reviewed by:
Carissa M. Soto,
Naval Research Laboratory, USA
Ousmane Boukar,
International Institute of Tropical
Agriculture (IITA), Nigeria
Ana Campa,
Servicio Regional de Investigación y
Desarrollo Agroalimentario, Spain
Deidre Fourie,
ARC-Grain Crops Institute,
South Africa

***Correspondence:**
Imade Y. Nsa,
Department of Microbiology, University
of Lagos, Akoka, Yaba, Lagos, Nigeria
imadee@hotmail.com;
insa@unilag.edu.ng

Imade Y. Nsa[1*] *and Kehinde T. Kareem*[2]

[1] *Department of Microbiology, University of Lagos, Lagos, Nigeria,* [2] *Institute of Agricultural Research and Training, Moor Plantation, Obafemi Awolowo University, Ibadan, Nigeria*

This study was carried out to determine the effects of single infections and co-infections of three unrelated viruses on three cowpea cultivars (one commercial cowpea cultivar "White" and 2 IITA lines; IT81D-985 and TVu 76). The plants were inoculated with Cowpea aphid-borne mosaic virus (CABMV), genus *Potyvirus*, Cowpea mottle virus (CMeV), genus *Carmovirus* and Southern bean mosaic virus (SBMV), genus *Sobemovirus* singly and in mixture (double and triple) at 10, 20, and 30 days after planting (DAP). The treated plants were assessed for susceptibility to the viruses, growth, and yield. In all cases of infection, early inoculation resulted in higher disease severity compared with late infection. The virus treated cowpea plants were relatively shorter than buffer inoculated control plants except the IT81D-985 plants that were taller and produced more foliage. Single infections by CABMV, CMeV, and SBMV led to a complete loss of seeds in the three cowpea cultivars at 10 DAP; only cultivar White produced some seeds at 30 DAP. Double and triple virus infections led to a total loss of seeds in all three cowpea cultivars. None of the virus infected IITA lines produced any seeds except IT81D-985 plants co-infected with CABMV and SBMV at 30 DAP with a reduction of 80%. Overall, the commercial cultivar "White" was the least susceptible to the virus treatments and produced the most yield (flowers, pods, and seeds). CABMV was the most aggressive of these viruses and early single inoculations with this virus resulted in the premature death of some of the seedlings. The presence of the *Potyvirus*, CABMV in the double virus infections did not appear to increase disease severity or yield loss. There was no strong evidence for synergistic interactions between the viruses in the double virus mixtures.

Keywords: cowpea, severity, virus, yield, interaction

Introduction

Cowpea (*Vigna unguiculata* (L). Walp), accounts for a huge portion of the dietary protein of the people in sub-Saharan Africa (Li et al., 2001). Nigeria ranks first in cowpea production worldwide, and is responsible for about 61% of total world production of cowpea (IITA, 2009). Historically, cowpeas are known to produce a more satisfactory yield than most leguminous plants under a variety of climatic, soil, and cultural conditions. However, they are very susceptible to insect pests (Lephale et al., 2012) and many virus diseases (Karungi et al., 2000) that reduce yields. More than 140 viruses have been identified as naturally infecting cowpea (Hughes and Shoyinka, 2003) but only nine have been reported in Nigeria (Taiwo, 2003), and these are Cowpea aphid-borne mosaic virus (CABMV), Cowpea golden mosaic virus (CPGMV), Southern bean mosaic virus (SBMV), Sunhemp mosaic virus (SHMV), Blackeye mosaic virus (BICMV), Cucumber mosaic virus (CMV), Cowpea mottle virus (CMeV), Cowpea yellow mosaic virus (CPMV), and Cowpea mild mottle virus (CPMMV).

CABMV was first reported and described as a distinct *Potyvirus* infecting cowpea in Italy (Vidano and Conti, 1965; Lovisolo and Conti, 1966; Bashir et al., 2002), then in Nigeria in 1976 (Ladipo, 1976). It is of widespread in distribution in Nigeria and in other major cowpea growing regions of the world. It has also been reported from several African countries including, Kenya (Bock, 1973), Tanzania (Patel and Kuwite, 1982), Botswana (Burke et al., 1986), Uganda (Orawu et al., 2005). CABMV is the most important viral pathogen of cowpea in Nigeria, as far as pathogenic variation and yield losses are concerned (Raheja and Leleji, 1974; Thottappilly and Rossel, 1992; Shoyinka et al., 1997). It is readily transmissible by sap inoculation and by several aphids in a non-persistent manner (Vidano and Conti, 1965; Atiri, 1982; Atiri et al., 1984). It is also seed-borne in cowpea, but transmission is dependent upon cowpea cultivar and virus strain (Aboul-Ata et al., 1982; Gumedzoe, 1985). It has a wide experimental host range including many species in the Leguminosae, Chenopodiaceae, Curcubitaceae, Solanaceae, and Amaranthaceae (Lovisolo and Conti, 1966; Bos, 1970; Bock, 1973). Virus may induce local lesions or systemic infections depending on the host, infected plants show dark green vein–banding, leaf deformation, crinkling, blistering, and stunting (Bock and Conti, 1974). However, the severity of the symptoms depends on the interaction between the host cultivar and virus strain (Rossel and Thottappilly, 1985).

Field occurrence of SBMV on cowpea has also been reported from the world, U.S.A. (Kuhn, 1963), India (Singh and Singh, 1974), Ghana (Lamptey and Hamilton, 1974), Ivory Coast (Fauquet and Thouvenel, 1980), Senegal (Gaikward and Thottappilly, 1988), Togo (Gumedzoe et al., 1989), and Nigeria (Shoyinka et al., 1979). The systemic symptoms induced by SBMV include vein clearing, followed by a mild to severe mottling or coarse mosaic pattern (Shepherd and Fulton, 1962; Allen et al., 1981). It is transmissible by inoculation of sap, beetle, and is seed borne (Tremaine and Hamilton, 1983).

CMeV was first described from Nigeria, where it was isolated from Bambarra groundnut (*Vigna subterranea*) and cowpea, *V. unguiculata* (Robertson, 1963, 1966). In Nigeria, CMeV is commonly found in the southern rainforest and guinea savannah zones where most of the Bambara groundnut is grown (Rossel and Thottappilly, 1985). It has also been reported from other West African countries, Ivory Coast (Thouvenel, 1988), Togo (Gumedzoe et al., 1989), and the Republic of Benin (Thottappilly and Rossel, 1988). In tolerant cowpea varieties, the symptom of this virus consists basically of mottling; whereas in severe infections, CMeV may induce leaf distortion, reduction in leaf size and witches broom syndrome. CMeV is transmitted principally by a chrysomelid beetle vector, *Ootheca mutabilis* (Shoyinka et al., 1978) and by sap inoculation.

In Nigeria, there have been reports of single and multiple (double and triple) virus infections of cultivated cowpeas resulting in complete yield loss (Raheja and Leleji, 1974; Kareem and Taiwo, 2007; Taiwo et al., 2007). Mixed virus infection of plants often results in intensified symptom severity than those caused by each individual virus, and higher virus accumulation, a phenomenon referred to as synergy (Wang et al., 2002; Untiveros et al., 2007). Synergism occurs in mixed infections when a pair of the viruses is unrelated, with the overall effect of the viruses eliciting more severe disease symptoms in the host plant than those produced by each of the viruses separately (Cho et al., 2000; Murphy and Bowen, 2006). Examples of synergistic viral diseases have been recorded for many decades (Shi et al., 1996). The severity of the disease symptoms is also greatly increased if one of the infecting viruses is a member of the genus *Potyvirus* (Pio-Ribeiro et al., 1978). Nevertheless, not all combinations of unrelated viruses result in increased symptoms (Anjos et al., 1992). Shoyinka et al. (1978) reported that CMeV frequently occurred in mixed infections with Cowpea yellow mosaic virus (CYMV) or SBMV in cowpea fields located in southwestern Nigeria.

One method of detecting viruses is by biological properties (Naidu and Hughes, 2003). In this study, two International Institute of Tropical Agriculture (IITA) cowpea lines (IT81D-985, TVu 76) and a local commercial cultivar (White) were evaluated for symptom severity, growth defects, and yield losses as a result of single and mixed infections of CABMV, CMeV, and SBMV. We also ascertained whether the qualitative observation of increased symptom severity, reductions in growth and yield presumably attributed to synergistic interactions of the viruses could be further validated by quantitative methods. Serological tests were done to determine relative virus concentration.

Materials and Methods

Source and Cultivation of Cowpea Lines/Cultivars

The cowpea lines, TVu 76 and IT 81D-985 used in this study were obtained from the Germplasm Resource Unit of the International Institute of Tropical Agriculture (IITA), Ibadan and the commercial "White" cultivar was purchased from Sandgrouse Market, Lagos. The IT81D-985 is a medium-late maturing semi spreading cowpea line (Singh, 2011) and line TVu 76 is known to be susceptible to leafhoppers (Raman et al., 1980). The cowpea lines/cultivars were cultivated as described by Kareem and Taiwo (2007).

Source and Maintenance of Viruses

The viruses used in this research were Cowpea aphid-borne mosaic virus genus, *Potyvirus*; Southern bean mosaic virus, genus *Sobemovirus*; and Cowpea mottle virus, genus *Carmovirus*. The viruses were obtained from infected cowpea plants or tissue stored over calcium chloride at the IITA Ibadan. These viruses were maintained on the Ife Brown variety in the Greenhouse of the Botanical Garden of University of Lagos (Kareem and Taiwo, 2007).

Inoculation of Cowpea Cultivars with Viruses

Seedlings were mechanically inoculated at 10, 20, and 30 days after planting. Each pot was labeled according to the type of virus treatment, including negative controls, name of cultivar and date of inoculation. Test plants were dusted with carborundum (180 mesh) to increase entry of the virus into the hosts. Single virus inoculum was prepared by grinding virus-infected leaves with a sterile mortar and pestle at the rate of 1 g of tissue to 2 ml of buffer. However, mixed viral treatments were obtained by mixing saps from the relevant inocula at ratio 1:1 (V/V) just before inoculation. The pestle was used to rub sap over the upper leaf surface from base to the top of the leaf. After inoculation, the leaves were rinsed immediately with water to prevent the harmful effects of the sap. Inoculated plants were kept in the University of Lagos Greenhouse. The different treatments are listed below.

a. *Single virus inoculation*
   Each of the cultivars: TVu 76, IT81D-985 and commercial cultivar "White" was inoculated singly with CABMV, CMeV, or SBMV.
b. *Double virus inoculation*
   Each of the three cowpea lines/cultivar was inoculated with double virus mixtures i.e., CABMV + CMeV, CABMV + SBMV, and CMeV + SBMV.
c. *Triple virus inoculation*
   Each of the three cowpea lines was also inoculated with a combination of the three viruses, i.e., CABMV + SBMV + CMeV.
d. *Mock inoculation with buffer*
   Control plants of each of the cowpea lines/cultivars were inoculated with buffer only. The buffer was prepared by dissolving 17.4 g of di-potassium hydrogen orthophosphate (K_2HPO_4) in one liter (1l) distilled water (0.1 M solution); 0.1 M potassium di-hydrogen orthophosphate (KH_2PO_4) was added (3.4 g of KH_2PO_4 dissolved in 250 ml) to achieve a pH 7.5 (Walkey, 1985). The working concentration was 0.05 M.

Inoculation Procedure

a. *First stage inoculation*
   The first inoculation was done 10 days after planting (DAP). The sap from the different virus treatments above or buffer was rubbed onto the plant surface of the fully expanded leaves of each of the three cultivars. Symptoms were observed at 10, 20, and 30 days after inoculation (DAI).
b. *Second stage inoculation*
   The second inoculation was done 20 DAP. The sap from the different virus treatments or buffer was rubbed on the leaves of each of the three cultivars. Symptoms were observed at 10, 20, and 30 DAI.
c. *Third stage inoculation*
   The second inoculation was carried out 30 days after planting. Symptoms were observed at 10, 20, and 30 DAI. After each inoculation, plants were rinsed with water to reduce the effect of caborundum on leaves that may interfere with photosynthesis. Plants were kept in the greenhouse under ambient temperatures ranging from 25 to 28°C, watered every other day, and observed for symptom development.

Experimental Design

The experimental design used in the greenhouse was randomized complete block design (RCBD). There were three cowpea lines/cultivars, three replicates, three different stages of virus inoculation and eight virus treatments including the negative control ($3 \times 3 \times 3 \times 8$) making a total of 216 pots.

Effect of Virus Treatments on Disease Severity

The reaction of the cultivars to mechanical inoculation with the viruses (single and multiple) and buffer was determined by rating inoculated cowpea plants based on symptom development at 10, 20, and 30 DAI. The rating was done on a scale of 1 to 5, where 1, no symptoms; (resistant) 2, slight mosaic/mottle on leaves (susceptible); 3, moderate mosaic/mottle/blistering/curling of leaves (susceptible); 4, severe mosaic/mottle/blistering, leaf reduction (susceptible); 5, very severe symptoms leading to stunting, apical necrosis, and plant death.

Effects of Virus Treatments on Growth Traits

The effect of virus treatments and buffer on the plant height, number of leaves of White, IT81D-985 and TVu 76, was determined at 10 DAP 10 DAI and 30 DAP 10 DAI. The plant height was measured with a ruler in cm. and the leaves were counted. The percentage reduction in plant height, and number of leaves in virus-infected plants was calculated as:

$$\% \text{ reduction in plant height} = [(H_B - H_V)/H_B] \times 100$$

Where:
H_B = plant height of buffer inoculated plant
H_V = plant height of virus inoculated plant

$$\% \text{ reduction in leaf number} = [(L_B - L_V)/L_B] \times 100$$

Where:
L_B = number of leaves of buffer inoculated plant
L_V = number of leaves of virus inoculated plant

Effect of Virus Treatments on Yield Parameters of the Cowpea Varieties

The effect of virus treatments and buffer on the number of flowers, number of pods, pod length, seed number, and seed weight was determined. Pod length was measured with a ruler and the seeds were weighed on a Mettler -Toledo balance (Columbus, OH, U.S.A).

Determination of the Type of Biological Response among the Viruses

To determine the type of biological interactions existing between the viruses in double co-infections, we compared the growth traits of cowpea plants exposed to double virus treatments to the single virus inoculated plants. We calculated the degree of interactions of the viruses in the mixed virus treatments using the Abbott's formula (Abbott, 1925; Gisi, 1996).

$$\text{Cexp} = A + B - (AB/100)$$

Where Cexp is the expected level of disease, and A and B are corresponding responses due to infection by virus A and B, respectively, as observed in the experiment, AB is the response of the double virus infection (Murphy and Bowen, 2006). This was done for all the double infections, CABMV + CMeV, CABMV + SBMV, and CMeV + SBMV of the three cowpea varieties under investigation.

The means of the different growth traits were used to quantitatively determine the degree of the observed reduction. Plant height and leaf number were the two traits considered. Rather than using the strict mathematical definition of a synergistic interaction as the ratio of the observed response (Cobs) to the expected response (Cexp) if greater than 1.0, additive, if equal to 1 and antagonistic, if less than 1. We adopted the cut offs by Gisi et al. (1985), where a biological response can be categorized as follows: (1) synergistic, if the interaction of the ratio of the observed response (Cobs) to the expected response (Cexp) is greater than 1.5, (2) additive, between 0.5 and 1.5, and (3) antagonistic, less than 0.5.

Data Analysis

All the data obtained were analyzed using Statistical Package for Social Scientists (SPSS) version 16.0 and Duncan Multiple Range Test was used to compare means obtained at 5% level of probability (Little and Hills, 1972).

Effect of Virus Treatment on Relative Virus Concentration

Leaf samples from cultivar White, IT81D-985 and TVu 76 that received the different treatments were taken at 10 and 30 days after inoculation at the different stages of growth to determine virus concentration. The relative concentration of virus in single and mixed infections was determined serologically with antigen coated plate enzyme linked immunosorbent assay (ACP ELISA). The ELISA protocol was carried out as described as Taiwo et al. (2007).

Results

Symptomatology

The three cultivars/lines were susceptible to the three viruses used in this investigation. The inoculated leaves were observed for symptom development at different stages of growth after inoculation and recorded. The symptoms observed at 20 DAP were not intermediary or significantly different from those observed at 30 DAP (data not shown).

Symptom Severity of Virus Treatments on Cowpea (White)

The mixture of the triple viruses, CABMV + CMeV + SBMV induced the most symptoms. Its common symptoms were mosaic, mottling and especially apical necrosis that eventually led to plant death (**Figure 1A**). The triple virus treatments caused death at all stages of plant growth regardless of the time of inoculation. As for the double virus infections, only early inoculations (10 DAP) resulted in a severity score of 5 indicating that the plants died due to early infection (**Table 1**). Leaf deformation was observed in the mixture of CABMV and CMeV (**Figure 1B**). No symptoms were observed on the leaves of the buffer inoculated plants (**Figure 1C**).

For the single viruses, SBMV induced mostly mosaic and leaf curling symptoms but did not cause the death of any plants (**Figure 1D**). CABMV was the most aggressive of the viruses in this variety. Early inoculations with this virus resulted in premature death of some of the seedlings giving a severity score of 5 (**Table 1**). Its common symptoms included mosaic, vein - banding, chlorosis, blistering (**Figure 1E**) and stunting. Whereas, single infections by CMeV, induced mostly mottling and defoliation. The leaves of the buffer inoculated plants/controls

FIGURE 1 | Viral symptoms induced on the local commercial cultivar White. (A) Apical necrosis and defoliation induced by triple virus treatment, CABMV + CMeV + SBMV that eventually led to plant death. **(B)** Leaf deformation and mosaic symptoms induced by double virus infection, CABMV + CMeV. **(C)** Buffer inoculated healthy fully expanded trifoliate leaves. **(D)** Mottling, leaf reduction and crinkling induced by SBMV. **(E)** Blistering and leaf chlorosis induced by CABMV.

TABLE 1 | Severity of single and multiple virus treatments on cowpea varieties.

Virus treatment	DAP^{x}	10^{y}			20^{y}			30^{y}		
		White	IT81D-985	TVu 76	White	IT81D-985	TVu 76	White	IT81D-985	TVu 76
CABMV	10	3^{c}	3^{c}	4^{d}	2.5^{bc}	5^{e}	3^{c}	5^{e}	5^{e}	4^{d}
	30	2^{b}	2^{b}	1^{a}	2^{b}	3^{c}	1^{a}	2.5^{bc}	4^{d}	2^{b}
CMeV	10	2^{b}	2^{b}	4.5^{de}	3.5^{cd}	2^{b}	3.5^{cd}	3.5^{cd}	4^{d}	4.5^{de}
	30	1^{a}	2^{b}	2^{b}	2.5^{bc}	2^{b}	2^{b}	3.5^{cd}	3.5^{cd}	4.5^{de}
SBMV	10	2.5^{bc}	2^{b}	2^{b}	3^{c}	2^{b}	4^{d}	3.5^{cd}	5^{e}	4^{d}
	30	2^{b}	2^{b}	2^{b}	2^{b}	2^{b}	2^{b}	3^{c}	2^{b}	2^{b}
CABMV + CMeV	10	2.5^{bc}	4^{d}	4.5^{cd}	5^{e}	2^{b}	5^{e}	5^{e}	5^{e}	5^{e}
	30	2^{b}	2^{b}	2^{b}	2^{b}	2^{b}	5^{e}	5^{e}	4^{d}	5^{e}
CABMV + SBMV	10	2.5^{bc}	3.5^{cd}	4^{d}	5^{e}	3.5^{cd}	5^{e}	5^{e}	3.5^{cd}	5^{e}
	30	2^{b}	3^{c}	2^{b}	2^{b}	3^{c}	4^{d}	2.5^{bc}	3.5^{cd}	5^{e}
CMeV + SBMV	10	2.5^{bc}	2^{b}	3.5^{cd}	5^{e}	3^{c}	4.5^{de}	5^{e}	5^{e}	5^{e}
	30	2^{b}	2^{b}	2^{b}	2^{b}	2^{b}	2^{b}	2.5^{bc}	3^{c}	3^{c}
CABMV + CMeV + SBMV	10	4.5^{de}	4.5^{de}	4^{d}	4.5^{de}	3.5^{cd}	5^{e}	5^{e}	5^{e}	5^{e}
	30	2^{b}	2^{b}	2^{b}	2^{b}	2^{b}	3^{c}	5^{e}	5^{e}	3.5^{cd}
BUFFER (negative control)	10	1^{a}	1^{a}	1^{a}	1^{a}	1^{a}	1^{a}	1^{a}	1^{a}	1^{a}
	30	1^{a}	1^{a}	1^{a}	1^{a}	1^{a}	1^{a}	1^{a}	1^{a}	1^{a}

Means followed by the same letter in each column are not significantly different according to Duncan's multiple range Test at $P \leq 5\%$. ^{x} = age of plant at inoculation in days; ^{y} = disease severity was observed at 10, 20, and 30 days after inoculation at growth stages 10 and 30 DAP. CABMV, Cowpea aphid-borne mosaic virus; CMeV, Cowpea mottle virus; SBMV, Southern bean mosaic virus.

were healthy, remained fully expanded and symptomless; the other plant parts also showed no disease symptoms.

Symptom Severity of Virus Treatments on Line IT81D-985

The most severe symptom induced by the triple virus infection of CABMV + CMeV + SBMV on IT81D-985 was apical necrosis. The symptoms induced on its primary leaves were reddish necrotic lesions. Double virus treatments involving CABMV + CMeV and CMeV + SBMV led to the maximum severity score of 5 at 10 DAP, while early inoculation with CABMV + SBMV resulted in the highest severity score of 3.5 (**Table 2**).

In this variety, single inoculations with CABMV or CMeV caused death in some of the plants at 10 and 30 DAP, respectively. CABMV induced vein- banding, mosaic, internode shortening, chlorotic lesions and apical necrosis. In addition to the characteristic mottling of CMeV (**Figure 2**), reduction in size of leaves, defoliation, and apical necrosis were also observed. The general symptoms induced by SBMV on this line included mosaic, leaf reduction, and defoliation. Inoculations with each of the double virus mixtures caused the death of some of the plants. While early inoculation with CMeV + SBMV at 10 DAP resulted in the death of some plants, as well as severe mottling among other symptoms. Single inoculation with either CABMV or SBMV produced a severity score of 5 that caused death in the plants at 10 DAP 10 DAI and 10 DAP 30 DAI. The severity score of 2 was observed in the early- inoculated CMeV plants, severity worsened to a score of 4 at 10 DAP 30 DAI. Mottling was observed on the leaves of cowpea inoculated with CMeV (**Figure 2A**) but not death. The symptoms induced on its primary leaves were reddish necrotic lesions but like in the "White" variety, CABMV also induced chlorotic lesions.

Symptom Severity of Virus Treatments on Cowpea (TVu 76)

As observed in "White" and IT81D-985 varieties, triple infections resulted in chronic symptoms, most especially apical necrosis which eventually led to the death of some of the TVu 76 plants. Generally, all the virus treatments (whether single, double, or triple) induced defoliation, while the buffer inoculated plants remained healthy and symptomless.

Severity scores in plants with single virus treatments were significantly reduced at 30 DAP compared with 10 DAP. TVu 76 plants inoculated with CABMV only at 30 DAP, were tolerant at the initial stage (10 DAI) of infection but later showed mild symptoms with a score of 2 at 30 DAI. The most common symptom caused by CABMV was internode shortening. Single infections by CMeV induced the most symptoms including mottling and apical necrosis, while SBMV induced basal necrosis and leaf curling among other symptoms. The severity scores of CABMV + CMeV and CABMV + SBMV at 10 DAP were high (between 4 and 5) causing death of some plants (**Table 1**). At 30 DAP, CMeV + SBMV induced less symptoms than CABMV + CMeV and CABMV + SBMV treatments and did not cause plant death. The buffer inoculated plants had a score of 1 implying that the plants remained healthy and symptomless.

CABMV was the Most Aggressive of Single Virus Infections in the White Cultivar but not in Line TVu 76

To determine the effects of single virus treatments to the different cowpea, we exposed each cowpea variety to single treatments of the 3 viruses—CABMV, CMeV, and SBMV. Generally, the entire virus treatments at 10 DAP caused a significant reduction in

TABLE 2 | The effects of single virus, double infections of CABMV + CMeV treatments and buffer on plant height and number of leaves in the White cultivar, and TVu 76.

Growth Parameter		Plant Height (cm)		Number of leaves	
		White	TVu 76	White	TVu 76
TREATMENT	DAP^{x}				
CABMV	10	19.19^{d}	24.69^{c}	3.44^{d}	14.11^{b}
	30	92.99^{ab}	46.84^{ab}	19.22^{ab}	16.67^{ab}
CMeV	10	27.09^{d}	16.3^{d}	4.33^{d}	5.25^{c}
	30	52.78^{bc}	36.12^{b}	13.67^{b}	12.22^{bc}
CABMV + CMeV	10	24.13^{d}	16.08^{d}	2.22^{d}	6.33^{c}
	30	98.64^{ab}	34.14^{b}	13.44^{b}	12.33^{bc}
BUFFER	10	50.09^{bc}	34.43^{b}	26.44^{a}	14.78^{b}
	30	102.44^{a}	54.49^{a}	27.11^{a}	18.33^{a}
Percent reduction in growth parameters of virus treated plants compared with mock inoculated plants					
CABMV	10	61.69	28.29	86.99	4.53
	30	9.22	14.04	29.10	9.06
CMeV	10	45.92	52.66	83.62	64.48
	30	48.48	33.71	49.58	33.33
CABMV + CMeV (Cobs)	10	51.83	53.30	91.60	57.17
	30	3.71	37.35	50.42	32.73
Cexp	10	79.28	66.05	97.87	71.93
	30	53.23	43.02	64.25	39.37
Cobs: Cexp	10	0.65	0.81	0.94	0.79
	30	0.07	0.87	0.78	0.83

DAP^{x}, Growth parameters were measured 10 days after inoculation at 10 and 30 DAP. The age of the plant at 10 DAP was 20 days old, at 30 DAP, the age of the plant was 40 days old. CABMV, Cowpea aphid-borne mosaic virus; CMeV, Cowpea mottle virus. Each value is the mean of 3 replicates. In each column, means followed by the same letter are not significantly different ($P = 0.05$) according to Duncan's multiple range tests. Cexp = A + B − AB/100, A stands for CABMV alone and B for CMeV alone. If Cobs: Cexp <0.5, interactions are antagonistic, Cobs: Cexp 0.5–1.5, interactions are additive, Cobs: Cexp >1.5, interactions are synergistic.

TABLE 3 | The effects of single virus, double infections of CABMV + SBMV treatments and buffer on plant height and number of leaves in the White cultivar and TVu 76.

Growth Parameter		Plant Height (cm)		Number of leaves	
		White	TVu 76	White	TVu 76
TREATMENT	DAP^{x}				
CABMV	10	19.19^{d}	24.69^{c}	3.44^{d}	14.11^{b}
	30	92.99^{ab}	46.84^{b}	19.22^{ab}	16.67^{ab}
SBMV	10	21.43^{d}	16.07^{d}	8.11^{c}	6.33^{c}
	30	51.92^{bc}	34.13^{b}	23.67^{ab}	12.33^{bc}
CABMV + SBMV	10	26.24^{d}	22.15^{c}	4.89^{d}	2.67^{d}
	30	43.11^{c}	34.44^{b}	15.67^{b}	4.22^{cd}
BUFFER	10	50.09^{bc}	34.43^{b}	26.44^{a}	14.78^{b}
	30	102.44^{a}	54.49^{a}	27.11^{a}	18.33^{a}
Percent reduction in growth parameters of virus treated plants compared with mock inoculated plants					
CABMV	10	61.69	28.29	86.99	4.53
	30	9.22	14.04	29.10	9.06
SBMV	10	57.22	53.33	69.33	57.17
	30	49.32	37.36	12.69	32.73
CABMV + SBMV (Cobs)	10	47.61	35.67	81.51	81.94
	30	57.92	36.80	42.20	76.98
Cexp	10	83.61	66.53	96.01	59.11
	30	53.99	46.16	38.10	38.83
Cobs: Cexp	10	0.57	0.54	0.85	1.39
	30	1.07	0.80	1.11	1.98

DAP^{x},- Growth parameters were measured 10 days after inoculation at 10 and 30 DAP. The age of the plant at 10 DAP was 20 days old, at 30 DAP, the age of the plant was 40 days old. CABMV, Cowpea aphid borne mosaic virus; SBMV, Southern bean mosaic virus. Each value is the mean of 3 replicates. In each column, means followed by the same letters are not significantly different ($P = 0.05$) according to Duncan's multiple range tests. Cexp = A + B − AB/100, A stands for CABMV alone and B for SBMV alone. If Cobs: Cexp < 0.5, interactions are antagonistic, Cobs: Cexp 0.5–1.5, interactions are additive, Cobs: Cexp > 1.5, interactions are synergistic.

FIGURE 2 | (A) Mottling and leaf reduction induced on line IT81D-985 by CMeV. **(B)** Symptomless buffer inoculated trifoliate leaf.

plant height and leaf number in comparison with the controls (**Tables 2–4**). We observed that in the White cultivar, CABMV was the most aggressive but caused the least reduction in vegetative growth in TVu 76 (**Table 2**). In the single infection of the cowpea cultivar "White," CABMV infected plants were the shortest with an average height of 19.19 cm, while the average height of the healthy was 50.09 cm. compared to the healthy's average of 26 leaves. In TVu 76, single infections with CMeV and SBMV caused similar percentage reductions in plant height and leaf number at 10 and 30 DAP (**Table 4**).

The Presence of a Potyvirus (CABMV) had no Significant Effect on Severity of Growth Reduction in the Double Infections

On exposure to the different double virus combinations, CABMV + CMeV, CABMV + SBMV, CMeV + SBMV, there were no significant differences in the response of the White and TVu 76 varieties to the 3 types of double infections (**Tables 2–4**).

In the White variety at 10 DAP, although the mixture of CMeV and SBMV caused the most stunting with an average height of 19.48 cm, there was no significant difference in the average height and leaf number of the plants that received CABMV + CMeV or CABMV + SBMV or CMeV + SBMV. However, there were significant differences in these parameters when the plants received the virus treatment at 30 DAP.

All the TVu 76 plants that received virus treatments at 10 and 30 DAP were significantly shorter than the buffer inoculated plants. There was no significant differences in the effects of the 3 types of double virus infections on plant height at 10 and 30 DAP.

Virus Treatments Promoted Increase in Foliage and Plant Height in the IITA Line IT81D-985

Unexpectedly, some of the virus treated plants were taller and produced more leaves than the control plants. For the plants inoculated at 10 DAP, those that received the following treatments (i) CMeV (ii) CABMV + SBMV (iii) CABMV + CMeV were taller than the control plants and also produced more leaves. Considering those treated at 30 DAP, all the plants that received the virus treatments produced more leaves and were taller than the controls except those that were treated with SBMV alone (**Table 5**).

TABLE 4 | The effects of single virus, double infections of CMeV + SBMV treatments and buffer on plant height and number of leaves in the White cultivar and line TVu 76.

Growth Parameter		Plant Height (cm)		Number of leaves	
		White	TVu 76	White	TVu 76
TREATMENT	**DAP**^{x}				
CMeV	10	27.09^{d}	16.3^{d}	4.33^{d}	5.25^{c}
	30	52.78^{bc}	36.12^{b}	13.67^{b}	12.22^{bc}
SBMV	10	21.43^{d}	16.07^{d}	8.11^{c}	6.33^{c}
	30	51.92^{bc}	34.13^{b}	23.67^{ab}	12.33^{bc}
CMeV + SBMV	10	19.48^{d}	18.2^{d}	4.33^{d}	4.5^{cd}
	30	87.48^{b}	34.00^{b}	20.56^{ab}	16.56^{ab}
BUFFER	10	50.09^{bc}	34.43^{b}	26.44^{a}	14.78^{b}
	30	102.44^{a}	54.49^{a}	27.11^{a}	18.33^{a}
Percent reduction in growth parameters of virus treated plants compared with mock inoculated plants					
CMeV	10	45.92	52.66	83.62	64.48
	30	48.48	33.71	49.58	33.33
SBMV	10	57.22	53.33	69.33	57.17
	30	49.32	37.36	12.69	32.73
CMeV + SBMV (Cobs)	10	61.11	47.14	83.62	69.55
	30	14.60	37.60	24.16	9.66
Cexp	10	76.86	77.90	94.98	84.79
	30	73.89	58.48	55.97	55.16
Cobs: Cexp	10	0.80	0.61	0.88	0.82
	30	0.20	0.64	0.43	0.18

DAP^{x}*, Growth parameters were measured 10 days after inoculation at 10 and 30 DAP. The age of the plant at 10 DAP was 20 days old, at 30 DAP, the age of the plant was 40 days old. CMeV, Cowpea mottle virus; SBMV, Southern bean mosaic virus. Each value is the mean of 3 replicates. In each column, means followed by the same letter are not significantly different (P = 0.05) according to Duncan's multiple range tests. If Cobs: Cexp <0.5, interactions are antagonistic, Cobs: Cexp 0.5–1.5, interactions are additive, obs: Cexp >1.5, interactions are synergistic.*

Co-infection with Triple Viruses did not have a Greater Impact than Double Virus Infections on Growth Traits

To assess if the extent of damage caused by mixed infections could be greater if there were more than 2 viruses involved, we looked at the effects of the triple virus treatment CABMV + CMeV + SBMV on the three cowpea varieties (**Table 6**). In all the cultivars/lines, the triple virus treated plants were significantly shorter than the control plants and produced less leaves. In White, there was no difference between the effects of the double and triple virus infections. However, the double infections were more aggressive than the triple virus infections in TVu 76 (**Tables 2–4, 6**).

None of the Early Virus Inoculated Plants (10 DAP) Produced any Flowers, Pods, or Seeds

In all the three varieties, none of the virus treated plants produced any flowers, pods or seeds at 10 DAP. However, the plants inoculated with buffer only, produced flowers and seeds. For those inoculated at 30 DAP, all the virus treated cultivar White plants produced an average of 1–4 flowers compared to the 7 flowers produced by the healthy. CABMV + SBMV, as well as CABMV + CMeV + SBMV treated ones produced the least—an average of 1 flower. All of the infected plants produced pods and seeds except those that received the CABMV + CMeV + SBMV and CABMV + SBMV treatments.

The pods from the CABMV + CMEV treated plants were the shortest at 2.13 cm whereas the average length of the pods from the negative controls was 9.27 cm. CMeV treated plants

TABLE 5 | The effect of single virus, double virus infections and buffer on plant height and leaf number in IT 81D-985.

Treatment	DAP^{x}	Plant height (cm)	Leaf number
CABMV	10	8.47^{d}	4.33^{d}
	30	50.80^{ab}	13.89^{b}
CMEV	10	26.33^{c}	10.56^{b}
	30	61.24^{a}	19.33^{a}
SBMV	10	6.17^{d}	7.44^{c}
	30	21.22^{c}	17.89^{a}
CABMV + CMeV	10	21.87^{c}	8.78^{c}
	30	36.98^{b}	11.56^{b}
CABMV + SBMV	10	23.42^{c}	13.78^{b}
	30	50.94^{b}	15.56^{a}
CMEV + SBMV	10	13.41^{c}	3.78^{d}
	30	75.77^{ab}	18.33^{a}
Buffer	10	19.33^{c}	8.00^{d}
	30	34.71^{b}	8.56^{c}

DAP^{x}*, Growth parameters were measured 10 days after inoculation at 10 and 30 DAP. The age of the plant at 10 DAP was 20 days old, at 30 DAP, the age of the plant was 40 days old. CABMV, Cowpea aphid borne mosaic virus. CMeV, Cowpea mottle virus; SBMV, Southern bean mosaic virus. Each value is the mean of 3 replicates. In each column, means followed by the same letter are not significantly different (P = 0.05) according to Duncan's multiple range tests.*

TABLE 6 | The effect of co-infection with a mixture of the three viruses/buffer and age of plant at inoculation on growth parameters on commercial cultivar White and IITA lines IT81D-985 and TVu 76.

Virus Treatment	DAP^x	PLANT HEIGHT			LEAF NUMBER		
		White	IT81D-985	TVu 76	White	IT81D-985	TVu 76
CABMV + CMeV + SBMV	10	27.15^d	13.82^c	21.17^c	3.22^d	3.89^d	10.111^{bc}
	30	58^{bc}	40.07^{ab}	36.12^b	12^b	6.89^c	3.11^d
BUFFER	10	50.09^{bc}	19.33^c	34.43^b	26.44^a	8^c	14.78^b
	30	102.44^a	34.71^b	54.49^a	27.11^a	8.56^c	18.33^a

DAP^x, Growth parameters were measured 10 days after inoculation at 10 and 30 DAP. The age of the plant at 10 DAP was 20 days old, at 30 DAP, the age of the plant was 40 days old. CABMV, Cowpea aphid borne mosaic virus; CMeV, Cowpea mottle virus; SBMV, Southern bean mosaic virus. Each value is the mean of 3 replicates. In each column, means followed by the same letter are not significantly different (P = 0.05) according to Duncan's multiple range tests.

had the least yield of approximately 1 seed, while the CABMV treated plants as well as SBMV treated plants had the same average yield of 3 seeds. CMeV + SBMV treated plants had the most yields for the double infections with an average seed number of 2. Reduction in seed number ranged from 47 to 89%. The average unit weight of the seeds produced from the virus treated plants was less than that produced by the control plants (0.15 g) except in those inoculated with CMeV + SBMV. CABMV treated plants produced the smallest seeds (0.09 g), while the average weight of seeds from CMeV + SBMV treated ones was 0.21 g.

For the IT81D-985 inoculated at 30 DAP, only those treated with CABMV singly, CABMV + CMeV and CABMV + SBMV and the healthy produced flowers. Of the virus treated plants that produced flowers, only those treated with CABMV + SBMV produced pods, which were one-third the length of those produced by the healthy. The average number of seeds they produced was 2 compared to about 5 produced by the control plants. The average unit weight of seeds from the healthy plants was at least 4 times the weight of the seeds from infected plant (**Table 7**).

Of the TVu 76 plants that received virus treatments at 30 DAP, only the plants infected singly with CABMV or CMeV produced flowers. All the others did not produce any flower or pod. None of the virus treated TVU plants produced any seeds. The control plants produced flowers, pods and seeds (**Table 7**).

Co-infection with Triple Viruses Resulted Almost in a Total Loss in Yield Loss

The triple virus infections resulted in a total loss of flowers, pods, and seeds except for the one flower produced by cultivar White. With regards to yield loss, the triple virus infections caused the most damage.

Relative Concentration of the Viruses in the Cowpea Lines

The ELISA test was also used to establish the virus concentration in the cowpea cultivar White, IITA lines- IT81D-985 and TVu 76 infected singly by CABMV, CMEV, and SBMV and in mixtures. The samples were obtained from the inoculated leaf samples at 10 and 30 DAP. The ELISA results showed that the cowpea cultivar White and TVu 76 were susceptible to the each of the viruses treatments (Taiwo et al., 2007) and the mock inoculated plants were virus free.

For IITA line, IT 81D-985, only SBMV was detected at reasonable concentrations in the applicable treatments. In the single virus treatments, the absorbance values of CABMV and CMeV were 0.1805 and 0.155; these values were not significantly different from their negative controls of 0.1355 and 0.113, respectively. In the double infections and triple infections, the virus titres of CABMV and CMeV were also not significant to be considered positive. For the sample to be considered positive for virus presence, it must have at least twice the absorbance value of negative control. The optical density readings are shown in **Figure 3**.

Lack of Strong Evidence for Synergistic Interactions among the Unrelated Viruses in the Mixed Infection

Based on our preliminary statistical analysis, the effects on the growth parameters did not suggest synergistic interactions. To validate our observation on the absence of synergism, we quantified using the Abbott's formula and set the boundaries for the type of response according to Gisi et al. (1985) instead of using the strict mathematical cut offs of synergism as being defined as Cobs: Cexp > 1 (Gisi, 1996). From the calculations, the reductions in the vegetative traits are not indicative of synergism in White and TVu 76 except the reduction in leaf number by CABMV + SBMV in TVu 76 (**Table 3**). The dual infections in IT81D-985 resulted in an increase in plant height and number of leaves; therefore, we could not make any calculations for this line (**Table 5**). The data is more indicative of an additive relationship between these viruses in White and TVu 76 (**Tables 2–4**).

Discussion

Natural infections of cowpea grown in Nigeria by Cowpea aphid-borne mosaic virus (CABMV), Southern bean mosaic virus (SBMV), and Cowpea mottle virus (CMeV) and others have been reported by Shoyinka et al. (1997). Comparative studies of single and mixed unrelated virus infections of cowpea have been done (Pio-Ribeiro et al., 1978; Owolabi et al., 1988; Shoyinka et al., 1997; Martin et al., 2004; Orawu et al., 2005; Kareem and Taiwo, 2007; Taiwo et al., 2007; Akinjogunla et al., 2008).

TABLE 7 | The effect of virus/buffer treatments and age of plant at inoculation on yield parameters on commercial cultivar White and IITA lines IT81D-985 and TVu 76.

Virus Treatment	*X*	Flower number			Pod number			Pod length			Seed number/pod			Seed weight (g)		
		White	IT81D-985	TVu 76	White	IT81D-985	TVu 76	White	IT81D-985	TVu 76	White	IT81D-985	TVu 76	White	IT81D-985	TVu 76
CABMV	10	0	0	0	0	0	0	0	0	0	0	0	0	0	0	0
	30	2.67^{c}	0.33^{b}	2.33^{b}	0.67^{c}	0	0	5.2^{b}	0	0	3^{ab}	0	0	0.09^{c}	0	0
CMeV	10	0	0	0	0	0	0	0	0	0	0	0	0	0	0	0
	30	2.33^{c}	0	1.67^{b}	1.33^{bc}	0	0	6.33^{b}	0	0	0.67^{c}	0	0	0.11^{b}	0	0
SBMV	10	0	0	0	0	0	0	0	0	0	0	0	0	0	0	0
	30	3.3^{b}	0	0	2.67^{b}	0	0	7.03^{b}	0	0	3^{ab}	0	0	0.09^{c}	0	0
CABMV +	10	0	0	0	0	0	0	0	0	0	0	0	0	0	0	0
CMeV	30	3.67^{b}	0.33^{b}	0	1.67^{bc}	0	0	2.13^{c}	0	0	1.33^{b}	0	0	0.05^{c}	0	0
CABMV +	10	0	0	0	0	0	0	0	0	0	0	0	0	0	0	0
SBMV	30	1^{cd}	3^{ab}	0	0	0.67^{b}	0	0	3.57^{b}	0	0	2^{b}	0	0	0.04^{b}	0
CMeV +	10	0	0	0	0	0	0	0	0	0	0	0	0	0	0	0
SBMV	30	4^{b}	0	0	3.3^{b}	0	0	5.57^{b}	0	0	1.87^{b}	0	0	0.21^{a}	0	0
CABMV + CMeV +	10	0	0	0	0	0	0	0	0	0	0	0	0	0		0
SBMV	30	1^{cd}	0	0	0	0	0	0	0	0	0	0	0	0	0	0
BUFFER	10	6.33^{a}	4^{a}	7.33^{a}	5.67^{ab}	3.3^{a}	1.00^{b}	10.03^{a}	6.27^{b}	8.43^{b}	4.67^{a}	4.67	5	0.15^{b}	0.1768	0.1403^{b}
	30	7^{a}	3.67^{a}	2^{b}	6.3^{a}	2.67^{a}	1.33^{a}	9.27^{a}	10.1^{a}	9.23^{a}	4	4.67	5	0.15^{b}	0.178^{a}	0.1423^{a}

DAP^{x}, Growth parameters were measured 10 days after inoculation at 10 and 30 DAP. The age of the plant at 10 DAP was 20 days old, at 30 DAP, the age of the plant was 40 days old. CABMV, Cowpea aphid borne mosaic virus. CMeV, Cowpea mottle virus; SBMV, Southern bean mosaic virus. Each value is the mean of 3 replicates. In each column, means followed. Each value is the mean of 3 replicates. In each column, means followed same letter are not significantly different ($P = 0.05$), and all zero values carry the same letter according to Duncan's multiple range tests.

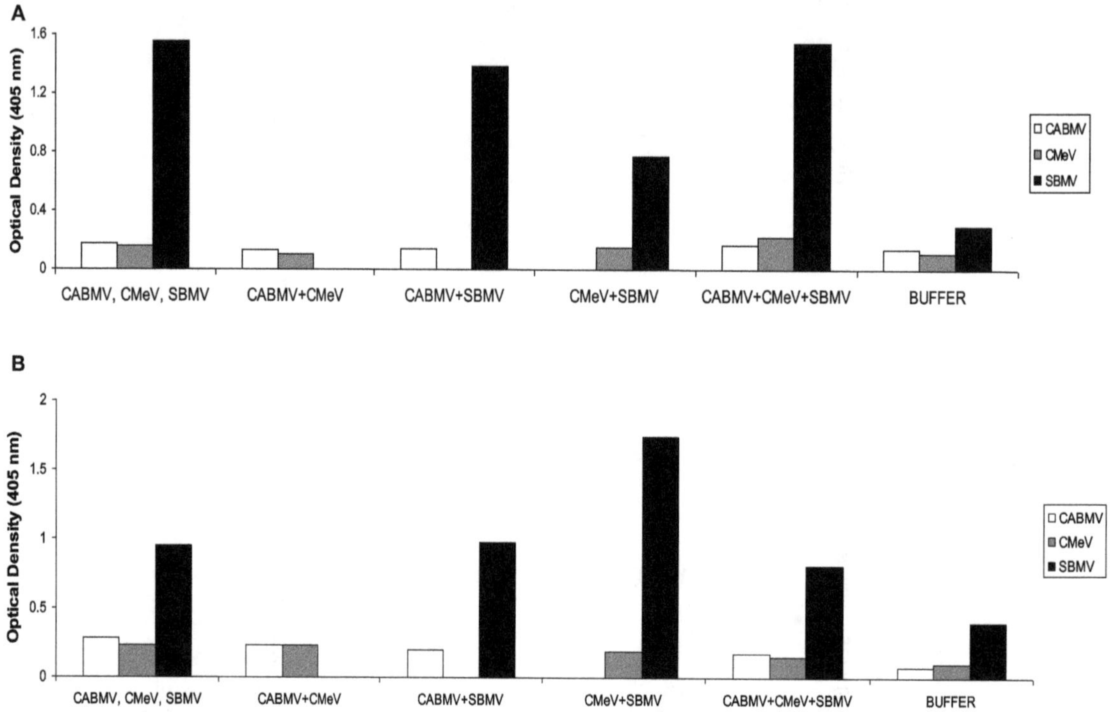

FIGURE 3 | Optical density readings (absorbance values) representing relative virus concentrations in leaf samples from IITA cowpea line IT81D-985 inoculated with single and mixed viruses at (A) 10 DAP (B) 30 DAP. The virus titres were determined by ACP ELISA. Values are averages of samples taken at 10 and 30 days after inoculation.

This study showed that the single and multiple virus infections have significant effects on the three Cowpea varieties investigated. The three Cowpea varieties- Commercial cultivar "White," IITA Lines IT 81D-985 and TVu 76 were susceptible to the single and mixed infections of Cowpea aphid-borne mosaic virus (CABMV), Southern bean mosaic virus (SBMV), and Cowpea mottle virus (CMeV). The common symptoms induced by the virus infections were mosaic, mottling, vein banding, chlorosis, and stunting. The investigation confirmed that triple infections in the three varieties produced more severe visual symptoms than the double or single infections. Severity scores ranging from 2 to 5 were recorded for the cowpea cultivars infected with the viruses. The high scores might be due to the fact that cowpea is the primary host of these viruses. The severity of the virus diseases varied depending on the cowpea variety, the type of treatment and age at the onset of virus treatment. The severity of virus infection was higher at 10 DAP than at 30 DAP; the severity also increased from 10 DAI to 30 DAI. In line with this, Uyemoto et al. (1981) reported that early infection of plants by viruses results in a more drastic response than infections at an advanced stage. Kareem and Akinjogunla (2008) had also previously reported that the infection of cowpea with three unrelated viruses resulted in increased symptom severity at early stage of plant growth. The reason for this is that at an early stage of growth, cultivars might not have built up enough defense mechanism to combat diseases.

Generally, all the virus treatments caused some degree of reduction in vegetative growth and total yield. The virus-infected plants were shorter than buffer inoculated plants due to shorter internodes with fewer leaves. However, in IT81D-985, an anomaly was observed as some of the virus treated plants produced more leaves than the control and were taller. In spite of this, there was hardly any yield from this cultivar. This means that the increase in the number of leaves and plant height was due to the reaction to the virus presence. On the other hand, the most common symptom in TVu 76 was defoliation. There was also hardly any yield from the infected TVu 76 plants. The commercial cultivar, White was the least susceptible in terms of yield. The different responses could be due to the varying tolerance level of the cultivars. Our results show that virus diseases can damage an entire crop, leaves, stems, flowers, and seeds in these varieties under investigation.

The effect of CABMV and CMeV on "White" and TVu 76 led to at least 60% reduction in leaf number and ultimately caused the withering of all the leaves inoculated at 10 days after planting. Our results also did not show that the double virus infections had a more severe effect than the single virus infections. This supports the report of Anjos et al. (1992) that not all the combinations of unrelated viruses result in increased symptoms. In some cases, the single virus had a more devastating effect on the crop than double infections involving that same virus. For example, the IT81D-985 plants doubly infected with CABMV and SBMV were the only ones that produced seeds. Wells and Deba (1961) reported a similar occurrence of 100% loss in yield in cowpea due to the single infection of Cowpea mosaic virus (CpMV).

Raheja and Leleji (1974) had also reported that CABMV caused a complete loss of yield of commercial cowpea.

Infections at 10 DAP with CABMV in "White," IT81D-985 and TVu 76 resulted in the death of some of the plants but not in those infected at 30 DAP. Of the 3 Cowpea varieties, the contrast in the impact of early and late infections on yield was most pronounced in the "White" variety. In the White variety, with the exception of plants that received the mixture of CABMV and SBMV, all the plants infected at 30 days after planting produced seeds unlike those infected at 10 DAP (**Table 7**). The results also correlate with previous information that the most severe effect on yield occurs as a result of early infection (Gay and Winstead, 1970; Kareem and Taiwo, 2007). This trend was also observed in the growth traits, as the heights of some "White" plants infected at 30 DAP were not significantly different from the height of the control plants. This compares favorably with the results of investigation by Owolabi et al. (1988), who reported that Cultivar "Ife Brown" plants inoculated with CpMV alone and in combination with BlCMV at the initiation of flowering showed no significant difference in yield compared with the control. A similar trend was also observed in TVu 76. Despite the observation that the vegetative growth in IT81D-985 was an aberration from the other two, some of the IT81D-985 plants showed this trend.

It was also observed that the response of the cowpea varieties to virus infections differed from one variety to another. While the major symptom of virus treated TVu 76 was defoliation, the response of the IT81D-985 plants to infection was increased leaf production. IT81D-985 is therefore sturdier than TVu 76 in response to virus infections. It is also significant that IT81D-985 was able to overcome infection by CABMV + SBMV and produced seeds, whereas, the White variety had no yield with the same treatment. The White variety shows similar characteristics to another local variety, Oloyin, reported by Kareem and Taiwo (2007) in that it was the most resistant of the three cultivars/varieties and produced seeds at 30 DAP.

Our results for growth and yield parameters for uninfected (buffer inoculated) IT81D-985 plants compare favorably with a previous report by Ekpo et al. (2012). They had evaluated plant height and leaf number of IT 81D-985 among other cowpea lines. In their report, at 21 days, plant height of IT81D-985 was 18.4 ± 1.38 cm and at 14 days, the number of leaves was 8.42 ± 0.2, our average value at 20 days old for negative controls were 19.33 cm and 8 for plant height and number of leaves, respectively. However, at advanced stages of plant growth, our findings were different from Ekpo et al. (2012). At 42 days old, they recorded an average plant height of 89.8 ± 13.72 cm and 18.25 ± 0.32 leaves, whereas at 40 days old, our healthy IT81D-985 plants measured an average of 34.19 cm and produced an average of 9 leaves. For yield parameters, the IT81D-985 plants inoculated with buffer at 10 DAP and 30 DAP produced pods with an average length of 6.27 and 10.1 cm, respectively, bearing similarity to the average pod length of 9.1 ± 2.26 cm recorded by Ekpo et al. (2012). Although our IT81D-985 plants inoculated with the buffer at 30 DAP did not grow as tall as expected, we ruled out the possibility that these group of plants did not grow optimally because serological evaluation showed that they were virus free and they had a good yield.

Serological detection of the three viruses in all the treatments for White and TVu 76 validated that the symptoms and reductions in growth traits were due to virus presence. However, in line IT81D-985 virus treated plants, ELISA results showed substantial accumulation of SBMV but not significant concentrations of CABMV or CMeV. Based on serological analysis, line IT81D-985 appeared resistant to CABMV and CMeV. In spite of the symptomatic IT81D-985 leaf samples having low titres of these two viruses, and more vegetative growth, there was still a significant reduction in yield.

Although Shoyinka et al. (1978) reported that the response of plants most times to mixed infections is more of synergism, and Pio-Ribeiro et al. (1978) reported that the symptoms induced by mixed infection in California Black eye by cucumber mosaic virus and BICMV were more severe and distinct from the relatively mild symptoms caused by either of the viruses, it is clear from our results that the interaction of the viruses in the double virus infections could not be suggestive of synergism, because the overall effect of these mixed treatments was not stronger than the sum of the effects caused by each of the viruses separately in all the cowpea varieties. The quantitative analysis of the growth parameters provided evidence for an additive relationship in the double virus infections in cultivar White and line TVu 76. With the unexpected results of increase in plant height and number of leaves of virus treated IT81D-985, in comparison with the healthy plants, we could not assess this line for synergistic interactions of viruses as in cultivar White and TVu 76.

It has been well documented for decades that diseases caused by viruses have been responsible for great damage, causing serious losses in cowpea crop yield in several countries (Lima et al., 1979; Damiri et al., 2013). It is also evident that our observations here, agree with earlier findings of Lana and Adegbola (1977), who reported that economic injury to cowpea due to virus infection depends on three factors- virus isolate, tolerance/resistance of the infected Cowpea cultivar and most importantly the age of the host plant at the time of infection. In summary, the results obtained confirm that plant virus infections cause the decline of cowpea plants, lower product quality, and ultimately result in crop losses.

The basic approaches for the control of viruses have not been overly successful because of non-curable nature of viral infections, lack of resistant cultivars with good agronomic quality and diversity of viruses and their natural vectors (Lecoq, 1998). The implications of these results are that there is a need to intensify efforts in developing advanced cowpea breeding lines/cultivars with multiple resistance to economically important viruses. With few virus control strategies available, pathogen-derived resistance is a promising tool in viral disease management. Since its first demonstration by Powell-Abel et al. (1986), it has been used with success against several other virus groups (Hull and Davies, 1992; Bau et al., 2003). It offers great potential to develop in a short time plants with highly durable resistance and broad efficacy (Kaniewski and Lawson, 1998). IITA reported that one of its 2003 milestones was the initiation of field testing of transgenic cowpea plants. Cowpea lines resistant to CABMV have been identified (Dhanasekar and Reddy, 2015).

The benefits of increased cowpea production include improved nutrition for humans and livestock, improved soil properties and substantial opportunities for greater income. The control of these viruses therefore is crucial to sustainable cowpea production most especially in sub-Saharan Africa.

Acknowledgments

This research was carried out as part of a Master's thesis by the first author under the supervision of Professor Monilola A. Taiwo at the University of Lagos. We thank Mr. Oye for excellent technical assistance in the greenhouse, and Chidiadi C. Nwachukwu for his assistance with the initial sorting of the data.

References

Abbott, W. S. (1925). A method of computing the effectiveness of an insecticide. *J. Econ. Entomol.* 18, 265–267. doi: 10.1093/jee/18.2.265a

Aboul-Ata, A. E., Allen, D. J., Thottappilly, G., and Rossel, H. W. (1982). Variation in rate of seed transmission of cowpea aphid-borne mosaic virus in cowpea. *Trop. Grain Legume Bull.* 25, 2–7.

Akinjogunla, O. J., Taiwo, M. A., and Kareem, K. T. (2008). Immunological and molecular diagnostic methods for detection of viruses infecting cowpea (*Vigna unguiculata*). *Afr. J. Biotechnol.* 7, 2099–2103. doi: 10.5897/AJB08.120

Allen, D. J., Anno-Nyako, F. O., Ochieng, R. S., and Ratinam, M. (1981). Beetle transmission of cowpea mottle and southern bean mosaic viruses in West Africa. *Trop. Agric. (Trinidad)* 58, 171–175.

Anjos, J. R., Jarlfors, U., and Ghabrial, S. A. (1992). Soybean mosaic virus, Potyvirus enhances the titre of two comoviruses in dually infected soybean plants. *Phytopathology* 82, 1022–1027. doi: 10.1094/Phyto-82-1022

Atiri, G. I. (1982). *Virus-Vector-Host Relationship of Cowpea Aphid-Borne Mosaic Virus in Cowpea (Vigna Unguiculata (L). Walp).* Ph.D. Thesis, University of Ibadan.

Atiri, G. I., Ekpo, E. J. A., and Thottappilly, G. (1984). The effect of aphid-resistance in cowpea infestation and development of *Aphis craccivora* and the transmission of cowpea aphid-borne mosaic virus. *Ann. Appl. Biol.* 104, 339–346. doi: 10.1111/j.1744-7348.1984.tb05619.x

Bashir, M., Ahmad, Z., and Ghafoor, A. (2002). Cowpea aphid borne mosaic virus-a review. *Int. J. Pest Manage.* 48, 155–168. doi: 10.1080/09670870110118722

Bau, H. J., Cheng, Y. H., Yang, J. S., and Yeh, S. D. (2003). Broad-spectrum resistance to different geographic strains of Papaya ringspot virus in coat protein transgenic papaya. *Phytopathology* 93, 112–120. doi: 10.1094/PHYTO.2003.93.1.112

Bock, K. R. (1973). East African strains of cowpea aphid-borne mosaic virus. *Ann. Appl. Biol.* 74, 75–83.

Bock, K. R., and Conti, M. (1974). "Cowpea aphid-borne mosaic virus," in *CMI/AAB Description of Plant Viruses,* No. 134 (Surrey: Kew).

Bos, L. (1970). The identification of three new viruses isolated from *Wisteria* and *Pisum* in the Netherlands, and the problem of variation within the potato virus Y group. *Neth. J. Plant Pathol.* 76, 8–46. doi: 10.1007/BF01976763

Burke, D. W., Ditshipi, P., and DeMooy, C. J. (1986). Virus diseases of cowpeas in dry land and irrigated plots in Botswana. *Plant Dis.* 70, 801.

Cho, J. D., Kim, J. S., Choi, H. S., La, Y. J., and Kim, K. S. (2000). Ultrastructural aspects of mixed infections of watermelon mosaic *Potyvirus* isolated from pumpkin and cucumber green mottle mosaic tobamovirus from watermelon. *Plant Pathol. J.* 16, 216–221.

Damiri, D., Al-Shahwan, I. M., Al-Saleh, O. A., Abdalla, O. A., and Amer, M. A. (2013). Identification and Characterization of cowpea aphid-borne mosaic virus isolates in Saudi Arabia. *J. Plant Pathol.* 95, 79–85. doi: 10.4454/JPP.V95I1.019

Dhanasekar, P., and Reddy, K. (2015). Serological screening of cowpea genotypes for resistance against Cowpea Aphid Borne Mosaic virus using DAS-ELISA. *Asian J. Plant Pathol.* 9, 83–90. doi: 10.3923/ajppaj.2015.83.90

Ekpo, I. A., Agbor, R. B., Osuagwu, A. N., Okpako, E. C., and Ekanem, B. E. (2012). Evaluation of eight cowpea (*Vigna unguiculata,* L. Walp) spcies for yield and associated traits. *Int. J. Pure Appl. Sci. Technol.* 12, 1–7.

Fauquet, C., and Thouvenel, J. C. (1980). *Plant Viral Diseases in the Ivory Coast, Documentation Techniques, No. 46.* Reedition 1987. (Paris: ORSTOM).

Gaikward, D. G., and Thottappilly, D. (1988). Occurrence of southern bean mosaic virus in Senegal. *J. Phytopathol.* 121, 366–369.

Gay, D. J., and Winstead, E. E. (1970). Seedborne viruses and fungi from southern pea grown in eight states. *Plant Dis. Rep.* 54, 243–245.

Gisi, U. (1996). Synergistic interactions of fungicides in mixture. *Phytopathology* 86, 1273–1279.

Gisi, U., Binder, H., and Rimbach, E. (1985). Synergistic interactions of fungicides with different modes of action. *Trans. Br. Mycol. Soc.* 85, 299–306. doi: 10.1016/S0007-1536(85)80192-3

Gumedzoe, M. Y. (1985). *Studies of Variability of the Cowpea Aphid-Borne Mosaic Virus (CAMBV) Complex in Nigeria.* Ph.D. Thesis. University of Laval, Quebec, QC.

Gumedzoe, M. Y., Sunu, D. Y., and Thottappilly, G. (1989). "Inventaire des principaux viroses du niébé," in *Papier Presente à la Conférence Biennale de l'Association Ouest Africaine, Cotonou le 10–15, Septembre.*

Hughes, J. A., and Shoyinka, S.A. (2003). "Overview of viruses of legumes other than groundnut in Africa," in *Plant Virology in Sub-Sahara Africa,* eds J. A. Hughes and J. Odu (Proceedings of a conference organized by IITA Ibadan, Nigeria: International Institute of Tropical Agriculture), 553–68.

Hull, R., and Davies, J. W. (1992). Approaches to non-conventional control of plant virus disease. *Crit. Rev. Plant Sci.* 11, 17–33. doi: 10.1080/07352689209382328

International Institute of Tropical Agriculture (IITA) (2009). *IITA Cereals and Legume Systems.* Available online at: www.iita.org/cms/details/cowpeaproject_details.asp

Kaniewski, W., and Lawson, C. (1998). "Coat protein and replicase-mediated resistance to plant viruses," in *Plant Virus Disease Control,* eds A. Hadidi, R. K. Khartepal, and K. Koagenezang (APS press), 65–78.

Kareem, K. T., and Akinjogunla, O. J. (2008). "Severity of infection of three unrelated viruses on cowpea cultivars," *Proceedings of the Third Conference on Science and National Development,* (Abeokuta), 102–107.

Kareem, K. T., and Taiwo, M. A. (2007). Interaction of viruses in cowpea: effect on growth and yield parameters. *Virol. J.* 4:15. doi: 10.1186/1743-422X-4-15

Karungi, J., Adipala, E., Ogenga-Latigo, M. W., Kyamanywa, S., and Oyobo, N. (2000). Pest management in cowpea. Part 1. Influence of planting time and planting density on cowpea field pests infestation in eastern Uganda. *Crop Prot.* 19, 231–236. doi: 10.1016/S0261-2194(00)00013-2

Kuhn, C. W. (1963). Field occurrence and properties of cowpea strain of southern bean mosaic virus. *Phytopathology* 53, 732–733.

Ladipo, J. L. (1976). A vein-banding strain of cowpea aphid-borne mosaic virus in Nigeria. *Niger. J. Sci.* 10, 77–78.

Lamptey, P. N., and Hamilton, R. I. (1974). A new cowpea strain of southern bean mosaic virus from Ghana. *Phytopathology* 64, 1100–1104. doi: 10.1094/Phyto-64-1100

Lana, A. O., and Adegbola, M. O. K. (1977). Important virus diseases in West African crops. *CMI Rev. Plant Pathol.* 56, 854–855.

Lecoq, H. (1998). "Control of plant virus diseases by cross protection," in *Plant Virus Disease Control,* eds A. Hadidi, R. K. Khartepal, and K. Koagenezang (St. Paul, MN: APS press), 33–40.

Lephale, S., Addo-Bediako, A., and Ayodele, V. (2012). Susceptibility of seven cowpea cultivars (*Vigna unguiculata*) to cowpea beetle (*Callosobruchus maculates*). *Agric. Sci. Res. J.* 2, 65–69.

Li, C.-D., Fatokun, C. A., Ubi, B., Singh, B. B., and Scoles, G. J. (2001). Determining genetic similarities and relationships among cowpea breeding lines and cultivars by microsatellite markers. *Crop Sci.* 41, 189–197. doi: 10.2135/cropsci2001.411189x

Lima, J. A. A., Purcifull, D. E., and Hiebert, E. (1979). Purification, partial characterization and serology of blackeye cowpea mosaic virus. *Phytopathology* 69, 1252–1258. doi: 10.1094/Phyto-69-1252

Little, T. M., and Hills, F. J. (1972). "Transformations," *Statistical Methods in Agricultural Research,* (Davis, CA: University of California), 103–120.

Lovisolo, O., and Conti, M. (1966). Identification of an aphid-transmitted cowpea mosaic virus. *Neth. J. Plant Pathol.* 72, 265–269. doi: 10.1007/BF02650216

Martin, E. M., Cho, J. D., Kim, S. C., Goeke, S. C., Kim, K. S., and Gergerich, R. C. (2004). Novel cytopathological structures induced by mixed infection of unrelated plant viruses. *Phytopathology* 94, 111–119. doi: 10.1094/PHYTO.2004.94.1.111

Murphy, J. F., and Bowen, K. L. (2006). Synergistic disease in pepper caused by the mixed infection of *Cucumber mosaic virus* and *Pepper mottle virus. Phytopathology* 96, 240–247. doi: 10.1094/PHYTO-96-0240

Naidu, R. A., and Hughes, J. A. (2003). "Methods for the detection of plant viral diseases," in *Plant Virology in Sub-Sahara Africa*, eds J. A. Hughes and J. Odu (Proceedings of a conference organized by IITA Ibadan, Nigeria: International Institute of Tropical Agriculture), 233–260.

Orawu, M., Melis, R., De Milliano, W., Laing, M., and Adipala, E. (2005). Occurrence and prevalence of cowpea virus diseases in Uganda. *Afr. Crop Sci. Proc.* 7, 1279–1283.

Owolabi, A. T., Taiwo, M. A., and Mabadeje, S. A. (1988). Effects of single and mixed inoculations with blackeye cowpea mosaic virus on two Nigerian cowpea cultivars. *Niger. J. Basic Appl. Sci.* 2, 25–33.

Patel, P. N., and Kuwite, C. (1982). Prevalence of cowpea aphid-borne mosaic virus and two strains of cowpea mosaic virus in Tanzania. *Indian Phytopathol.* 35, 467–472.

Pio-Ribeiro, G., Wyatt, S. D., and Kuhn, C. W. (1978). Cowpea stunt: a disease caused by the synergistic interaction of two viruses. *Phytopathology* 68, 1260–1265. doi: 10.1094/Phyto-68-1260

Powell-Abel, P., Nelson, R. S., De, B., Hoffmann, N., Rogers, S. G., Fraley, R. T., et al. (1986). Delay of disease development in transgenic plants that express the tobacco mosaic virus coat protein gene. *Science* 232, 738–742. doi: 10.1126/science.3457472

Raheja, A. K., and Leleji, O. I. (1974). An aphid-borne virus disease of irrigated cowpea in northern Nigeria. *Plant Dis. Rep.* 58, 1080–1084.

Raman, K. V., Singh, S. S., and van Emden, H. F. (1980). Mechanisms of resistance to leafhopper damage in cowpea. *Entomol. Soc. Am.* 73, 484–488. doi: 10.1093/jee/73.4.484

Robertson, D. G. (1963). "Cowpea virus research in Nigeria," in *Proceedings of the First Nigerian Grain Legume Conference. Samaru, Zaria.* Institute of Agricultural Research, Ahmadu Bello University, Zaria, Nigeria.

Robertson, D. G. (1966). *Seed-Borne Viruses of Cowpea in Nigeria.* B.Sc. Thesis, University of Oxford, UK.

Rossel, H. W., and Thottappilly, G. (1985). *Virus Diseases of Important Food Crops in Tropical Africa.* (Ibadan: IITA Publication Series), 61.

Shepherd, R. J., and Fulton, R. W. (1962). Identity of a seed-borne virus of cowpea. *Phytopathology* 52, 489–493.

Shi, X. M., Pruss, G., Ge, X., Sriskanda, V., and Vance, V. B. (1996). *Identification of Plant Viral Synergism Genes.* Doctoral dissertation, University of South Carolina.

Shoyinka, S. A., Bozarth, R. F., Reese, J., and Okunsanya, B. O. (1979). Field occurrence and identification of southern bean mosaic virus (cowpea strain) in Nigeria. *Turrialba* 29, 111–116.

Shoyinka, S. A., Bozarth, R. F., Reese, J., and Rossel, H. W. (1978). Cowpea mottle virus: a seed-borne virus with distinctive properties infecting cowpea in Nigeria. *Phytopathology* 68, 693–699. doi: 10.1094/Phyto-68-693

Shoyinka, S. A., Thottappilly, G., Adebayo, G. G., and Anno-Nyako, F. O. (1997). Survey on cowpea virus incidence and distribution in Nigeria. *Int. J. Pest Manage.* 43, 127–132. doi: 10.1080/096708797228816

Singh, B. B. (2011). *Cowpea Genetics and Breeding - A Historical Perspective.* Brasil: IV Reunião de Biofort Fiação.

Singh, R., and Singh, R. (1974). Natural infection of sickle senna (*Cassia tora* L.) and cowpea (*Vigna sinensis savi*) plants by some new strains of southern bean mosaic virus. *Partugalia Data Biol.* 13, 87–89.

Taiwo, M. A. (2003). "Viruses infecting legumes in Nigeria: case history," in *Plant Virology in Sub-Saharan Africa,* eds J. A. Hughes and B. Odu (Proceedings of Plant Virology Ibadan, Nigeria: IITA), 364–378.

Taiwo, M. A., Kareem, K. T., Nsa, I. Y., and Hughes, J. A. (2007). Cowpea viruses: effect of single and mixed infections on symptomatology and virus concentration. *Virol. J.* 4:95. doi: 10.1186/1743-422X-4-95

Thottappilly, G., and Rossel, H. W. (1988). Occurrence of cowpea mottle virus and other viruses (cowpea yellow mosaic virus, southern bean mosaic virus) in cowpea. *FAO Plant Prot. Bull.* 36, 184–185.

Thottappilly, G., and Rossel, H. W. (1992). Virus of cowpea in tropical Africa. *Trop. Pest Manage.* 38, 337–348. doi: 10.1080/09670879209371724

Thouvenel, J. C. (1988). A serious disease caused by cowpea mottle virus in Ivory Coast. *Plant Dis.* 72, 363. doi: 10.1094/PD-72-0363F

Tremaine, J. H., and Hamilton, R. I. (1983). "Southern bean mosaic virus," in *CMI/AAB Descriptions of Plant Viruses*, No. 274 (Surrey: Kew).

Untiveros, M., Fuentes, S., and Salazar, L. F. (2007). Synergistic interaction of Sweet potato chlorotic stunt virus (Crinivirus) with carla-, cucumo-, ipomo-, and potyviruses infecting sweet potato. *Plant Dis.* 91, 669–676. doi: 10.1094/PDIS-91-6-0669

Uyemoto, J. K., Claflin, L. E., Wilson, D. L., and Rayney, R. J. (1981). Maize chlorotic mottle and maize dwarf mosaic viruses: effect of single and double inoculations on symptomatology and yield. *Plant Dis.* 65, 9–41. doi: 10.1094/PD-65-39

Vidano, C., and Conti, M. (1965). Transmissione con afidi d'un 'cowpea mosaic virus' isolato da *Vigna sinensis* Endl. *Ann. Accad. Sci. Torino (Italia)*, 99, 1041–1050.

Walkey, D. G. A. (1985). *Applied Plant Virology.* (New York, NY: John Wiley & Sons), 329.

Wang, Y., Gaba, V., Yang, J., Palukaitis, P., and Gal-on, A. (2002). Characterization of synergy between Cucumber mosaic virus and potyviruses in cucurbit hosts. *Phytopathology* 92, 51–58. doi: 10.1094/PHYTO.2002.92.1.51

Wells, D. G., and Deba, R. (1961). Sources of resistance to cowpea yellow mosaic virus. *Plant Dis. Rep.* 45, 878–881.

Conflict of Interest Statement: The authors declare that the research was conducted in the absence of any commercial or financial relationships that could be construed as a potential conflict of interest.

14

Influence of *Rhizoctonia solani* and *Trichoderma* spp. in Growth of Bean (*Phaseolus vulgaris* L.) and in the Induction of Plant Defense-related Genes

Edited by:
Antonio M. De Ron,
Spanish National Research Council - Misión Biológica de Galicia, Spain

Reviewed by:
Cristina Cabaleiro Sobrino,
Universidade de Santiago de Compostela, Spain
Eleonora Cominelli,
Istituto di Biologia e Biotecnologia Agraria, Consiglio Nazionale delle Ricerche, Italy

***Correspondence:**
Pedro A. Casquero,
Research Group of Engineering and Sustainable Agriculture, Natural Resources Institute, University of León, Av. Portugal 41, 24071 León, Spain
pedro-casquero@unileon.es

Sara Mayo[1], *Santiago Gutiérrez*[2], *Monica G. Malmierca*[2], *Alicia Lorenzana*[1], *M. Piedad Campelo*[1], *Rosa Hermosa*[3] *and Pedro A. Casquero*[1*]

[1] Research Group of Engineering and Sustainable Agriculture, Natural Resources Institute, University of León, León, Spain, [2] Area of Microbiology, University School of Agricultural Engineers, University of León, Ponferrada, Spain, [3] Department of Microbiology and Genetics, Spanish-Portuguese Centre for Agricultural Research, University of Salamanca, Salamanca, Spain

Many *Trichoderma* species are well-known for their ability to promote plant growth and defense. We study how the interaction of bean plants with *R. solani* and/or *Trichoderma* affect the plants growth and the level of expression of defense-related genes. *Trichoderma* isolates were evaluated *in vitro* for their potential to antagonize *R. solani*. Bioassays were performed in climatic chambers and development of the plants was evaluated. The effect of *Trichoderma* treatment and/or *R. solani* infection on the expression of bean defense-related genes was analyzed by real-time PCR and the production of ergosterol and squalene was quantified. *In vitro* growth inhibition of *R. solani* was between 86 and 58%. In *in vivo* assays, the bean plants treated with *Trichoderma harzianum* T019 always had an increased size respect to control and the plants treated with this isolate did not decrease their size in presence of *R. solani*. The interaction of plants with *R. solani* and/or *Trichoderma* affects the level of expression of seven defense-related genes. Squalene and ergosterol production differences were found among the *Trichoderma* isolates, T019 showing the highest values for both compounds. *T. harzianum* T019 shows a positive effect on the level of resistance of bean plants to *R. solani*. This strain induces the expression of plant defense-related genes and produces a higher level of ergosterol, indicating its ability to grow at a higher rate in the soil, which would explain its positive effects on plant growth and defense in the presence of the pathogen.

Keywords: antifungal activity, defense-related genes, qPCR, ergosterol, squalene

Introduction

The common bean (*Phaseolus vulgaris* L.) is the third most important food legume crop worldwide, surpassed only by the soybean [*Glycine max* (L.) Merr.] and peanut (*Arachis hypogea* L.). Among the southern countries of the European Union, Spain together with Italy and Greece are the main common bean producers. León, a province located at the northwest of Spain, is the main producer province by quantity and quality, with almost 45% of Spanish production in 2014. Socio-economic conditions of León province enabled possible the maintenance of local varieties in traditional cropping systems, which are based in small-scale farms (Casquero et al., 2006). The high quality of this legume has been awarded with a Protected Geographic Indication (PGI) (EC Reg. n.256/2010 published on 26 March 2010, OJEU L880/17). In the last few years, however dry bean production has gone through difficulties due to relatively low yields (mainly caused by fungus, virus, and bacteria) and insufficient income for growers.

Root rots are the main diseases caused by soil fungi having their incidence on bean yield. *Rhizoctonia solani* JG Kühn [Teleomorph: *Thanatephorus cucumeris* (AB Frank) Donk] is the main root rot in León, being detected in 91.8% of affected plants in an evaluation of its occurrence in bean plants (Valenciano et al., 2006). Plant infection occurs through wounds or by a coating of an organ with mycelium, which tears the cuticle and penetrates the epidermis. This pathogen is more aggressive at temperatures between 15 and 18°C and in moist soils. It is a necrotrophic pathogen, distributed worldwide (Guerrero-González et al., 2011). *R. solani* is one of the root and hypocotyl pathogen that causes most economic losses worldwide.

Trichoderma (Teleomorph: *Hypocrea*) is a fungal genus that is found in the soil. It is a secondary fast growing opportunistic invasive, which produces large numbers of spores, enzymes able to degrade the fungal cell wall (chitinases, glucanases, and proteases) and compounds with antimicrobial activity. Many *Trichoderma* species are also well known as biocontrol agents (BCA) of important phytopathogenic fungi. The primary mechanisms of biocontrol used by *Trichoderma* in direct confrontation with pathogenic fungi are the mycoparasitism (Papavizas, 1985) antibiosis, and competition for nutrients with the pathogen (Harman and Kubicek, 1998).

Many *Trichoderma* species colonize the root surface and cause substantial changes in plant metabolism (Harman et al., 2004). The physical interaction between *Trichoderma* and plants is limited to the first cell layer of the epidermis and the root bark. This symbiotic relationship would thus protect plants against pathogens. *Trichoderma* induces the expression of genes involved in defense response and promotes plant growth, root development and nutrient availability (Hermosa et al., 2012). During the *Trichoderma*-plant interaction various classes of metabolites could induce resistance such as proteins with enzymatic activity, low molecular weight compounds, related to the fungal or the plant cell wall, originated by the enzymatic activity of *Trichoderma* (Woo et al., 2006; Woo and Lorito, 2007) and other secondary metabolites that trigger plant defense mechanisms against the pathogen (Hermosa et al., 2012; Malmierca et al., 2014), by inducing the expression of pathogenesis-related (PR) proteins that reduce the diseases symptoms. Thus, when the plant contacts with a pathogen it is activated a mechanism of systemic acquired resistance (SAR). However, when they interact with a non-pathogen organism the plants activated a mechanism for induced systemic resistance (ISR) (Hermosa et al., 2013; Mukherjee et al., 2013).

Squalene is a polyunsaturated terpene that is an intermediate in the ergosterol biosynthetic pathway, which has an essential function in the fungal cell structure. The levels of squalene will influence the level of ergosterol biosynthesis (Garaiová et al., 2013). In addition to its structural function, and as a result of its importance in fungal development, ergosterol is also able to activate the expression of a number of defense genes and could increase the resistance of plants against pathogens (Lochman and Mikeš, 2005).

In this work 23 *Trichoderma* isolates were collected from bean fields. These isolates were used to study their effect on the growth of bean plants, and also in the defense response of plants against the phytopathogen *R. solani*. Thus, parameters as plant growth in the presence of the pathogen and/or the different *Trichoderma* isolates were evaluated, also analyzing the level of expression of defense-related genes in plants treated with the selected *Trichoderma* isolate.

Materials and Methods

Trichoderma and *R. solani* Isolates and Culture Collections

The present study was conducted with twenty-three isolates of *Trichoderma* (**Table 1**) collected from the production area of the Protected Geographical Indication (PGI), called "Alubia La Bañeza—León," without any genetic manipulation and three isolates from other collections. The *Trichoderma* isolates were stored in the collection "Pathogens and Antagonists of the Laboratory Diagnosis of Pests and Diseases" (PALDPD, University of León, León, Spain). *R. solani* R43 was also collected from plants of the same PGI and selected by its high virulence (**Table 1**).

In vitro Antifungal Assays

Trichoderma isolates were evaluated for their *in vitro* potential to antagonize the plant pathogenic fungus *R. solani* using two different tests. For all tests, plugs of 7 mm diameter collected from the edge of growing fungal colonies were used to inoculate potato dextrose agar medium (PDA) in sterile Petri dishes of 9 cm diameter. The dishes were incubated in the dark at 22°C for 7 days.

The aim of these tests was to study the percentage of *R. solani* growth inhibition caused by the different *Trichoderma* isolates.

The antifungal assay on membranes was used to quantify the ability of the *Trichoderma* isolates to produce metabolites and/or enzymes with inhibitory activity against *R. solani*. The surface of Petri dishes containing PDA medium was overlaid with a sterile cellophane membrane. *Trichoderma* plugs, extracted from PDA dishes grown for 7 days at 22°C, were placed in the center of the dish with the cellophane sheet, containing PDA medium, and incubated for 48 h at 22°C. Then, the cellophane membranes along with the mycelia of *Trichoderma* isolates were removed and *R. solani* plugs were placed in the same plates. Growth of *R. solani* was recorded after 72 h to calculate the percentage of pathogen growth inhibition (**Figure 1A**). Control PDA plates of *R. solani*, where *Trichoderma* spp. had not been previously grown, were also prepared in the same conditions as above.

TABLE 1 | *Trichoderma* and *Rhizoctonia* strains used in this study.

Lab. Code	Culture collection***	Localization/Received as	Identified as/References
T001*	PAULET20	Bustillo del Páramo (León)	*T. harzianum*
T002*	PAULET21	Roperuelos del Páramo (León)	*Trichoderma* spp.
T003*	PAULET22	Bercianos del Páramo (León)	*Trichoderma* spp.
T004*	PAULET23	Riego de la Vega (León)	*T. gamsii*
T005*	PAULET24	Valderrey (León)	*T. longibrachiatum*
T006*	PAULET25	San Esteban de Nogales (León)	*Trichoderma* spp.
T007*	PAULET26	Villamejil (León)	*Trichoderma* spp.
T008*	PAULET27	Fresno de la Vega (León)	*T. citrinoviride*
T009	PAULET28	Bustillo del Páramo (León)	*Trichoderma* spp.
T010	PAULET29	Bustillo del Páramo (León)	*T. harzianum*
T011*	PAULET30	San Pedro de Bercianos (León)	*T. harzianum*
T012*	PAULET31	Quintana del Castillo (León)	*T. harzianum*
T013*	PAULET32	Quintana del Castillo (León)	*T. atroviride*
T014	PAULET33	Santa Marina del Rey (León)	*Trichoderma* spp.
T015*	PAULET34	San Cristobal de la Polantera (León)	*T. harzianum*
T016	PAULET35	Bustillo del Páramo (León)	*Trichoderma* spp.
T017	PAULET36	Bustillo del Páramo (León)	*Trichoderma* spp.
T018	PAULET37	Urdiales del Páramo (León)	*Trichoderma* spp.
T019*	PAULET38	Carrizo de la Ribera (León)	*T. harzianum*
T020	PAULET39	Soto de la Vega (León)	*T. harzianum*
T021	PAULET40	Pozuelo del Páramo (León)	*T. harzianum*
T022	PAULET41	Villaornate (León)	*Trichoderma* spp.
T023	PAULET42	Cabreros del Río (León)	*Trichoderma* spp.
T024*	IMI 352941	*T. atroviride*	Hermosa et al., 2000
T025*	NBT 59	*T. virens*	Hermosa et al., 2004
T34**	CECT 2413	*T. harzianum*	Kullnig et al., 2001
R43	PAULER006	Santa María del Páramo (León)	*Rhizoctonia solani*

**Strains included in in vivo assays.*

***Strain only used in ergosterol and squalene assays.*

****All PAULE strains are in Pathogens and Antagonists of the Laboratory Diagnosis of Pests and Diseases (PALDPD) Collection, University of León, León, Spain; IMI, CABI Bioscience (Egham); NBT, Newbiotechnic S.A. (Seville); CECT, Spanish Type Culture Collection, Burjassot, Spain.*

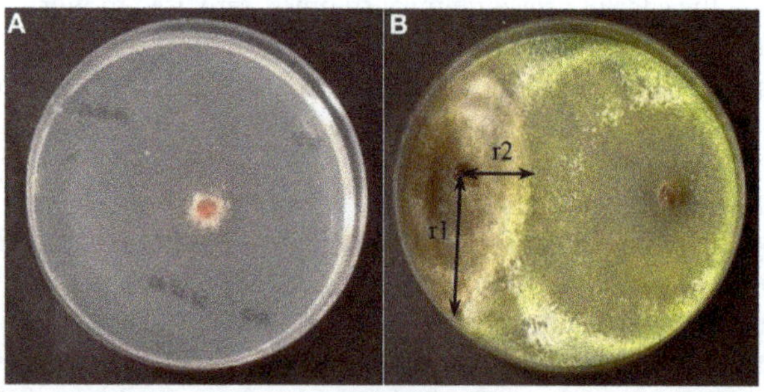

FIGURE 1 | (A) Inhibition of *R. solani* growth by metabolites of Trichoderma spp in assay on membranes **(B)** Growth of Trichoderma (right) and *R. solani* (left) in direct confrontation assays. Parameters to calculate the percentage of inhibition of growth in direct confrontation (ID) % ID = [(r1−r2)/r1] x 100.

The percentage of inhibition (IM) was calculated after 3 days of growth of *R. solani* in this medium using the formula %IM = [(C−T)/C]x100 (C: diameter of the *R. solani* control; T: diameter of *R. solani* after being exposed to the metabolites of *Trichoderma* spp.). Experiments were performed with four replicates. The results were compared by analysis of variance (ANOVA) and Fisher least significant difference (LSD) tests using SAS (SAS Institute Inc., 2004, Cary, NC, USA).

The direct confrontation assays were used to verify the ability of *Trichoderma* spp. to overgrow the pathogen. Each *Trichoderma* isolate was grown in dual culture with *R. solani* R43. The isolates were placed 5.5 cm apart on the same plate and incubated at 22°C for 5 days. Experiments were performed with four replicates. The parameters were measured after 5 days: r1 (distance between the pathogen sowing point and furthest point of the colony) and r2 (distance between the pathogen sowing point and the edge of the colony) from where *R. solani* and *Trichoderma* mycelia came into contact. Thus, the percentage of inhibition in the direct confrontation assay (ID) was calculated by the formula: %ID =[(r1−r2)/r1]x100 (**Figure 1B**). Inhibition of *R. solani* growth was compared by analysis of variance (ANOVA) and Fisher least significant difference (LSD) tests using SAS. (SAS Institute Inc., 2004, Cary, NC, USA).

In vivo Assay of the Antifungal Activity

These assays were only performed with those *Trichoderma* isolates that gave percentages of inhibition greater that 40% in membrane assays or 20% in direct confrontation assays, and that were able to sporulate on PDA medium.

The bioassays were performed in climatic chambers with 32 treatments as follow: 15 *in vitro* selected *Trichoderma* isolates against *R. solani* (R43) in order to test the antagonistic activity (RT0-number of *Trichoderma* isolate); 15 *in vitro* selected *Trichoderma* isolates in order to test their effect on plant (CT0-number of *Trichoderma* isolate); one control with *R. solani* (RC); and one control without fungi (CC). Thirty pots were used per treatment, with polypropylene pots (1 liter capacity) with substrate (80% white peat, 20% black peat and 5.5 pH). Each pot was watered with 250 ml of water prior inoculation. *R. solani* R43 was inoculated by surface irrigation with 50 ml per pot of a suspension of triturated micromicete culture of this pathogen using five Petri dishes (18 ml of PDA per dish) per liter of water. For control inoculation, only PDA medium was used without any pathogen. Pots were kept in a growth chamber for 8 days at 25°C (16 h) and 16°C (8 h), 60% relative humidity (RH) in the dark.

Trichoderma isolates were inoculated on PDA medium to grow in dark conditions (25°C) for 1 week. After that, they were exposed to light in order to induce the formation of spores. Spores suspensions were prepared at a final concentration of 2×10^7 spores/ml. Bean seeds of "Canela" variety were surface sterilized (sodium hypochlorite 1% for 3 min and distilled water for 6 min). Then, they were coated with a spore suspension of each *Trichoderma* isolate. The seeds were submerged in the spore suspension (45 seeds per 20 ml spores suspension) and they were dried in a flow chamber for 12 h. Coated seeds were sown after 8 days of the inoculation of *R. solani* R43. The culture was maintained for 45 days with a photoperiod of 16 h light, 25°C/16°C (day/night), 60% RH and brightness of 3500 lux. Irrigations were performed every 4 days with tap water (about 250 ml/pot). On the 2nd–4th week a nutrient solution was added (Rigaud and Puppo, 1975). Plants were removed after 45 days from sowing, tissues with symptoms were placed in PDA medium, incubating the plates at 22°C for 5 days and identifying the fungus for fulfill Koch's postulates. The next parameters were evaluated in removed plants after 45 days from sowing: wet weight and dry weight (72 h in an oven, 82°C) of the aerial part and root system.

The data were transformed by the formula $\sqrt{x + 0.5}$ and they were compared by analysis of variance (ANOVA) and Fisher least significant difference (LSD) tests using SAS (SAS Institute Inc., 2004, Cary, NC, USA).

Nucleic Acid Extraction and Manipulation

Genomic DNA from those isolates showing positive and negative phenotypic effects on bean plants, were extracted by growing the *Trichoderma* isolates in PDB medium (potato dextrose broth). Mycelia were then recovered by filtration, washed with 0.9% NaCl and dried on absorbent filter paper. The procedure for fungal genomic DNA isolation was performed as previously described (Cardoza et al., 2006).

PCR Amplification, Sequencing and DNA Analysis

The amplification of the ITS regions of the nuclear rDNA gene cluster and an approximately 0.56 kb fragment of the *tef1* (translation elongation factor 1-α) gene were carried out with the primer pairs ITS1/ITS4 and EF1-728F/EF1-LLErev, respectively, as described previously (Hermosa et al., 2004). The PCR products were purified from agarose gels using the NucleoSpin Extract II Kit (Macherey-Nagel, Düren, Germany), according to the manufacturer's protocol. PCR fragments were sequenced in an ABI 377 Prism Sequencer (Applied Biosystems, Foster City, CA).

The *Trichoderma* ITS and *tef1* sequences obtained in this work were analyzed using the online interactive key (available from http://www.isth.info/tools/blast/index.php) (Druzhinina et al., 2005).

Analysis of Expression of Bean Defense-related Genes

Three bean leaves from 45 day-old plants of each treatment were randomly collected and stored at −80°C until use. Leaves were detached from plants inoculated with *Trichoderma* isolate showing positive phenotypic results in the *in vivo* test. Leaves were then reduced to a fine powder in a mortar under liquid nitrogen. Plant RNA isolation were performed as previously described (Malmierca et al., 2013).

cDNA were synthesized using 1 µg total RNA and a Reverse Transcription System with an Oligo$(dT)_{15}$ as the primer (Promega, Madison, WT). cDNA were quantified using a Nanodrop 2000 (Thermo Scientific, Wilmington, DE) and used for further studies.

Real Time-PCR Analysis

In order to analyze the effect of *Trichoderma* treatment and/or *R. solani* infection of bean plants, oligonucleotides corresponding to seven defense-related genes were designed based on their available sequences (**Table 2**). *PR1, PR2, PR3,* and *PR4,* which encode for pathogenesis related proteins related to the salicylate (SA) pathway; *CH5b, CH1* encoding for related to the jasmonate/ethylene pathway (JA/ET), and *PAL* involved in the phenylpropanoid pathway, were selected to be analyzed in the present study. *α-actin* (Upchurch and Ramirez, 2010; Guerrero-González et al., 2011) and *PvEF1α* (this work) encoding genes were used as reference (housekeeping genes) for comparative analysis. The qPCR reactions were carried out using Step One Plus™ (Applied Biosystems, Foster City, CA). The reactions were performed in a total volume of 20 µl: 10 µl Power SYBR® Green PCR Master Mix (Applied Biosystems, USA), 0.4 µl Forward Primer 10 µM, 0.4 µl Reverse Primer 10 µM, 5 µl cDNA, and H_2O to 20 µl. The REST 2009© software (Pfaffl et al., 2002) was used to calculate the relative expression ratio and the significance of the differences between the gene expression levels. For each primer pair used in this work, we performed a standard curve with 320, 160, 80, 40, 20, and 10 ng cDNA to determine the PCR amplification efficiency (E value). Each measurement was made in triplicate (Malmierca et al., 2013).

TABLE 2 | Oligonucleotides designed for Real-Time PCR analysis.

Gene	Function gene	GenBank Accession number	Oligonucleotide name	Oligonucleotide sequence (5′-3′)	References
PvEF1α	Elongation Factor 1	EF660340.1	EF1α-F	CGGGTATGCTGGTGACTTTT	This work
			EF1α-R	CACGCTTGAGATCCTTGACA	
α-actin	Actine	U60500.1	α-actin-F	GAGCTATGAATTGCCTGATGG	This work
			α-actin-R	CGTTTCATGAATTCCAGTAGC	
PvCH5b	Chitinase	FE897014.1	CH5b-F	CAGCCAAAGGCTTCTACACC	This work
			CH5b-R	TTGTTTCGTGAGACGTTTGC	
PvPR1	Pathogenesis related 1	HO864272.1	PR1-F	TGGTCCTAACGGAGGATCAC	This work
			PR1-R	TGGCTTTTCCAGCTTTGAGT	
PvPR2	Pathogenesis related 2	HO864270.1	PR2-F	GTGAAGGACGCCGATAACAT	This work
			PR2-R	ACTGAGTTTGGGGTCGATTG	
PvPR4	Pathogenesis related 4	HO864354.1	PR4-F	CGCAGTGAGTGCATATTGCT	This work
			PR4-R	TGTTTGTCACCCTCAAGCAC	
PvCH1	Chitinase				Pereira et al., 2014
PvPR3	Pathogenesis related 3				Pereira et al., 2014
PvPAL	Phenylalanine ammonia-lyase				Pereira et al., 2014

Quantification of Ergosterol and Squalene

The *Trichoderma* selected strain was inoculated in 100 ml of CM medium (0.5% malt extract, 0.5% yeast extract, and 0.5% glucose) with 10^6 spores/ml, and incubated 24 h at 28°C. Then 20 ml from the previous cultures were inoculated on 100 ml potato dextrose broth (PDB medium) and were incubated as before during 24–96 h. The mycelia were filtered through nytal filters (30 μm diameter) and the liquid removed by drying between filter papers. The dry weight of the fungal pellet was calculated. Total intracellular sterols were extracted and ergosterol and squalene content were quantified as previously reported other authors (Cardoza et al., 2007; Ghimire et al., 2009). All measurements were made in duplicate in the *Trichoderma* selected isolate and a strain used as control of the same species of the selected isolate. The results were compared by analysis of variance (ANOVA) and Fisher least significant difference (LSD) tests using SAS (SAS Institute Inc., 2004, Cary, NC, USA).

Results

Analysis of the *in vitro* Antagonistic Activity of *Trichoderma* Isolates with *R. solani*

The first test to determine the *in vitro* antifungal ability of the different *Trichoderma* isolates was based on their ability to produce metabolites that may inhibit the growth of *R. solani* (**Table 3**, **Figure 1A**). *Trichoderma* isolates T003, T004, T006, T020, T022, T012, T013, T025, T016, T007, T024, T005, and T010 inhibited *R. solani* growth by more than 75%, with the highest inhibition produced by T003, T004, T006, T020, and T022 (86.70%). T019, T008, T002, T021, T001, T018, and T023 showed a remarkable inhibition (75–40%). Finally, T015, T014, T017, T011, and T009 inhibited *R. solani* growth by less than 40%, and T009 showed the lowest percentage (15.82%).

T021 was the *Trichoderma* isolate showing the highest percentage of inhibition (72.77%) in the direct confrontation assays (**Table 3**, **Figure 1B**), whereas T009 showed the lowest inhibition values (14.63%). The inhibition percentages detected for the other *Trichoderma* isolates ranged from 47.13 to 30.55%.

Analysis of the *in vivo* Antagonistic Activity of *Trichoderma* Isolates with *R. solani*

The results of the *in vitro* membrane assays and direct confrontation assays against *R. solani*, showed above, were used to select the isolates that would be used for the *in vivo* analysis. Thus, fifteen *Trichoderma* isolates (marked with asterisk in **Tables 1**, **3**) were tested since they were able to sporulate in PDA medium and showed a percentage of inhibition of *R. solani* growth higher than 40% in the membrane assays, and/or 20% in the direct confrontation assays.

On plants removed after 45 days from sowing for Koch's postulates, *R. solani* was found and *Trichoderma* isolates were also present in the medium.

When dry aerial parts (**Figure 2**) were analyzed, plants treated with CT019 had the greatest weight of the aerial part, being significantly different from the control (CC). In the RT019 treatment, plants did not show significant differences in comparison with the control (CC), what is indicative of a biocontrol effect by T019 isolate.

In the case of the root system (**Figure 3**), the situation was similar to that observed in the aerial parts. CT019 treated plants were not significantly different in dry weight, although it was always greater than the weight of the control plants (CC). If the pathogen was present in the soil with same *Trichoderma* isolate, RT019 treatment, in the case of wet weight, there were no significant differences between control (CC) and control pathogen (RC) (data not shown).

Based on these results, the *Trichoderma* isolate T019 was selected for further studies since it showed the best positive effects on plant phenotype among all the analyzed isolates (**Figure 4**).

Molecular Identification of *Trichoderma* Isolates

Those *Trichoderma* isolates able to sporulate in PDA medium, showing a percentage of inhibition higher than

TABLE 3 | *In vitro* antifungal activity of *Trichoderma* strains against *R. solani* R43.

Strain	Inhibition in growth assay on membranes (% ± Standard error) (1)		Strain	Inhibition in growth assay on direct confrontation (% ± Standard error) (2)	
T003*	86.70 ± 0.15	a	T021	72.77 ± 4.49	a
T004*	86.70 ± 0.15	a	T004*	47.13 ± 0.91	b
T006*	86.70 ± 0.15	a	T013*	43.49 ± 1.41	bc
T020	86.70 ± 0.15	a	T011*	42.57 ± 2.00	bcd
T022	86.70 ± 0.15	a	T014	41.99 ± 1.33	cde
T012*	86.67 ± 0.14	a	T023	41.57 ± 1.40	cdef
T013*	85.04 ± 1.04	a	T018	41.01 ± 1.93	cdefg
T025*	83.97 ± 1.97	a	T020	39.21 ± 1.41	cdefgh
T016	82.29 ± 2.33	a	T006*	38.68 ± 0.59	defghi
T007*	76.77 ± 2.52	b	T017	37.42 ± 1.61	efghij
T024*	76.22 ± 2.32	bc	T016	37.10 ± 0.63	fghij
T005*	75.65 ± 2.23	bc	T002*	36.44 ± 0.29	ghij
T010	75.14 ± 3.30	bc	T024*	36.43 ± 3.71	hij
T019*	72.36 ± 2.01	bcd	T007*	36.34 ± 1.30	hij
T008*	70.61 ± 1.82	cde	T005*	36.16 ± 0.67	hij
T002*	66.75 ± 1.18	def	T019*	35.09 ± 1.01	hijk
T021	65.98 ± 0.77	ef	T022	35.05 ± 0.34	hijk
T001*	65.51 ± 1.75	f	T008*	35.04 ± 0.89	hijk
T018	54.12 ± 2.35	g	T012*	34.59 ± 0.98	ijk
T023	48.26 ± 3.10	h	T025*	33.28 ± 4.24	kjl
T015*	39.99 ± 0.82	i	T010	30.55 ± 1.32	kl
T014	38.82 ± 2.96	ij	T015*	28.84 ± 0.85	lm
T017	34.10 ± 0.88	j	T001*	24.66 ± 0.74	m
T011*	21.40 ± 2.66	k	T003*	19.85 ± 2.97	n
T009	15.82 ± 1.40	l	T009	14.63 ± 0.83	o

Values in the same column followed by different letters are significantly different (Fisher's LSD. $p < 0.05$).

(1) Growth assay on membranes (effect of Trichoderma isolates on percent growth inhibition of R. solani. using cellophane membranes. (2) in direct confrontation assays (percentage of inhibition of R43 growth when grown confronted with Trichoderma strains during 5 days in PDA medium plates).

**Trichoderma isolates selected for in vivo experiments because they were able to sporulate in PDA medium and showed a percentage of inhibition higher than 40% in the membrane assays. and/or 20% in the direct confrontation assays.*

40% in the membrane assays, and/or 20% in the direct confrontation assays, were identified. ITS1 region of rDNA and a fragment of the translation elongation factor 1 (*tef1*) were amplified and sequenced for 12 isolates, and both nucleotide sequences were used for identification at species level. Eight out of these 12 isolates were identified as *T. harzianum*, and *T. atroviride*, *T. gamsii*, *T. longibrachiatum* and *T. citrinoviride* species were represented by one isolate (**Table 1**).

Effect of *Trichoderma* Treatment and/or *R. solani* Infection in the Expression of Bean Defense Related Genes

The amplification efficiencies of the oligo-pairs (**Table 2**) were: *α-actin* 1.150, *PvEF1α* 0.903, *CH5b* 0.883, *CH1* 1.098, *PR1* 1.094, *PR2* 1.048, *PR3* 0.947, *PR4* 0.922, and *PAL* 0.962.

α-actin and *PvEF1α* were used as housekeeping genes to determine the relative expression level of the other genes analyzed in the present work. *Trichoderma* T019 strain was selected, based on its positive effects on bean phenotype with and without *R. solani* infection. The results included in **Figure 5** showed that: (i) *R. solani* down-regulated the expression of all the *P. vulgaris* defense-related genes analyzed (**Figure 5A**), raising values of expression ranging between 0.099 ($p = 0.029$) for *PR2* and 0.397 ($p = 0.045$) for *CH1*; (ii) In plants treated with T019 compared with control plants (bean plants not treated with *Trichoderma* nor infected with *R. solani*) only the *CH5b* and *PR2* were significantly up-regulated, raising comparative expression values of 1.495 ($p = 0.000$) and 24.492 ($p = 0.000$), respectively (**Figure 5B**). Finally, (iii) treatment with T019 and infection with *R. solani* significantly up-regulated the expression ratio of the analyzed genes, except *PAL*, in comparison with plants only infected with *R. solani*, with values ranging from 1.420 ($p = 0.000$) to 42.975 ($p = 0.000$) for *CH1* and *PR4* respectively (**Figure 5C**).

A non-remarkable significant up-regulation of the defense-related gene expression was observed when expression ratios of the selected genes in plants treated with *T. harzianum* T019 and infected with *R. solani* were compared with those from plants untreated or uninfected (data not shown).

Production of Ergosterol and Squalene by the *Trichoderma* Selected Strains

Strain T019 (*T. harzianum*) was selected for this analysis because it was those producing the most remarkable positive phenotypic effect on beans (see above). *T. harzianum* T34 (**Table 1**) was used as control strain for comparative purposes because it is a well-known strain, widely characterized (Kullnig et al., 2001).

Production of ergosterol at 24 h of growth did not show significant differences between the strains *T. harzianum* (T34 and T019) (**Table 4**). When the production of ergosterol was analyzed at 96 h, T019 produced significantly higher amounts than its respective control (T34).

Regarding the squalene production at 24 h, *T. harzianum* (T34 and T019) strains did not show significant differences. However, at 96 h of growth, T019 produced amounts significantly higher than its control.

Discussion

Most of the *Trichoderma* isolates studied in the present work have been able to control the growth of *R. solani*. The percentage of growth inhibition in the direct confrontation assays raised values up to 72.77% for T021, but most of the isolates were between 47 and 24%. By contrast, in the antifungal assays on membranes, those percentages ranged mostly between 86 and 58%. In the assays conducted by Campelo et al. (2010) the percentage of inhibition in membrane assays using *T. virens* T59 (NBT59) and *T. atroviride* T11 (IMI352941) range between 100 and 84.7%, respectively. The differences observed in the different *in vitro* assays might be due to the variability of genotypes, with differences in growth, sporulation and in their environmental

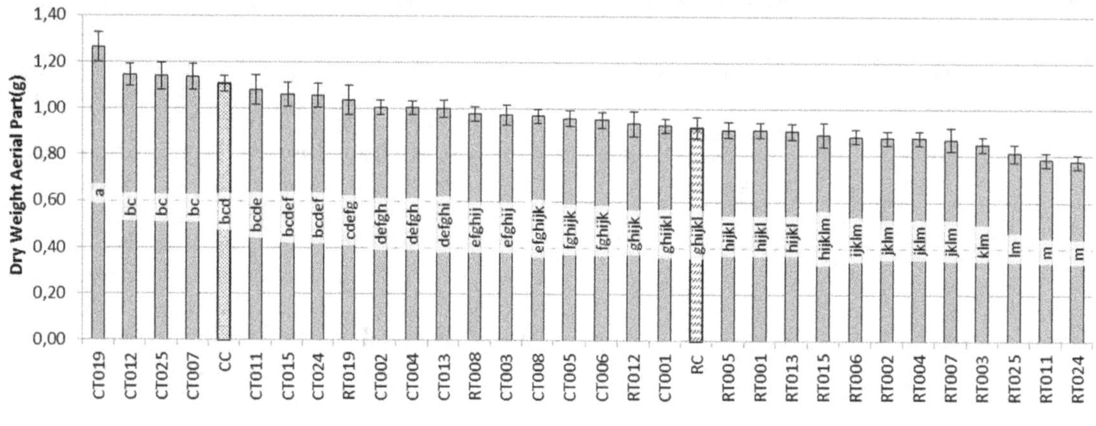

FIGURE 2 | Evaluation of the dry weight (g) of the aerial parts of bean plants grown during 45 days after sowing with 60 replicates. [*Trichoderma* isolates without pathogen (CT0-number of *Trichoderma* isolate), *Trichoderma* isolates with *R. solani* (RT0-number of *Trichoderma* isolate), *R. solani* control (RC) and control without fungus (CC)].

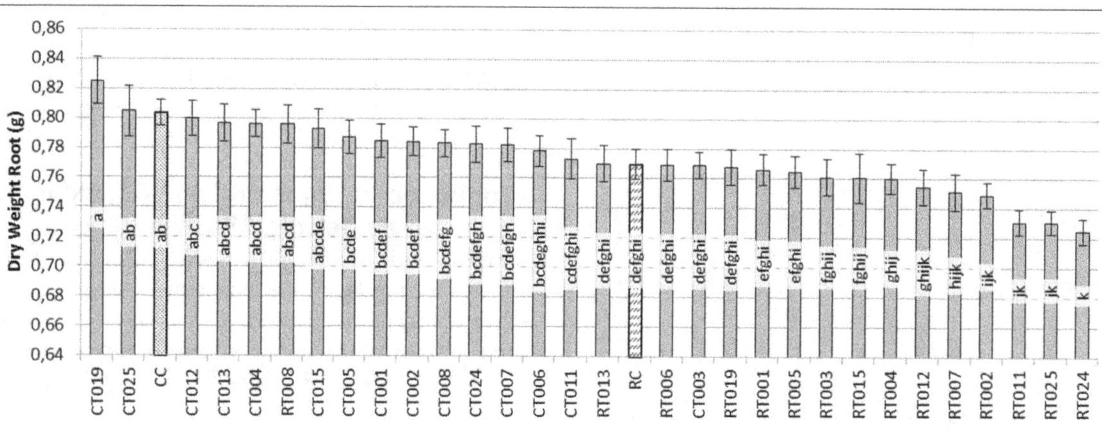

FIGURE 3 | Evaluation of the dry weight (g) of the root system of bean plants grown for 45 days after sowing with 60 replicates. [*Trichoderma* isolates without pathogen (CT0-number of *Trichoderma* isolate), *Trichoderma* isolates with *R. solani* (RT0-number of *Trichoderma* isolate), *R. solani* control (RC) and control without fungus (CC)].

biological interactions as a consequence of the prevalence of different modes of action on each of the analyzed strains (Ruano-Rosa et al., 2010).

The *in vitro* antagonistic activity of the *Trichoderma* isolates is an indication of their *in vivo* biological activity against phytopathogenic fungi. However, it has been found that the *in vitro* antifungal activity of *Trichoderma* does not correlate in a direct way with its *in vivo* ability to control diseases caused by phytopathogenic fungi, since many other factors influence that activity (Anees et al., 2010).

Regarding the *in vivo* assays, bean plants that had been in contact with some *Trichoderma* isolate had an increased size when the pathogen was not present. Pereira et al. (2014) also observed that *T. harzianum* was able to promote the growth of common bean plants in comparison to plants grown in its absence. However, in the present work only T019 treated plants did not decrease their size in the presence of *R. solani*. These data are in agreement with those from Yedidia and coworkers reporting a much stronger effect on cucumber plants treated with *T. harzianum*, which increased by 75% the length of the root, 95% aerial parts, 80% dry weight and 80% the size of the blade relative to the untreated control (Yedidia et al., 2001). *T. harzianum* was able to promote the growth of common bean plants in comparison to plants grown in its absence (Pereira et al., 2014).

However, in the present work opposite results were observed to those previously reported (Tello et al., 1985), indicating that infection with *R. solani* not always resulted in a reduction of the bean plant size, which was explained as a result of the activation of the natural plant-defense mechanisms (Cardoso and Echandi, 1987) that would lead to an enhanced development of the plants when this pathogen was present in the soil. In the present study we have observed that treatment with *R. solani* resulted in a down-regulation of all the bean defense-related genes analyzed, indicating a certain compensation effect between the intensity of plant defense response and plant growth (Hermosa et al., 2013).

FIGURE 4 | Differences in growth of bean plants. CC plant control; RC plant with *R. solani*, CT019 plant with *Trichoderma* T019; RT019 plant with *R. solani* and *Trichoderma* T019.

Little is known about the effect of *Trichoderma* treatment and/or infection with *R. solani* in the expression of bean defense-related gene expression. As indicated above, interaction of the plant with *R. solani* during 45 days of growth caused the repression of the seven defense-related genes studied (*CH5b, CH1, PR1, PR2, PR3, PR4*, and *PAL1*), as a mechanism to overcome the plant defense response and thus facilitating the progression of the infection process within the plant. Other authors have observed similar results in assays with tomato plants infected with *Pseudomonas syringae* pv. *tomato*, where a repression of *PR1* and *PR4* was shown, as an indication of the decrease in the plant self-defense mechanism, then facilitating disease progression (Zhao et al., 2003).

Expression of *P. vulgaris* defense-related genes was analyzed in leaves, even when the interaction with the pathogen *R. solani* is initially produced at the root level, to determine if the signals generated in roots as result of this interaction are able to systematically stimulate the bean defense along long distance from seed to leaf. When the interaction of plant with T019 was analyzed, *PR2* was up-regulated, *PR4* was slightly restrained but *PR1* and *PR3* were not affected. Pereira and coworkers observed that *T. harzianum* also seems to potentiate common bean response against the phytopathogenic fungus *R. solani*, as shown by the increase in the levels of *glu1* and *pod3* in the double treatment in comparison to that obtained for plants in the presence of *R. solani* alone (Pereira et al., 2014). Other studies showed that there was an increase in expression *PR1* at 16 h of interaction (Guerrero-González et al., 2011). Rivière and coworkers suggested that *PR1* was down-regulated by the β-1,3-glucanases (PR2 and PR3) (Rivière et al., 2008). In the present study, after 45 days of the inoculation of the biocontrol isolate we observed a higher expression of *PR2*, which could be due to an alteration of plant defense responses at these longer times in comparison with shorter ones assayed in the previous reported works. This would result in a higher β-1,3-glucanase activity in the cell wall that would increase oligosaccharides released by its action, then acting as elicitors of the plant defense response and/or of the fungal secondary metabolism (Druzhinina et al., 2011).

In the case of plants inoculated with *Trichoderma* and seeded in soil infected with *R. solani*, when compared with plants only infected with *R. solani*, thus excluding the effect due exclusively to this pathogen, a significant up-regulation of all the analyzed genes was observed, except *PAL*, indicating that in presence of *R. solani* several mechanisms are induced in *Trichoderma* that potentiate its ability to elicit plant defense-responses. In previous works, during the interaction of *Solanum tuberosum* with *R. solani* it was observed that after 120 h from the initial infection and damage to the first floor, new outbreaks produce less symptoms, suggesting that the plant might be prepared to defend and avoid disease progression (Lehtonen et al., 2008). Other studies have shown that the *PR4* was overexpressed in eggplants treated with BCAs suggesting that these agents promoted defensive reactions within the plant (Angelopoulou et al., 2014). In other assays with pepper plants an increase in the accumulation of PR1 and chitinases was also observed when these plants were previously inoculated with a strain of *Fusarium oxysporum* and subsequently exposed to *Verticillium dahliae* (Veloso and Díaz, 2012), emphasizing their importance in the response to pathogen infection and to abiotic stresses (Jung and Hwang, 2000). Thus, it could be inferred that the effect of *Trichoderma* would therefore modulate the response of the plant and prevent the suppression of defense genes caused by *R. solani*. However, the effect of the different *Trichoderma* strains in the different plants would indicate the existence of great differences between them.

In the case of *CH5b*, related to JA pathway, when *R. solani* was in contact with the plant for 45 days of development, the expression of this gene was down-regulated in comparison with non-infected plants. When bean plants were inoculated with *Trichoderma*, its expression was slightly induced. However, if the antagonist and the pathogen were present in the medium, the effect caused by *R. solani* was the overexpression of this gene. In previous works (Broglie et al., 1991) it was shown that *CH5b*, responsible for chitinase production, was overexpressed in *Nicotiana tabacum* and *Brassica napus* infected with *R. solani*, and their disease symptoms were reduced. Benhamou et al. (1993) found that the expression of this gene caused a reduction of disease in *B. napus* infected with the same pathogen, which correlates with similar results observed in strawberry plants inoculated with *Botrytis cinerea* (Vellicce et al., 2006). In the present study we observed that the contact of the plant with the pathogen during a period of 45 days resulted in a significant repression of this gene, similarly to what it was observed for the expression of *CH1*, which thereby facilitates disease progression. However, in the presence of *Trichoderma* a similar effect to that previously described for SA-related genes was observed, so facilitating a prevention of disease progression.

In the case of the phenylalanine ammonia lyase (PAL), this is a key enzyme in the metabolic pathway of phenylpropanoid compounds by catalyzing the amino acid L-phenylalanine

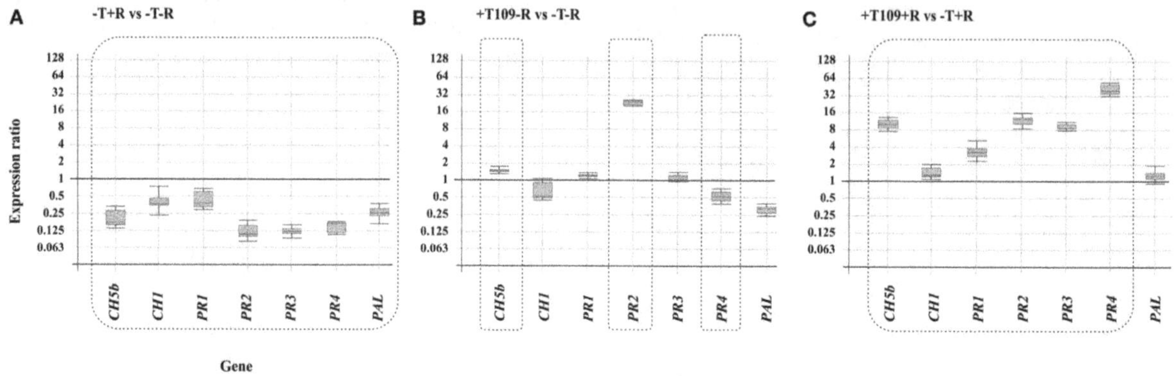

Gene	Expression ratio	p(H1)
CH5b	0.168*	0.029
CH1	0.397*	0.045
PR1	0.364*	0.038
PR2	0.099*	0.029
PR3	0.122*	0.000
PR4	0.127*	0.000
PAL	0.256*	0.000

Gene	Expression ratio	p(H1)
CH5b	1.495*	0.000
CH1	0.636	0.087
PR1	1.127	0.065
PR2	24.492*	0.000
PR3	1.173	0.152
PR4	0.524*	0.036
PAL	0.316	0.081

Gene	Expression ratio	p(H1)
CH5b	10.410*	0.000
CH1	1.420*	0.000
PR1	3.457*	0.000
PR2	12.267*	0.000
PR3	8.896*	0.000
PR4	42.975*	0.000
PAL	1.305	0.174

FIGURE 5 | Expression of *CH5b, CH1, PR1, PR2, PR3, PR4,* and *PAL* genes in comparison with *α-actin* and *PvEF-1α* reference genes. The comparisons and statistical analysis were performed using the REST2009©program (Pfaffl et al., 2002). The numeric data are illustrated at the bottom of the figure and those showing differences statistically significant ($p < 0.05$) are indicated with an asterisk and they are boxed in the graphic representation. Comparison of the gene expression of the bean defense-related genes **(A)** in plants infected with *R. solani* R43 vs. control plants. **(B)** In plants treated with *Trichoderma* T019 vs. control plants **(C)** In plants infected with *R. solani* R43 and treated with *Trichoderma* T019 vs. plants infected with *R. solani* R43.

TABLE 4 | Ergosterol and squalene quantification from 24 and 96 h samples of the selected *Trichoderma* strains selected for this study.

Trichoderma spp.	Ergosterol (mg E/g d wt)		Squalene (mg S/g d wt)	
	24 h	96 h	24 h	96 h
T34	16.197 ± 1.019 a	8.970 ± 0.388 b	0.317 ± 0.020 a	0.052 ± 0.049 b
T019	14.336 ± 2.459 a	12.502 ± 3.568 a	0.303 ± 0.041 a	0.348 ± 0.103 a

Values in the same column followed by different letters are significantly different (Fisher's LSD, $p < 0.05$).

deamination, giving rise to trans-cinnamic acid and ammonia. The trans-cinnamic acid is used for the synthesis of various phenolic compounds, which are precursors in the synthesis of esters, coumarins, flavonoids, and lignin. The production of this enzyme is controlled during plant growth, but is also induced in cells neighboring the infection site and, besides infection, various environmental stimuli such as injury, heavy metal contamination, light and growth regulators (Rahman and Punja, 2005). In the present work, we observed that this gene was down-regulated or not affected in all the conditions assayed, indicating that in the analyzed strain, PAL activity would be more focused in the plant response against environmental stress than in the response against pathogens, which contrast with the role of this enzyme described in pepper, where it is involved in the positive regulation of SA-dependent defense signaling (Kim and Hwang, 2014).

Regarding squalene and ergosterol production, in this study it was found that there were differences among the isolates. Ergosterol is a sterol found in the fungal membrane, which although considered by the plant as a PAMP (pathogen associated molecular patterns) (Nürnberger et al., 2004) triggers a series of reactions (Cervone et al., 1997) which would cause an activation of the genes of plant defense (Rossard et al., 2010). Squalene is a precursor of the ergosterol biosynthetic pathway (Malmierca et al., 2013) located in the cellular membranes or accumulated as droplets in the cytoplasm whose physiological function, apart from acting as an ergosterol precursor, still remains unclear, but it might play some role in the elicitation of plant defense responses. Thus, in this assay, an increased production of these compounds by *Trichoderma* would result in the induction of defense genes in the bean, then the plant could grow better under a pathogen presence in the soil.

Conclusions

Trichoderma isolates inhibit the development of *R. solani* under *in vitro* conditions, when they grow in Petri dishes. *In vivo* conditions, the bean plants that had been in contact with the *Trichoderma* isolates always had an increased size when the pathogen was not present. When *R. solani* was present in the soil, the development of the bean plants was significantly reduced. However, bean plants treated with strain T019 did not decrease their size in the presence of *R. solani.*

The interaction of bean plants with *R. solani* caused, after 45 days of growth, the down-regulation of the seven defense-related genes studied (*CH5b, CH1, PR1, PR2, PR3, PR4, PAL*) as a mechanism to overcome the plant defense response and thus facilitating the progression of the infection process within the plant.

An increased production of ergosterol and squalene by *Trichoderma* resulted in the induction of defense genes in the bean plants. In this way, the plant would grow better under a pathogen presence in the soil.

T. harzianum T019 shows a positive effect on the level of resistance of bean plants to *R. solani.* This strain induces the expression of plant defense-related genes and produces a higher level of ergosterol, indicating its ability to grow at a higher rate in the soil, which would explain its positive effects on plant growth and defense in the presence of the pathogen.

Acknowledgments

The grant awarded to Sara Mayo (FPU12/00677) by the Ministry of Education, Culture and Sport (Spain) according the Resolution of April 25, 2012 (BOE of 10 May), as amended by Resolution of May 18, 2012 (BOE of 22 May), and by Resolution of October 31, 2012 (BOE of 12 November) and granted in the Resolution of December 20, 2012. Ministry of Economy and Competitiveness for National project "Farnesol as self-regulatory molecule in *Trichoderma.* Tyrosol and farnesol signaling in *Trichoderma*-bean interaction" (AGL2012-40041-C02-02). Junta de Castilla y León, Consejería de Educación for the project "Effect of terpenes and physiologically related compounds produced by *Trichoderma parareesei* in the development of common bean (*Phaseolus vulgaris* L.) and in defensive responses of this plant" (LE228U14). J. Alvarez for his technical support.

References

Anees, M., Tronsmo, A., Edel-Hermann, V., Hjeljord, L. G., Héraud, C., and Steinberg, C. (2010). Characterization of field isolates of *Trichoderma* antagonistic against *Rhizoctonia solani. Fungal Biol.* 114, 691–701. doi: 10.1016/j.funbio.2010.05.007

Angelopoulou, D. J., Naska, E. J., Paplomatas, E. J., and Tjamos, S. E. (2014). Biological control agents (BCAs) of verticillium wilt: influence of application rates and delivery method on plant protection, triggering of host defence mechanisms and rhizosphere populations of BCAs. *Plant Pathol.* 63, 1062–1069. doi: 10.1111/ppa.12198

Benhamou, N., Broglie, K., Chet, I., and Broglie, R. (1993). Cytology of infection of 35S-bean chitinase transgenic canola plants by *Rhizoctonia solani*: cytochemical aspects of chitin breakdown *in vivo. Plant J.* 4, 295–305. doi: 10.1046/j.1365-313X.1993.04020295.x

Broglie, K., Chet, I., Holliday, M., Cressman, R., Biddle, P., Knowlton, S., et al. (1991). Transgenic plants with enhanced resistance to the fungal pathogen *Rhizoctonia solani. Science* 254, 1194–1197.

Campelo, P., Cardoza, R., Lorenzana, A., Hermosa, M., Monte, E., Reinoso, B., et al. (2010). Biological control of phytopathogenic fungi in bean (*Phaseolus vulgaris* L.) with *Trichoderma atroviride* and *Trichoderma virens.* [Abstract]. *Bean Improv. Coop. Colo.* 53, 114–115.

Cardoso, J., and Echandi, E. (1987). Biological control of Rhizoctonia root rot of snap bean with binucleate *Rhizoctonia*-like fungi. *Plant Dis.* 71, 167–170. doi: 10.1094/PD-71-0167

Cardoza, R. E., Hermosa, M. R., Vizcaíno, J. A., González, F., Llobell, A., Monte, E., et al. (2007). Partial silencing of a hydroxy-methylglutaryl-CoA reductase-encoding gene in *Trichoderma harzianum* CECT 2413 results in a lower level of resistance to lovastatin and lower antifungal activity. *Fungal Genet. Biol.* 44, 269–283. doi: 10.1016/j.fgb.2006.11.013

Cardoza, R. E., Vizcaíno, J. A., Hermosa, M. R., Sousa, S., González, F. J., Llobell, A., et al. (2006). Cloning and characterization of the erg1 gene of *Trichoderma harzianum*: effect of the erg1 silencing on ergosterol biosynthesis and resistance to terbinafine. *Fungal Genet. Biol.* 43, 164–178. doi: 10.1016/j.frb.2005.11.002

Casquero, P. A., Lema, M., Santalla, M., and De Ron, A. M. (2006). Performance of common bean (*Phaseolus vulgaris* L.) landraces from Spain in the Atlantic and Mediterranean environments. *Genet. Res. Crop Evol.* 53, 1021–1032. doi: 10.1007/s10722-004-7794-1

Cervone, F., Castoria, R., Leckie, F., and De Lorenzo, G. (1997). "Perception of fungal elicitors and signal transduction," in *Signal Transduction in Plants*, ed P. Aducci (Basel, Boston, Berlin: Birkhäuser), 153–177.

Druzhinina, I. S., Kopchinskiy, A. G., Komon, M., Bissett, J., Szakacs, G., and Kubicek, C. P. (2005). An oligonucleotide barcode for species identification in *Trichoderma* and *Hypocrea. Fungal Genet. Biol.* 42, 813–828. doi: 10.1016/j.fgb.2005.06.007

Druzhinina, I. S., Seidl-Seiboth, V., Herrera-Estrella, A., Horwitz, B. A., Kenerley, C. M., Monte, E., et al. (2011). *Trichoderma*: the genomics of opportunistic success. *Nat. Rev. Microbiol.* 9, 749–759. doi: 10.1038/nrmicro2637

Garaiová, M., Zambojová, V., Šimová, Z., Griac, P., and Hapala, I. (2013). Squalene epoxidase as a target for manipulation of squalene levels in the yeast *Saccharomyces cerevisiae. FEMS Yeast Res.* 14, 310–323. doi: 10.1111/1567-1364.12107

Ghimire, G. P., Hei, C. L., and Jae, K. S. (2009). Improved squalene production via modulation of the methylerythritol 4-phosphate pathway and heterologous expression of genes from *Streptomyces peucetius* ATCC 27952 in *Escherichia coli. Appl. Environ. Microbiol.* 75, 7291–7293. doi: 10.1128/AEM.01402-09

Guerrero-González, M. L., Rodríguez-Kessler, M., Rodríguez-Guerra, R., González-Chavira, M., Simpson, J., Sanchez, F., et al. (2011). Differential expression of *Phaseolus vulgaris* genes induced during the interaction with *Rhizoctonia solani. Plant Cell Rep.* 30, 1465–1473. doi: 10.1007/s00299-011-1055-5

Harman, G. E., Howell, C. R., Viterbo, A., Chet, I., and Lorito, M. (2004). *Trichoderma* species-opportunistic, avirulent plant symbionts. *Nat. Rev. Microbiol.* 2, 43–56. doi: 10.1038/nrmicro797

Harman, G. E., and Kubicek, C. P. (1998). *Trichoderma and Gliocladium: Enzymes, Biological Control and Commercial Applications.* Vol. 2, London: CRC Press.

Hermosa, M. R., Grondona, I., Iturriaga, E. A., Díaz-Mínguez, J. M., Castro, C., Monte, E., et al. (2000). Molecular characterization and identification of biocontrol isolates of *Trichoderma* spp. *Appl. Environ. Microbiol.* 66, 1890–1898. doi: 10.1128/AEM.66.5.1890-1898.2000

Hermosa, M. R., Keck, E., Chamorro, I., Rubio, B., Sanz, L., Vizcaíno, J. A., et al. (2004). Genetic diversity shown in *Trichoderma* biocontrol isolates. *Mycol. Res.* 108, 897–906. doi: 10.1017/S0953756204000358

Hermosa, R., Belén Rubio, M., Cardoza, R. E., Nicolás, C., Monte, E., and Gutiérrez, S. (2013). The contribution of *Trichoderma* to balancing the costs of plant growth and defense. *Int. Microbiol.* 16, 69–80. doi: 10.2436/20.1501.01.181

Hermosa, R., Viterbo, A., Chet, I., and Monte, E. (2012). Plant-beneficial effects of *Trichoderma* and of its genes. *Microbiol* 158, 17–25. doi: 10.1099/mic.0.052274-0

Jung, H. W., and Hwang, B. K. (2000). Pepper gene encoding a basic ß-1,3-glucanase is differentially expressed in pepper tissues upon pathogen infection and ethephon or methyl jasmonate treatment. *Plant Sci.* 159, 97–106. doi: 10.1016/S0168-9452(00)00334-4

Kim, D. S., and Hwang, B. K. (2014). An important role of the pepper phenylalanine ammonia-lyase gene (PAL1) in salicylic acid-dependent signalling of the defence response to microbial pathogens. *J. Exp. Bot.* 65, 2295–2306. doi: 10.1093/jxb/eru109

Kullnig, C., Krupica, T., Woo, S. L., Mach, R. L., Rey, M., Benítez, T., et al. (2001). Confusion abounds over identities of *Trichoderma* biocontrol isolates. *Mycol. Res.* 105, 770–772. doi: 10.1017/S0953756201229967

Lehtonen, M. J., Somervuo, P., and Valkonen, J. P. T. (2008). Infection with *Rhizoctonia solani* induces defense genes and systemic resistance in potato sprouts grown without light. *Phytopathol* 98, 1190–1198. doi: 10.1094/PHYTO-98-11-1190

Lochman, J., and Mikeš, V. (2005). Activation of different defence-related genes expression by ergosterol. *FEBS J.* 272, 470. doi: 10.1111/j.1742-4658.2005.4739_12.x

Malmierca, M. G., Barua, J., Mccormick, S. P., Izquierdo-Bueno, I., Cardoza, R. E., Alexander, N. J., et al. (2014). Novel aspinolide production by *Trichoderma arundinaceum* with a potential role in *Botrytis cinerea* antagonistic activity and plant defence priming. *Environ. Microbiol.* 17, 1103–1118. doi: 10.1111/1462-2920.12514

Malmierca, M. G., Cardoza, R. E., Alexander, N. J., McCormick, S. P., Collado, I. G., Hermosa, R., et al. (2013). Relevance of trichothecenes in fungal physiology: disruption of tri5 in *Trichoderma arundinaceum*. *Fungal Genet. Biol.* 53, 22–33. doi: 10.1016/j.fgb.2013.02.001

Mukherjee, P. K., Horwitz, B. A., Herrera-Estrella, A., Schmoll, M., and Kenerley, C. M. (2013). *Trichoderma* research in the genome era. *Annu. Rev. Phytopathol.* 51, 105–129. doi: 10.1146/annurev-phyto-082712-102353

Nürnberger, T., Brunner, F., Kemmerling, B., and Piater, L. (2004). Innate immunity in plants and animals: Striking similarities and obvious differences. *Immunol. Rev.* 198, 249–266. doi: 10.1111/j.0105-2896.2004.0119.x

Papavizas, G. (1985). *Trichoderma* and *Gliocladium*: biology, ecology, and potential for biocontrol. *Annu. Rev. Phytopathol.* 23, 23–54. doi: 10.1146/annurev.py.23.090185.000323

Pereira, J. L., Queiroz, R. M. L., Charneaum, S. O., Felix, C. R., Ricart, C. A. O., Lopes Da Silva, F., et al. (2014). Analysis of *Phaseolus vulgaris* response to its association with *Trichoderma harzianum* (ALL-42) in the presence or absence of the phytopathogenic fungi *Rhizoctonia solani* and *Fusarium solani*. *PLoS ONE* 9:e98234. doi: 10.1371/journal.pone.0098234

Pfaffl, M. W., Horgan, G. W., and Dempfle, L. (2002). Relative expression software tool (REST) for group-wise comparison and statistical analysis of relative expression results in real-time PCR. *Nucleic Acids Res.* 30:e36. doi: 10.1093/nar/30.9.e36

Rahman, M., and Punja, Z. K. (2005). Biochemistry of ginseng root tissues affected by rusty root symptoms. *Plant Physiol. Biochem.* 43, 1103–1114. doi: 10.1016/j.plaphy.2005.09.004

Rigaud, J. R., and Puppo, A. (1975). Indole 3 acetic acid catabolism by soybean bacteroids. *J. Gen. Microbiol.* 88, 223–228. doi: 10.1099/00221287-88-2-223

Rivière, M., Marais, A., Ponchet, M., Willats, W., and Galiana, E. (2008). Silencing of acidic pathogenesis-related PR-1 genes increases extracellular β-(1→3)-glucanase activity at the onset of tobacco defence reactions. *J. Exp. Bot.* 59, 1225–1239. doi: 10.1093/jxb/ern044

Rossard, S., Roblin, G., and Atanassova, R. (2010). Ergosterol triggers characteristic elicitation steps in *Beta vulgaris* leaf tissues. *J. Exp. Bot.* 61, 1807–1816. doi: 10.1093/jxb/erq047

Ruano-Rosa, D., del Moral-Navarrete, L., and Lopez-Herrera, C. J. (2010). Selection of *Trichoderma* spp. isolates antagonistic to Rosellinia necatrix. *Span. J. Agric. Res.* 8, 1084–1097. doi: 10.5424/sjar/2010084-1403

Tello, J., Lacasa, A., and Molina, R. (1985). Una nota fitopatológica sobre el complejo parasitario del pie de la judía (*Phaseolus vulgaris*). *ITEA* 61, 57–69.

Upchurch, R. G., and Ramirez, M. E. (2010). Defense-related gene expression in soybean leaves and seeds inoculated with *Cercospora kikuchii* and *Diaporthe phaseolorum* var. meridionalis. *Physiol. Mol. Plant Pathol.* 75, 64–70. doi: 10.1016/j.pmpp.2010.08.007

Valenciano, J. B., Casquero, P. A., Boto, J. A., and Marcelo, V. (2006). Evaluation of the occurrence of root rots on bean plants (*Phaseolus vulgaris*) using different sowing methods and with different techniques of pesticide application. *N.Z. J. Crop. Horticul. Sci.* 34, 291–298. doi: 10.1080/01140671.2006.9514419

Vellicce, G. R., Ricci, J. C. D., Hernández, L., and Castagnaro, A. P. (2006). Enhanced resistance to *Botrytis cinerea* mediated by the transgenic expression of the chitinase gene ch5B in strawberry. *Transgenic Res.* 15, 57–68. doi: 10.1007/s11248-005-2543-6

Veloso, J., and Díaz, J. (2012). *Fusarium oxysporum* Fo47 confers protection to pepper plants against *Verticillium dahliae* and *Phytophthora capsici*, and induces the expression of defence genes. *Plant Pathol.* 61, 281–288. doi: 10.1111/j.1365-3059.2011.02516.x

Woo, S. L., and Lorito, M. (2007). "Exploiting the interactions between fungal antagonists, pathogens and the plant for biocontrol," in *Novel Biotechnologies for Biocontrol Agent Enhancement and Management*, eds M. Vurro and J. Gressel (Dordrecht: Springer), 107–130.

Woo, S., Scala, F., Ruocco, M., and Lorito, M. (2006). The molecular biology of the interactions between *Trichoderma* spp., phytopathogenic fungi, and plants. *Phytopathol* 96, 181–185. doi: 10.1094/PHYTO-96-018

Yedidia, I., Srivastva, A. K., Kapulnik, Y., and Chet, I. (2001). Effect of *Trichoderma harzianum* on microelement concentrations and increased growth of cucumber plants. *Plant Soil* 235, 235–242. doi: 10.1023/A:1011990013955

Zhao, Y., Thilmony, R., Bender, C. L., Schaller, A., He, S. Y., and Howe, G. A. (2003). Virulence systems of *Pseudomonas syringae* pv. tomato promote bacterial speck disease in tomato by targeting the jasmonate signaling pathway. *Plant J.* 36, 485–499. doi: 10.1046/j.1365-313X.2003.01895.x

Conflict of Interest Statement: The authors declare that the research was conducted in the absence of any commercial or financial relationships that could be construed as a potential conflict of interest.

15

Proteomics and Metabolomics: Two Emerging Areas for Legume Improvement

Edited by:
Francesca Sparvoli,
Institute of Agricultural Biology and Biotechnology, Italy

Reviewed by:
Hao Peng,
Washington State University, USA
Luca Espen,
Università degli Studi di Milano, Italy

***Correspondence:**
Rajeev K. Varshney
r.k.varshney@cgiar.org

Abirami Ramalingam*[1]*, Himabindu Kudapa*[1]*, Lekha T. Pazhamala*[1]*, Wolfram Weckwerth*[2] *and Rajeev K. Varshney*[1,3]

[1] International Crops Research Institute for the Semi-Arid Tropics (ICRISAT), Hyderabad, India, [2] Department of Ecogenomics and Systems Biology, University of Vienna, Vienna, Austria, [3] School of Plant Biology and Institute of Agriculture, The University of Western Australia, Crawley, WA, Australia

The crop legumes such as chickpea, common bean, cowpea, peanut, pigeonpea, soybean, etc. are important sources of nutrition and contribute to a significant amount of biological nitrogen fixation (>20 million tons of fixed nitrogen) in agriculture. However, the production of legumes is constrained due to abiotic and biotic stresses. It is therefore imperative to understand the molecular mechanisms of plant response to different stresses and identify key candidate genes regulating tolerance which can be deployed in breeding programs. The information obtained from transcriptomics has facilitated the identification of candidate genes for the given trait of interest and utilizing them in crop breeding programs to improve stress tolerance. However, the mechanisms of stress tolerance are complex due to the influence of multi-genes and post-transcriptional regulations. Furthermore, stress conditions greatly affect gene expression which in turn causes modifications in the composition of plant proteomes and metabolomes. Therefore, functional genomics involving various proteomics and metabolomics approaches have been obligatory for understanding plant stress tolerance. These approaches have also been found useful to unravel different pathways related to plant and seed development as well as symbiosis. Proteome and metabolome profiling using high-throughput based systems have been extensively applied in the model legume species, *Medicago truncatula* and *Lotus japonicus*, as well as in the model crop legume, soybean, to examine stress signaling pathways, cellular and developmental processes and nodule symbiosis. Moreover, the availability of protein reference maps as well as proteomics and metabolomics databases greatly support research and understanding of various biological processes in legumes. Protein-protein interaction techniques, particularly the yeast two-hybrid system have been advantageous for studying symbiosis and stress signaling in legumes. In this review, several studies on proteomics and metabolomics in model and crop legumes have been discussed. Additionally, applications of advanced proteomics and metabolomics approaches have also been included in this review for future applications in legume research. The integration of these "omics" approaches will greatly support the identification of accurate biomarkers in legume smart breeding programs.

Keywords: abiotic and biotic stresses, developmental process, functional genomics, signaling pathways, stress tolerance

INTRODUCTION

The legume crops such as chickpea (*Cicer arietinum*), common bean (*Phaseolus vulgaris*), cowpea (*Vigna unguiculata*), faba bean (*Vicia faba*), lentil (*Lens culinaris*), lupin (*Lupinus luteus*), mungbean (*Vigna radiata*), pea (*Pisum sativum*), peanut (*Arachis hypogaea*), pigeonpea (*Cajanus cajan*), and soybean (*Glycine max*) have greatly contributed in providing nutrition, food security, and environmental sustainability (Graham and Vance, 2003; Varshney et al., 2013a). However, a majority of them are grown in marginal environments, subjected to abiotic (e.g., drought, heat, cold, salinity, waterlogging, heavy metal toxicities etc.) and biotic stresses (e.g., anthracnose, bean rust, bacterial blight, *Fusarium* wilt etc.), thus limiting their productivity (Dita et al., 2006; Varshney and Tuberosa, 2013; Rodziewicz et al., 2014). Moreover, these environmental conditions severely affect rhizobia-legume symbiosis, which contributes to ~45% of nitrogen required for agriculture (Karmakar et al., 2015). Therefore, for the development of superior varieties with enhanced stress tolerance, it is extremely important to understand stress response mechanisms in legumes which include changes in gene expression, and the consequent variations in the transcriptome, proteome and metabolome.

The availability of high throughput and cost effective next generation sequencing (NGS) platforms as well as high throughput genotyping technologies, have facilitated the generation of massive genomic data for model as well as crop legumes. These platforms have been vital in producing the genome sequence assemblies for the following legumes: adzuki bean (Kang et al., 2015; Yang et al., 2015), chickpea (Varshney et al., 2013b), common bean (Schmutz et al., 2014), Lotus (Sato et al., 2008), Medicago (Young et al., 2011), mung bean (Kang et al., 2014), pigeonpea (Varshney et al., 2012), and soybean (Schmutz et al., 2010). Furthermore, whole genome-resequencing data are also becoming readily available for mining superior alleles (Varshney et al., 2009; Lam et al., 2010). Similarly, transcriptomics/gene expression studies, using a range of platforms, have been valuable for identifying candidate genes associated with tolerance/resistance to different stresses as well as several agronomic traits (Kudapa et al., 2013; Campbell et al., 2014; Brasileiro et al., 2015).

Although QTLs/candidate genes/ alleles have been used in breeding programmes, it has been observed that structural/ expression variation identified at the genetic level are not always translated into the "predicted" phenotype. Additionally, mechanisms involved in stress tolerance can be complicated, e.g., the involvement of metabolites, multigenes and post-translational modifications (PTM) which cannot be investigated by genomics or transcriptomics approaches (Mazzucotelli et al., 2008; Weckwerth, 2011a). In this context, proteomics and metabolomics are promising approaches to enhance our understanding of functional molecules on specific aspects of multigene families and PTMs, instead of analyzing the genetic code (DNA) or transcript (RNA) abundance which may not correlate with their corresponding proteins (Weckwerth, 2011b; Hossain et al., 2012).

Proteomics, defined as the high-throughput study of proteins, has taken the lead in plant biological research and stress responses, especially due to the increasing number of plant genomes being sequenced and released (Pandey and Mann, 2000; Weckwerth, 2011b; Jorrín-Novo et al., 2015). In addition, advancements in mass spectrometry (MS), quantitative methods and bioinformatics approaches have allowed comprehensive identification, quantitation, validation and characterization for a wide range of proteins from specific organ/tissue/cells (Glinski and Weckwerth, 2006). The information obtained through these advanced approaches are valuable for deciphering protein structure and complex mechanisms such as enzymatic and regulatory functions of proteins coded by specific genes (Bachi and Bonaldi, 2008; Wienkoop et al., 2010a; Nanjo et al., 2011; Abdallah et al., 2012). Furthermore, proteomics approaches provide valuable information such as protein levels associated with stress tolerance, the modifications in proteomes under stress that link transcriptomics and metabolomics analyses as well as the role of expressed genes in the functionally translated regions of the genome linked to traits of interest (Kosová et al., 2011). Many proteomics based publications, especially related to plant development and other biological phenomenon such as symbiosis in legumes are available in the model legumes and *Arabidopsis thaliana*, as well as in some crop plants such as rice (*Oryza sativa*), wheat (*Triticum aestivum*), maize (*Zea mays*), soybean, tomato (*Solanum lycopersicum*) and tobacco (*Nicotiana tabacum*) (Jorrín-Novo et al., 2015).

In addition to proteomics, metabolomics is another important approach of functional genomics in which the identification and quantitation of metabolomes (collection of metabolites or small molecules) within a cell, tissue or organism produced through cellular metabolism, connects the cellular biochemical activity with the phenotype (Weckwerth, 2003). Major plant metabolomics approaches include metabolic fingerprinting, metabolite profiling and targeted analysis (Fiehn et al., 2000; Halket et al., 2005; Shulaev, 2006). Depending on the objective of study, different metabolomics approaches or a combination of approaches are applied. Furthermore, the use of MS, bioinformatics tools and softwares, allows metabolites to be measured quickly, simultaneously, in large numbers from a small amount of sample, which are spatially localized within the biological material (Bhalla et al., 2005; Patti et al., 2012). Since metabolites are closer to the phenotype, they reflect gene expressions and different regulatory processes more accurately and metabolomics is a powerful tool to study plant molecular phenotypes in response to stresses (Scherling et al., 2010; Arbona et al., 2013; Doerfler et al., 2013). For example, under abiotic stress conditions, the plant metabolism is affected due to factors such as inhibition of metabolic enzymes, shortage of substrate, extreme demand for specific compounds and a combination of these factors. Thus, the plant undergoes metabolic reprogramming to adapt to the predominant stress conditions through the production of anti-stress components such as compatible solutes, antioxidants and stress-responsive proteins (Wienkoop et al., 2008a; Obata and Fernie, 2012; Doerfler et al., 2014). In crop breeding programs, there has been great interest in using metabolites as selection biomarkers, since metabolites integrate the complex interaction between genotype and the environment (Fernie and Schauer, 2009).

With proteomics and metabolomics emerging as cutting edge functional biology disciplines for understanding plant adaptation mechanisms to stresses at cellular and developmental stages in different plant systems, there has been great interest in applying the knowledge to understand responses in different crop plants. These approaches, integrated with information obtained from genomics data, allow accurate identification of candidate genes and pathways involved in important agronomic traits that can be applied in crop breeding programs (Langridge and Fleury, 2011).

In this article different proteomics and metabolomics tools available and the use of these approaches in legumes especially to understand stress response mechanism, cellular and developmental processes and symbiosis, have been discussed. The assemblage of information and resources available for these aspects of legume proteomics and metabolomics will facilitate our understanding and utilization of these resources for legume crop improvement programs.

PROTEOMICS APPROACHES AND CHALLENGES FOR LEGUME RESEARCH

Proteomics approaches have been implemented according to the objective of the study and are based on certain criteria such as (1) descriptive proteomics for classification of proteins, (2) comparative proteomics for comparison of protein profiles (genotypes, cells, organs, developmental stages, stress response, etc., (3) PTMs that determine how proteins are modified, (4) Protein-protein interactions (PPI) for identifying protein complexes and interacting networks, (5) proteinomics for studying protein structure and functional groups, and (6) translational proteomics that involves transfer of methodology and knowledge for crop improvement (Jorrín-Novo et al., 2015). Data obtained through above mentioned approaches together with bioinformatics, provide significant information on biological processes and stress tolerance mechanisms that can be applied in crop breeding programs (Salekdeh and Komatsu, 2007; Nanjo et al., 2011; Hu et al., 2015).

Furthermore, the comprehensive analysis using differential proteomics in complex systems has demanded development of new technologies to study the cell proteome. Each approach has advantages and disadvantages from both the conceptual and the methodological perspectives. Several approaches are being used to study proteomics in plants and these include electrophoresis and/or chromatography combined with chemical or metabolic labeling and MS. In the following section, we have discussed the gel-based system which is significant for differential expression profiling for studying legume stress responses, and the advantages of the technologically more advanced gel-free, quantitative systems. The application of these approaches for protein profiling in model and crop legumes for understanding stress responses have been addressed with successful examples where proteomics approaches have been utilized for the advancement of crop legumes research. In the interest of channeling the usage of more advanced approaches for legume crop improvement research, the application of more advanced approaches applied in other plant systems, have also been addressed.

Gel-Based System for Protein Differential Expression Profiling

Gel-based systems have been widely applied for protein differential expression analysis which involves the use of two-dimensional gel electrophoresis (2D-GE) or two-dimensional difference gel electrophoresis (2D-DIGE) for protein separation coupled with MS applications for the identification and quantitation of proteins (Kosová et al., 2011; Subramanian and Smith, 2013). The 2D-GE method has an advantage of resolving and visualizing thousands of spots corresponding to different molecular forms of same proteins. However, it has low sensitivity and is not suited for protein quantitation (Marouga et al., 2005; Abdallah et al., 2012). It was found that the use of fluorescent dyes in place of Coomassie can considerably increase the sensitivity and compatibility for MS analyses (Vanderschuren et al., 2013). These factors have led to the usage of a more refined gel electrophoresis method, called the 2D-DIGE which is based on pre-labeling of protein mixtures with fluorescence. Thus, 2D-DIGE is a much more sensitive method, which allows protein detection at sub-picomolar levels for highly precise quantitative proteomic studies (Marouga et al., 2005). Few studies have been reported in legumes, which employed 2D-DIGE for monitoring early response to symbiosis and pathogenesis, such as in the root proteome in Medicago (Schenkluhn et al., 2010) and parasitic infections in pea (Castillejo et al., 2012). The advantages and disadvantages of 2D-GE and 2D-DIGE have been reviewed before (see Griffin and Aebersold, 2001; Marouga et al., 2005; Abdallah et al., 2012).

For identification of proteins separated by 2D-GE, MS based techniques were found to have high accuracy, resolution and sensitivity but the sample preparation seemed to be laborious (van Wijk, 2001; Subramanian and Smith, 2013). Earlier, the usage of techniques such as matrix-assisted laser desorption ionization time-of-flight (MALDI-TOF) MS and electrospray ionization (ESI) MS with tandem MS (MS/MS) overcame the slower and less sensitive methods (Griffin and Aebersold, 2001; Griffin et al., 2001; van Wijk, 2001). In particular, the application of tandem MS methods provided sequence information for more peptides accurately for the identification of proteins and PTMs (Pandey and Mann, 2000; Bachi and Bonaldi, 2008). The use of MALDI ionization with two TOF analysers, (MALDI TOF/TOF) and liquid chromatography tandem MS (LC-MS/MS) have been fairly new technologies for protein identification that enhanced the pipeline for *de novo* assembly. The availability of these tools, together with the genome sequences of more plants and accessibility of sequence databases, intensified the identification of protein spots in different plants (see Jorrín-Novo et al., 2015). The most commonly used pipeline for the identification of proteins involves comparing the MS/MS spectra to the reference databases (Romero-Rodríguez et al., 2014). Although SEQUEST, MASCOT and X!Tandem are the widely used database search programs for protein identification, MassMatrix is also suited for tandem MS data (Xu and Freitas, 2009; Senkler and Braun, 2012).

Gel-Free Proteomics Tools and Approaches for Quantitation

Shot-gun proteomics is a gel-free "bottom-up" strategy, in which complex peptide fractions produced after proteolytic digestion of proteins are analyzed by LC-MS/MS. The protein identification rates are enhanced using different peptide or protein fractionation strategies (Abdallah et al., 2012). This approach allows high throughput and comprehensive investigation, by providing an overview of the organelle or cell-type proteomes (Glinski and Weckwerth, 2006). For large-scale proteomics, a multidimensional chromatography technique called multidimensional protein identification technology (MudPIT) has been used to separate complex protein samples prior to MS analyses to increase the proteome coverage and dynamic range (Washburn et al., 2001; Bachi and Bonaldi, 2008).

MS based protein quantitation strategies may include untargeted quantitation, (database dependent or database independent protein identification) or targeted absolute quantitation, which are important for biomarker discovery and protein stoichiometries in protein complexes (Wienkoop et al., 2008b; Weckwerth, 2011b; Schmidt and Urlaub, 2012). Relative quantitation is an unbiased, large-scale screening strategy, which is useful to detect proteins involved in unknown regulatory processes, protein modifications and mechanisms in systems biology. MS-based methods for protein quantitation can be label (stable-isotope *in vivo* or *in vitro* label) or non-label based (see Bachi and Bonaldi, 2008; Abdallah et al., 2012). Label based proteomics for relative quantitation consists of chemical labeling and metabolic labeling. The iTRAQ (isobaric tags for relative and absolute quantitation), is a chemical labeling approach and has been applied to study salinity stress tolerance in the phloem sap proteome of *Cucumis sativus* (Fan et al., 2015), leaf proteome of *Vitis vinifera* to understand heat stress response (Liu et al., 2014) and the root proteome of rice to investigate aluminum stress response (Wang et al., 2014). The platforms available for relative quantitation include Mass Accuracy Precursor Alignment (MAPA), ProtMax, MASS WESTERN, and PROMEX (Hoehenwarter et al., 2008; Weckwerth et al., 2014). Recently, a rapid shot-gun LC-MS approach for relative quantitation called full-scan (FS) selective peptide extraction (Selpex) was successfully used for generating a reference library of targeted peptides from leaf tissue of Medicago (Castillejo et al., 2014).

Selective reaction monitoring (SRM) or multiple reaction monitoring (MRM), which involves the use of triple-quadrupole (QQQ) MS is a quantitative, targeted and label-free approach, that complements untargeted shotgun methods due to its reliable quantitation of low abundance proteins in complex mixtures (Lange et al., 2008; Wienkoop et al., 2010b; Schmidt and Urlaub, 2012). This approach has been useful to measure a predetermined set of proteins that could constitute cellular networks or candidate biomarkers across samples consistent and reproducibly (Wienkoop et al., 2010b; Picotti and Aebersold, 2012). In Medicago, the MRM approach was applied for the absolute quantitation of sucrose synthase isoforms and N-metabolism enzymes in symbiotic root nodules as well as in the analysis of nodule metabolism under drought stress (Wienkoop et al., 2008b; Larrainzar et al., 2009).

Protein Profiling in Legumes

Proteomic studies have been carried out predominantly in Medicago to investigate stress tolerance, seed physiology, plant growth, development and symbiosis which has immense significance in agricultural research (Colditz and Braun, 2010; Jorrín-Novo et al., 2015). A majority of studies have been carried out to investigate abiotic stress responses, e.g., drought tolerance in shoots, leaves and roots using gel-based as well as non-gel based approaches (**Table 1**). For instance, 26 differentially expressed proteins were identified using 2D-GE and ESI-LC-MS/MS approaches from leaf samples subjected to drought stress (Aranjuelo et al., 2011). The study revealed the regulation process involved in the synthesis of amino acids pertaining to osmoregulation. Using non-gel based LC-MS/MS approaches, root nodule proteome and symbiotic nitrogen fixation was studied under drought stress in Medicago (Larrainzar et al., 2007, 2009). The former study (Larrainzar et al., 2007) not only identified a large number of nodule proteins, but also grouped the drought responsive proteins from plant and bacteroid which was helpful in understanding the mechanism in which each symbiotic member responded to drought stress. The later study (Larrainzar et al., 2009) was important for the relative quantification of root nodule proteins and absolute quantification of a key enzyme in sucrose metabolism to understand the regulation of nitrogen fixation under drought stress. In order to understand the roles of diverse LEA proteins in developing seeds of Medicago, differential profiling was effectively used to identify proteins that may have a role in desiccation tolerance and seed longevity (Chatelain et al., 2012). The study of arbuscular mycorrhizal symbiosis in Medicago roots identified differential accumulation of 96 membrane proteins which could have a role during symbiosis when compared between the mycorrhizal and non-mycorrhizal roots. The study has also identified proteins that stimulated changes in membrane trafficking (Abdallah et al., 2014).

In addition, there is a significant contribution of proteomics for studying abiotic stress in soybean at subcellular, organ and whole plant levels. The methodologies applied and the major discoveries in these studies have been reviewed in a number of articles (see Hossain et al., 2013; Hossain and Komatsu, 2014a,b). A few important examples of proteomics approaches used for analyzing soybean abiotic stress are described here. For instance, osmotic stress in the plasma membrane of soybean seedling was studied using both gel-based and nanoLC MS/MS approaches (Nouri and Komatsu, 2010). While the former technique identified four up-regulated and eight down-regulated protein spots, the latter approach, identified 11 up-regulated and 75 down-regulated proteins, of which 7 were identified in both the studies. In dissecting the mechanism of Cadmium (Cd) uptake and distribution in soybean, 2D-GE and LC-MS/MS approaches were employed on contrasting soybean cultivars to identify 13 and 11 differentially expressed proteins, respectively.

TABLE 1 | Key studies on protein differential expression analysis in response to various stress in some model and crop legumes.

Legume	Cell/Tissue/ Organ	Stress	Method	Protein differential expression and identification		References
				Differentially expressed proteins	Function	
Medicago	Shoots	Salt, drought	nanoESI-LC-MS/MS	–	Protein regulation, photosystem (PS11)	Staudinger et al., 2012
	Shoots	Cadmium	2D-GE, MALDI-TOF-MS	17	Photosynthesis, chaperones	Aloui et al., 2011
	Leaves	Drought	2D-GE, ESI-LC-MS/MS	26	Metabolism, energy, protein storage	Aranjuelo et al., 2011
	Nodules	Drought	LC-MS/MS	16	Nodule plant and bacteoid protein	Larrainzar et al., 2007
	Nodules	Drought	LC-MS/MS	–	Sucrose synthase, symbiotic nitrogen fixation	Larrainzar et al., 2009
Chickpea	Microsomal fraction from aerial tissue	Dehydration	2D-GE, MALDI-TOF-TOF	184	Photosynthesis, transport, metabolism	Jaiswal et al., 2014
	Leaves	Cold	2D-GE with MALDI-TOF/TOF and/or with LC-MS/MS	70	Defense, signal transduction, storage	Heidarvand and Maali-Amiri, 2013
	ECM	Dehydration	2D-GE, LC-ESI-MS/MS	81	Cellular function	Bhushan et al., 2011
	ECM	Dehydration	2D-GE, ESI-Q-TOF-MS/MS	134	Cell wall modification, signal transduction, metabolism, defense	Bhushan et al., 2007
	ECM	Dehydration	2D-GE, ESI-TOF-MS	147	Molecular chaperones, cell signaling	Pandey et al., 2008
Common bean	Leaves	Drought	2D-DIGE, LC-MS/MS	130	Metabolism, photosynthesis, protein synthesis, proteolysis, defense	Zadražnik et al., 2013
Green gram	Roots	Cadmium	2D-GE and MALDI-TOF MS	23	Nutrient metabolism	Muneer et al., 2014
Peanut	Mature seeds	Water deficit	LC MS/MS	93	Glycolysis, sucrose and starch, fatty acid metabolism	Kottapalli et al., 2013
	Leaves	Water deficit	2D-GE, MALDI-TOF-MS, Q-TOF-MS/MS	79	Photosynthesis, signal transduction, energy, metabolism	Kottapalli et al., 2009
Pea	Seeds	Osmotic	2D-GE and MALDI-TOF-MS	230	Glycolysis, signal transduction, detoxification	Brosowska-Arendt et al., 2014
	Roots	Drought	2D-GE, MALDI-TOF/TOF and LC-ESI-QTOF	18	Flavonoid and sulfur metabolism	Irar et al., 2014
	Roots	Salinity	2D-GE, ESI-Q-TOF MS/MS	35	Defense, stress related	Kav et al., 2004
Soybean	Roots	Cold, osmotic	2D-GE, LC/nanoESI-MS	59	Signal transduction, secondary metabolism, defense, energy, protein synthesis, development, translocation,storage	Swigonska and Weidner, 2013
	Seedlings	Flooding	2D-GE, nano-LC-MS/MS	168	Metabolism, transportation, localization, Isoflavone reductase	Khatoon et al., 2012
	Leaves	Salinity	2D-GE, MALDI-TOF-TOF-MS	91	Stress related, proteolysis, protein biosynthesis, photosynthesis	Ma et al., 2012
	Leaves, hypocotyls and roots	Drought	2D-GE, nanoLC-MS/MS	57	Stress related, defense	Mohammadi et al., 2012
	Roots	Flooding	2D-GE, MALDI-TOF- MS, nanoLC MS/MS	70	Transportation, localization, storage, metabolism, cell wall modification, programmed cell death	Salavati et al., 2012
	Developing seeds	High temperature, humidity	2D-GE, MALDI-TOF-MS	42	Signal transduction, protein biosynthesis, photosynthesis, protein folding, defense, metabolism, regulation, secondary metabolite biosynthesis	Wang et al., 2012a
	Leaves	Fungus	2D-GE, MALDI-TOF-TOF-MS	41	Defense, carbohydrate metabolism, energy	Wang et al., 2012b

(Continued)

TABLE 1 | Continued

Legume	Cell/Tissue/ Organ	Stress	Method	Protein differential expression and identification		References
				Differentially expressed proteins	Function	
	Plasma membrane	Osmotic	2D-GE, nano-LC-MS/MS	96	Transport	Nouri and Komatsu, 2010
	Endoplasmic reticulum	Flooding	2D-GE and BN-PAGE, nano-LC-MS/MS	~50	Heat shock proteins, chaperonins	Komatsu et al., 2011
	Microsomal proteins	Cadmium	2D-GE, nanoLC-MS/MS	13 and 11	Stress, protein biosynthesis	Ahsan et al., 2010a
	Leaves, stems and roots	High temperature	MALDI-TOF-MS, nanoLC-MS/MS and protein sequencing	54, 35, 61	Defense photosynthesis, secondary metabolism, protein biosynthesis	Ahsan et al., 2010b
	Roots	Waterlogging	2D-GE, MALDI-TOF- MS, ESI-MS/MS	24	Signal transduction, programmed cell death, homeostasis and metabolism	Alam et al., 2010
	Hypocotyls, roots	Salt	2D-GE ESI-Q/TOF-MS/MS	>20	LEA protein, protease inhibitor	Aghaei et al., 2009
	Seedlings	Aluminum	2D-GE, MALDI-TOF-MS	39	Defense, signal transduction, protein folding, transport	Zhen et al., 2007

In this study, highly up-regulated proteins associated with lignin biosynthesis indicated that xylem lignification may be preventing the translocation of Cd (Ahsan et al., 2010a). For deciphering the mechanism of heat tolerance in soybean seedlings, 2D-GE, MALDI-TOF-MS, LC-MS/MS and protein sequencing were applied in which 54, 35 and 61 proteins were differentially expressed respectively in leaves, stems and roots, respectively (Ahsan et al., 2010b). The study showed that heat shock proteins (HSPs) and antioxidant defense related proteins were induced and identified different proteins involved in tissue specific and common defense mechanisms.

In other legume crops, there are reports of significant discoveries being made for understanding abiotic stress responses through protein profiling. In the case of chickpea, drought is one of the most important abiotic stresses that severely affect its productivity, for which efforts have been made to dissect the genetic basis of tolerance (Thudi et al., 2014; Varshney et al., 2014). Toward understanding proteomic response to dehydration stress in chickpea, a few protein profiling studies have been made available (**Table 1**). For instance, the investigation of a dehydration-responsive microsomal proteome with 2D-GE and MALDI-TOF/TOF identified 184 proteins that showed significant differential expression (Jaiswal et al., 2014). This study was significant in identifying a novel component involved in dehydration signaling called CaSUN1. In a different study, using 2D-GE with MALDI-TOF/TOF and LC-MS/MS, proteomic changes were identified at early stage of cold stress in chickpea leaves (Heidarvand and Maali-Amiri, 2013). The analysis indicated that energy resources and primary metabolites respond by recreating a new homeostasis in preparation for long-term cold stress adaptation. In an earlier study, 81 dehydration responsive proteins were identified from profiling analysis of the chickpea organellar proteome in two contrasting genotypes, with 2D-GE and LC-ESI-MS/MS (Bhushan et al., 2011). The study proposed that cell wall restructuring and the control of reactive oxygen species were mainly responsible for better adaptability to the stress.

A few studies have also been conducted in pea through protein profiling for understanding abiotic stress response and are being described here. For example, 2D-GE, MALDI-TOF/TOF, and LC-ESI-QTOF was employed to dissect the signaling pathway leading to inhibition of biological nitrogen fixation under drought stress (Irar et al., 2014). The study successfully identified 18 nodule proteins regulated by both pea and rhizobium genomes under drought stress and unraveling the SNF regulation machinery in nodules. In a different study, changes in protein accumulation in germinating seeds under different osmotic conditions were monitored with 2D-GE and MALDI-TOF-MS (Brosowska-Arendt et al., 2014). This work showed that under optimal conditions, proteins associated to glycolysis, fatty acids synthesis and detoxification under osmotic stress considerably decreased, while proteins involved in signal transduction and protection were accumulated. Similarly, investigation of salinity stress response in pea with 2D-GE and ESI-Q-TOF MS/MS approach revealed significant differential expression of 35 proteins, of which 10 pathogenesis-related (PR) proteins was reported for the first time to be involved in salinity stress response, possibly involving a new signal transduction pathway (Kav et al., 2004).

Protein Reference Maps in Legumes

Protein reference maps have been developed for model and crop legumes, which identify as many proteins as possible in a particular tissue or cell cultures at a given point of time. This provides useful insights into important plant processes, such as stress tolerance, nutrient uptake and symbiotic association with rhizobia (Brechenmacher et al., 2009). These resources will allow further studies for efficient proteomics applications for the crop legumes, especially through the classification and characterization of proteins related to development and stress tolerance important for unambiguous candidate gene identifications. Reference maps have been established in Medicago for cell suspension cultures, roots, leaves, stems, flowers and pods (Watson et al., 2003). A protein reference map generated for cell suspension culture of Medicago consisted of 1367 proteins with 907 unique accessions, which could identify a complete tricarboxylic acid cycle, a nearly complete glycolytic pathway, a partial ubiquitin pathway, enzymes involved in secondary metabolism through functional annotations (Lei et al., 2005). Other specialized proteome maps generated include Medicago mitochondria (Dubinin et al., 2011); cell wall (Gokulakannan and Niehaus, 2010); embryonic cell cultures generated from single protoplasts (Imin et al., 2004) and roots (Mathesius et al., 2011).

Lotus proteome reference maps for nodules and roots consist of 780 and 790 protein spots identified with 2D-GE, with 45% of the corresponding unique gene accessions common in both the tissues (Dam et al., 2014). The study showed that PTMs were more prominent in nodules rather than in roots. In addition to this, higher levels of proteins such as pathogen-related 10, heat shock, and redox processes related were found in the nodules prior to nitrogen fixation and nodulin related proteins were prevalent in mature nitrogen fixing nodules. In Lotus proteome reference maps were also generated for pod and seed development (Nautrup-Pedersen et al., 2010). Similar reference maps were also developed in crop legumes such as soybean and peanut. In soybean, a comprehensive proteome reference map was generated with 5702 proteins identified for a single root hair cell (Brechenmacher et al., 2012). A previous reference map in root hair cells identified 1492 proteins (Brechenmacher et al., 2009). Other proteome maps generated in soybean includes root apex and differentiated root zone (Mathesius et al., 2011); seed filling (Hajduch et al., 2005), leaf (Xu et al., 2006), and response to pathogen invasion (Mithöfer et al., 2002). Similarly, leaf proteome reference map was developed in peanut (Katam et al., 2010). In chickpea, reference maps were generated for understanding the complexity of plant nuclear proteins (Pandey et al., 2006) and membrane proteins (Jaiswal et al., 2012). In the case of pea, reference maps have been developed for vegetative tissues (Schiltz et al., 2004) and mature seeds (Bourgeois et al., 2009).

Proteome reference maps generated through gel-based approaches consist of gel images in which selected "spots" were linked with protein identity information with arrows and numbers (Senkler and Braun, 2012). However, web-based resources with interactive features of the reference maps have also been made available. Examples of these resources include the "Seed Proteome," the "Rice proteome database," and the "Arabidopsis seed proteome." Recently, for constructing proteome reference map, a software tool called "GelMap" was developed (Rode et al., 2011). This tool was used for generating a 3D GelMap of Arabidopsis complex 1 in which its unique proteins constituents has been defined (Peters et al., 2013).

Legume Databases for Proteomics Analyses

Various databases have been developed that store a large resource of plant proteins from the proteome reference maps of legumes and other plants, such as the plant proteomics database PROMEX (Hummel et al., 2007; Wienkoop et al., 2012). A continuously growing database, PROMEX consists of 116,364 tryptic peptide product ion spectra entries of 48,218 different peptide sequence entries from Lotus, Medicago, common bean and soybean as well as other plants such as Arabidopsis, rice, etc. The database could be searched for whole experiments with an experimental ID, meta-information and single proteins and their corresponding peptide reference spectra. Furthermore, new LC-MS/MS analyses can be searched against this spectral library (http://promex.pph.univie.ac.at/promex/). LegProt (http://bioinfo.noble.org/manuscript-support/legumedb) is a legume specific protein database consisting of amino acid sequences translated from predicted gene models and 6-frame translations of tentative consensus sequences expressed sequence tags (ESTs) (Lei et al., 2011). The ProteomeXchange contains the dataset of seed phosphoproteins from Lotus (PXD000053, http://proteomecentral.proteomexchange.org/cgi/GetDataset?ID=PXD00053) and has been valuable for understanding the regulatory mechanisms of seed germination in legumes (Ino et al., 2014). This dataset contains a total of 721 phosphopeptides from 343 phosphoproteins in cotyledons and 931 phosphopeptides from 473 phosphoproteins in hypocotyls.

The Medicago PhosphoProtein Database (MPPD, http://phospho.Medicago.wisc.edu), contains 3457 unique phosphopeptides with 3404 non-redundant sites of phosphorylation on 829 proteins. This database represents the most comprehensive Medicago phosphorylation data, which allows browsing of identified proteins, searching proteins of interests, in addition to conducting BLAST searches of the database using peptide sequences and phosphorylation motifs as queries (Rose et al., 2012a). The Soybean Proteome Database (SPD, http://proteome.dc.affrc.go.jp/Soybean/) consists of a repository of functional analysis of abiotic stresses (flooding, drought, and salt). In total, it consists of 23 reference maps of soybean and proteins collected from several organs, tissues, and organelles of soybean (Ohyanagi et al., 2012). Recently, another database for storage, allergen, and anti-nutritional proteins from cultivated soybean called Soybean Protein Database (SoyProDB; http://bioinformatics.towson.edu/Soybean_Seed_Proteins_2D_Gel_DB/Home.aspx) has also been developed (Tavakolan et al., 2013). In the case of yellow lupin (*Lupinus luteus* L.), a seed-protein catalog has been developed. In this study, 736 proteins corresponding to 152 unique proteins have been deposited in the WORLD-2DPAGE repository (http://world-2dpage.expasy.org/repository/0066/; Ogura et al., 2014).

Post-Translational Modifications (PTMs) in Legumes

PTMs are required for the functionality of proteins that regulate processes and the subcellular localization that could be analyzed using proteomics approaches (Pandey and Mann,

2000; Seo and Lee, 2004; Downes and Vierstra, 2005). Although mostly identified through MS approaches, PTM analyses are not straightforward as protein identifications, the reason being peptide analyses do not show the expected molecular mass and therefore more protein samples are required (Pandey and Mann, 2000). However, advancement in phosphoproteomics procedures, e.g., the enrichment of phosphopeptides using immobilized metal affinity chromatography (IMAC) or proteins with aluminum hydroxide or titanium dioxide called metal oxide affinity chromatography (MOAC) (Wolschin et al., 2005) have improved the identification efficiency (Ndassa et al., 2006; Chen et al., 2010; Hoehenwarter et al., 2013; Beckers et al., 2014). Among the PTMs, phosphorylation is the major post-translational regulatory processes in all eukaryotes followed by ubiquitin and SUMO (Small Ubiquitin-like MOdifier) conjugations (Mazzucotelli et al., 2008).

The root proteome of the Medicago genotype, Jemalong A173457, revealed unique phosphopeptides that covers 3404 non-redundant sites of *in vivo* phosphorylation on 829 proteins (Grimsrud et al., 2010). The large scale phosphoproteomic study identified multiple sites of phosphorylation on a number of crucial proteins in rhizobial symbiosis initiation such as SICKLE, NUCLEOPORIN133 and INTERACTING PROTEIN OF DM13. Further, the rapid Nod-factors (NF) induced changes in the phosphorylation levels of 13,506 phosphosites in 7739 proteins was recently measured for rhizobia-legume symbiosis in Medicago which was found useful for quantifying phosphorylation, specifically associated with NF-signaling (Rose et al., 2012b).

In chickpea, a differential phosphoproteomic study in response to dehydration stress identified 91 phosphoproteins that are likely to be involved in cell defense, photosynthesis, photorespiration, molecular chaperones and ion transport. The study also identified multiple sites of phosphorylation in key regulatory and functional proteins (Subba et al., 2013). Recently, a nucleus-specific phosphoproteome map of 107 identified phosphoproteins was constructed in chickpea, which identified a collection of phosphoproteins involved in many cellular functions such as protein folding, signaling, gene regulation, DNA replication/repair/modification, metabolism, etc. (Kumar et al., 2014). In an earlier study, the PTMs of αAI (α-amylase inhibitor) were compared among transgenic pea and chickpea expressing αAI from common bean, with the processed form of the protein from several bean varieties. The αAI proteins displayed microheterogeneity due to differences in glycan addition frequency, variation in glycan processing and differences in C-terminal exopeptidase activity (Campbell et al., 2011). Thus, PTMs in common beans were also investigated on three seed defensive proteins αAI-1, αAI-2, and arcelin-5 (Young et al., 1999). The data showed that the proteolytic cleavage is required for the activation of the proteins, which resulted in the loss of the terminal Asn residue in αAI-1, and a minimum of seven residues from the C-termini of all three proteins. Additionally, a significant difference in the glycosylation patterns of αAI-1 and αAI-2 has been reported, although the proteins showed high sequence homology. Similarly in pea, the presence of multiple trypsin inhibitors (TI) isoforms were attributed to PTMs and particularly post-translational processing at the C-terminus during the desiccation stage of seed development resulted in the appearance of TI isoforms in pea (Domoney et al., 1995).

Histone modifications and histone variants are known to be vital for various biological processes. In this context, variants of soybean histone, H3 and H4 and their PTMs were reported, which revealed several distinct variants of soybean histone and their modifications that were different from Arabidopsis (Wu et al., 2009). The study thus, provides important biological information toward understanding histone modifications and their functional relevance in legumes.

This area of research possesses great promise for legume improvement as it is supported by advanced proteomics technologies, in particular, developments in the strategies for detection and selective isolation of proteins with known function. As discussed above functional properties of proteins are often regulated by PTMs of proteins and the numerous techniques developed can be applied to the global identification of PTMs and their processing sites in legumes.

PPI Approaches and Application in Legume Research

PPIs provide functional knowledge about proteins by analysing the interacting partners and interactomics, which is the large-scale study of PPI networks using high throughput methods. This provides valuable insights for understanding cellular function, metabolism and signaling mechanisms (Braun et al., 2013; Stasi et al., 2015). Interactomes (maps of PPI) are constructed on the basis of experimental data and computational prediction of interactions (Stasi et al., 2015). Approaches used in PPI analyses and mapping can be genetic, biochemical or proteomics-based. The strategies involve direct interactions with binary methods such as yeast two-hybrid (Y2H) or split ubiquitin, analyses of protein complexes with co-immunoprecipitation or affinity purification followed by MS (AP-MS), imaging and database analyses (Pandey and Mann, 2000; Fukao, 2012; Braun et al., 2013; Stasi et al., 2015). The identification of minor PPIs are also essential for studying transient interactions and proteins of low-abundance, which may be functionally relevant (Fukao, 2012). In recent years, many proteins have been identified with advances in MS technologies and large-scale proteomics, however, the consequence of this is the increased number of false positive protein identifications. Therefore, for the accurate appraisal of PPI identification, the inclusion of independent experimental validation is required (Fukao, 2012). Y2H system is sensitive in detecting transient and unstable interactions, and is suitable for PPI mapping and generating high-throughput data with a fine resolution to understand cellular process at the systemic level. However, it has limitations with incidences of false-positives and false-negative interactions. The PPIs are usually validated with techniques such as pull-down assays, co-immunoprecipitation, *in-situ* hybridization (von Mering et al., 2002; Parrish et al., 2006; Brückner et al., 2009).

Although limited PPI based studies have been undertaken in legumes, the studies involving stress signaling networks, nodule formation and symbiosis have provided valuable information in understanding these processes. The Y2H system has been mostly used in model legumes to study symbiosis and nodulation mechanisms and in the crop legumes mainly to study stress responses (**Table 2**). For example in pea, using the Y2H system and validation using *in planta* co-immunoprecipitation, the PPI observed among the heterotrimeric G proteins and GPCR protein were shown to be relevant in the salt and heat stress signaling pathways (Misra et al., 2007). In another study, the exposure to salt and cold simultaneously stimulated the expression of genes encoding a protein kinase (*PsCIPK*) and a calcineurin B-like protein (*PsCBL*) in pea (Mahajan et al., 2006). Although immunofluorescence and confocal microscopy showed that the PsCBL was localized in the cytosol and PsCIPK in the outer membrane, Y2H analysis indicated that both the protein products interacted and this was supported with Western blots. In chickpea, a few candidate interactors were identified for the 1R-MYB for drought tolerance with the Y2H system (Ramalingam et al., 2015).

Bimolecular fluorescence complementation (BiFC) uses the yellow fluorescent protein (YFP), split into two-non-overlapping N-terminal (YN) and C-terminal (YC) fragments, where each fragment is cloned in-frame to the gene of interest, allowing the expression of fusion proteins. PPI detected *in planta* causes the detection of yellow fluorescence, which is not found for non-interacting pairs or non-fused YN/YC (Bracha-Drori et al., 2004). Although identification of PPI by screening a cDNA library is usually performed in yeast, BiFC technology was applied to screen an Arabidopsis cDNA library against a bait protein *in planta* since subcellular compartmentation and protein modifications differ between plant and yeast cells (Lee et al., 2012).

A challenging area in crops research would be the development of a proteome-wide PPI maps to understand the complex biological pathways and cellular networks. A high-throughput, Y2H system (HTP-YTH), suitable for mapping PPIs was described by Fang et al. (2002) to screen PPI in plants. This system involves a yeast gap-repair cloning and a selectivity that reduces false positives and negative clones with automation in laboratory procedures. This system has been used to study the defense signal transduction pathway in rice, where more than 100 genes were selected as "baits" for HTP-YTH screening in which many known and novel PPIs were identified. In Arabidopsis, the first binary PPI map for the interactome network of plants was developed (Arabidopsis Interactome

TABLE 2 | PPI analyses in some model and crop legume using Y2H and validated with other approaches.

Legume	Stress/Condition	Interacting components	Method of confirmation	References
Lotus	Nodule development	LjNSP2 homodimers	β-galactosidase assay	Murakami et al., 2013
	Symbiotic signaling	SINA4 and SYMRK	BiFC	Den Herder et al., 2012
	Nodule development	CASTOR homodimer, POLLUX homodimer	BiFC	Charpentier et al., 2008
		SIP1 and SymRK	Pull down assay	Zhu et al., 2008
	Cell growth and differentiation	LjRac and LjRacGAP1	Affinity chromatography	Borg et al., 1999
Medicago	Nodulation signaling	RAM1 and NSP2	BiFC	Gobbato et al., 2012
		NSP1 and NSP2	BiFC	Hirsch et al., 2009
		DMI3 and IPD3	BiFC	Messinese et al., 2007
Chickpea	Salt	CaCIPK6 and NtCBL3	-	Tripathi et al., 2009
Cowpea	Osmotic and heat	VuDRIP and VuDREB2A	Antibiotic and X-α-Gal	Sadhukhan et al., 2014
Mungbean	Osmotic stress	VrUBC1 and AtVBP1	BiFC	Chung et al., 2013
Pea	Salt and heat	Gα subunit with the G_β subunit and phospholipase C at the calcium-binding domain	Co-immunoprecipitation	Misra et al., 2007
	Salt and cold	PsCIPK and PsCBL	β-galactosidase assay, Far-western blotting	Mahajan et al., 2006
Soybean	Cold	SCOF-1 and SGBF-1	β-galactosidase assay	Kim et al., 2001
	ABA, drought, cold, salt	GmMYB76 homodimers, GmMYB76 and GmMYB177, GmZIP46 homodimer, GmZIP46 and GmZIP62, GmZIP46 and GmMYB76.	β-galactosidase assay	Liao et al., 2008a
			β-galactosidase assay	Liao et al., 2008b
	Salt and heat	GmGBP1 with R2R3 domain of GmGAMYB1 in	X-Gal assay	Zhang et al., 2013

Mapping Consortium, 2011). The construction of these maps in legumes should facilitate system biology approaches and will greatly benefit legume crop improvement programs.

METABOLOMICS APPROACHES IN LEGUMES

Plants synthesize specialized metabolites that define the biochemical phenotype of a cell or tissue and can be viewed as the end products of gene expression (Sumner et al., 2003). Quantitative and qualitative measurements of cellular metabolites provide a broad view of the biochemical status of an organism that could be used to monitor and assess gene function (Fiehn et al., 2000). Furthermore, metabolomics contributes significantly to the study of stress biology by identifying different compounds such as by-products of stress metabolism, stress signal transduction molecules, molecules that are part of plant acclimation process etc. (Weckwerth, 2003; Larrainzar et al., 2009). The identified metabolic compounds could be further studied by direct measurement or correlating with the changes in transcriptome and proteome expression and confirmed by mutant analysis. In this section, recent applications of metabolomics approaches in the area of legume development, symbiosis and stress response are discussed.

Metabolomics is a rapidly developing technology and at present metabolic fingerprinting and metabolite profiling approaches are being used. For wide coverage of the vast range of metabolites present, several analytical techniques involving separation and detection are implemented (Scherling et al., 2010; Doerfler et al., 2013, 2014). The separation technique is selective for certain groups of metabolites which includes gas chromatography (GC) for volatile and primary metabolites such as sugars and amino acids (Weckwerth, 2011a), LC for mainly secondary metabolites (Scherling et al., 2010; Weckwerth, 2011a), capillary electrophoresis (CE) for ionic metabolites to be separated (Soga et al., 2003; Soga, 2007) and ultra-performance liquid chromatography (UPLC). UPLC is a powerful technique which has high resolution, sensitivity and throughput than conventional high performance liquid chromatography (HPLC). The MS analysers have been commonly used for metabolite profiling, particularly those that provide accurate mass measurements such as FTICR_MS, Orbitrap-MS or TOF-MS due to its fast scan times with improved deconvolution, run times and high mass accuracy (Scherling et al., 2010; Weckwerth, 2010, 2011a; Obata and Fernie, 2012; Arbona et al., 2013; Doerfler et al., 2013, 2014). GC-MS has been widely used in plant metabolomics research and electron impact (EI) supports strong interfacing of GC with MS which allows fragmentation patterns to be highly reproducible. Major metabolomics approaches and their applications in legumes are described below and are summarized in **Table 3**.

Metabolite Profiling

Metabolite profiling is the simultaneous measurement of all or a set of metabolites in a given sample. Several analytical techniques such as nuclear magnetic resonance (NMR), GC-MS, LC-MS, capillary electrophoresis–MS (CE-MS) and Fourier transform infrared (FT-IR) spectroscopy have been reported for analysing the data from metabolite profiling (Sumner et al., 2003; Weckwerth, 2003; Shulaev, 2006). The advantages and disadvantages of each technique for metabolite profiling have been discussed previously (Fiehn et al., 2000; Roessner et al., 2000; Sumner et al., 2003; Weckwerth, 2003; Weckwerth et al., 2004a,b; Shulaev, 2006). The use of metabolic profiling has been limited in crop legumes but this approach has been successfully demonstrated in model legumes. For example, in Medicago, untargeted quantitative MS approach was used to profile metabolites treated with rhizobial Nod factors to study the metabolic changes between the symbionts (Zhang et al., 2013). The study showed decrease in concentration of (9)-HODE class of oxylipins upon Nod factor treatment *in planta* and together with jasmonic acid inhibited Nod factor signaling. This suggests an important role for the oxylipin pathway in Nod factor signaling in symbiosis. In a different study, under early salt and drought stress conditions, the involvement of certain metabolites in nutritional priming through symbiotic interaction of nodulated plants and N-fertilized Medicago has been reported (Staudinger et al., 2012).

Similarly, to understand drought acclimation in model and forage legumes, a comprehensive and progressive reprogramming of metabolic pathways were suggested for increased water stress in Lotus (Sanchez et al., 2012). Using GC coupled to electron impact ionization (EI)-TOF-MS (GC-EI-TOF-MS), this study reported gradual increase in most of the soluble molecules profiled. In addition, comparative metabolomics between the Lotus species showed the presence of metabolites that were conserved and unique in response to drought stress. Metabolite profiling using a combination of ionomic and GC-MS was conducted for the shoots of extremophile Lotus species, adapted to highly saline coastal regions and was compared with that of cultivated glycophytic grassland forage Lotus species, to understand salt tolerance mechanisms. The extremophile Lotus species was identified to have higher salt levels with a differential rearrangement of shoot nutrient levels upon salt exposure (Sanchez et al., 2011). In a similar study, the accumulation of alanine under anoxic conditions was examined in Lotus, which is highly tolerant to water logging (Rocha et al., 2010). High accumulation of succinate, alanine and the direct co-substrates for alanine synthesis, glutamate and gamma aminobutyric acid (GABA) in the roots of Lotus during water logging was reported. Whereas, majority of amino acids that are derived from TCA cycle intermediate were found to be decreased, which support earlier findings that the metabolic equilibriums are expected to drive the metabolic flux from glycolysis, via alanine synthesis and oxoglutarate to succinate, which prevents the accumulation of pyruvate activating fermentation leading to ATP production by succinyl- CoA ligase. In another study in Lotus, GC-TOF-MS (primary metabolism) and LC-Fourier-Transformation-MS (secondary metabolism) was applied to study plant/plant competition responses in a large biodiversity experiment (Scherling et al., 2010). Significant effects in Lotus were not associated with primary metabolism (sugars, amino acids, organic acids) but with the secondary metabolism. A

TABLE 3 | Application of metabolomics approaches in stress, development and symbiotic processes in some legumes.

Legume	Stress/Condition, Tissue	Methods	Metabolites	References
Lotus	Drought, shoots	GC-EI-TOF-MS	Serine, proline, arabitol	Sanchez et al., 2012
	Salt, shoots	GC-MS	Citric acid, malic acid	Sanchez et al., 2011
	Biodiversity and plant/plant competition, leaves	GC-TOF-MS and LC-Orbitrap-MS	Especially secondary metabolites, Flavonoids	Scherling et al., 2010
Medicago	Metabolic reprogramming, roots	GC-MS, LC-MS	Flavonoids, triterpenoid, starch hydrolysis metabolites	Watson et al., 2015
	Symbiosis, roots	MSI(MALDI)-MS imaging	Organic acids, amino acids, sugars.	Ye et al., 2013
		LC-ESI-TOF-MS	Oxylipins	Zhang et al., 2012
	Flavonoid profiling, roots	LC-MS	Flavonoids	Staszków et al., 2011
	Arbuscular mycorhizal symbiosis, roots	GC-MS, HPLC, LC-MS	Amino acids, fatty acids, isoflavonoids	Schliemann et al., 2008
	Signaling pathway roots	HPLC coupled to UV photodiode array and ESI ion-trap MS(HPLC-PDA-ESI-ITMS)	Phenylpropanoid, isoflavonoid	Farag et al., 2007
	Biotic, abiotic signaling pathway, roots	GC-MS, LC-MS	Primary metabolites, amino acids, organic acids, carbohydrate, shikimic acids, saponins	Broeckling et al., 2005
	Microbial, roots	Reversed-phase HPLC-PDA-ESI-MS (HPLC-PDA-ESI-MS)	Saponins	Huhman and Sumner, 2002
Pea	Drought, leaves	NMR	Amino acids	Charlton et al., 2008
Soybean	Aphid infection, leaves	CE-TOF-MS	Flavonoids, alkaloids, amino acids	Sato et al., 2013
	Defense, cotyledon	LC-MS, NMR	Prenylated-isoflavones	Cheng et al., 2011
	Flooding, roots, hypocotyls	CE-MS	Succinate, citrate, pyruvate	Komatsu et al., 2011
	Symbiosis, root hairs	GC-MS, UPLC-QTOF-MS	Isoflavonoids, fatty acids,carboxylic acids	Brechenmacher et al., 2010
	Metabolic profiling, leaves	GC-MS	Sugars, organic acids, fatty acids	Benkeblia et al., 2007
	GM and isogenic	CE–TOF-MS	Amino acids	García-Villalba et al., 2008
	Salt stress, seeds	HPLC-UV-ESI-MS	Isoflavonoids, saponins	Wu et al., 2008

significant gradient of several putative flavonoids structures showed a high correlation to increasing biodiversity in the close environment of the individual plant (Scherling et al., 2010).

As described above, plants show a variety of metabolic responses against varied abiotic stresses. With the availability of genomics platforms, scientists are now able to opt for metabolomics for studying metabolites involving non-biased approaches. However, it would be important to study if there are any common metabolic responses associated with all the abiotic stresses or if the responses are specific to the stress. In this context, Komatsu et al. (2011) identified 81 metabolites related to the mitochondria under flooding stress in roots and hypocotyls of soybean using CE-MS which showed that the TCA-cycle-related metabolites, glycolysis related metabolites, GABA, pyruvate, NAD, NADH and amino acids increased, while ATP decreased. This kind of accumulation/increase in GABA and amino acids was also reported in Lotus unlike TCA cycle related metabolites were found to be low in Lotus and high in soybean. These studies support that integrative analysis is required on the metabolite accumulation within the context of understanding metabolic responses to stress.

Phosphorous is an essential component of energy metabolism, signaling molecules, and structural macromolecules. Therefore, studies have been conducted to understand the role of phosphorus in stress response metabolite profiling of common bean roots and nodules under P starvation (Hernández et al., 2007). Increase in the levels of most of the amino acids and several sugars was reported in P-stressed roots. It was suggested that the accumulation of sugars may be partitioned preferentially to P-stressed roots to support the expression of P stress-induced genes. On the other hand reduced amounts of organic acids are reported in P-starved roots that likely reflect exudation of these metabolites from the roots into the rhizosphere (Hernández et al., 2007). The metabolic response of P-starved nodules is in contrast to that of P-starved roots in common bean. It was reported that amino acids, N-containing metabolites and sugars were decreased, while organic acids were accumulated in P-deficient nodules (Hernández et al., 2009). Such contrasting response may be due to the suppression of N supply from fixed N_2 under environmental limitations such as P-starvation in nodules. Recently, metabolite profiling study in chickpea revealed 49 primary metabolites in contrasting salt stress responsive cultivars (Dias et al., 2015).

One drawback with the metabolite profiling is that the data obtained is not sufficient to determine the regulation mechanisms of the pathways of interest. In order to overcome this issue, integrated analysis of metabolomics data together with that of proteomic and transcriptomic data need to be performed. For example in Medicago, an integrated metabolomics and transcriptomic approach was found useful to study metabolic reprogramming of the border cells in roots through cumulative and pathway specific datasets (Watson et al., 2015). This integrative approach showed that there were significant differences in the levels of phytohormone, supported by variation in lipoxygenases and auxin responsive transcripts in the border cells and root tips. Additionally, this approach identified metabolic resources for growth and development redirected to the border cells for the accumulation of specialized metabolites that were defense and symbiosis related. Similarly, Komatsu et al. (2011) used an integrative proteomic and metabolomics approach which was useful in identifying the expression and regulation of components linked to flooding stress in soybean seedlings. Larrainzar et al. (2009) integrated GC-TOF-MS metabolite profiling with untargeted and targeted proteomics to reveal nodule metabolic responses under drought stress and recovery by re-watering in Medicago. Metabolite profiling was able to show a highly pronounced reprogramming of metabolism during drought response and the ability of the nodules to recover completely after re-watering.

Targeted and Untargeted Metabolite Analysis

Metabolites are analyzed either using a targeted or untargeted method (Patti et al., 2012). The targeted method is used to analyse a specified set of metabolites which targets one or more pathways of interest and involves the setup of selected reaction monitoring methods of the standard compounds of the metabolite of interest, followed by the extraction of metabolites from the sample and analysis. The data obtained provides quantitation based on standardized methods for the metabolites of interest. This method has been widely used to follow the dynamics of a limited number of metabolites known to be involved in a particular stress and also for comparative metabolite profiling of a large number of known metabolites. For example, in a single chromatographic run, highly parallel targeted assays based on SRM can be used for sensitive simultaneous analysis of over 100 metabolites (see Bajad and Shulaev, 2007). Alternatively, quantitative profiling may provide *in vivo* enrichment of metabolites with stable isotopes like C-13 and N-15. This can be possible only by growing plants or plant cells in liquid media containing N-15-labeled inorganic nitrogen sources or C-13-labeled carbon dioxide or glucose (Hegeman et al., 2007; Huege et al., 2007). Metabolic labeling combined with MS has been successfully used for quantitative metabolic profiling in microorganisms (Mashego et al., 2004; Lafaye et al., 2005; Wu et al., 2005).

On the other hand, the untargeted metabolite profiling is often used for global and broader applications, e.g., understanding cellular metabolism. In this approach, metabolites are isolated from samples followed by the LC-MS analysis. The data obtained is processed using bioinformatics and values for peaks of interest are searched against metabolite databases for possible identity. These metabolite identities are confirmed using tandem MS (MS/MS) data and retention time data compared with standard components. Untargeted metabolite profiling in Lotus demonstrated a major and reproducible change of the metabolic phenotype in the course of salt acclimatization, which was most evident for amino acid, sugars and organic acid metabolism (Sanchez et al., 2008). Accumulation of amino acids and other nitrogen-containing compounds is a remarkable biochemical feature of almost all plant stress responses reported so far. However, the main disadvantage of untargeted profiling is that it is a semi-quantitative method and provides relative concentration data based on the same "surrogate" internal standard. These semi-quantitative data have to be further validated using targeted quantitative assays.

Metabolic Fingerprinting

Metabolic fingerprinting is mainly used to identify metabolic signatures, for example, finding patterns associated with a particular stress response without precise quantitation/identification of different metabolites in the given sample. Features specific to a fingerprint can be identified using a variety of pattern recognition and multivariate statistical analysis (e.g., principal component analysis (PCA), self-organizing maps (SOMs) and hierarchical clustering, discriminant function analysis (DFA), ANOVA etc.) on the data (Sumner et al., 2003). Different analytical techniques, including NMR spectroscopy (Krishnan et al., 2005), MS (Goodacre et al., 2003), and FT-IR (Johnson et al., 2003) may be used to perform metabolic finger printing. Of these, MS is advantages over NMR spectroscopy because of the low sensitivity of NMR, which makes it difficult to detect low abundance cellular metabolites. On contrary, MS has high resolving power compared to NMR, providing higher sensitivity and lower detection limit but generates more complex spectrum because of its results in the form of discriminant ions which remains as a challenge for data validation. Moreover, a larger subset of metabolites associated with the phenotype can be identified using MS.

There have been few studies that reported the application of these approaches in legumes. Metabolic finger printing has been utilized to study drought in seven different model and forage species of the Lotus genus. Analysis using PCA of the metabolite features regardless metabolite identification status yielded sample "fingerprints," which classified primarily according to the genotype. Infrared (IR) spectroscopy has been used to obtain a snapshot of the sample metabolome (typically low-molecular-weight compounds) at a given time. This study identified metabolic compounds responsible for rapid fermentation for the efficient conservation of forage proteins (Johnson et al., 2004). In another study, NMR-based approach has been used for metabolic fingerprinting of 21 grass and legume cultivars. Applying PCA, variation between cultivars and the magnitude of changes in the metabolic fingerprint between the spring growth and the second regrowth was elucidated in the study. Furthermore, variation in metabolic compounds such as malic acid, choline, and glucose was reported due to seasonal change (Bertram et al., 2010).

Metabolomics Databases for Legumes

As with transcriptomics and proteomics data, metabolomics approaches also generate huge datasets that require specialized data mining and bioinformatics tools. It is imperative to integrate functional genomics data to comprehensively study biological components, using a systemic approach, e.g., through the mathematical modeling of biological systems. Metabolomics data handling, mining and analysis etc. have been improved tremendously, due to advances in bioinformatics tools. In this scenario, several databases have been developed for plant metabolomics data analysis. For example, a metabolic pathway reconstruction was used to generate a pathway database for Medicago called MedicCyc (http://www.noble.org/MedicCyc/) which features more than 250 pathways with related genes, enzymes and metabolites (Urbanczyk-Wochniak and Sumner, 2007). The database contains Medicago specific pathways including isoflavonoid, lignin and triterpene saponin biosynthesis which were added or modified based on literature and in-house expertise. MedicCyc is designed to visualize functional genomics datasets from Medicago within the biological context of metabolic pathways and has been believed that this is best achieved through the visualization of data within the biological context of metabolic pathways in legumes. The pathways were engineered to enable the correlated visualization of integrated functional genomics data. Another database, Soybean Knowledge Base (SoyKB) (http://soykb.org) has been reported to be a comprehensive resource for soybean translational genomics and contains integrative information on soybean genomics, transcriptomics, proteomics and metabolomics (Joshi et al., 2014). This is a web resource that would not only be useful for soybean translational genomics, but also for legume crop improvement programs. Also, a database for plant metabolomics, PlantMetabolomics.org (PM) (http://www.plantmetabolomics.org) was developed (Bais et al., 2010). This database represents metabolomics data generated from Arabidopsis through an integration of experiments compiled from different platforms with visualization tools. PlantMetabolomics has been widely used for exploring, visualizing and downloading plant metabolomics data and well-annotated metabolomics datasets which is useful for establishing metabolomics as a functional genomics tool in legumes.

Metabolomics for Crop Breeding

There is growing interest in using metabolites as selection markers in crop breeding programs, because metabolite biomarkers have been linked with strong environmentally-controlled traits (Steinfath et al., 2010). Mapping and metabolomics genome wide association studies (mGWAS) have been conducted to develop "metabotypes" using metabotype quantitative trait locus (mQTL) (association of genomic markers and metabolic markers) which enabled the associations between metabolic concentrations and genetic polymorphisms. Overall, plant metabolomics has benefited from a rich array of pre-existing methodological approaches and bioanalytical knowledge for the characterization of the chemically diverse classes of metabolites. However, speedy progress in the application of these approaches in legumes will be quite useful for legume improvement.

SUMMARY AND FUTURE OUTLOOK

The development in the area of proteomics and metabolomics has enhanced the power of "omics" with the possibility of studying at different levels of plant regulations, namely transcriptome, proteome and the metabolome (Shulaev et al., 2008). Plant growth, development and stress responses are not straightforward to be able to understand by just looking into one or two level(s), which has been done traditionally. For instance, to understand the molecular basis of stress physiology and biochemistry, the quantitative correlation of different protein groups and classes of metabolites with stress levels are required for the identification of candidate bio-markers (Weckwerth, 2011b; Rodziewicz et al., 2014). Moreover, with the recent developments in the technologies, availability of legume genomics and protein databases, it would be possible to have efficient and high throughput identification of stress related proteins in crop legumes (Hiremath et al., 2011; Kosová et al., 2011; Varshney et al., 2013a; Hossain and Komatsu, 2014a,b).

Legume crops are widely cultivated in the semi-arid tropics where various abiotic and biotic stresses pose severe threats to the productivity. The crops are subjected to not only one stress, but also a combination of stresses at a given point of time under natural field conditions. To sustain legume cultivation under these environments, crop improvement programs requires resourceful methods such as integrated "omics" approaches to understand stress responses at the molecular level such as cellular mechanisms, signaling pathways, biochemical processes. Such studies have largely been initiated in model plants such as Arabidopsis. For instance, transcriptomics and the metabolomics data revealed a different pattern of defense response in Arabidopsis subjected to a combination of drought and heat stress. It was found that proline (Pro) was accumulated in response to drought stress, but during the combination of stresses, Pro was replaced by sucrose as osmoprotectants. Similar studies needs to be undertaken in legume crops such as chickpea, pigeonpea and peanut, which are of agronomic value especially in the semi-arid tropics.

In this review, the available proteomics and metabolomics resources for legume research and their applications for further our understanding of the stress biology in model as well as crop legumes have been provided (**Figure 1**). A large number of protein reference maps for various tissues are already available for model legume crops that will greatly support comparative proteomics approaches. There is a great need to generate proteome maps from crop legumes that would greatly help in precise understanding of various cellular processes and signaling pathways. These resources would improve the functional annotations of the gene models in identifying novel ORFs as in the case of Arabidopsis (Castellana et al., 2008) and also validates the existing annotation (Agrawal et al., 2012, 2013; Dash et al., 2015; Walley and Briggs, 2015). Moreover, it is now possible to uncover the regulation of

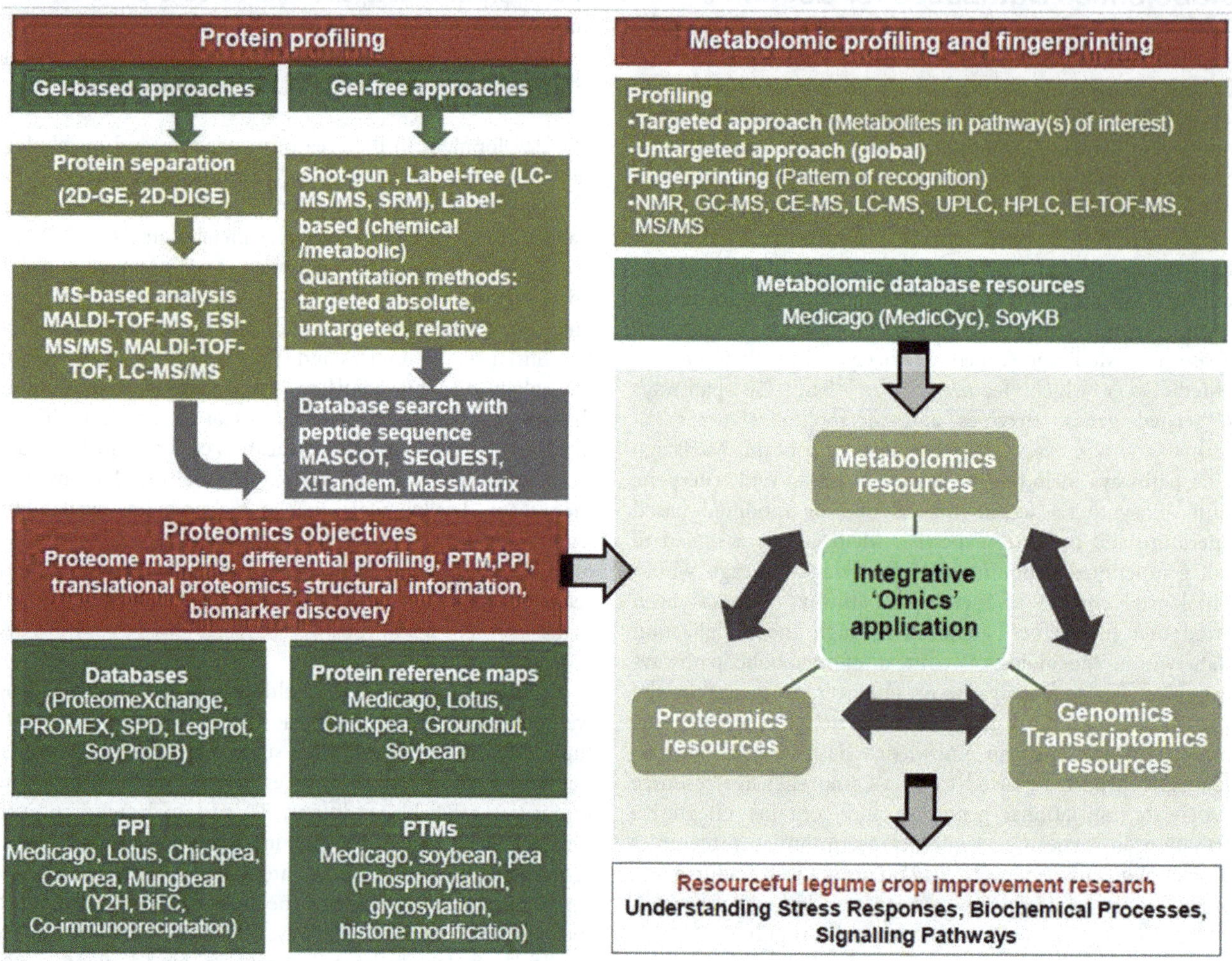

FIGURE 1 | Significant advances in proteomic and metabolomics for integrative "omics" approaches in legume crop research.

these processes by studying the post-translational modifications. Further, huge amount of information on metabolomics data related to stress responses in legumes have been generated which includes a large number of metabolic pathways under stress conditions. The integrative analyses of data through genomics, transcriptomics, proteomics and metabolomics will be important for a system biology approach and for efficient application in legume crop improvement. At this point, it would be worthwhile to caution that combining and integrating omics data together is challenging, because of the difficulty in correlating the data due to the differences in the time for quenching metabolism. In addition, some errors can be introduced due to the nature of the samples used and the lack of appropriate mathematical models that would allow identification of various biochemical and signaling pathways involved in stress response. Therefore, there is a need to advance at all the fronts, i.e., sampling strategy, data generation, data analysis, etc. in different disciplines so that precise system biology approach can be deployed to understand molecular basis of different traits related to plant biology as well as breeding applications.

ACKNOWLEDGMENTS

This work has been undertaken as part of the CGIAR Research Program on Grain Legumes. ICRISAT is a member of CGIAR Consortium.

REFERENCES

Abdallah, C., Dumas-Gaudot, E., Renaut, J., and Sergeant, K. (2012). Gel-based and gel-free quantitative proteomics approaches at a glance. *Int. J. Plant Genomics* 2012:494572. doi: 10.1155/2012/494572

Abdallah, C., Valot, B., Guillier, C., Mounier, A., Balliau, T., Zivy, M., et al. (2014). The membrane proteome of *Medicago truncatula* roots displays qualitative and quantitative changes in response to arbuscular mycorrhizal symbiosis. *J. Proteomics* 108, 354–368. doi: 10.1016/j.jprot.2014.05.028

Aghaei, K., Ehsanpour, A. A., Shah, A. H., and Komatsu, S. (2009). Proteome analysis of soybean hypocotyl and root under salt stress. *Amino Acids* 36, 91–98. doi: 10.1007/s00726-008-0036-7

Agrawal, G. K., Pedreschi, R., Barkla, B. J., Bindschedler, L. V., Cramer, R., Sarkar, A., et al. (2012). Translational plant proteomics: A perspective. *J. Proteomics* 75, 4588–4601. doi: 10.1016/j.jprot.2012.03.055

Agrawal, G. K., Sarkar, A., Righetti, P. G., Pedreschi, R., Carpentier, S., Wang, T., et al. (2013). A decade of plant proteomics and mass spectrometry: Translation of technical advancements to food security and safety issues. *Mass Spectrom. Rev.* 32, 335–365. doi: 10.1002/mas.21365

Ahsan, N., Donnart, T., Nouri, M. Z., and Komatsu, S. (2010b). Tissue-specific defense and thermo-adaptive mechanisms of soybean seedlings under heat stress revealed by proteomic approach. *J. Proteome Res.* 9, 4189–4204. doi: 10.1021/pr100504j

Ahsan, N., Nanjo, Y., Sawada, H., Kohno, Y., and Komatsu, S. (2010a). Ozone stress-induced proteomic changes in leaf total soluble and chloroplast proteins of soybean reveal that carbon allocation is involved in adaptation in the early developmental stage. *Proteomics* 10, 2605–2619. doi: 10.1002/pmic.201000180

Alam, I., Lee, D. G., Kim, K. H., Park, C. H., Sharmin, S. A., Lee, H., et al. (2010). Proteome analysis of soybean roots under water logging stress at an early vegetative stage. *J. Biosci.* 35, 49–62. doi: 10.1007/s12038-010-0007-5

Aloui, A., Recorbet, G., Robert, F., Schoefs, B., Bertrand, M., Henry, C., et al. (2011). Arbuscular mycorrhizal symbiosis elicits shoot proteome changes that are modified during cadmium stress alleviation in *Medicago truncatula*. *BMC Plant Biol.* 11:75. doi: 10.1186/1471-2229-11-75

Arabidopsis Interactome Mapping Consortium (2011). Evidence for network evolution in an *Arabidopsis interactome map*. *Science* 333, 601–607. doi: 10.1126/science.1203877

Aranjuelo, I., Molero, G., Erice, G., Avice, J. C., and Nogués, S. (2011). Plant physiology and proteomics reveals the leaf response to drought in alfalfa (*Medicago sativa* L.). *J. Exp. Bot.* 62, 111–123. doi: 10.1093/jxb/erq249

Arbona, V., Manzi, M., de Ollas, C., and Gómez-Cadenas, A. (2013). Metabolomics as a tool to investigate abiotic stress tolerance in plants. *Int. J. Mol. Sci.* 14, 4885–4911. doi: 10.3390/ijms14034885

Bachi, A., and Bonaldi, T. (2008). Quantitative proteomics as a new piece of the systems biology puzzle. *J. Proteomics* 71, 357–367. doi: 10.1016/j.jprot.2008.07.001

Bais, P., Moon, S. M., He, K., Leitao, R., Dreher, K., Walk, T., et al. (2010). PlantMetabolomics. org: A web portal for plant metabolomics experiments. *Plant Physiol.* 152, 1807–1816. doi: 10.1104/pp.109.151027

Bajad, S., and Shulaev, V. (2007). Highly-parallel metabolomics approaches using LC-MS for pharmaceutical and environmental analysis. *Trends Analyt. Chem.* 26, 625–636. doi: 10.1016/j.trac.2007.02.009

Beckers, G. J., Hoehenwarter, W., Röhrig, H., Conrath, U., and Weckwerth, W. (2014). Tandem metal-oxide affinity chromatography for enhanced depth of phosphoproteome analysis. *Methods Mol. Biol.* 1072, 621–632. doi: 10.1007/978-1-62703-631-3_42

Benkeblia, N., Shinano, T., and Osaki, M. (2007). Metabolite profiling and assessment of metabolome compartmentation of soybean leaves using non-aqueous fractionation and GC-MS analysis. *Metabolomics* 3, 297–305. doi: 10.1007/s11306-007-0078-y

Bertram, H. C., Weisbjerg, M. R., Jensen, C. S., Pedersen, M. G., Didion, T., Petersen, B. O., et al. (2010). Seasonal changes in the metabolic fingerprint of 21 grass and legume cultivars studied by nuclear magnetic resonance-based metabolomics. *J. Agric. Food Chem.* 58, 4336–4341. doi: 10.1021/jf904321p

Bhalla, R., Narasimhan, K., and Swarup, S. (2005). Metabolomics and its role in understanding cellular responses in plants. *Plant Cell Rep.* 24, 562–571. doi: 10.1007/s00299-005-0054-9

Bhushan, D., Jaiswal, D. K., Ray, D., Basu, D., Datta, A., Chakraborty, S., et al. (2011). Dehydration-responsive reversible and irreversible changes in the extracellular matrix: Comparative proteomics of chickpea genotypes with contrasting tolerance. *J. Proteome Res.* 10, 2027–2046. doi: 10.1021/pr200010f

Bhushan, D., Pandey, A., Choudhary, M. K., Datta, A., Chakraborty, S., and Chakraborty, N. (2007). Comparative proteomics analysis of differentially expressed proteins in chickpea extracellular matrix during dehydration stress. *Mol. Cell. Proteomics* 6, 1868–1884. doi: 10.1074/mcp.M700015-MCP200

Borg, S., Pødenphant, L., Jensen, T. J., and Poulsen, C. (1999). Plant cell growth and differentiation may involve GAP regulation of Rac activity. *FEBS Lett.* 453, 341–345. doi: 10.1016/S0014-5793(99)00750-4

Bourgeois, M., Jacquin, F., Savois, V., Sommerer, N., Labas, V., Henry, C., et al. (2009). Dissecting the proteome of pea mature seeds reveals the phenotypic plasticity of seed protein composition. *Proteomics* 9, 254–271. doi: 10.1002/pmic.200700903

Bracha-Drori, K., Shichrur, K., Katz, A., Oliva, M., Angelovici, R., Yalovsky, S., et al. (2004). Detection of protein–protein interactions in plants using bimolecular fluorescence complementation. *Plant J.* 40, 419–427. doi: 10.1111/j.1365-313X.2004.02206.x

Brasileiro, A. C., Morgante, C. V., Araujo, A. C., Leal-Bertioli, S. C., Silva, A. K., Martins, A. C., et al. (2015). Transcriptome profiling of wild *Arachis* from water-limited environments uncovers drought tolerance candidate genes. *Plant Mol. Biol. Rep.* 1–17. doi: 10.1007/s11105-015-0882-x

Braun, P., Aubourg, S., Van Leene, J., De Jaeger, G., and Lurin, C. (2013). Plant protein interactomes. *Annu. Rev. Plant Biol.* 64, 161–187. doi: 10.1146/annurev-arplant-050312-120140

Brechenmacher, L., Lee, J., Sachdev, S., Song, Z., Nguyen, T. H. N., Joshi, T., et al. (2009). Establishment of a protein reference map for soybean root hair cells. *Plant Physiol.* 149, 670–682. doi: 10.1104/pp.108.131649

Brechenmacher, L., Lei, Z., Libault, M., Findley, S., Sugawara, M., Sadowsky, M. J., et al. (2010). Soybean metabolites regulated in root hairs in response to the symbiotic bacterium *Bradyrhizobium japonicum*. *Plant Physiol.* 153, 1808–1822. doi: 10.1104/pp.110.157800

Brechenmacher, L., Nguyen, T. H. N., Hixson, K., Libault, M., Aldrich, J., Pasa-Tolic, L., et al. (2012). Identification of soybean proteins from a single cell type: The root hair. *Proteomics* 12, 3365–3373. doi: 10.1002/pmic.201200160

Broeckling, C. D., Huhman, D. V., Farag, M. A., Smith, J. T., May, G. D., Mendes, P., et al. (2005). Metabolic profiling of *Medicago truncatula* cell cultures reveals the effects of biotic and abiotic elicitors on metabolism. *J. Exp. Bot.* 56, 323–336. doi: 10.1093/jxb/eri058

Brosowska-Arendt, W., Gallardo, K., Sommerer, N., and Weidner, S. (2014). Changes in the proteome of pea (*Pisum sativum* L.) seeds germinating under optimal and osmotic stress conditions and subjected to post-stress recovery. *Acta Physiol. Plant* 36, 795–807. doi: 10.1007/s11738-013-1458-8

Brückner, A., Polge, C., Lentze, N., Auerbach, D., and Schlattner, U. (2009). Yeast two-hybrid, a powerful tool for systems biology. *Int. J. Mol. Sci.* 18, 2763–2788. doi: 10.3390/ijms10062763

Campbell, P. M., Reiner, D., Moore, A. E., Lee, R. Y., Epstein, M. M., and Higgins, T. J. V. (2011). Comparison of the α-amylase inhibitor-1 from common bean (*Phaseolus vulgaris*) varieties and transgenic expression in other legumes post-translational modifications and immunogenicity. *J. Agric. Food Chem.* 59, 6047–6054. doi: 10.1021/jf200456j

Campbell, R., Pont, S. D., Morris, J. A., McKenzie, G., Sharma, S. K., Hedley, P. E., et al. (2014). Genome-wide QTL and bulked transcriptomic analysis reveals new candidate genes for the control of tuber carotenoid content in potato (*Solanum tuberosum* L.). *Theor. Appl. Genet.* 127, 1917–1933. doi: 10.1007/s00122-014-2349-0

Castellana, N. E., Payne, S. H., Shen, Z., Stanke, M., Bafna, V., and Briggs, S. P. (2008). Discovery and revision of Arabidopsis genes by proteogenomics. *Proc. Natl. Acad. Sci. U.S.A.* 105, 21034–21038. doi: 10.1073/pnas.0811066106

Castillejo, M. Á., Fernández-Aparicio, M., and Rubiales, D. (2012). Proteomic analysis by two-dimensional differential in gel electrophoresis (2D DIGE) of the early response of *Pisum sativum* to *Orobanche crenata*. *J. Exp. Bot.* 63, 107–119. doi: 10.1093/jxb/err246

Castillejo, M. A., Staudinger, C., Egelhofer, V., and Wienkoop, S. (2014). *Medicago truncatula* proteomics for systems biology: Novel rapid shotgun LC-MS approach for relative quantification based on full-scan selective peptide extraction (Selpex). *Methods Mol. Biol.* 1072, 303–313. doi: 10.1007/978-1-62703-631-3_22

Charlton, A. J., Donarski, J. A., Harrison, M., Jones, S. A., Godward, J., Oehlschlager, S., et al. (2008). Responses of the pea (*Pisum sativum* L.) leaf metabolome to drought stress assessed by nuclear magnetic resonance spectroscopy. *Metabolomics* 4, 312–327. doi: 10.1007/s11306-008-0128-0

Charpentier, M., Bredemeier, R., Wanner, G., Takeda, N., Schleiff, E., and Parniske, M. (2008). *Lotus japonicus* CASTOR and POLLUX are ion channels essential for perinuclear calcium spiking in legume root endosymbiosis. *Plant Cell* 20, 3467–3479. doi: 10.1105/tpc.108.063255

Chatelain, E., Hundertmark, M., Leprince, O., Gall, S. L., Satour, P., Deligny-Penninck, S., et al. (2012). Temporal profiling of the heat-stable proteome during late maturation of *Medicago truncatula* seeds identifies a restricted subset of late embryogenesis abundant proteins associated with longevity. *Plant Cell Environ.* 35, 1440–1455. doi: 10.1111/j.1365-3040.2012.02501.x

Chen, Y., Hoehenwarter, W., and Weckwerth, W. (2010). Comparative analysis of phytohormone−responsive phosphoproteins in *Arabidopsis thaliana* using TiO2−phosphopeptide enrichment and mass accuracy precursor alignment. *Plant J.* 63, 1–17. doi: 10.1111/j.1365-313x.2010.04218.x

Cheng, J., Yuan, C., and Graham, T. L. (2011). Potential defense-related prenylated isoflavones in lactofen-induced soybean. *Phytochem.* 72, 875–881. doi: 10.1016/j.phytochem.2011.03.010

Chung, E., Cho, C. W., So, H. A., Kang, J. S., Chung, Y. S., and Lee, J. H. (2013). Overexpression of VrUBC1, a mung bean E2 ubiquitin-conjugating enzyme, enhances osmotic stress tolerance in *Arabidopsis. PLoS ONE* 8:e66056. doi: 10.1371/journal.pone.0066056

Colditz, F., and Braun, H. P. (2010). *Medicago truncatula* proteomics. *J. Proteomics* 73, 1974–1985. doi: 10.1016/j.jprot.2010.07.004

Dam, S., Dyrlund, T. F., Ussatjuk, A., Jochimsen, B., Nielsen, K., Goffard, N., et al. (2014). Proteome reference maps of the *Lotus japonicus* nodule and root. *Proteomics* 14, 230–240. doi: 10.1002/pmic.201300353

Dash, S., Campbell, J. D., Cannon, E. K. S., Cleary, A. M., Huang, W., Kalberer, S. R., et al. (2015). Legume information system (LegumeInfo. org): A key component of a set of federated data resources for the legume family. *Nucleic Acids Res.* gkv1159. doi: 10.1093/nar/gkv1159

Den Herder, G., Yoshida, S., Antolín-Llovera, M., Ried, M. K., and Parniske, M. (2012). *Lotus japonicus* E3 ligase SEVEN IN ABSENTIA4 destabilizes the symbiosis receptor-like kinase SYMRK and negatively regulates rhizobial infection. *Plant Cell* 24, 1691–1707. doi: 10.1105/tpc.110.082248

Dias, D. A., Hill, C. B., Jayasinghe, N. S., Atieno, J., Sutton, T., and Roessner, U. (2015). Quantitative profiling of polar primary metabolites of two chickpea cultivars with contrasting responses to salinity. *J. Chromatogr. B* 1000, 1–13. doi: 10.1016/j.jchromb.2015.07.002

Dita, M. A., Rispail, N., Prats, E., Rubiales, D., and Singh, K. B. (2006). Biotechnology approaches to overcome biotic and abiotic stress constraints in legumes. *Euphytica* 147, 1–24. doi: 10.1007/s10681-006-6156-9

Doerfler, H., Lyon, D., Nägele, T., Sun, X., Fragner, L., Hadacek, F., et al. (2013). Granger causality in integrated GC–MS and LC–MS metabolomics data reveals the interface of primary and secondary metabolism. *Metabolomics* 9, 564–574. doi: 10.1007/s11306-012-0470-0

Doerfler, H., Sun, X., Wang, L., Engelmeier, D., Lyon, D., and Weckwerth, W. (2014). mzGroupAnalyzer-Predicting pathways and novel chemical structures from untargeted high-throughput metabolomics data. *PLoS ONE* 9:e96188. doi: 10.1371/journal.pone.0096188

Domoney, C., Welham, T., Sidebottom, C., and Firmin, J. L. (1995). Multiple isoforms of *Pisum* trypsin inhibitors result from modification of two primary gene products. *FEBS Lett.* 360, 15–20. doi: 10.1016/0014-5793(95)00070-P

Downes, B., and Vierstra, R. D. (2005). Post-translational regulation in plants employing a diverse set of polypeptide tags. *Biochem. Soc. Trans.* 33, 393–400. doi: 10.1042/BST0330393

Dubinin, J., Braun, H. P., Schmitz, U., and Colditz, F. (2011). The mitochondrial proteome of the model legume *Medicago truncatula. Biochim. Biophys. Acta* 1814, 1658–1668. doi: 10.1016/j.bbapap.2011.08.008

Fan, H., Xu, Y., Du, C., and Wu, X. (2015). Phloem sap proteome studied by iTRAQ provides integrated insight into salinity response mechanisms in cucumber plants. *J. Proteomics* 125, 54–67. doi: 10.1016/j.jprot.2015.05.001

Fang, Y., Macool, D. J., Xue, Z., Heppard, E. P., Hainey, C. F., Tingey, S. V., et al. (2002). Development of a high-throughput yeast two-hybrid screening system to study protein-protein interactions in plants. *Mol. Genet. Genomics* 267, 142–153. doi: 10.1007/s00438-002-0656-7

Farag, M. A., Huhman, D. V., Lei, Z., and Sumner, L. W. (2007). Metabolic profiling and systematic identification of flavonoids and isoflavonoids in roots and cell suspension cultures of *Medicago truncatula* using HPLC-UV-ESI-MS and GC-MS. *Phytochemistry* 68, 342–354. doi: 10.1016/j.phytochem.2006.10.023

Fernie, A. R., and Schauer, N. (2009). Metabolomics-assisted breeding: A viable option for crop improvement? *Trends Genet.* 25, 39–48. doi: 10.1016/j.tig.2008.10.010

Fiehn, O., Kopka, J., Dörmann, P., Altmann, T., Trethewey, R. N., and Willmitzer, L. (2000). Metabolite profiling for plant functional genomics. *Nat. Biotechnol.* 18, 1157–1161. doi: 10.1038/81137

Fukao, Y. (2012). Protein-protein interactions in plants. *Plant Cell Physiol.* 53, 617–625. doi: 10.1093/pcp/pcs026

García-Villalba, R., León, C., Dinelli, G., Segura-Carretero, A., Fernández-Gutiérrez, A., Garcia-Cañas, V., et al. (2008). Comparative metabolomic study of transgenic versus conventional soybean using capillary electrophoresis–time-of-flight mass spectrometry. *J. Chromatogr. A* 1195, 164–173. doi: 10.1016/j.chroma.2008.05.018

Glinski, M., and Weckwerth, W. (2006). The role of mass spectrometry in plant systems biology. *Mass Spectrom. Rev.* 25, 173–214. doi: 10.1002/mas.20063

Gobbato, E., Marsh, J. F., Vernié, T., Wang, E., Maillet, F., Kim, J., et al. (2012). A GRAS-type transcription factor with a specific function in mycorrhizal signaling. *Curr. Biol.* 22, 2236–2241. doi: 10.1016/j.cub.2012.09.044

Gokulakannan, G. G., and Niehaus, K. (2010). Characterization of the *'Medicago truncatula'* cell wall proteome in cell suspension culture upon elicitation and suppression of plant defense. *J. Plant Physiol.* 167, 1533–1541. doi: 10.1016/j.jplph.2010.06.023

Goodacre, R., York, E. V., Heald, J. K., and Scott, I. M. (2003). Chemometric discrimination of unfractionated plant extracts analyzed by electrospray mass spectrometry. *Phytochemistry* 62, 859–863. doi: 10.1016/S0031-9422(02)00718-5

Graham, P. H., and Vance, C. P. (2003). Legumes: Importance and constraints to greater use. *Plant Physiol.* 131, 872–877. doi: 10.1104/pp.017004

Griffin, T. J., and Aebersold, R. (2001). Advances in proteome analysis by mass spectrometry. *J. Biol. Chem.* 276, 45497–45500. doi: 10.1074/jbc.R100014200

Griffin, T. J., Gygi, S. P., Rist, B., Aebersold, R., Loboda, A., Jilkine, A., et al. (2001). Quantitative proteomic analysis using a MALDI quadrupole time-of-flight mass spectrometer. *Anal. Chem.* 73, 978–986. doi: 10.1021/ac001169y

Grimsrud, P. A., den Os, D., Wenger, C. D., Swaney, D. L., Schwartz, D., Sussman, M. R., et al. (2010). Large-scale phosphoprotein analysis in *Medicago truncatula* roots provides insight into *in vivo* kinase activity in legumes. *Plant Physiol.* 152, 19–28. doi: 10.1104/pp.109.149625

Hajduch, M., Ganapathy, A., Stein, J. W., and Thelen, J. J. (2005). A systematic proteomic study of seed-filling in soybean: Establishment of high resolution two-dimensional reference maps, expression profiles, and an interactive proteome database. *Plant Physiol.* 137, 1397–1419. doi: 10.1104/pp.104.056614

Halket, J. M., Waterman, D., Przyborowska, A. M., Patel, R. K., Fraser, P. D., and Bramley, P. M. (2005). Chemical derivatization and mass spectral libraries in metabolic profiling by GC/MS and LC/ MS/MS. *J. Exp. Bot.* 56, 219–243. doi: 10.1093/jxb/eri069

Hegeman, A. D., Schulte, C. F., Cui, Q., Lewis, I. A, Huttlin, E. L., Eghbalnia, H., et al. (2007). Stable isotope assisted assignment of elemental compositions for metabolomics. *Anal. Chem.* 79, 6912–6921. doi: 10.1021/ac070346t

Heidarvand, L., and Maali-Amiri, R. (2013). Physio-biochemical and proteome analysis of chickpea in early phases of cold stress. *J. Plant Physiol.* 170, 459–469. doi: 10.1016/j.jplph.2012.11.021

Hernández, G., Ramírez, M., Valdés-López, O., Goffard, N., Weiller, G., Aparicio-Fabre, R., et al. (2009). Global changes in the transcript and metabolic profiles during symbiotic nitrogen fixation in phosphorus-stressed common bean plants. *Plant Physiol.* 151, 1221–1238. doi: 10.1104/pp.109.143842

Hernández, G., Ramírez, M., Valdés-López, O., Tesfaye, M., Graham, M. A., Czechowski, T., et al. (2007). Phosphorus stress in common bean: Root transcript and metabolic responses. *Plant Physiol.* 144, 752–767. doi: 10.1104/pp.107.096958

Hiremath, P. J., Farmer, A., Cannon, S. B., Woodward, J., Kudapa, H., Tuteja, R., et al. (2011). Large−scale transcriptome analysis in chickpea (*Cicer arietinum* L.), an orphan legume crop of the semi−arid tropics of Asia and Africa. *Plant Biotechnol. J.* 9, 922–931. doi: 10.1111/j.1467-7652.2011.00625.x

Hirsch, S., Kim, J., Muñoz, A., Heckmann, A. B., Downie, J. A., and Oldroyd, G. E. (2009). GRAS proteins form a DNA binding complex to induce gene expression during nodulation signaling in *Medicago truncatula. Plant Cell* 21, 545–557. doi: 10.1105/tpc.108.064501

Hoehenwarter, W., Thomas, M., Nukarinen, E., Egelhofer, V., Röhrig, H., Weckwerth, W., et al. (2013). Identification of novel *in vivo* MAP kinase substrates in *Arabidopsis thaliana* through use of tandem metal oxide affinity chromatography. *Mol. Cell Proteomics* 12, 369–380. doi: 10.1074/mcp.M112.020560

Hoehenwarter, W., van Dongen, J. T., Wienkoop, S., Steinfath, M., Hummel, J., Erban, A., et al. (2008). A rapid approach for phenotype-screening and database independent detection of cSNP/protein polymorphism using mass accuracy precursor alignment. *Proteomics* 8, 4214–4225. doi: 10.1002/pmic.200701047

Hossain, Z., Khatoon, A., and Komatsu, S. (2013). Soybean proteomics for unravelling abiotic stress response mechanism. *J. Proteome Res.* 12, 4670–4684. doi: 10.1021/pr400604b

Hossain, Z., and Komatsu, S. (2014a). Potentiality of soybean proteomics in untying the mechanism of flood and drought stress tolerance. *Proteomes* 2, 107–127. doi: 10.3390/proteomes2010107

Hossain, Z., and Komatsu, S. (2014b). Soybean proteomics. *Methods Mol. Biol.* 1072, 315–331. doi: 10.1007/978-1-62703-631-3_23

Hossain, Z., Nouri, M. Z., and Komatsu, S. (2012). Plant cell organelle proteomics in response to abiotic stress. *J. Proteome Res.* 11, 37–48. doi: 10.1021/pr200863r

Hu, J., Rampitsch, C., and Bykova, N. V. (2015). Advances in plant proteomics toward improvement of crop productivity and stress resistance. *Front. Plant Sci.* 6:209. doi: 10.3389/fpls.2015.00209

Huege, J., Sulpice, R., Gibon, Y., Lisec, J., Koehl, K., and Kopka, J. (2007). GC-EI-TOF-MS analysis of *in vivo* carbon-partitioning into soluble metabolite pools of higher plants by monitoring isotope dilution after 13CO2 labelling. *Phytochemistry* 68, 2258–2272. doi: 10.1016/j.phytochem.2007.03.026

Huhman, D. V., and Sumner, L. W. (2002). Metabolic profiling of saponins in *Medicago sativa* and *Medicago truncatula* using HPLC coupled to an electrospray ion-trap mass spectrometer. *Phytochemistry* 59, 347–360. doi: 10.1016/S0031-9422(01)00432-0

Hummel, J., Niemann, M., Wienkoop, S., Schulze, W., Steinhauser, D., Selbig, J., et al. (2007). ProMEX: A mass spectral reference database for proteins and protein phosphorylation sites. *BMC Bioinformatics* 8:216. doi: 10.1186/1471-2105-8-216

Imin, N., De-Jong, F., Mathesius, U., Van Noorden, G., Saeed, N. A., Wang, X. D., et al. (2004). Proteome reference maps of *Medicago truncatula* embryogenic cell cultures generated from single protoplasts. *Proteomics* 4, 1883–1896. doi: 10.1002/pmic.200300803

Ino, Y., Ishikawa, A., Nomura, A., Kajiwara, H., Harada, K., and Hirano, H. (2014). Phosphoproteome analysis of *Lotus japonicus* seeds. *Proteomics* 14, 116–120. doi: 10.1002/pmic.201300237

Irar, S., González, E. M., Arrese-Igor, C., and Marino, D. (2014). A proteomic approach reveals new actors of nodule response to drought in split-root grown pea plants. *Physiol. Plant* 152, 634–645. doi: 10.1111/ppl.12214

Jaiswal, D. K., Mishra, P., Subba, P., Rathi, D., Chakraborty, S., and Chakraborty, N. (2014). Membrane-associated proteomics of chickpea identifies Sad1/UNC-84 protein (CaSUN1), a novel component of dehydration signaling. *Sci. Rep.* 4, 4177. doi: 10.1038/srep04177

Jaiswal, D. K., Ray, D., Subba, P., Mishra, P., Gayali, S., Datta, A., et al. (2012). Proteomic analysis reveals the diversity and complexity of membrane proteins in chickpea (*Cicer arietinum* L.). *Proteome Sci.* 10, 59–5956. doi: 10.1186/1477-5956-10-59

Johnson, H. E., Broadhurst, D., Goodacre, R., and Smith, A. R. (2003). Metabolic fingerprinting of salt-stressed tomatoes. *Phytochemistry* 62, 919–928. doi: 10.1016/S0031-9422(02)00722-7

Johnson, H. E., Broadhurst, D., Kell, D. B., Theodorou, M. K., Merry, R. J., and Griffith, G. W. (2004). High-throughput metabolic fingerprinting of legume silage fermentations via Fourier transform infrared spectroscopy and chemometrics. *Appl. Environ. Microbiol.* 70, 1583–1592. doi: 10.1128/AEM.70.3.1583-1592.2004

Jorrín-Novo, J. V., Pascual, J., Sánchez-Lucas, R., Romero-Rodríguez, M. C., Rodríguez-Ortega, M. J., Lenz, C., et al. (2015). Fourteen years of plant proteomics reflected in Proteomics: Moving from model species and 2DE–based approaches to orphan species and gel-free platforms. *Proteomics* 15, 1089–1112. doi: 10.1002/pmic.201400349

Joshi, T., Fitzpatrick, M. R., Chen, S., Liu, Y., Zhang, H., Endacott, R. Z., et al. (2014). Soybean knowledge base (SoyKB): A web resource for integration of soybean translational genomics and molecular breeding. *Nucleic Acids Res.* 42, D1245–D1252. doi: 10.1093/nar/gkt905

Kang, Y. J., Kim, S. K., Kim, M. Y., Lestari, P., Kim, K. H., Ha, B. K., et al. (2014). Genome sequence of mungbean and insights into evolution within *Vigna* species. *Nat. Commun.* 5, 5443. doi: 10.1038/ncomms6443

Kang, Y. J., Satyawan, D., Shim, S., Lee, T., Lee, J., Hwang, W. J., et al. (2015). Draft genome sequence of adzuki bean, *Vigna angularis*. *Sci. Rep.* 5:8069. doi: 10.1038/srep08069

Karmakar, K., Rana, A., Rajwar, A., Sahgal, M., and Johri, B. N. (2015). "Legume-rhizobia symbiosis under stress," in *Plant Microbes Symbiosis: Applied Facets*, ed N. K. Arora (New Delhi; Heidelberg: Springer), 241–258.

Katam, R., Basha, S. M., Suravajhala, P., and Pechan, T. (2010). Analysis of peanut leaf proteome. *J. Proteome Res.* 9, 2236–2254. doi: 10.1021/pr901009n

Kav, N. N. V., Srivastava, S., Goonewardene, L., and Blade, S. F. (2004). Proteome–level changes in the roots of *Pisum sativum* in response to salinity. *Ann. Appl. Biol.* 145, 217–230. doi: 10.1111/j.1744-7348.2004.tb00378.x

Khatoon, A., Rehman, S., Hiraga, S., Makino, T., and Komatsu, S. (2012). Organ-specific proteomics analysis for identification of response mechanism in soybean seedlings under flooding stress. *J. Proteomics* 75, 5706–5723. doi: 10.1016/j.jprot.2012.07.031

Kim, J. C., Lee, S. H., Cheong, Y. H., Yoo, C. M., Lee, S. I., Chun, H. J., et al. (2001). A novel cold–inducible zinc finger protein from soybean, SCOF–1, enhances cold tolerance in transgenic plants. *Plant J.* 25, 247–259. doi: 10.1046/j.1365-313x.2001.00947.x

Komatsu, S., Yamamoto, A., Nakamura, T., Nouri, M. Z., Nanjo, Y., Nishizawa, K., et al. (2011). Comprehensive analysis of mitochondria in roots and hypocotyls of soybean under flooding stress using proteomics and metabolomics techniques. *J. Proteome Res.* 10, 3993–4004. doi: 10.1021/pr2001918

Kosová, K., Vítámvás, P., Prášil, I. T., and Renaut, J. (2011). Plant proteome changes under abiotic stress–Contribution of proteomics studies to understanding plant stress response. *J. Proteomics* 74, 1301–1322. doi: 10.1016/j.jprot.2011.02.006

Kottapalli, K. R., Rakwal, R., Shibato, J., Burow, G., Tissue, D., Burke, J., et al. (2009). Physiology and proteomics of the water-deficit stress response in three contrasting peanut genotypes. *Plant Cell Environ.* 32, 380–407. doi: 10.1111/j.1365-3040.2009.01933.x

Kottapalli, K. R., Zabet-Moghaddam, M., Rowland, D., Faircloth, W., Mirzaei, M., Haynes, P. A., et al. (2013). Shotgun label-free quantitative proteomics of water-deficit-stressed midmature peanut (*Arachis hypogaea* L.) seed. *J. Proteome Res.* 12, 5048–5057. doi: 10.1021/pr400936d

Krishnan, P., Kruger, N. J., and Ratcliffe, R. G. (2005). Metabolite fingerprinting and profiling in plants using NMR. *J. Exp. Bot.* 56, 255–265. doi: 10.1093/jxb/eri010

Kudapa, H., Ramalingam, A., Nayakoti, S., Chen, X., Zhuang, W. J., Liang, X., et al. (2013). Functional genomics to study stress responses in crop legumes: Progress and prospects. *Funct. Plant Biol.* 40, 1221–1233. doi: 10.1071/FP13191

Kumar, R., Kumar, A., Subba, P., Gayali, S., Barua, P., Chakraborty, S., et al. (2014). Nuclear phosphoproteome of developing chickpea seedlings (*Cicer arietinum* L.) and protein-kinase interaction network. *J. Proteomics* 105, 58–73. doi: 10.1016/j.jprot.2014.04.002

Lafaye, A., Labarre, J., Tabet, J. C., Ezan, E., and Junot, C. (2005). Liquid chromatography-mass spectrometry and 15N metabolic labeling for quantitative metabolic profiling. *Anal. Chem.* 77, 2026–2033. doi: 10.1021/ac048657g

Lam, H. M., Xu, X., Liu, X., Chen, W., Yang, G., Wong, F., et al. (2010). Resequencing of 31 wild and cultivated soybean genomes identifies patterns of genetic diversity and selection. *Nat. Genet.* 42, 1053–1059. doi: 10.1038/ng.715

Lange, V., Picotti, P., Domon, B., and Aebersold, R. (2008). Selected reaction monitoring for quantitative proteomics: A tutorial. *Mol. Sys. Biol.* 4, 222. doi: 10.1038/msb.2008.61

Langridge, P., and Fleury, D. (2011). Making the most of 'omics' for crop breeding. *Trends Biotechnol.* 29, 33–40. doi: 10.1016/j.tibtech.2010.09.006

Larrainzar, E., Wienkoop, S., Scherling, C., Kempa, S., Ladrera, R., Arrese-Igor, C., et al. (2009). Carbon metabolism and bacteroid functioning are involved in the regulation of nitrogen fixation in *Medicago truncatula* under drought and recovery. *Mol. Plant Microbe Interact.* 22, 1565–1576. doi: 10.1094/MPMI-22-12-1565

Larrainzar, E., Wienkoop, S., Weckwerth, W., Ladrera, R., Arrese-Igor, C., and González, E. M. (2007). *Medicago truncatula* root nodule proteome analysis reveals differential plant and bacteroid responses to drought stress. *Plant Physiol.* 144, 1495–1507. doi: 10.1104/pp.107.101618

Lee, L. Y., Wu, F. H., Hsu, C. T., Shen, S. C., Yeh, H. Y., Liao, D. C., et al. (2012). Screening a cDNA library for protein-protein interactions directly in planta. *Plant Cell* 24, 1746–1759. doi: 10.1105/tpc.112.097998

Lei, Z., Dai, X., Watson, B. S., Zhao, P. X., and Sumner, L. W. (2011). A legume specific protein database (LegProt) improves the number of identified peptides, confidence scores and overall protein identification success rates for legume proteomics. *Phytochemistry* 72, 1020–1027. doi: 10.1016/j.phytochem.2011.01.026

Lei, Z., Elmer, A. M., Watson, B. S., Dixon, R. A., Mendes, P. J., and Sumner, L. W. (2005). A two-dimensional electrophoresis proteomic reference map and systematic identification of 1367 proteins from a cell suspension culture of the model legume *Medicago truncatula*. *Mol. Cell Proteomics* 4, 1812–1825. doi: 10.1074/mcp.D500005-MCP200

Liao, Y., Zou, H. F., Wang, H. W., Zhang, W. K., Ma, B., Zhang, J. S., et al. (2008a). Soybean GmMYB76, GmMYB92, and GmMYB177 genes confer stress tolerance in transgenic *Arabidopsis* plants. *Cell Res.* 18, 1047–1060. doi: 10.1038/cr.2008.280

Liao, Y., Zou, H. F., Wei, W., Hao, Y. J., Tian, A. G., Huang, J., et al. (2008b). Soybean GmbZIP44, GmbZIP62 and GmbZIP78 genes function as negative regulator of ABA signaling and confer salt and freezing tolerance in transgenic *Arabidopsis*. *Planta* 228, 225–240. doi: 10.1007/s00425-008-0731-3

Liu, G. T., Ma, L., Duan, W., Wang, B. C., Li, J. H., Xu, H. G., et al. (2014). Differential proteomic analysis of grapevine leaves by iTRAQ reveals responses to heat stress and subsequent recovery. *BMC Plant Biol.* 14:110. doi: 10.1186/1471-2229-14-110

Ma, H., Song, L., Shu, Y., Wang, S., Niu, J., Wang, Z., et al. (2012). Comparative proteomic analysis of seedling leaves of different salt tolerant soybean genotypes. *J. Proteomics* 75, 1529–1546. doi: 10.1016/j.jprot.2011.11.026

Mahajan, S., Sopory, S. K., and Tuteja, N. (2006). Cloning and characterization of CBL–CIPK signalling components from a legume (*Pisum sativum*). *FEBS J.* 273, 907–925. doi: 10.1111/j.1742-4658.2006.05111.x

Marouga, R., David, S., and Hawkins, E. (2005).The development of the DIGE system: 2D fluorescence difference gel analysis technology *Anal. Bioanal. Chem.* 382, 669–678. doi: 10.1007/s00216-005-3126-3

Mashego, M. R., Wu, L., Van Dam, J. C., Ras, C., Vinke, J. L., Van Winden, W. A., et al. (2004). MIRACLE: Mass isotopomer ratio analysis of U-13C-labeled extracts. A new method for accurate quantification of changes in concentrations of intracellular metabolites. *Biotechnol. Bioeng.* 85, 620–628. doi: 10.1002/bit.10907

Mathesius, U., Djordjevic, M. A., Oakes, M., Goffard, N., Haerizadeh, F., Weiller, G. F., et al. (2011). Comparative proteomic profiles of the soybean (*Glycine max*) root apex and differentiated root zone. *Proteomics* 11, 1707–1719. doi: 10.1002/pmic.201000619

Mazzucotelli, E., Mastrangelo, A. M., Crosatti, C., Guerra, D., Stanca, A. M., and Cattivelli, L. (2008). Abiotic stress response in plants: When post-transcriptional and post-translational regulations control transcription. *Plant Sci.* 174, 420–431. doi: 10.1016/j.plantsci.2008.02.005

Messinese, E., Mun, J. H., Yeun, L. H., Jayaraman, D. K., Rougé, P., Barre, A., et al. (2007). A novel nuclear protein interacts with the symbiotic DMI3 calcium- and calmodulin-dependent protein kinase of *Medicago truncatula*. *Mol. Plant Microbe Interact.* 20, 912–921. doi: 10.1094/MPMI-20-8-0912

Misra, S., Wu, Y., Venkataraman, G., Sopory, S. K., and Tuteja, N. (2007). Heterotrimeric G-protein complex and G-protein-coupled receptor from a legume (*Pisum sativum*): Role in salinity and heat stress and cross-talk with phospholipase C. *Plant J.* 51, 656–669. doi: 10.1111/j.1365-313X.2007.03169.x

Mithöfer, A., Müller, B., Wanner, G., and Eichacker, L. A. (2002). Identification of defence-related cell wall proteins in *Phytophthora sojae*-infected soybean roots by ESI-MS/MS. *Mol. Plant Pathol.* 3, 163–166. doi: 10.1046/j.1364-3703.2002.00109.x

Mohammadi, P., Moieni, A., Hiraga, S., and Komatsu, S. (2012). Organ-specific proteomic analysis of drought-stressed soybean seedlings. *J. Proteomics* 75, 1906–1923. doi: 10.1016/j.jprot.2011.12.041

Muneer, S., Hakeem, K. R., Mohamed, R., and Lee, J. H. (2014). Cadmium toxicity induced alterations in the root proteome of green gram in contrasting response towards iron supplement. *Int. J. Mol. Sci.* 15, 6343–6355. doi: 10.3390/ijms15046343

Murakami, Y., Imaizumi-Anraku, H., Kouchi, H., Kawagushi, M., and Kawasaki, S. (2013). The transcription activation and homodimerization of *Lotus japonicus* Nod factor Signaling Pathway2 protein. *Plant Signal Behav.* 8:e26457. doi: 10.4161/psb.26457

Nanjo, Y., Skultety, L., Uváčková, L. U., Klubicová, K., Hajduch, M., and Komatsu, S. (2011). Mass spectrometry-based analysis of proteomic changes in the root tips of flooded soybean seedlings. *J. Proteome Res.* 11, 372–385. doi: 10.1021/pr200701y

Nautrup-Pedersen, G., Dam, S., Laursen, B. S., Siegumfeldt, A. L., Nielsen, K., Goffard, N., et al. (2010). Proteome analysis of pod and seed development in the model legume *Lotus japonicus*. *J. Proteome Res.* 9, 5715–5726. doi: 10.1021/pr100511u

Ndassa, Y. M., Orsi, C., Marto, J. A., Chen, S., and Ross, M. M. (2006). Improved immobilized metal affinity chromatography for large-scale phosphoproteomics applications. *J. Proteome Res.* 5, 2789–2799. doi: 10.1021/pr0602803

Nouri, M. Z., and Komatsu, S. (2010). Comparative analysis of soybean plasma membrane proteins under osmotic stress using gel–based and LC MS/MS–based proteomics approaches. *Proteomics* 10, 1930–1945. doi: 10.1002/pmic.200900632

Obata, T., and Fernie, A. R. (2012). The use of metabolomics to dissect plant responses to abiotic stresses. *Cell. Mol. Life Sci.* 69, 3225–3243. doi: 10.1007/s00018-012-1091-5

Ohyanagi, H., Sakata, K., and Komatsu, S. (2012). Soybean Proteome Database 2012: Update on the comprehensive data repository for soybean proteomics. *Front. Plant Sci.* 3:110. doi: 10.3389/fpls.2012.00110

Ogura, T., Ogihara, J., Sunairi, M., Takeishi, H., Aizawa, T., Olivos-Trujillo, M. R., et al. (2014). Proteomic characterization of seeds from yellow lupin (*Lupinus luteus* L.). *Proteomics* 14, 1543–1546. doi: 10.1002/pmic.201300511

Pandey, A., Chakraborty, S., Datta, A., and Chakraborty, N. (2008). Proteomics approach to identify dehydration responsive nuclear proteins from chickpea (*Cicer arietinum* L.). *Mol. Cell. Proteomics* 7, 88–107. doi: 10.1074/mcp.M700314-MCP200

Pandey, A., Choudhary, M. K., Bhushan, D., Chattopadhyay, A., Chakraborty, S., Datta, A., et al. (2006). The nuclear proteome of chickpea (*Cicer arietinum* L.) reveals predicted and unexpected proteins. *J. Proteome Res.* 5, 3301–3311. doi: 10.1021/pr060147a

Pandey, A., and Mann, M. (2000). Proteomics to study genes and genomes. *Nature* 405, 837–846. doi: 10.1038/35015709

Parrish, J. R., Gulyas, K. D., and Finley, R. L. (2006). Yeast two-hybrid contributions to interactome mapping. *Curr. Opin. Plant Biol.* 17, 387–393. doi: 10.1016/j.copbio.2006.06.006

Patti, G. J., Yanes, O., and Siuzdak, G. (2012). Innovation: Metabolomics: The apogee of the omics trilogy. *Nat. Rev. Mol. Cell Biol.* 13, 263–269. doi: 10.1038/nrm3314

Peters, K., Belt, K., and Braun, H. P. (2013). 3D gel map of *Arabidopsis* complex I. *Front. Plant Sci.* 4:153. doi: 10.3389/fpls.2013.00153

Picotti, P., and Aebersold, R. (2012). Selected reaction monitoring-based proteomics: Workflows, potential, pitfalls and future directions. *Nat. Methods* 9, 555–566. doi: 10.1038/nmeth.2015

Ramalingam, A., Kudapa, H., Pazhamala, L. T., Garg, V., and Varshney, R. K. (2015). Gene expression and Yeast two-hybrid studies of a 1R-MYB transcription factor mediating drought stress response in root tissues of chickpea (*Cicer arietinum* L.). *Front. Plant Sci.* 6:1117. doi: 10.3389/fpls.2015.01117

Rocha, M., Licausi, F., Araújo, W. L., Nunes-Nesi, A., Sodek, L., Fernie, A. R., et al. (2010). Glycolysis and the tricarboxylic acid cycle are linked by alanine aminotransferase during hypoxia induced by waterlogging of *Lotus japonicus*. *Plant Physiol.* 152, 1501–1513. doi: 10.1104/pp.109.150045

Rode, C., Senkler, M., Klodmann, J., Winkelmann, T., and Braun, H. P. (2011). GelMap - a novel software tool for building and presenting proteome reference maps. *J. Proteomics* 74, 2214–2219. doi: 10.1016/j.jprot.2011.06.017

Rodziewicz, P., Swarcewicz, B., Chmielewska, K., Wojakowska, A., and Stobiecki, M. (2014). Influence of abiotic stresses on plant proteome and metabolome changes. *Acta Physiol. Plant* 36, 1–19. doi: 10.1007/s11738-013-1402-y

Roessner, U., Wagner, C., Kopka, J., Trethewey, R. N., and Willmitzer, L. (2000). Technical advance: Simultaneous analysis of metabolites in potato tuber by gas chromatography-mass spectrometry. *Plant J.* 23, 131–142. doi: 10.1046/j.1365-313x.2000.00774.x

Romero-Rodríguez, M. C., Pascual, J., Valledor, L., and Jorrín-Novo, J. (2014). Improving the quality of protein identification in non-model species. Characterization of Quercus ilex seed and *Pinus radiata* needle proteomes by using SEQUEST and custom databases. *J. Proteomics* 105, 85–91. doi: 10.1016/j.jprot.2014.01.027

Rose, C. M., Venkateshwaran, M., Grimsrud, P. A., Westphall, M. S., Sussman, M. R., Coon, J. J., et al. (2012a). *Medicago* PhosphoProtein Database: A repository for *Medicago truncatula* phosphoprotein data. *Front. Plant Sci.* 3:122. doi: 10.3389/fpls.2012.00122

Rose, C. M., Venkateshwaran, M., Volkening, J. D., Grimsrud, P. A., Maeda, J., Bailey, D. J., et al. (2012b). Rapid phosphoproteomic and transcriptomic changes in the rhizobia-legume symbiosis. *Mol. Cell. Proteomics* 11, 724–744. doi: 10.1074/mcp.M112.019208

Sadhukhan, A., Panda, S. K., and Sahoo, L. (2014). The cowpea RING ubiquitin ligase VuDRIP interacts with transcription factor VuDREB2A for regulating abiotic stress responses. *Plant Physiol. Biochem.* 83, 51–56. doi: 10.1016/j.plaphy.2014.07.007

Salavati, A., Khatoon, A., Nanjo, Y., and Komatsu, S. (2012). Analysis of proteomic changes in roots of soybean seedlings during recovery after flooding. *J. Proteomics* 75, 878–893. doi: 10.1016/j.jprot.2011.10.002

Salekdeh, G. H., and Komatsu, S. (2007). Crop proteomics: Aim at sustainable agriculture of tomorrow. *Proteomics* 7, 2976–2996. doi: 10.1002/pmic.200700181

Sanchez, D. H., Lippold, F., Redestig, H., Hannah, M. A., Erban, A., Krämer, U., et al. (2008). Integrative functional genomics of salt acclimatization in the model legume *Lotus japonicus*. *Plant J.* 53, 973–987. doi: 10.1111/j.1365-313X.2007.03381.x

Sanchez, D. H., Pieckenstain, F. L., Escaray, F., Erban, A., Kraemer, U. T. E., Udvardi, M. K., et al. (2011). Comparative ionomics and metabolomics in extremophile and glycophytic *Lotus* species under salt stress challenge the metabolic pre-adaptation hypothesis. *Plant Cell Environ.* 34, 605–617. doi: 10.1111/j.1365-3040.2010.02266.x

Sanchez, D. H., Schwabe, F., Erban, A., Udvardi, M. K., and Kopka, J. (2012). Comparative metabolomics of drought acclimation in model and forage legumes. *Plant Cell Environ.* 35, 136–149. doi: 10.1111/j.1365-3040.2011.02423.x

Sato, D., Akashi, H., Sugimoto, M., Tomita, M., and Soga, T. (2013). Metabolomic profiling of the response of susceptible and resistant soybean strains to foxglove aphid, *Aulacorthum solani* Kaltenbach. *J. Chromatogr. B* 925, 95–103. doi: 10.1016/j.jchromb.2013.02.036

Sato, S., Nakamura, Y., Kaneko, T., Asamizu, E., Kato, T., Nakao, M., et al. (2008). Genome structure of the legume, *Lotus japonicus*. *DNA Res.* 15, 227–239. doi: 10.1093/dnares/dsn008

Schenkluhn, L., Hohnjec, N., Niehaus, K., Schmitz, U., and Colditz, F. (2010). Differential gel electrophoresis (DIGE) to quantitatively monitor early symbiosis-and pathogenesis-induced changes of the *Medicago truncatula* root proteome. *J. Proteomics* 73, 753–768. doi: 10.1016/j.jprot.2009.10.009

Scherling, C., Roscher, C., Giavalisco, P., Schulze, E. D., and Weckwerth, W. (2010). Metabolomics unravel contrasting effects of biodiversity on the performance of individual plant species. *PLoS ONE* 5:e12569. doi: 10.1371/journal.pone.0012569

Schiltz, S., Gallardo, K., Huart, M., Negroni, L., Sommerer, N., and Burstin, J. (2004). Proteome reference maps of vegetative tissues in pea. An investigation of nitrogen mobilization from leaves during seed filling. *Plant Physiol.* 135, 2241–2260. doi: 10.1104/pp.104.041947

Schliemann, W., Ammer, C., and Strack, D. (2008). Metabolite profiling of mycorrhizal roots of *Medicago truncatula*. *Phytochemistry* 69, 112–146. doi: 10.1016/j.phytochem.2007.06.032

Schmidt, C., and Urlaub, H. (2012). Absolute quantification of proteins using standard peptides and multiple reaction monitoring. *Methods Mol. Biol.* 893, 249–265. doi: 10.1007/978-1-61779-885-6_17

Schmutz, J., Cannon, S. B., Schlueter, J., Ma, J., Mitros, T., Nelson, W., et al. (2010). Genome sequence of the palaeopolyploid soybean. *Nature* 463, 178–183. doi: 10.1038/nature08670

Schmutz, J., McClean, P. E., Mamidi, S., Wu, G. A., Cannon, S. B., Grimwood, J., et al. (2014). A reference genome for common bean and genome-wide analysis of dual domestications. *Nat. Genet.* 46, 707–713. doi: 10.1038/ng.3008

Senkler, M., and Braun, H. P. (2012). Functional annotation of 2D protein maps: The GelMap portal. *Front. Plant Sci.* 14:87. doi: 10.3389/fpls.2012.00087

Seo, J., and Lee, K. J. (2004). Post-translational modifications and their biological functions: Proteomic analysis and systematic approaches. *J. Biochem. Mol. Biol.* 37, 35–44. doi: 10.5483/BMBRep.2004.37.1.035

Shulaev, V. (2006). Metabolomics technology and bioinformatics. *Brief Bioinform.* 7, 128–139. doi: 10.1093/bib/bbl012

Shulaev, V., Cortes, D., Miller, G., and Mittler, R. (2008). Metabolomics for plant stress response. *Physiol. Plant.* 132, 199–208. doi: 10.1111/j.1399-3054.2007.01025.x

Soga, T. (2007). Capillary electrophoresis-mass spectrometry for metabolomics. *Methods Mol. Biol.* 358, 129–137. doi: 10.1007/978-1-59745-244-1_8

Soga, T., Ohashi, Y., Ueno, Y., Naraoka, H., Tomita, M., and Nishioka, T. (2003). Quantitative metabolome analysis using capillary electrophoresis mass spectrometry. *J. Proteome Res.* 2, 488–494. doi: 10.1021/pr034020m

Stasi, M., De Luca, M., and Bucci, C. (2015). Two-hybrid-based systems: Powerful tools for investigation of membrane traffic machineries. *J. Biotechnol.* 202, 105–117. doi: 10.1016/j.jbiotec.2014.12.007

Staszków, A., Swarcewicz, B., Banasiak, J., Muth, D., Jasiński, M., and Stobiecki, M. (2011). LC/MS profiling of flavonoid glycoconjugates isolated from hairy roots, suspension root cell cultures and seedling roots of *Medicago truncatula*. *Metabolomics* 7, 604–613. doi: 10.1007/s11306-011-0287-2

Staudinger, C., Mehmeti, V., Turetschek, R., Lyon, D., Egelhofer, V., and Wienkoop, S. (2012). Possible role of nutritional priming for early salt and drought stress responses in *Medicago truncatula*. *Front. Plant Sci.* 3:285. doi: 10.3389/fpls.2012.00285

Steinfath, M., Strehmel, N., Peters, R., Schauer, N., Groth, D., Hummel, J., et al. (2010). Discovering plant metabolic biomarkers for phenotype prediction using an untargeted approach. *Plant Biotechnol. J.* 8, 900–911. doi: 10.1111/j.1467-7652.2010.00516.x

Subba, P., Barua, P., Kumar, R., Datta, A., Soni, K. K., Chakraborty, S., et al. (2013). Phosphoproteomic dynamics of chickpea (*Cicer arietinum* L.) reveals shared and distinct components of dehydration response. *J. Proteome Res.* 12, 5025–5047. doi: 10.1021/pr400628j

Subramanian, S., and Smith, D. L. (2013). "A proteomics approach to study soybean and its symbiont *Bradyrhizobium japonicum*- a review," in *A Comprehensive Survey of International Soybean Research—Genetics, Physiology, Agronomy and Nitrogen Relationships*, ed J. Board (Rijeka: INTECH Open Access Publisher), 978–953.

Sumner, L.W., Mendes, P., and Dixon, R. A. (2003). Plant metabolomics: Large-scale phytochemistry in the functional genomics era. *Phytochemistry* 62, 817–836. doi: 10.1016/S0031-9422(02)00708-2

Swigonska, S., and Weidner, S. (2013). Proteomic analysis of response to long-term continuous stress in roots of germinating soybean seeds. *J. Plant Physiol.* 170, 470–479. doi: 10.1016/j.jplph.2012.11.020

Tavakolan, M., Alkharouf, N. W., Khan, F. H., and Natarajan, S. (2013). SoyProDB: A database for the identification of soybean seed proteins. *Bioinformation* 9, 165–167. doi: 10.6026/97320630009165

Thudi, M., Upadhyaya, H. D., Rathore, A., Gaur, P. M., Krishnamurthy, L., Roorkiwal, M., et al. (2014). Genetic dissection of drought and heat tolerance in chickpea through genome-wide and candidate gene-based association mapping approaches. *PLoS ONE* 9:e96758. doi: 10.1371/journal.pone.0096758

Tripathi, V., Parasuraman, B., Laxmi, A., and Chattopadhyay, D. (2009). CIPK6, a CBL-interacting protein kinase is required for development and salt tolerance in plants. *Plant J.* 58, 778–790. doi: 10.1111/j.1365-313X.2009.03812.x

Urbanczyk-Wochniak, E., and Sumner, L. W. (2007). MedicCyc: A biochemical pathway database for *Medicago truncatula*. *Bioinformatics* 23, 1418–1423. doi: 10.1093/bioinformatics/btm040

Vanderschuren, H., Lentz, E., Zainuddin, I., and Gruissem, W. (2013). Proteomics of model and crop plant species: Status, current limitations and strategic advances for crop improvement. *J. Proteomics.* 93, 5–19. doi: 10.1016/j.jprot.2013.05.036

van Wijk, K. J. (2001). Challenges and prospects of plant proteomics. *Plant Physiol.* 126, 501–508. doi: 10.1104/pp.126.2.501

Varshney, R. K., Chen, W., Li, Y., Bharti, A. K., Saxena, R. K., Schlueter, J. A., et al. (2012). Draft genome sequence of pigeonpea (*Cajanus cajan*), an orphan legume crop of resource-poor farmers. *Nat. Biotechnol.* 30, 83–89. doi: 10.1038/nbt.2022

Varshney, R. K., Close, T. J., Singh, N. K., Hoisington, D. A., and Cook, D. R. (2009). Orphan legume crops enter the genomics era! *Curr. Opin. Plant Biol.* 12, 202–210. doi: 10.1016/j.pbi.2008.12.004

Varshney, R. K., Mohan, S. M., Gaur, P. M., Gangarao, N. V., Pandey, M. K., Bohra, A., et al. (2013a). Achievements and prospects of genomics-assisted breeding in three legume crops of the semi-arid tropics. *Biotechnol. Adv.* 31, 1120–1134. doi: 10.1016/j.biotechadv.2013.01.001

Varshney, R. K., Song, C., Saxena, R. K., Azam, S., Yu, S., Sharpe, A. G., et al. (2013b). Draft genome sequence of chickpea (*Cicer arietinum*) provides a resource for trait improvement. *Nat. Biotechnol.* 31, 240–246. doi: 10.1038/nbt.2491

Varshney, R. K., Thudi, M., Nayak, S. N., Gaur, P. M., Kashiwagi, J., Krishnamurthy, L., et al. (2014). Genetic dissection of drought tolerance in chickpea (*Cicer arietinum* L.). *Theor. Appl. Genet.* 127, 445–462. doi: 10.1007/s00122-013-2230-6

Varshney, R. K., and Tuberosa, R. (2013). "Translational genomics in crop breeding for biotic stress resistance: An introduction" in *Translational Genomics for Crop Breeding: Biotic Stress*, Vol. 1, eds R. K. Varshney and R. Tuberosa (Chichester, UK: John Wiley & Sons), 1–9.

von Mering, C., Krause, R., Snel, B., Cornell, M., Oliver, S. G., Fields, S., et al. (2002). Comparative assessment of large-scale data sets of protein–protein interactions. *Nature* 417, 399–403. doi: 10.1038/nature750

Walley, J. W., and Briggs, S. P. (2015). Dual use of peptide mass spectra: Protein atlas and genome annotation. *Curr. Plant Biol.* 2, 21–24. doi: 10.1016/j.cpb.2015.02.001

Wang, L., Ma, H., Song, L., Shu, Y., and Gu, W. (2012a). Comparative proteomics analysis reveals the mechanism of pre-harvest seed deterioration of soybean under high temperature and humidity stress. *J. Proteomics* 75, 2109–2127. doi: 10.1016/j.jprot.2012.01.007

Wang, Y., Yuan, X., Hu, H., Liu, Y., Sun, W., Shan, Z., et al. (2012b). Proteomic analysis of differentially expressed proteins in resistant soybean leaves after *Phakopsora pachyrhizi* infection. *J. Phytopathol.* 160, 554–560. doi: 10.1111/j.1439-0434.2012.01949.x

Wang, Z. Q., Xu, X. Y., Gong, Q. Q., Xie, C., Fan, W., Yang, J. L., et al. (2014). Root proteome of rice studied by iTRAQ provides integrated insight into aluminum stress tolerance mechanisms in plants. *J. Proteomics* 98, 189–205. doi: 10.1016/j.jprot.2013.12.023

Washburn, M. P., Wolters, D., and Yates, J. R. (2001). Large-scale analysis of the yeast proteome by multidimensional protein identification technology. *Nat. Biotechnol.* 19, 242–247. doi: 10.1038/85686

Watson, B. S., Asirvatham, V. S., Wang, L., and Sumner, L. W. (2003). Mapping the proteome of barrel medic (*Medicago truncatula*). *Plant Physiol.* 131, 1104–1123. doi: 10.1104/pp.102.019034

Watson, B. S., Bedair, M. F., Urbanczyk-Wochniak, E., Huhman, D. V., Yang, D. S., Allen, S. N., et al. (2015). Integrated metabolomics and transcriptomics reveal enhanced specialized metabolism in *Medicago truncatula* root Border Cells. *Plant Physiol.* 167, 1699–1716. doi: 10.1104/pp.114.253054

Weckwerth, W. (2003). Metabolomics in systems biology. *Annu Rev Plant Biol.* 54, 669–689. doi: 10.1146/annurev.arplant.54.031902.135014

Weckwerth, W. (2010). Metabolomics: An integral technique in systems biology. *Bioanalysis* 2, 829–836. doi: 10.4155/bio.09.192

Weckwerth, W. (2011a). Unpredictability of metabolism- the key role of metabolomics science in combination with next-generation genome sequencing. *Anal. Bioanal. Chem.* 400, 1967–1978. doi: 10.1007/s00216-011-4948-9

Weckwerth, W. (2011b). Green systems biology- From single genomes, proteomes and metabolomes to ecosystems research and biotechnology. *J. Proteomics* 75, 284–305. doi: 10.1016/j.jprot.2011.07.010

Weckwerth, W., Loureiro, M. E., Wenzel, K., and Fiehn, O. (2004a). Differential metabolic networks unravel the effects of silent plant phenotypes. *Proc. Natl. Acad. Sci. U.S.A.* 101, 7809–7814. doi: 10.1073/pnas.0303415101

Weckwerth, W., Wenzel, K., and Fiehn, O. (2004b). Process for the integrated extraction, identification and quantification of metabolites, proteins and RNA to reveal their co–regulation in biochemical networks. *Proteomics* 4, 78–83. doi: 10.1002/pmic.200200500

Weckwerth, W., Wienkoop, S., Hoehenwarter, W., Egelhofer, V., and Sun, X. (2014). "From proteomics to systems biology: MAPA, MASS WESTERN, PROMEX, and COVAIN as a user-oriented platform," in *Plant Proteomics: Methods and Protocols*, eds J. V. Jorrin-Novo, S. Komatsu, W. Weckwerth, and S. Wienkoop (New York, NY: Springer, LLC), 15–27.

Wienkoop, S., Baginsky, S., and Weckwerth, W. (2010a). *Arabidopsis thaliana* as a model organism for plant proteome research. *J. Proteomics* 73, 2239–2248. doi: 10.1016/j.jprot.2010.07.012

Wienkoop, S., Larrainzar, E., Glinski, M., González, E. M., Arrese-Igor, C., and Weckwerth, W. (2008b). Absolute quantification of *Medicago truncatula* sucrose synthase isoforms and N-metabolism enzymes in symbiotic root nodules and the detection of novel nodule phosphoproteins by mass spectrometry. *J. Exp. Bot.* 59, 3307–3315. doi: 10.1093/jxb/ern182

Wienkoop, S., Morgenthal, K., Wolschin, F., Scholz, M., Selbig, J., and Weckwerth, W. (2008a). Integration of metabolomic and proteomic phenotypes analysis of data covariance dissects starch and RFO metabolism from low and high temperature compensation response in *Arabidopsis thaliana*. *Mol. Cell. Proteomics* 7, 1725–1736. doi: 10.1074/mcp.M700273-MCP200

Wienkoop, S., Staudinger, C., Hoehenwarter, W., Weckwerth, W., and Egelhofer, V. (2012). ProMEX- a mass spectral reference database for plant proteomics. *Front. Plant. Sci.* 3:125. doi: 10.3389/fpls.2012.00125

Wienkoop, S., Weiß, J., May, P., Kempa, S., Irgang, S., Recuenco-Munoz, L., et al. (2010b). Targeted proteomics for *Chlamydomonas reinhardtii* combined with rapid subcellular protein fractionation, metabolomics and metabolic flux analyses. *Mol Biosyst.* 6, 1018–1031. doi: 10.1039/b920913a

Wolschin, F., Wienkoop, S., and Weckwerth, W. (2005). Enrichment of phosphorylated proteins and peptides from complex mixtures using metal oxide/hydroxide affinity chromatography (MOAC). *Proteomics* 5, 4389–4397. doi: 10.1002/pmic.200402049

Wu, L., Mashego, M. R., van Dam, J. C., Proell, A. M., Vinke, J. L., Ras, C., et al. (2005). Quantitative analysis of the microbial metabolome by isotope dilution mass spectrometry using uniformly 13C-labeled cell extracts as internal standards. *Anal. Biochem.* 336, 164–171. doi: 10.1016/j.ab.2004.09.001

Wu, T., Yuan, T., Tsai, S. N., Wang, C., Sun, S. M., Lam, H. M., et al. (2009). Mass spectrometry analysis of the variants of histone H3 and H4 of soybean and their post-translational modifications. *BMC Plant Biol.* 9:98. doi: 10.1186/1471-2229-9-98

Wu, W., Zhang, Q., Zhu, Y., Lam, H. M., Cai, Z., and Guo, D. (2008). Comparative metabolic profiling reveals secondary metabolites correlated with soybean salt tolerance. *J. Agric. Food Chem.* 56, 11132–11138. doi: 10.1021/jf8024024

Xu, C., Garrett, W. M., Sullivan, J., Caperna, T. J., and Natarajan, S. (2006). Separation and identification of soybean leaf proteins by two-dimensional gel electrophoresis and mass spectrometry. *Phytochemistry* 67, 2431–2440. doi: 10.1016/j.phytochem.2006.09.002

Xu, H., and Freitas, M. A. (2009). MassMatrix: A database search program for rapid characterization of proteins and peptides from tandem mass spectrometry data. *Proteomics* 9, 1548–1555. doi: 10.1002/pmic.200700322

Yang, K., Tian, Z., Chen, C., Luo, L., Zhao, B., Wang, Z. et al. (2015). Genome sequencing of adzuki bean (*Vigna angularis*) provides insight into high starch and low fat accumulation and domestication. *Proc. Natl. Acad. Sci. U.S.A.* 112, 13213–13218. doi: 10.1073/pnas.1420949112

Ye, H., Gemperline, E., Venkateshwaran, M., Chen, R., Delaux, P. M., Howes-Podoll, M., et al. (2013). MALDI mass spectrometry-assisted molecular imaging of metabolites during nitrogen fixation in the *Medicago truncatula–Sinorhizobium meliloti* symbiosis. *Plant J.* 75, 130–145. doi: 10.1111/tpj.12191

Young, N. D., Debellé, F., Oldroyd, G. E., Geurts, R., Cannon, S. B., Udvardi, M. K., et al. (2011). The *Medicago* genome provides insight into the evolution of rhizobial symbioses. *Nature* 480, 520–524. doi: 10.1038/nature10625

Young, N. M., Thibault, P., Watson, D. C., and Chrispeels, M. J. (1999). Post-translational processing of two K-amylase inhibitors and an arcelin from the common bean, *Phaseolus vulgaris*. *FEBS Lett.* 446, 203–206. doi: 10.1016/S0014-5793(99)00212-4

Zadražnik, T., Hollung, K., Egge-Jacobsen, W., Megliè, V., and Šuštar-Vozliè, J. (2013). Differential proteomic analysis of drought stress response in leaves of common bean (*Phaseolus vulgaris* L.). *J. Proteomics* 78, 254–272. doi: 10.1016/j.jprot.2012.09.021

Zhang, N., Venkateshwaran, M., Boersma, M., Harms, A., Howes-Podoll, M., den Os, D., et al. (2012). Metabolomic profiling reveals suppression of oxylipin biosynthesis during the early stages of legume–rhizobia symbiosis. *FEBS Lett.* 586, 3150–3158. doi: 10.1016/j.febslet.2012.06.046

Zhang, Y., Zhao, L., Li, H., Gao, Y., Li, Y., Wu, X., et al. (2013). GmGBP1, a homolog of human ski interacting protein in soybean, regulates flowering and stress tolerance in *Arabidopsis*. *BMC Plant Biol.* 13:21. doi: 10.1186/1471-2229-13-21

Zhen, Y., Qi, J. L., Wang, S. S., Su, J., Xu, G. H., Zhang, M. S., et al. (2007). Comparative proteome analysis of differentially expressed proteins induced by Al toxicity in soybean. *Physiol. Plant* 131, 542–554. doi: 10.1111/j.1399-3054.2007.00979.x

Zhu, H., Chen, T., Zhu, M., Fang, Q., Kang, H., Hong, Z., et al. (2008). A novel ARID DNA-binding protein interacts with SymRK and is expressed during early nodule development in *Lotus japonicus*. *Plant Physiol.* 148, 337–347. doi: 10.1104/pp.108.119164

Conflict of Interest Statement: The authors declare that the research was conducted in the absence of any commercial or financial relationships that could be construed as a potential conflict of interest.

16

Variations in Protein Concentration and Nitrogen Sources in Different Positions of Grain in Wheat

Edited by:
Antonio M. De Ron,
Spanish National Research Council (CSIC) - Misión Biológica de Galicia, Spain

Reviewed by:
Francisco Perez-Alfocea,
Consejo Superior de Investigaciones Científicas, Spain
Wanda Waaterworth,
University of Leeds, UK

***Correspondence:**
Qin Zhou
qinzhou@njau.edu.cn;
Dong Jiang
jiangd@njau.edu.cn

Xiangnan Li[1,2], Longjing Zhou[1], Fulai Liu[2], Qin Zhou[1], Jian Cai[1], Xiao Wang[1], Tingbo Dai[1], Weixing Cao[1] and Dong Jiang[1*]*

[1] National Engineering and Technology Center for Information Agriculture/Key Laboratory of Crop Physiology and Ecology in Southern China, Ministry of Agriculture, Nanjing Agricultural University, Nanjing, China, [2] Department of Plant and Environmental Sciences, Faculty of Science, University of Copenhagen, Taastrup, Denmark

The distribution patterns of total protein and protein components in different layers of wheat grain were investigated using the pearling technique, and the sources of different protein components and pearling fractions were identified using ^{15}N isotope tracing methods. It was found that N absorbed from jointing to anthesis (JA) and remobilized to the grain after anthesis was the principal source of grain N, especially in the outer layer. For albumin and globulin, the amount of N absorbed during different stages all showed a decreasing trend from the surface layer to the center part. Whereas, for globulin and glutenin, the N absorbed after anthesis accounted for the main part indicating that for storage protein, the utilization of N assimilated after anthesis is greater than that of the stored N assimilated before anthesis. It is concluded that manipulation of the N application rate during different growth stages could be an effective approach to modulate the distribution of protein fractions in pearled grains for specific end-uses.

Keywords: protein, pearling, wheat, nitrogen, isotope

INTRODUCTION

The grain protein concentration, which to a great extent determines the quality of pasta and bread, is one of the major pricing factors for wheat trading, and is an important nutritional factor for human health (Masclaux-Daubresse et al., 2008). The protein concentration is closely related to the nitrogen (N) content and dry mass in wheat grains and can be separated into the fractions of albumin, gliadin, globulin, and glutenin according to its solubility in different solvents (Malik et al., 2013). Grain N is derived from N that is absorbed after anthesis and from remobilized N that is stored in vegetative organs assimilated before anthesis. Both N sources support the synthesis of the storage proteins in grain (Dupont and Altenbach, 2003), and the former contributes approximately 60–95% to the grain N (Palta and Fillery, 1995). On the contrary, approximately 90% of carbon in grain is acquired by the concurrent photo-assimilates after anthesis (Zhang et al., 2011). In an extreme case, when plants are grown without N supply after the emergence of flag leaf, the remobilization of N from leaves, glumes, stems, and roots contributed 40, 23, 23, and 16% to grain N, respectively (Simpson et al., 1983).

In many studies, the contribution of the pre-anthesis assimilated N to grain N was calculated by the difference in the N stored in vegetative organs at anthesis and maturity (Arduini et al., 2006; Ercoli et al., 2008). Similarly, the contribution of post-anthesis absorbed N to grain N was calculated by the difference between the N in the mature grain and the remobilized N assimilated pre-anthesis

(Shi et al., 2012a). However, errors might exist using this method, due to the loss of senescence organs (especially for leaves, which present the highest N concentration and account for a large share of the remobilized N), nutrient leaches, and the fact that the post-anthesis absorbed N might be retained in vegetative organs such as sheathes and stems, which usually depends on the N supply level of the soil and the N fertilization level (Shi et al., 2012a). To overcome these problems, the ^{15}N isotope tracing technique is used to discriminate the N that is absorbed at different growth stages and between the concurrent absorbed N and the remobilized N (Kichey et al., 2007). It has been shown that the contribution of N assimilated before anthesis to grain N is largely dependent on the genotypes, soil N status, field management practices such as fertilization and irrigation, and environmental conditions (Ferrise et al., 2010). However, the contributions of N that is absorbed during different growing periods and the N that is remobilized from different vegetative organs to the mature grains have not been well-quantified.

Pearling (debranning) is a new milling technique in the milling and baking industry, where the bran layers are removed sequentially by friction and abrasion operations with a type of device that is a modified rice polisher (Beta et al., 2005). The advantages of pearling before roller milling include improving the semolina yield and the quality of the durum wheat (Dexter and Wood, 1996), enhancing the bread-baking quality (Mousia et al., 2004), depressing the xylanase activity in both whole meal and flour (Gys et al., 2004), and decreasing the α-amylase activity in the flour of pre-harvest sprouted grains of wheat (Hareland, 2003). At the same time, pearling provides us an effective way to investigate the distribution patterns of both the chemical components and quality of flour that are located at different positions in wheat grain, which is essential for separating the milling fractions for specific products (Liu et al., 2007). The distribution patterns of ash, iron, and zinc and the phenolic compounds, with antioxidant activity, have been illustrated in different pearling fractions (Fares et al., 1996; Beta et al., 2005; Liu et al., 2007, 2008). It has also been reported that the flour obtained from pearled wheat was significantly different from the unpearled wheat in terms of flour particle size distribution, percentage of damaged starch granules, starch gelatinization temperature, moisture, and concentrations in the chemical compositions, such as ash and protein (Mousia et al., 2004). In addition, a clear gradient in the protein concentration has been found across the starchy endosperm by microscopy, which was higher in the sub-aleurone cells and lower in the central endosperm cells (Tosi et al., 2011). Moreover, the concentrations of the protein components, such as the gluten proteins, were reported to be expressed in gradient along the different parts of the endosperm (Tosi et al., 2009). However, little is known about the concentration gradient of proteins and protein fractions in different layers of wheat grain and the N sources of those proteins. In this study, the distribution patterns and concentration gradient of the protein and protein components in wheat grain were investigated using the pearling technique, and the ^{15}N isotope tracing technique was used to discriminate the N sources of different protein components in different pearling fractions of grain. The results will be essential in understanding the mechanisms of grain quality formation and exploring better N fertilization protocols for super quality wheat production.

MATERIALS AND METHODS

Experimental Design and Materials

Seeds of winter wheat (*Triticum aestivum* L., cv. Yangmai 16) were sown into 40-L pots (45 cm in length, 30 cm in both width and height, with a valve near the bottom) filled with 16 kg deionized quartz sand. Eight seeds were sown in each pot. A Hoagland nutrient solution that contained 3.0 mM NH_4NO_3 ($^{15}NH_4$ $^{15}NO_3$, 5.0% in ^{15}N abundance), 1.0 mM KH_2PO_4, 2.0 mM KCl, 2.0 mM $CaSO_4$, 1.0 mM $MgSO_4$, 1.0 mM NaCl, 2.3 μM H_3BO_3, 0.46 μM $MnSO_4$, 0.038 μM $ZnSO_4$, 0.016 μM $CuSO_4$, 0.006 μM H_2MoO_4, 89.3 μM $FeSO_4$, and 43.1 μM Na_2EDTA was used to provide nutrients to the plants. Depending on the winter wheat growth stages, four treatments were established: (1) NH_4NO_3 was replaced with $^{15}NH_4^{15}NO_3$ during the period of emergence (November 14th 2011) to jointing (March 12th 2012; EJ); (2) NH_4NO_3 was replaced with $^{15}NH_4$ $^{15}NO_3$ during the period of jointing-anthesis (April 12th 2012; JA); (3) NH_4NO_3 was replaced with $^{15}NH_4$ $^{15}NO_3$ during the period of anthesis-maturity (2nd June 2012; AM); and (4) NH_4NO_3 was applied during the whole growth season as the control (NN). The wheat plants were grown in a greenhouse to prevent any rainfall. Nine liters of nutrient solution with or without $^{15}NH_4$ $^{15}NO_3$ were applied to each pot every 7 days. The experiment was a completely randomized block design, with three replicates per treatment. Each replicate included six pots. Uniform tillers that flowered on the same day were tagged. All plants in one pot were harvested and separated into the root, stem, leaf, and spike (chaff and grain at maturity) for each replicate at anthesis and maturity. The samples were then dried at 80°C until a constant weight was reached. The pots left were harvested at maturity to record the grain yield.

Approximately 10 g of wheat grain was used for analysis of the whole grain N concentration and the concentrations of total protein and protein components. Following the methods of Liu et al. (2007), the wheat grain was pearled into eight fractions (P_1–P_8) from the surface layer to the center part, with two rice polishers (JNMJ7 and JNMJ6, Taizhou Grain Industry Instrument Corp, China). The pearling fractions were classified as P_1 (0–10%), P_2 (10–20%), P_3 (20–30%), P_4 (30–40%), P_5 (40–50%), P_6 (50–60%), P_7 (60–70%), and P_8 (70–80%). The pearling residue (20% of the whole grain) of each sample was ground in a stainless grinder (DJ-04B, Shanghai Dianjiu Machinery Manufactory, China) and designated as P_9 (80–100%).

Chemical Analysis

The protein concentration in the whole wheat grain and the albumin, globulin, glutenin, and gliadin fractions were determined by the micro-Kjeldhl distillation method of AACC 46-13.01 with some modifications (Li et al., 2012).

The natural abundance and atom% of ^{15}N in the plant samples were detected using a pattern ZHT-O2 mass spectrometer following the method by Shi et al. (2012b) at the Agroforestry Academy, Chemical Institute of Hebei Province, China.

Calculation Methods

The percentage of N that was derived from fertilizer-N (Ndff, %) was calculated by the following Equation (1) (Cookson et al., 2001; Shi et al., 2012b):

$$\text{Ndff}\,(\%) = \frac{c-b}{a-b} \times 100 \quad (1)$$

where a is the atom% of ^{15}N in the fertilizer, b is the atom% of ^{15}N in the unfertilized plants, and c is the atom % of ^{15}N in the fertilized plants.

The accumulation of N that was assimilated by the plants from fertilizer-N was calculated following Equations (2) and (3):

$$\text{Plant total N accumulation (mg/plant)} = \frac{\text{plant dry weight} \times \text{N concentration}}{100} \times 1000 \quad (2)$$

$$\text{N from fertilizer N (mg/plant)} = \frac{(2) \times \text{Ndff}_{\text{plant}}}{100} \quad (3)$$

The N distribution was calculated as the ratio of the ^{15}N accumulation amount (mg) in a given organ to the total ^{15}N accumulation in the whole plant (mg).

The remobilization of stored N from a given organ to grain (RAN) was calculated by subtracting the N at maturity from the N at anthesis in this organ. The translocation efficiency of the stored N in a given organ (TEN) was calculated following Equation (4):

$$\text{TEN}\,(\%) = \frac{\text{RAN}}{\text{N amonut at anthesis}} \times 100 \quad (4)$$

The contribution of RAN to grain N (CRAN) was calculated following Equation (5):

$$\text{CRAN}\,(\%) = \frac{\text{RAN}}{\text{N amount in grain at maturity}} \times 100 \quad (5)$$

Statistical Analysis

All of the data were subjected to one-way ANOVA using SPSS version 20.0 for windows (IBM SPSS Statistics, Chicago, IL, USA). The Duncan's multiple range test was used to check the significance of the difference between treatments.

RESULTS

Distribution of Assimilated N before Anthesis in the Organs at Anthesis

The percentages of N derived from fertilizer-N (Ndff, %) in all organs (root, stem, leaf, and spike) were different between during EJ (Emergence-Jointing) and JA (Jointing-Anthesis; **Figure 1A**). The values of Ndff in the root and leaf during EJ were significantly higher than those during JA, while an opposite trend was found in the Ndff in the stem and spike. The accumulation of N during EJ in the root and leaf was 11.1 and 10.9% higher than that during JA, while the N accumulation during EJ in the stem and spike was 14.3 and 29.1% lower than that during JA (**Figure 1B**). The distribution of N in the root and leaf that was absorbed during EJ was significantly higher than that during JA ($P < 0.05$;

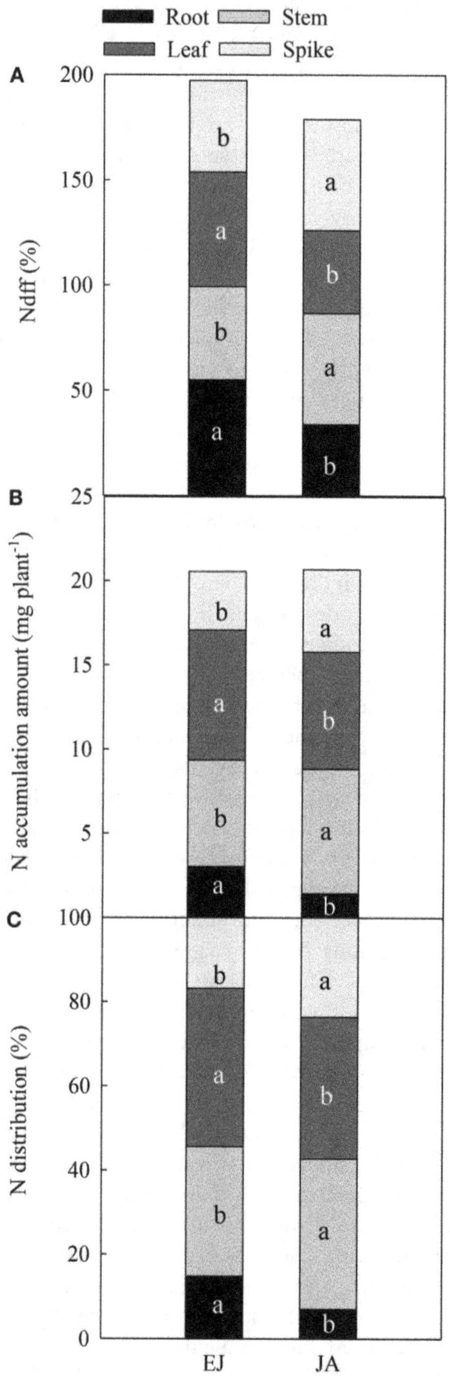

FIGURE 1 | Percentage of N derived from fertilizer-N (Ndff, %) (A), accumulation amount (B), and contribution of absorbed N (C) during periods from emergence to jointing (EJ), and from jointing to anthesis (JA). Different small letters for the same organ (root, stem, leaf, and spike) indicate significant difference at 0.05 level between different growth periods. Sheath was included in stem.

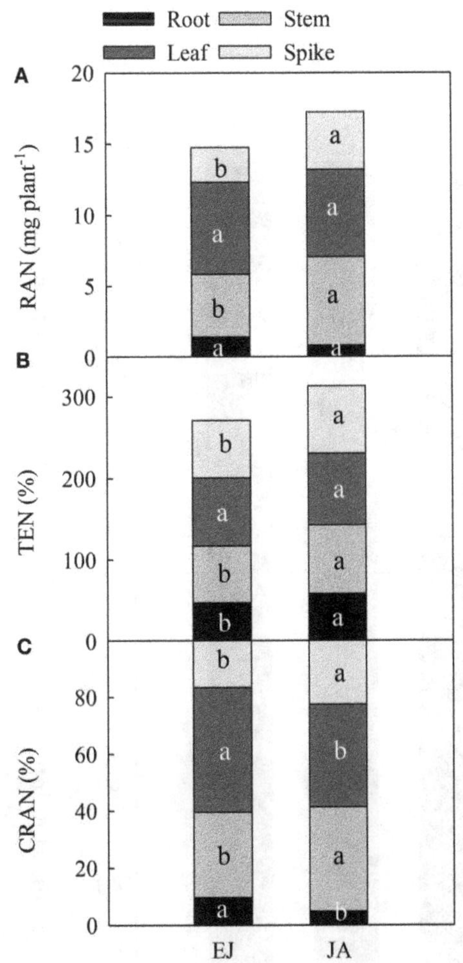

FIGURE 2 | Redistribution amount of N absorbed during periods from emergence to jointing (EJ) and from jointing to anthesis (JA) into grains (RAN) (A), translocation efficiency of the stored N in a given organ (TEN) (B), and contribution of RAN to grain N (CRAN) (C). Different small letters for the same organ (root, stem, leaf and spike) indicate significant difference at 0.05 level between different growth periods. Sheath was included in stem.

Figure 1C). On the contrary, the distribution of N in the stem and spike that was absorbed during EJ was significantly lower than that during JA.

Remobilization of the Stored N before Anthesis to the Grain

The remobilization of N (RAN) stored during EJ in the stem and spike to the grain was significantly lower than the RAN stored during JA (**Figure 2A**). However, for the leaf and root, no significant difference was found in RAN stored during EJ and JA. The translocation efficiency of N stored during EJ in the stem, spike and root (TEN) was significantly lower than TEN stored during JA (**Figure 2B**). The contribution of RAN (CRAN) stored in the leaf and root during EJ to grain was higher than the CRAN stored during JA (**Figure 2C**).

Distribution of Assimilated N at Different Periods in Organs at Maturity

The Ndff during EJ was the highest among all of the growth periods (**Figure 3A**). In the chaff and grain, the Ndff during EJ and JA was significantly higher than that during AM ($P < 0.05$). In the stem and leaf, the Ndff during JA was significantly higher than those during AM, but it was lower than that during EJ ($P < 0.001$). In the root, The Ndff during JA was significantly lower than that during AM (Anthesis-Maturity; $P < 0.001$).

There was no difference in dry weight of plants among these treatments at maturity (**Figure 3B**). The N accumulation in the chaff (including the spike axis and kernel husks) during EJ and JA were higher than that during AM (**Figure 3C**). The highest amount of N accumulation in the stem and leaf was found during EJ, followed by during JA, and the lowest was during AM. However, the highest N accumulation in the root was observed during EJ, followed by during AM, and the lowest was during JA. In addition, the distribution of N absorbed during AM was higher than that absorbed during EJ and JA (**Figure 3D**).

Distribution of Assimilated N at Different Periods in the Pearling Fractions of the Grain

Across the pearling fractions in the grain, the highest Ndff during EJ was found in the P_2 fraction, whereas the lowest Ndff was in the fractions P_6 and P_9 (**Table 1**). The Ndff during JA showed a decreasing trend from the surface layer to the center part, where the highest Ndff were found in the fractions P_1, P_2, and P_3. However, the Ndff during AM showed an opposite trend: the highest Ndff-values were in P_7 and P_8. The highest accumulation of N was found in the P_9 fraction in all these treatments. In addition, the average amount of N accumulation in the P_1–P_4 fractions was higher than in the P_5–P_9 fractions. In each pearling fraction the N was mostly accumulated during JA. The distribution of N assimilated during EJ, JA, and AM increased from the surface layer to the center part (except for P_1 and P_9).

Distribution of the Assimilated N in the Protein Fractions of Pearled Grains

The contribution of assimilated N during JA to the protein and protein fractions in each layer was the highest, except for the glutenin concentration in the P_1 fraction (**Figure 4**). For the total protein, the contribution of N assimilated during EJ was higher than that during AM in the fractions P_1–P_6; however, it was opposite in the fractions P_7–P_9. For albumin, the contribution of N assimilated during EJ was significantly higher than during AM in the fractions P_1–P_6 and P_9, and the contribution of N assimilated during EJ to the globulin was significantly higher than that during AM in all of the fractions. For gliadin, the N assimilated during EJ contributed more than that during AM in the fractions P_1–P_3, whereas the opposite was true in the fractions P_4–P_9. In the fractions of P_2–P_8, N assimilated during AM contributed more to glutenin than that during EJ; however,

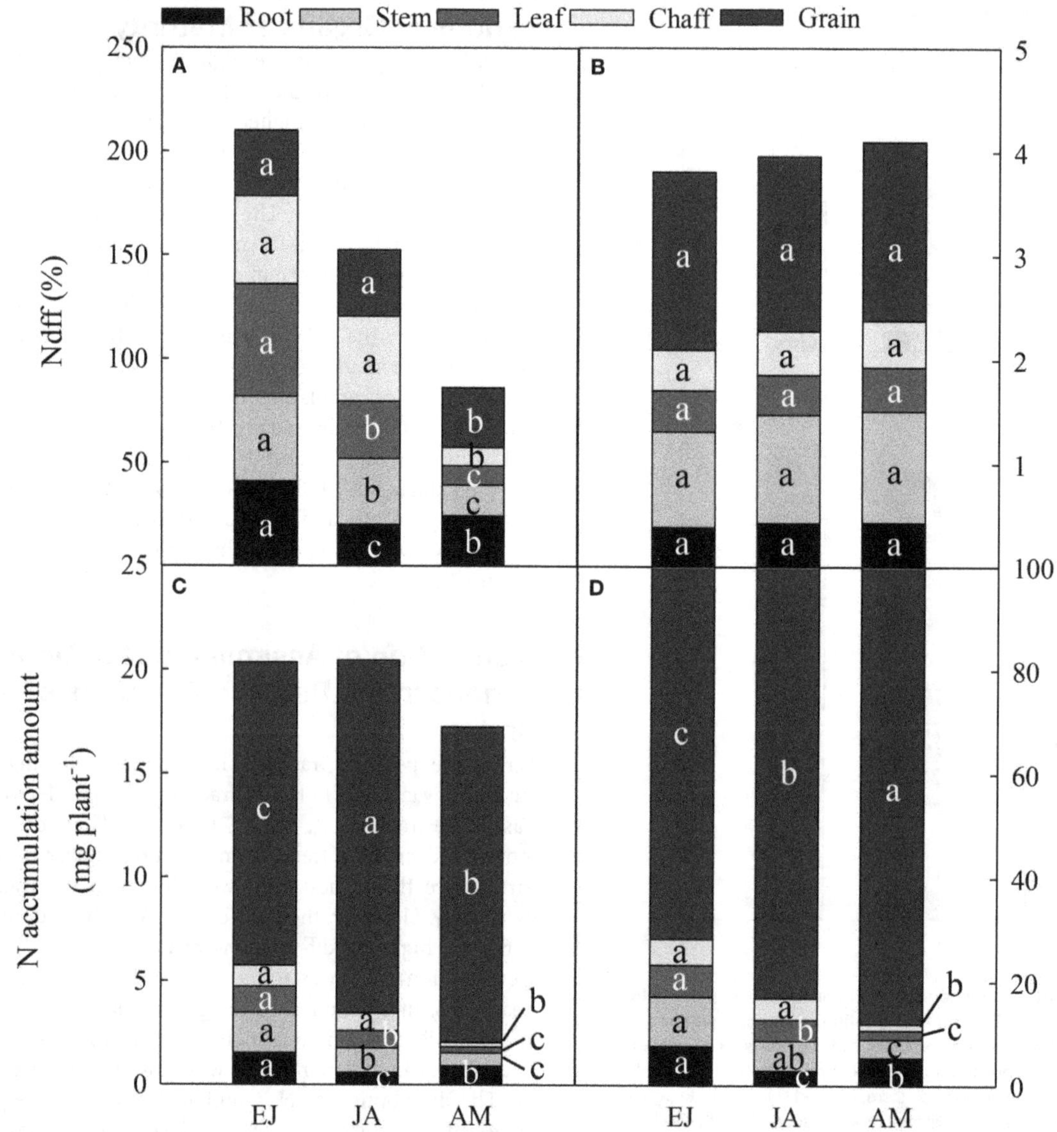

FIGURE 3 | Percentage of N derived from fertilizer-N (Ndff, %) (A), dry weight in different vegetative organs at maturity (B), accumulation amount (C), and contribution (D) of absorbed N during periods from emergence to jointing (EJ), from jointing to anthesis (JA) and from anthesis to maturity (AM). Different small letters for the same organ (root, stem, leaf, chaff, and grain) indicate significant difference at 0.05 level between different growth periods. The chaff includes spike axis and kernel husks. Sheath was included in stem.

in fraction P_1, the N contribution was found to be in the order: during AM> during JA> during EJ.

Across the pearling fractions, the distribution of N assimilated during EJ, JA, and AM in albumin and globulin was decreased from the surface layer to the center part (**Figure 5**). In albumin, the distribution of N assimilated during JA was higher than that during the other two periods in P_{1-4}. Also, a decreased trend from P_2 to P_8 was found in both gliadin and glutenin. For gliadin, the distribution of N assimilated during JA was higher than those during the other two periods in P_{1-4}, and in P_{5-9} the distribution of N assimilated during AM was the highest among these three periods. For glutenin, in P_{1-4} the distribution of N assimilated during AM was the highest among the three periods, and in P_{5-9} the distribution of N assimilated during JA and AM was higher than that during EJ.

DISCUSSION

Distribution and Remobilization of N Stored at Different Periods

The N utilization of plants involves several processes, such as uptake, assimilation, translocation, and remobilization (Masclaux-Daubresse et al., 2008). Our study illustrated the spatial and temporal N distribution in the grains, using ^{15}N isotope tracing technique. During vegetative growth stages,

TABLE 1 | Percentage of ^{15}N derived from fertilizer-N (Ndff, %), accumulation and distribution of absorbed N during different growing periods in different pearling fractions of wheat grain.

Pearling fractions	Ndff (%)			N accumulation amount (mg plant^{-1})			N contribution (%)			N distribution (%)		
	EJ	JA	AM	EJ	JA	AM	EJ	JA	AM	EJ	JA	AM
P_1	31.14c	31.59a	25.10f	1.82b	2.16c	1.67d	8.87b	10.43c	9.66d	12.35c	12.67c	11.68d
P_2	31.70a	31.70a	26.65e	2.10a	2.31b	1.86b	10.23a	11.16b	10.81b	14.25b	13.56b	13.06b
P_3	31.40b	31.58a	27.08d	1.84b	2.13c	1.76c	8.97b	10.30c	10.19c	12.50c	12.51c	12.31c
P_4	31.05c	31.31b	27.36c	1.62c	1.83d	1.54e	7.89c	8.84d	8.92e	10.98d	10.74d	10.78e
P_5	31.00c	31.37b	27.80b	1.46d	1.69e	1.45f	7.13d	8.19e	8.43f	9.93e	9.95e	10.19f
P_6	30.23e	31.01c	27.87b	1.43d	1.62e	1.41f	6.96d	7.85e	8.17f	9.70e	9.53e	9.87f
P_7	30.40d	31.04c	28.32a	1.20e	1.42f	1.23g	5.83e	6.85f	7.15g	8.12f	8.32f	8.64g
P_8	30.38de	31.09c	28.28a	1.10f	1.25g	1.14h	5.34f	6.03g	6.62h	7.43g	7.32g	7.99h
P_9	30.23e	31.10c	27.27c	2.18a	2.62a	2.21a	10.59a	12.68a	12.81a	14.74a	15.40a	15.48a
P_{1-4}	31.32	31.55	26.55	1.85	2.11	1.71	8.99	10.18	9.90	12.52	12.37	11.96
P_{5-9}	30.45	31.12	27.91	1.43	1.24	5.98	7.20	8.32	8.42	30.45	31.12	27.91
P_{5-9}/P_{1-4}	0.97	0.99	1.05	0.68	0.73	0.66	0.73	0.66	0.68	0.97	0.99	1.05

P_1–P_9, pearling fraction 1–9; P_{1-4}, mean value of P_1–P_4; P_{5-9}, mean value of P_5–P_9; P_{5-9}/P_{1-4}, ratio of P_{5-9} to P_{1-4}. EJ, emergence-jointing; JA, jointing-anthesis; AM, anthesis-maturity. The small letters in the same column indicate significant difference at 0.05 level among different pearling fractions under given treatment. The mean value of N accumulation amount, N contribution and N distribution in fractions of P_{5-9} were the total values divided by 6, respectively, since the P_{5-9} fractions accounted for 60% of the total grain mass.

the leaves have the highest N accumulation of any component because the photosynthetic apparatus is the greatest sink of N in cereal plants (Flood and Martin, 2001). In this study, the highest N accumulation was found in the leaves at anthesis, which mostly accumulated before jointing (**Figure 1**). However, the N that accumulated in the stem during JA was higher than that accumulated during EJ because of the rapid stem extension after jointing (**Figure 1**). During the initial phase of the grain growth, N accumulation continues in the stem and leaf, while the grain turn to be the major sink of N (Ferrise et al., 2010). In agreement with the study of Bertheloot et al. (2008) with bread wheat, here the N concentration in all of the organs started to decrease after anthesis with a rate that was proportional to their N concentration (**Figure 2**). The leaf and stem have nearly equal contributions to the grain N, but the main part of remobilized N, e.g., amino acids (lysine and proline) and small peptides, from these two organs was accumulated at different stages. For example, in leaf the remobilization of N stored during EJ was higher than the N stored during JA, whereas in the stems the remobilization of N stored during JA was higher than the N stored during EJ.

Rapid accumulation of N in grain after anthesis resulted in more than 70% of the total N being in the grains at maturity, which was mainly attributed to the remobilization of N assimilated during JA (**Figure 3**). In addition, the main part of the N was stored temporarily in the stems among all of the vegetative tissues after anthesis (**Figure 3**). Simpson et al. (1983) reported that the leaves, glumes, stems, and roots contribute 40, 23, 23, and 16%, respectively, to the daily rate of grain N accumulation in wheat at mid-grain filling. Here, the contribution of N stored during EJ was 43.8, 29.8, 9.5, and 16.5% from the leaves, stem, roots, and spike, respectively, to the grain. The N stored during JA in the leaves, stem, roots and spike contributed 36.1, 36.5, 5.0, and 23.7%, respectively, to the grain (**Figure 2**). This finding indicates that the contribution of N stored in the stems to the grain N increased during later vegetative growth stages before anthesis.

The Ndff during EJ was significantly higher than that during JA and AM in all of the organs, except for in the chaff and grain (**Figure 3**), suggesting that N assimilated during EJ is significantly involved in the functioning of meristematic tissue. It has been reported that up to 75% of the reduced N in cereal leaves is located in mesophyll cells, mostly as Rubisco (Bertheloot et al., 2008). On the other hand, the N absorbed and stored during JA and AM was mainly involved in the N metabolism as mobile constituents, which is remobilized to the growing grain during the reproductive stage (Bertheloot et al., 2008). Consistent with this, here higher accumulations and distributions of N stored during JA and AM than during EJ in mature grain were found.

Distribution of N Stored at Different Periods in the Protein Fractions of Pearled Grains

Wheat grain consists of the embryo, starchy endosperm, and bran. Among these three distinct parts, the bran comprises the pericarp (fruit coat), testa (seed coat), and outer endosperm (aleurone layer), which was included in fraction P_1 in the present study (Mousia et al., 2004). The starchy endosperm consists of an outer aleurone layer and inner columns of starchy endosperm cells, which are packed with most of the starch and protein, accounting for approximately 80% of the grain mass (Tosi et al., 2011). During milling, the germ and bran are removed, leaving the starchy endosperm as the principal contributor to white flour (Dupont and Altenbach, 2003). Typically, the protein concentration is relatively low in the cells near the endosperm cavity and increases in an outward radial direction

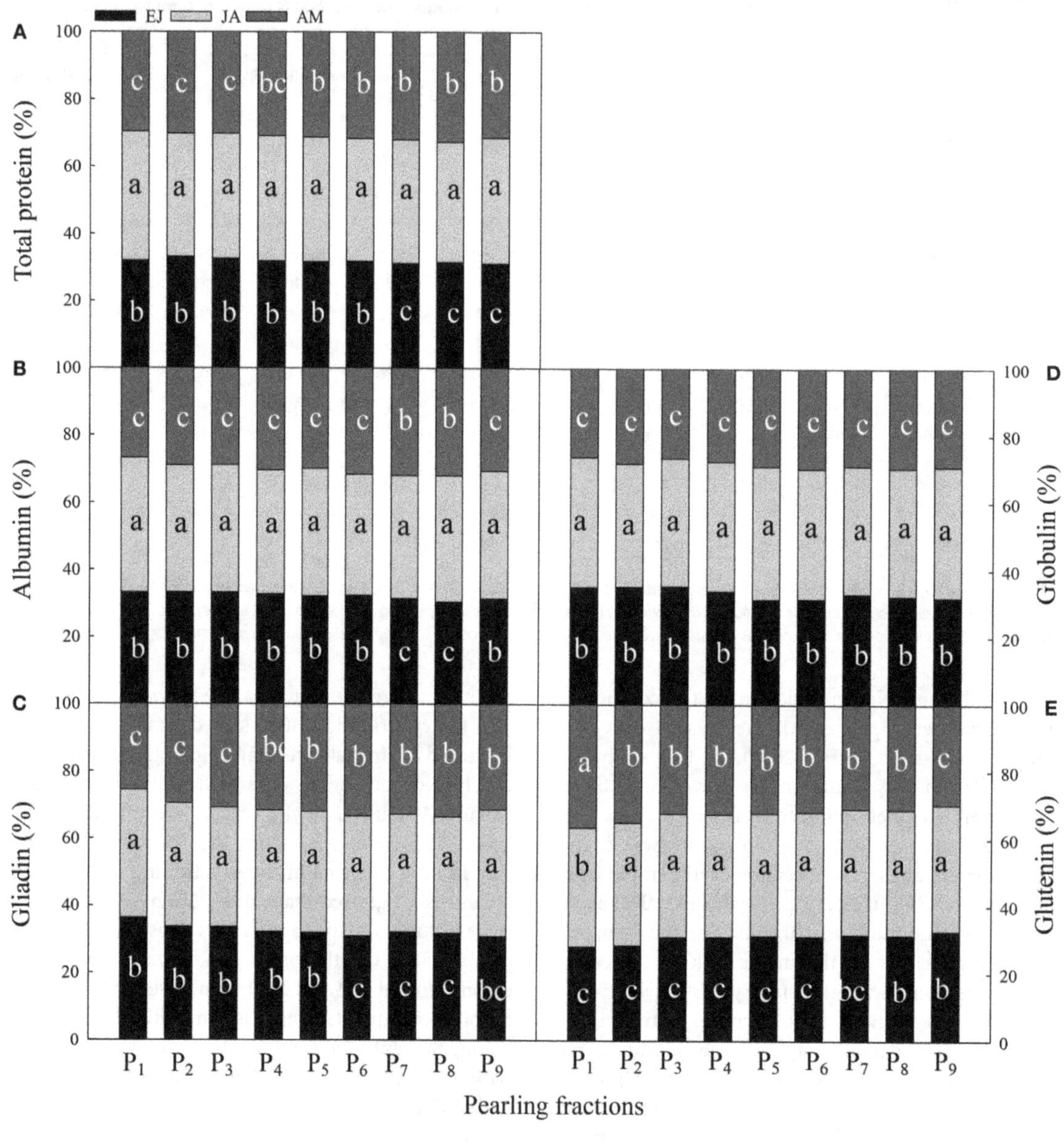

FIGURE 4 | Contribution of absorbed N druing periods from emergence to jointing (EJ), from jointing to anthesis (JA) and from anthesis to maturity (AM) to the total protein (A) and protein components (albumin, B; globulin, C; gliadin, D; glutenin, E) of different pearling fractions of wheat grain. P1–P9, pearling fraction 1–9. Different small letters for the same period (EJ, from emergence stage to jointing stage; JA, from jointing to anthesis stage; AM, from anthesis to maturity stage) indicate significant difference at 0.05 level between pearling fractions.

in a mature endosperm (Farrand and Hinton, 1974; Tosi et al., 2011). In the present study, the N accumulation also showed an increased trend from P_8 to P_2, where the endosperm is located (**Table 1**). Although it has been well-known that the gradients in the protein distribution are present in cereal grains, the underlying mechanisms by which the gradient was established during grain filling is still unclear. Tosi et al. (2011) hypothesized that the gradients have a developmental basis because both the quantitative and qualitative gradients of the protein distribution appear to follow the radial pattern of cell development in the endosperm. They concluded that the gradients in the gluten protein composition are related to the origin of the subaleurone cells, which are different from other starchy endosperm cells that derived from the re-differentiation of aleurone cells, but it could also be related to the signals produced by the maternal tissue that affect specific domains of the gluten protein gene promoters (Tosi et al., 2011).

It should be noted that the Ndff during EJ and JA increased from the inner starchy endosperm to the outer layer with a peak value in P_2, while it was decreased in P_1 (which corresponds to the outer layers, aleurone, and subaleurone; **Table 1**). However, the Ndff during AM decreased from P_8 (inner layer) to P_1. The

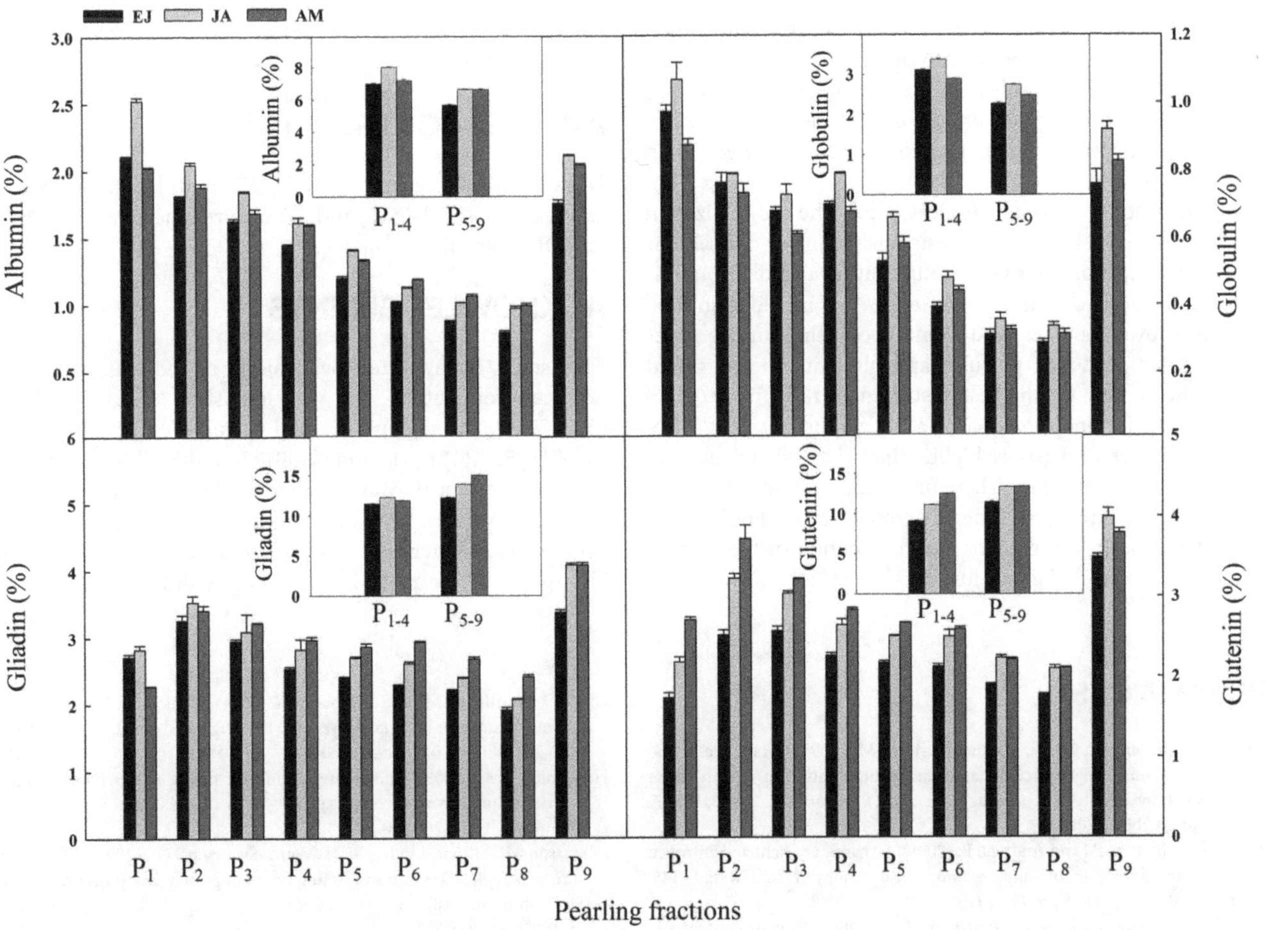

FIGURE 5 | Distribution of absorbed N druing periods of emergence-jointing (EJ), jointing-anthesis (JA), and anthesis-maturity (AM) in the protein components of different pearling fractions of wheat grain. P_1–P_9, pearling fraction 1-9; P_{1-4}, mean value of P_1–P_4; P_{5-9}, mean value of P_5–P_9.

varied trends in Ndff during the different growing stages could be explained by the fact that two separate pathways may operate in the developing wheat grain (Tosi et al., 2009). It could be hypothesized that the remobilization of N that stored before anthesis depends on the apoplast transport through the aleurone layer to the endosperm cells, and the N assimilated after anthesis is transferred following the radial direction in the endosperm cells. Nevertheless, the establishment of these gradients should be further studied considering the biological process of the grain development.

The contributions of N assimilated during different periods to the grain protein fractions varied across all of the layers, although the largest contribution was found from N assimilated during JA in each layer (**Figure 4**). For the albumin and gliadin, the contribution of N assimilated during EJ and JA declined from the outer layer to the inner layer, whereas the N contribution assimilated after anthesis showed a reversed trend. For globulin, the contribution of N assimilated during EJ decreased, whereas that during JA and AM increased from the outer layer to the inner layer. For glutenin, the contribution of N assimilated before anthesis increased, whereas the N contribution assimilated after anthesis decreased from the outer layer to the inner layer. The different contributions of N after anthesis to the protein fractions could be related to the synthesis process and transport pathway of each protein fraction during grain development (Zhu et al., 2005). The main grain structural proteins are synthesized during early grain growth, and later, the storage proteins, including gliadin and glutenin, are accumulated (Stone and Nicolas, 1996). The gliadin is a compound accumulated relatively late but fast, and part of the structural proteins can be converted into storage proteins during the latter grain-filling stage (Stone and Nicolas, 1996).

The distributions of N in grain assimilated during different periods across all of the layers showed different trends (**Figure 5**). For example, for albumin and gliadin, the N assimilated during JA had a high distribution ratio in the outer layer, whereas the contribution of N assimilated after anthesis increased faster than that before anthesis. This finding indicates that the utilization efficiency of N assimilated during JA is higher for structural proteins and greater amount of structural proteins were converted to storage proteins in the inner layers. However, for globulin and glutenin, the N assimilated after anthesis had the highest distribution ratio in most of the layers indicating that the utilization efficiency of those N is higher than that assimilated before anthesis for storage proteins, especially in the inner layers. This result is in accordance with the finding that the increased N application rate after anthesis can enhance the protein concentration in the endosperm of wheat grain (Johansson et al., 2004).

In conclusion, our results indicated that the remobilization of N stored from jointing to anthesis was the major contributor to the grain N, especially in the outer layer. The distributions of N assimilated during EJ, JA, and AM in albumin and globulin showed a decreased trend from the surface layer to the center part. For globulin and glutenin, the N stored after anthesis had the highest distribution ratio in most of the layers, which indicates that the utilization efficiency of N assimilated after anthesis is higher than that before anthesis for the storage proteins. Therefore, modification in the N application rate during the different growth stages should be an effective approach to regulating the distribution of the protein fractions in the pearled grains for specific end-use. However, the establishment of these gradients should be further studied while considering the biological processes that are involved in the grain development.

AUTHOR CONTRIBUTIONS

DJ and QZ designed research; XL and LZ performed the experiment; XW, JC, XL, and LZ analyzed the data; LF, TD, WC, and DJ revised the manuscript.

ACKNOWLEDGMENTS

This study is supported by projects of the National Natural Science Foundation for Distinguished Young Scientists (31325020), the National Natural Science Foundation of China (31471445, 31171484, and 31401326), the China Agriculture Research System (CARS-03), Jiangsu Collaborative Innovation Center for Modern Crop Production (JCIC-MCP), Collaborative Innovation Center of Gene Resources, and the National Non-profit Program by Ministry of Agriculture (2014039).

REFERENCES

Arduini, I., Masoni, A., Ercoli, L., and Mariotti, M. (2006). Grain yield, and dry matter and nitrogen accumulation and remobilization in durum wheat as affected by variety and seeding rate. *Eur. J. Agron.* 25, 309–318. doi: 10.1016/j.eja.2006.06.009

Bertheloot, J., Martre, P., and Andrieu, B. (2008). Dynamics of light and nitrogen distribution during grain filling within wheat canopy. *Plant Physiol.* 148, 1707–1720. doi: 10.1104/pp.108.124156

Beta, T., Nam, S., Dexter, J. E., and Sapirstein, H. D. (2005). Phenolic content and antioxidant activity of pearled wheat and roller-milled fractions. *Cereal Chem.* 82, 390–393. doi: 10.1094/CC-82-0390

Cookson, W., Rowarth, J., and Cameron, K. (2001). The fate of autumn-, late winter-and spring-applied nitrogen fertilizer in a perennial ryegrass (*Lolium perenne* L.) seed crop on a silt loam soil in Canterbury, New Zealand. *Agric. Ecosyst. Environ.* 84, 67–77. doi: 10.1016/S0167-8809(00)00196-1

Dexter, J., and Wood, P. (1996). Recent applications of debranning of wheat before milling. *Trends Food Sci. Technol.* 7, 35–41. doi: 10.1016/0924-2244(96)81326-4

Dupont, F. M., and Altenbach, S. B. (2003). Molecular and biochemical impacts of environmental factors on wheat grain development and protein synthesis. *J. Cereal Sci.* 38, 133–146. doi: 10.1016/S0733-5210(03)00030-4

Ercoli, L., Lulli, L., Mariotti, M., Masoni, A., and Arduini, I. (2008). Post-anthesis dry matter and nitrogen dynamics in durum wheat as affected by nitrogen supply and soil water availability. *Eur. J. Agron.* 28, 138–147. doi: 10.1016/j.eja.2007.06.002

Fares, C., Troccoli, A., and Di Fonzo, N. (1996). Use of friction debranning to evaluate ash distribution in Italian durum wheat cultivars. *Cereal Chem.* 73, 232–234.

Farrand, E., and Hinton, J. (1974). Study of relationships between wheat protein contents of two UK varieties and derived flour protein contents at varying extraction rates. II. Studies by hand-dissection of individual grains. *Cereal Chem.* 51, 56–65.

Ferrise, R., Triossi, A., Stratonovitch, P., Bindi, M., and Martre, P. (2010). Sowing date and nitrogen fertilization effects on dry matter and nitrogen dynamics for durum wheat: an experimental and simulation study. *Field Crops Res.* 117, 245–257. doi: 10.1016/j.fcr.2010.03.010

Flood, R. G., and Martin, P. J. (2001). Nitrogen accumulation and distribution at anthesis and maturity in ten wheats grown at three sites in north-western Victoria. *Aust. J. Exp. Agric.* 41, 533–540. doi: 10.1071/EA98025

Gys, W., Courtin, C. M., and Delcour, J. A. (2004). Reduction of xylanase activity in flour by debranning retards syruping in refrigerated doughs. *J. Cereal Sci.* 39, 371–377. doi: 10.1016/j.jcs.2004.01.003

Hareland, G. A. (2003). Effects of pearling on falling number and α-amylase activity of preharvest sprouted spring wheat. *Cereal Chem.* 80, 232–237. doi: 10.1094/CCHEM.2003.80.2.232

Johansson, E., Prieto-Linde, M. L., and Svensson, G. (2004). Influence of nitrogen application rate and timing on grain protein composition and gluten strength in Swedish wheat cultivars. *J. Plant Nutr. Soil Sci.* 167, 345–350. doi: 10.1002/jpln.200320332

Kichey, T., Hirel, B., Heumez, E., Dubois, F., and Le Gouis, J. (2007). In winter wheat (*Triticum aestivum* L.), post-anthesis nitrogen uptake and remobilisation to the grain correlates with agronomic traits and nitrogen physiological markers. *Field Crops Res.* 102, 22–32. doi: 10.1016/j.fcr.2007.01.002

Li, X., Cai, J., Li, H., Bo, Y., Liu, F., Jiang, D., et al. (2012). Effect of Shading from jointing to maturity on high molecular weight glutenin subunit accumulation and glutenin macropolymer concentration in grain of winter wheat. *J. Agron. Crop Sci.* 198, 68–79. doi: 10.1111/j.1439-037X.2011.00484.x

Liu, Z., Wang, H., Wang, X., Zhang, G., Chen, P., and Liu, D. (2007). Phytase activity, phytate, iron, and zinc contents in wheat pearling fractions and their variation across production locations. *J. Cereal Sci.* 45, 319–326. doi: 10.1016/j.jcs.2006.10.004

Liu, Z., Wang, H., Wang, X.-E., Xu, H., Gao, D., Zhang, G., et al. (2008). Effect of wheat pearling on flour phytase activity, phytic acid, iron, and zinc content. *LWT Food Sci. Technol.* 41, 521–527. doi: 10.1016/j.lwt.2007.04.001

Malik, A. H., Kuktaite, R., and Johansson, E. (2013). Combined effect of genetic and environmental factors on the accumulation ofproteins in the wheat grain and their relationship to bread-making quality. *J. Cereal Sci.* 57, 170–174. doi: 10.1016/j.jcs.2012.09.017

Masclaux-Daubresse, C., Reisdorf-Cren, M., and Orsel, M. (2008). Leaf nitrogen remobilisation for plant development and grain filling. *Plant Biol.* 10, 23–36. doi: 10.1111/j.1438-8677.2008.00097.x

Mousia, Z., Edherly, S., Pandiella, S. S., and Webb, C. (2004). Effect of wheat pearling on flour quality. *Food Res. Int.* 37, 449–459. doi: 10.1016/j.foodres.2004.02.012

Palta, J., and Fillery, I. (1995). N application increases pre-anthesis contribution of dry matter to grain yield in wheat grown on a duplex soil. *Crop Pasture Sci.* 46, 507–518. doi: 10.1071/AR9950507

Shi, Z., Jing, Q., Cai, J., Jiang, D., Cao, W., and Dai, T. (2012b). The fates of ^{15}N fertilizer in relation to root distributions of winter wheat under different N splits. *Eur. J. Agron.* 40, 86–93. doi: 10.1016/j.eja.2012.01.006

Shi, Z., Li, D., Jing, Q., Cai, J., Jiang, D., Cao, W., et al. (2012a). Effects of nitrogen applications on soil nitrogen balance and nitrogen utilization of winter wheat in a rice-wheat rotation. *Field Crops Res.* 127, 241–247. doi: 10.1016/j.fcr.2011.11.025

Simpson, R. J., Lambers, H., and Dalling, M. J. (1983). Nitrogen redistribution during grain growth in wheat (*Triticum aestivum* L.). *Plant Physiol.* 71, 7–14. doi: 10.1104/pp.71.1.7

Stone, P., and Nicolas, M. (1996). Varietal differences in mature protein composition of wheat resulted from different rates of polymer accumulation during grain filling. *Func. Plant Biol.* 23, 727–737. doi: 10.1071/pp9960727

Tosi, P., Gritsch, C. S., He, J., and Shewry, P. R. (2011). Distribution of gluten proteins in bread wheat (*Triticum aestivum*) grain. *Ann. Bot.* 108, 23–35. doi: 10.1093/aob/mcr098

Tosi, P., Parker, M., Gritsch, C. S., Carzaniga, R., Martin, B., and Shewry, P. R. (2009). Trafficking of storage proteins in developing grain of wheat. *J. Exp. Bot.* 60, 979–991. doi: 10.1093/jxb/ern346

Zhang, X., Jiang, D., Zheng, C., Dai, T., and Cao, W. (2011). Post-anthesis salt and combination of salt and waterlogging affect distributions of sugars, amino acids, Na^+ and K^+ in wheat. *J. Agron. Crop Sci.* 197, 31–39. doi: 10.1111/j.1439-037X.2010.00438.x

Zhu, X., Zhou, J., Feng, C., Guo, W., and Peng, Y. (2005). Differences of protein and its component accumulation in wheat for different end uses. *Acta Agron. Sin.* 31, 342–347.

Conflict of Interest Statement: The authors declare that the research was conducted in the absence of any commercial or financial relationships that could be construed as a potential conflict of interest.

17

Crossing Methods and Cultivation Conditions for Rapid Production of Segregating Populations in Three Grain Amaranth Species

Edited by:
Antonio M. De Ron,
Spanish National Research Council (CSIC) - Misión Biológica de Galicia (MBG), Spain

Reviewed by:
Johann Vollmann,
University of Natural Resources and Life Sciences, Austria
Peter Jeff Maughan,
Brigham Young University, USA
David M. Brenner,
Iowa State University, USA

***Correspondence:**
Markus G. Stetter
markus.stetter@uni-hohenheim.de

Markus G. Stetter*, Leo Zeitler, Adrian Steinhaus, Karoline Kroener, Michelle Biljecki and Karl J. Schmid

Institute of Plant Breeding, Seed Science and Population Genetics, University of Hohenheim, Stuttgart, Germany

Grain amaranths (*Amaranthus* spp.) have been cultivated for thousands of years in Central and South America. Their grains are of high nutritional value, but the low yield needs to be increased by selection of superior genotypes from genetically diverse breeding populations. Amaranths are adapted to harsh conditions and can be cultivated on marginal lands although little is known about their physiology. The development of controlled growing conditions and efficient crossing methods is important for research on and improvement of this ancient crop. Grain amaranth was domesticated in the Americas and is highly self-fertilizing with a large inflorescence consisting of thousands of very small flowers. We evaluated three different crossing methods (open pollination, hot water emasculation and hand emasculation) for their efficiency in amaranth and validated them with genetic markers. We identified cultivation conditions that allow an easy control of flowering time by day length manipulation and achieved flowering times of 4 weeks and generation times of 2 months. All three different crossing methods successfully produced hybrid F_1 offspring, but with different success rates. Open pollination had the lowest (10%) and hand emasculation the highest success rate (74%). Hot water emasculation showed an intermediate success rate (26%) with a maximum of 94% success. It is simple to perform and suitable for a more large-scale production of hybrids. We further evaluated 11 single nucleotide polymorphism (SNP) markers and found that they were sufficient to validate all crosses of the genotypes used in this study for intra- and interspecific hybridizations. Despite its very small flowers, crosses in amaranth can be carried out efficiently and evaluated with inexpensive SNP markers. Suitable growth conditions strongly reduce the generation time and allow the control of plant height, flowering time, and seed production. In combination, this enables the rapid production of segregating populations which makes amaranth an attractive model for basic plant research but also facilitates further the improvement of this ancient crop by plant breeding.

Keywords: amaranth, hybridization, hot water emasculation, hand emasculation, genetic resources, marker assisted breeding

1. INTRODUCTION

Ancient crops from the Americas such as quinoa (*Chenopodium quinoa* willd.) or amaranth (*Amaranthus* spp. L.) are a valuable addition to the human diet because of their high nutritional value. These pseudocereals have a high protein content and are rich in lysine and other essential amino acids that are limited in other grains (Vega-Gálvez et al., 2010; Rastogi and Shukla, 2013). In addition, these crops are well adapted to harsh environmental conditions and are therefore suitable for cultivation on marginal soils. Their yields are significantly lower than those of major crops due to a lack of plant breeding (Reta Alemayehu et al., 2015), but the presence of a high genetic and phenotypic diversity in these species indicates an excellent potential for breeding and variety development (Brenner et al., 2010).

Grain amaranth originated from Central and South America, where it was of great importance in pre-columbian agriculture until its cultivation strongly declined after the Spanish conquest (Sauer, 1967; Kauffman and Weber, 1990; Brenner et al., 2010). Three species of *Amaranthus* are cultivated for grain production: *A. caudatus* L., *A. cruentus* L., and *A. hypochondriacus* L. Amaranth expresses the C_4 carbon cycle, which is more common in grasses but rare in dicots. Despite a high genetic diversity (Stetter et al., 2015), breeding efforts in amaranth so far were limited to the selection of suitable genotypes from landraces. Amaranth is mainly self-pollinating and has numerous intricate flowers, which make crosses more difficult than in other crops. The ability to efficiently carry out crosses is an important requirement for plant research to understand genetic basis of relevant traits (Moose and Mumm, 2008; Olsen and Wendel, 2013). Crosses are equally important for plant breeding and are used to generate new genetic variation and to introgress exotic material into breeding populations.

In many crops, hybrid varieties are characterized by strongly increased yields (Duvick, 2001). The application of hybrid breeding in amaranth is also very promising, because a mid-parent heterosis of up to 88% has been reported (Lehmann et al., 1991). The ability to conduct crosses on a large scale with little effort is of central importance for the development and production of hybrid crop varieties. To use this potential in minor crops, an improvement of crossing methods is essential (Veerappan et al., 2014).

Several approaches for hybrid production are available, but for all methods the key step is to prevent self-fertilization by the male parent. This is either by using appropriate genetic self-incompatibility systems or by mechanical and chemical treatments that lead to male sterility. In several species, cytoplasmatic male sterility (CMS) systems prevent selfing of the female crossing partner (Laser and Lersten, 1972). To use CMS systems for breeding male sterile female parent and male parents with restorer genes are needed to allow seed production in the hybrid progeny. Additionally, a maintainer line is needed that allows multiplying the male sterile line without loosing the CMS. Male sterility has been reported in *A. hypochondriacus* but is not yet developed sufficiently to be used for breeding (Peters and Jain, 1987). Mechanical emasculation methods are efficient if the male and female flower are well separated on the plant (e.g., as in maize) because then male flowers can be removed without interfering with the female inflorescence. In other crops like tomato and *Medicago*, anthers are removed before pollen shedding (Veerappan et al., 2014). Another physical method is the heat treatment of the flowers of the female parent to destroy the pollen, for example by a hot water treatment. Here, the temperature is crucial, as differences by few degrees can influence the efficiency of the emasculation (Mukasa et al., 2007; García-Yzaguirre and Carreres, 2008; Otsuka et al., 2010). Chemical gametocides are used in hermaphrodite crops for which no CMS systems are available or are too costly, for example in wheat (Dotlacil and Apltauerová, 1978). The grain amaranth species have male and female flowers on the same inflorescence where several female flowers are arranged circularly around a male flower (**Figure 1**). The flowers are less than 1 mm in diameter, which makes mechanical emasculation difficult. For this reason other emasculation methods such as a hot water treatment may be more efficient.

Frequently, crossing methods are not completely reliable and require the validation of progeny. Phenotypic traits with a dominant-recessive inheritance can be used to identify successful crosses. In amaranth, traits such as seed or leaf color differ between genotypes and are available for validation (Kulakow et al., 1985). For phenotypic traits to be useful, however, parents need to differ in at least one trait and the male parent needs to express the dominant allele. In contrast, molecular markers allow an efficient and early evaluation of crosses without restricting the combination of parents, and cost-efficient PCR-based marker systems are available for this purpose (Maughan et al., 2011).

For model plants it is important to take specific requirements of development into account. Amaranth shows a strong photoperiod sensitivity and starts to flower under short day conditions (Brenner et al., 2010). A single plant has the potential

FIGURE 1 | Flower morphology. Inflorescence of *A. caudatus* consisting of flower clusters in which a male flower in the center is surrounded by several female flowers.

to produce several thousands of seeds and can therefore produce large populations. However, under field conditions amaranth plants are usually tall and require a significant amount of space for cultivation. If flowering time, plant size and seed production can be controlled in climate chambers and greenhouses, an efficient propagation of the plant may be possible.

In the work presented here we study the efficiency of three different crossing methods and describe environmental conditions in a controlled environment (growth chamber) to achieve efficient and rapid generation of progeny for genetic studies. We suggest a method for hybrid identification with cost efficient PCR- based markers. Subsequently, we apply our method to three species of amaranth to evaluate its potential for the wider application to species within the genus *Amaranthus*.

2. METHODS

2.1. Plant Material and Growth Conditions

The amaranth accessions for testing the three crossing methods were selected to comprise accessions with green seedlings as female parent and accessions with red seedlings as male parent. Additionally, amaranth varieties were used to verify hybridization and the use of genetic markers (**Table 1** and Table S2). Single seeds were planted in 7 × 7 cm pots in standard gardening soil. Plants were grown for 2 weeks under long day conditions (**Table 2**) before transferring them step-wise in weekly intervals to short day conditions (**Table 2**). This helped to synchronize flowering of different genotypes and spread workload for performing the crosses.

2.2. Crossing Methods

We evaluated three different methods for crossing wild (*A. hybr.* and *A. hybridus*) and cultivated species of amaranth (**Figure 2** and **Table 1**). The first method was open pollination by fixing the flowers of the female and male parent to each other and protecting them with a pollen proof bag (Sealed Air, Germany) from cross pollination by other plants. The second method was a warm water treatment of the inflorescence during flower initiation of the first emerging flowers (García-Yzaguirre and Carreres, 2008). Female flowers were dipped into a water bath of 45°C warm water for 10 min to emasculate the male flowers before proceeding as in the first method. The water treatment was repeated after 7 days. The third method was hand emasculation. For this approach, female flowers that were already open and all male flowers were removed from the inflorescence. The tip of the inflorescence was also removed to prevent the emergence of new flowers. The emasculation was repeated after 7 days and any flowers that developed later were removed. For all three methods plants were shaken daily to increase pollen dispersal and to assure cross-fertilization.

2.3. Success Evaluation and Statistical Analysis

Seeds of the female parent were harvested 4 weeks after crossing. For each cross 50 seeds were counted and planted in pots. Seedling color evaluation was performed 2 weeks after planting by counting green and red seedlings. The success rate was the

TABLE 1 | Parental genotypes for 11 KASP marker assays.

ID	Name	Species	AM17978	AM19584	AM19963	AM21336	AM21605	AM22029	AM22341	AM24451	AM24579	AM25548	AM26171
26	PI 642741	*A. caudatus*	FAM	HEX	HEX	HEX	HEX	HEX	FAM	FAM	-	HEX	HEX
34	PI 511679	*A. caudatus*	FAM	HEX	FAM	HEX	FAM	HEX	HEX	HEX	FAM	HEX	HEX
37	PI 649220	*A. caudatus*	FAM	HEX	FAM	HEX	FAM	HEX	FAM	FAM	FAM	HEX	HEX
117	PI 511684	*A. hybr.*	FAM	HEX	FAM	HEX	FAM	HEX	HEX	FAM	FAM	HEX	HEX
174	PI 649623	*A. hypochondriacus*	HEX	FAM	FAM	-	FAM	FAM	HEX	HEX	HEX	FAM	FAM
245	Baernkrafft	*A. cruentus*	FAM	FAM	FAM	FAM	FAM	HEX	HEX	HEX	HEX	FAM	HEX
246	C6	*A. cruentus*	FAM	HEX	FAM	FAM	FAM	HEX	HEX	HEX	HEX	FAM	HEX
247	Puerto Moutt	*A. cruentus*	FAM	FAM	FAM	FAM	FAM	HEX	HEX	HEX	HEX	FAM	HEX
248	Pastewny	*A. hybridus*	-	FAM	FAM	-	FAM	FAM	HEX	HEX	HEX	-	FAM
369	PI 511695	*A. caudatus*	FAM	HEX	HEX	HEX	HEX	HEX	FAM	FAM	FAM	HEX	HEX

HEX and FAM are the fluorescence dyes associated with each allele. Markers were tested in 10 individuals of which 8 were used for crosses.

ratio between red and green offspring. Data analysis to test the differences between methods and between crossing types was conducted with a Generalized Linear Model (GLM) with binomial variance and a logit link function that included the crossing method (α), the crossing type (β, Intra- and inter-specific) and the interaction as factors:

$$logit(\mu_{ij}) = log(\frac{\mu_{ij}}{1-\mu_{ij}}) = \eta_{ij} = \mu + \alpha_i + \beta_j + (\alpha\beta)_{ij}. \quad (1)$$

The calculation was done with the R statistical package version 3.2.0 using the `stats` library.

2.4. DNA Extraction

For genotyping the DNA was extracted with EconoSpin columns (Epoch Life Science Inc.) using 1% CTAB extraction buffer (Saghai-Maroof et al., 1984). Dry leaf samples were homogenized and incubated for 2 h at 50°C in 400 μl 1% CTAB extraction buffer and 4 μl Proteinase K. After addition of 300 μl Ammonium acetate (7.5 M) and 300 μl Ethanol (96%), the samples were centrifuged for 1 min at full speed. Then 800 μl of the supernatant were transferred on a EconoSpinÂő column placed in the collection tube and centrifuged for 1 min. The flow through was discarded. The columns were washed twice with wash buffers from Saghai-Maroof et al. (1984) before eluting DNA twice with 50 μl Tris-HCl (10 mM, pH 8).

2.5. Evaluation of Genetic Markers

Eleven KASP assays (LGC Berlin/Germany) were selected from Maughan et al. (2011) to validate crosses (Table S1). The assays were prepared with 5 μl DNA (10 ng/μl) and 5 μl genotyping mix and run on the LightCycler 480 Instrument II (Roche Life Science) with standard settings as given by the KASP manual (LGC Berlin/Germany) and analyzed using the LightCycler 480 Software. First, parental lines were evaluated to find polymorphic markers for each of the crosses. Later, these markers were used to validate the crosses. For a proof of concept we genotyped offspring that were evaluated before by their seedling color. Both offspring with green (selfed plants) and red (hybrids) were genotyped.

TABLE 2 | Growth conditions.

	Day length	Light intensity	Temp day	Temp night
Long day	16	150 mmol	35°C	30°C
Short day	8	150 mmol	30°C	25°C

Parameters for amaranth in growth chamber for long and short day conditions.

2.6. Additional Hybrid Production

The previously evaluated hand emasculation method was used to produce additional hybrids. Plants were grown as described above, but crossing partners were not restricted to different seedling colors. The success of the crosses was validated with SNP markers.

3. RESULTS AND DISCUSSION

3.1. Cultivation and Life Cycle

In the field the generation time of the three grain amaranth species is approximately 6 months and leads to very tall plants with thousands of flowers. To reduce the generation time, plant height and number of flowers, we cultivated the plants under short day conditions (8 h) and high temperature (30°C) which both induced early flowering 4 weeks after planting. Additionally, we controlled the initiation of flowering by transferring plants from long day (16 h, 35°C) conditions to short day conditions. Under long day conditions the plants displayed strong vegetative growth and did not flower within 10 weeks after planting, but started flowering approximately 14 days after a transfer to short day conditions. The step-wise transfer of plants from long to short-day conditions allows the production of plants in different flowering stages, which greatly facilitates synchronous flowering for crosses between genotypes that differ in their flowering time. This treatment is further useful to produce male parents that are able to shed large amounts of pollen when females parents start flowering. As soon as 4 weeks after flowering, mature seeds could be harvested. By employing these treatments, very short generation times can be achieved that allow up to six generations per year, which is comparable to the model plant *Arabidopsis thaliana*. In addition, plant height and seed number per plant can be controlled by adjusting growth conditions. Long day conditions lead to more vegetative growth, later flowering and more seeds, while short day conditions result in small early

FIGURE 2 | Crossing methods. Three crossing methods: **(A)** Hand emasculation by removal of male flowers from female plant. **(B)** Hot water emasculation by 10 min treatment with 45°C water bath. **(C)** Fixing male and female flower to each other for better pollen transfer.

TABLE 3 | Success rate of different crossing methods.

	Type	*N*	Mean (%)	SD (%)	Minimum (%)	Maximum (%)
Open pollination		7	10^{c}	5	4	18
	Intra-specific	3	11	3	8	14
	Inter-specific	4	10	6	4	18
Heat treatment		8	26^{b}	35	0	94
	Intra-specific	4	26	45	0	94
	Inter-specific	4	27	27	0	57
Hand emasculation		11	74^{a}	29	17	100
	Intra-specific	4	80	20	50	94
	Inter-specific	7	71	34	17	100

Success rates and standard deviation (SD) for different crossing methods based on seedling color of 50 offspring per sample. The mean was calculated on basis of four to seven crosses (N) per method and crossing type. Intra-specific crosses were performed with A. caudatus (PI 511679 × PI 649220) and inter-specific crosses between A. caudatus (PI 511679) and A. hybr. (PI 511684). A Generalized Linear Model (GLM) with binomial variance and a logit link function were used to analyze differences between methods. Different letters show significant differences between methods. There was no significant difference between intra- and inter- specific crosses.

flowers. This is useful for different applications, because crosses require only few flowers, whereas the resulting F_1 plants should produce larger amounts of seeds (e.g., for creating mapping populations).

3.2. Crossing Methods

We compared three crossing methods that included open pollination, hot water emasculation and hand emasculation. All three methods produced successful crosses, but the success rates and variances differed strongly between the methods (**Table 3**). The amount of seeds produced did not substantially differ between methods and the mother plants produced between 100 and 200 seeds. Open pollination between two plants under a single bag without emasculation of the female parent led to a mean success rate of 10% with a standard deviation (s.d.) of 0.05. The hot water treatment of the female parent led to a significantly increased success rate of 26% but with a very high deviation (s.d. = 0.35) and a minimal success rate of 0%. However, the maximal success after hot water treatment was 94%, which shows that the method has a high potential if the key conditions for a successful application can be identified. We sterilized flowers at 45°C and an adaptation of temperature may contribute to a higher rate of success. In other species (e.g., Acacia, buckwheat and rice), different temperatures change the efficiency of emasculation (Mukasa et al., 2007; García-Yzaguirre and Carreres, 2008; Otsuka et al., 2010). A temperature of 45°C for emasculation is rather high compared to other crops (García-Yzaguirre and Carreres, 2008), but not too high because the amaranth plants still set seeds after this treatment and a further optimization may be achieved by varying the length of the heat treatment. Overall, hot water emasculation works with amaranth and, if it can be further improved, is suitable for application in the field to large numbers of plants.

The most elaborated and time consuming method we evaluated was hand emasculation (**Figure 3**). The mean success rate of 74% was the highest of the three methods and the deviation (s.d. = 0.29) was lower than of the heat treatment.

FIGURE 3 | Hand crossing procedure. (A) Flower initiation of female plant. **(B)** Female plant prepared for crossing. Leaves near the flower are removed. **(C)** Male crossing partner with first open male flowers. **(D)** Male plant prepared for crossing. Leaves near the flower are removed for improved pollen exchange. **(E)** Female and male crossing partners attached to each other. **(F)** Crossing partners are isolated with pollen proof bag to avoid contamination by foreign pollen.

The minimum success was comparable to free pollination, but the maximum success was up to 100%. Hand emasculation is difficult because amaranth has many small flowers and each male flower sheds enough pollen to pollinate a whole plant. Therefore it is critical to remove all male flowers from the female parent before flower dehiscence. The deviation can be decreased by keeping only few flower clusters per plant. We also tested whether intra- and inter-specific crosses are different in their efficiency, but there was no significant difference between intra- and inter-specific crosses (**Table 3**). This shows that inter-specific hybridization is possible, but as the two species are closely related this might not be the case for distant member of the *Amaranthus* genus.

The comparison of the three methods shows that open pollination had low success rates, whereas heat treatment can be an effective and simple method for crosses if many seeds are required and simple morphological markers are available for the evaluation of offspring. Hand emasculation by well trained personnel shows the best performance. Since amaranth plants may produce thousands of seeds, a single successful cross can produce large F_2 populations, and the number of hand crosses needed can be kept low, which decreases the work load of the method substantially and makes it suitable for large projects.

3.3. Genetic Markers for Hybrid Identification

Since no crossing method provides a 100% success rates, unsuccessful crosses have to be excluded in early stages. Furthermore, crossing partners should not be limited by phenotypic differences in certain traits (e.g., different seedling color), but all possible combinations parents should be available. We therefore evaluated all accessions used in this study with 11 PCR-based SNP markers. The markers were the most polymorphic from a set of 411 KASP markers from Maughan et al. (2011). Each marker was polymorphic between at least two lines and each cross segregated at least for one marker (**Table 1**). After evaluating the parental lines, we selected suitable markers to evaluate crosses.

First, we investigated progeny which had already been evaluated by their seedling color, because we expected green

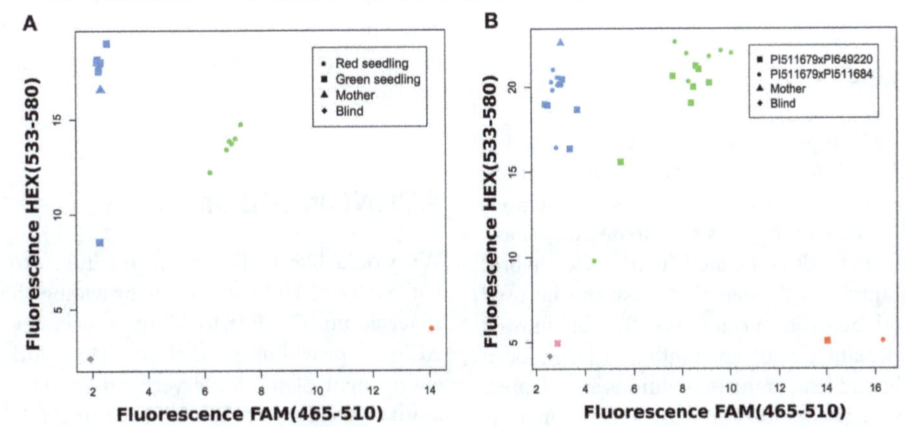

FIGURE 4 | SNP genotyping for known crosses. (A) Validation of PI511679xPI649220 with AM22341 and comparison with seedling color. **(B)** Validation of two crosses with AM24451.

TABLE 4 | Crosses of different amaranth varieties by hand emasculation and evaluation of success rates with SNP markers.

	ID (Mother)	ID (Father)	Marker	Genotyped	Selfings	Successful crosses	Failed assay
1	34	245	AM19584	7	0	7	0
2	34	245	AM19584	6	2	2	2
3	34	245	AM19584	7	2	4	1
4	34	248	AM19584	7	2	4	1
5	34	248	AM19584	7	0	6	1
6	34	248	AM19584	7	0	4	3
7	245	26	AM19584	6	4	2	0
8	245	26	AM19584	8	2	3	3
9	247	248	AM22029	10	1	9	0
10	247	248	AM22029	9	9	0	0
11	248	245	AM22029	6	6	0	0
12	248	245	AM22029	8	5	0	3
13	248	245	AM22029	7	4	0	3

All crosses are interspecific crosses between the three grain amaranths and/or their putative ancestors. The ID of mothers and fathers corresponds to ***Table 1****.*

seedlings to be homozygous for the maternal allele since the green allele is recessive, and red seedlings to be heterozygous. For example, the application of marker AM22341 in a cross of PI511679 × PI649220 showed that green seedlings were homozygous for the allele of parent PI511679 and red seedlings were heterozygous for both parental alleles (**Figure 4A**). Frequently, the same marker can be used in several crosses, which allows the evaluation of more than one cross simultaneously (**Figure 4B**). This strongly decreases the work load and the cost of the evaluation. When working with homozygous parental lines a single maker is sufficient to validate successful crosses.

To test the effectiveness of the hand crossing method and the validation with genetic markers, we produced hybrids between amaranth genotypes from different species and validated them with the marker system. Although not all crosses produced hybrids, for most crosses the number of hybrids produced was high and less than 10 offspring had to be evaluated per cross (**Table 4**).

4. CONCLUSIONS

Ancient and underutilized crops greatly benefit from the ongoing revolution in genomics. However, to utilize this information for the improvement of minor crops, efficient crossing methods which are the basis of breeding programs need to be established. We developed crossing methods and genetic markers for hybrid identification in amaranth and showed that these can be used for crosses within and between species. We further showed that the life cycle and plant size of amaranth can be reduced substantially when light and temperature conditions are adapted. For genetic and physiological studies a short generation time is advantageous, which is a common characteristic of model organisms for basic research. Under the conditions described here, generation times as short as those of *A. thaliana* are possible (Meyerowitz and Pruitt, 1985). Additionally, the amount of seeds can be controlled, which allows the production of large offspring populations for genetic mapping. Furthermore, amaranth has a relatively small genome (500 Mbp) with a reference sequence, and a large number of genotyped genebank accessions are available (Stetter et al., 2015; Clouse et al., 2016). Taken together, these resources and the possibility of interspecific crosses make the grain amaranth species a very suitable model organism for studying fundamental processes such as adaptation, speciation, heterosis, C_4 photosynthetic metabolism, or domestication. The ability to conduct crosses from genetically diverse material facilitates the establishment of advanced breeding programs and the selection of improved genotypes using current breeding methods such as genomic selection will improve the value of this minor crop for agricultural production.

AUTHOR CONTRIBUTIONS

KS and MS designed the experiments. LZ, AS, KK, and MB performed the crosses. MS performed the genotyping and analyzed the data. KS and MS wrote the manuscript. All authors read and approved the final manuscript.

ACKNOWLEDGMENTS

We would like to thank the Institute of Crop Science of the University of Hohenheim for providing the amaranth varieties (Baernkrafft, C6, Puerto Moutt and Pastewny) and the USDA-ARS for providing germplasm. This work was funded by an endowment of the Stifterverband der Deutschen Wissenschaft grant to KS (Project H1705701502016296). MS, LZ, AS, KK, and MB were supported by the Humboldt Reloaded project at the University of Hohenheim, Germany (BMBF 01PL11003).

REFERENCES

Brenner, D. M., Baltensperger, D. D., Kulakow, P. A., Lehmann, J. W., Myers, R. L., Slabbert, M. M., et al. (2010). "Genetic resources and breeding of *Amaranthus*," in *Plant Breeding Reviews*, Vol. 19 (New York, NY: John Wiley & Sons, Inc.), 227–285. doi: 10.1002/9780470650172.ch7

Clouse, J. W., Adhikary, D., Page, J. T., Ramaraj, T., Deyholos, M. K., Udall, J. A., et al. (2016). The amaranth genome: genome, transcriptome, and physical map assembly. *Plant Gen.* 9, 1–14. doi: 10.3835/plantgenome2015.07.0062

Dotlacil, L., and Apltauerová, M. (1978). Pollen sterility induced by ethrel and its utilization in hybridization of wheat. *Euphytica* 27, 353–360.

Duvick, D. N. (2001). Biotechnology in the 1930s: the development of hybrid maize. *Nat. Rev. Genetics* 2, 69–74. doi: 10.1038/35047587

García-Yzaguirre, A., and Carreres, R. (2008). Efficiency of different hybridization methods in single crosses of rice for pure line breeding. *Spanish J. Agricult. Res.* 6, 395–400. doi: 10.5424/sjar/2008063-332

Kauffman, C. S., and Weber, L. E. (1990). "Grain Amaranth," in *Advances in New Crops*, eds J. Janick and J. E. Simon (Portland: Timber Press), 127–139.

Kulakow, P., Hauptli, H., and Jain, S. (1985). Genetics of grain amaranths I. Mendelian analysis of six color characteristics. *J. Heredity* 76, 27–30.

Laser, K., and Lersten, N. (1972). Anatomy and cytology of microsporogenesis in cytoplasmic male sterile angiosperms. *Botan. Rev.* 38, 425–454.

Lehmann, J., Clark, R., and Frey, K. (1991). Biomass heterosis and combining ability in interspecific and intraspecific matings of grain amaranths. *Crop Sci.* 31, 1111–1116.

Maughan, P. J., Smith, S. M., Fairbanks, D. J., and Jellen, E. N. (2011). Development, characterization, and linkage mapping of single nucleotide polymorphisms in the grain amaranths (*Amaranthus* sp.). *Plant Gen. J.* 4, 92–101. doi: 10.3835/plantgenome2010.12.0027

Meyerowitz, E. M., and Pruitt, R. E. (1985). *Arabidopsis thaliana* and plant molecular genetics. *Science* 229, 1214–1218.

Moose, S. P., and Mumm, R. H. (2008). Molecular plant breeding as the foundation for 21st century crop improvement. *Plant Physiol.* 147, 969–977. doi: 10.1104/pp.108.118232

Mukasa, Y., Suzuki, T., and Honda, Y. (2007). Emasculation of Tartary buckwheat (*Fagopyrum tataricum* Gaertn.) using hot water. *Euphytica* 156, 319–326. doi: 10.1007/s10681-007-9377-7

Olsen, K. M., and Wendel, J. F. (2013). A bountiful harvest: genomic insights into crop domestication phenotypes. *Ann. Rev. Plant Biol.* 64, 47–70. doi: 10.1146/annurev-arplant-050312-120048

Otsuka, J., Yamaguchi, S., Chigira, O., and Kato, K. (2010). Application of hot water emasculation to Acacia auriculiformis for controlled pollination. *J. Forest Res.* 15, 210–216. doi: 10.1007/s10310-009-0174-0

Peters, I., and Jain, S. (1987). Genetics of grain amaranths III. Gene-cytoplasmic male sterility. *J. Heredity* 78, 251–256.

Rastogi, A., and Shukla, S. (2013). Amaranth: a new millennium crop of nutraceutical values. *Crit. Rev. Food Sci. Nutrit.* 53, 109–125. doi: 10.1080/10408398.2010.517876

Reta Alemayehu, F., Bendevis, M. A., and Jacobsen, S.-E. (2015). The potential for utilizing the seed crop Amaranth (*Amaranthus* spp.) in east Africa as an alternative crop to support food security and climate change mitigation. *J. Agron. Crop Sci.* 201, 321–329. doi: 10.1111/jac.12108

Saghai-Maroof, M. A., Soliman, K. M., Jorgensen, R. A., and Allard, R. W. (1984). Ribosomal DNA spacer-length polymorphisms in barley: mendelian inheritance, chromosomal location, and population dynamics. *Proc. Natl. Acad. Sci. U.S.A.* 81, 8014–8018.

Sauer, J. (1967). The grain amaranths and their relatives: a revised taxonomic and geographic survey. *Ann. Missouri Botan. Garden* 54, 103–137.

Stetter, M. G., Müller, T., and Schmid, K. (2015). Incomplete domestication of South American grain amaranth (*Amaranthus caudatus*) from its wild relatives. *bioRxiv*.

Veerappan, V., Kadel, K., Alexis, N., Scott, A., Kryvoruchko, I., Sinharoy, S., et al. (2014). Keel petal incision: a simple and efficient method for genetic crossing in *Medicago truncatula*. *Plant Methods* 10:11. doi: 10.1186/1746-4811-10-11

Vega-Gálvez, A., Miranda, M., Vergara, J., Uribe, E., Puente, L., and Martínez, E. A. (2010). Nutrition facts and functional potential of quinoa (*Chenopodium quinoa* willd.), an ancient Andean grain: a review. *J. Sci. Food Agricult.* 90, 2541–2547. doi: 10.1002/jsfa.4158

Conflict of Interest Statement: The authors declare that the research was conducted in the absence of any commercial or financial relationships that could be construed as a potential conflict of interest.

18

Development of a qPCR Strategy to Select Bean Genes Involved in Plant Defense Response and Regulated by the *Trichoderma velutinum–Rhizoctonia solani* Interaction

Edited by:
Jaime Prohens,
Polytechnic University of Valencia, Spain

Reviewed by:
Ana Paula Rodiño,
Misión Biológica de Galicia – CSIC, Spain
Eriston Vieira Gomes,
University of São Paulo, Brazil

***Correspondence:**
Pedro A. Casquero
pedro.casquero@unileon.es

Sara Mayo[1], Eleonora Cominelli[2], Francesca Sparvoli[2], Oscar González-López[1], Alvaro Rodríguez-González[1], Santiago Gutiérrez[3] and Pedro A. Casquero[1]*

[1] Research Group of Engineering and Sustainable Agriculture, Department of Agrarian Engineering and Sciences, Natural Resources Institute, University of León, León, Spain, [2] Institute of Agricultural Biology and Biotechnology, Consiglio Nazionale delle Ricerche, Milan, Italy, [3] Area of Microbiology, University School of Agricultural Engineers, University of León, Ponferrada, Spain

Bean production is affected by a wide diversity of fungal pathogens, among them *Rhizoctonia solani* is one of the most important. A strategy to control bean infectious diseases, mainly those caused by fungi, is based on the use of biocontrol agents (BCAs) that can reduce the negative effects of plant pathogens and also can promote positive responses in the plant. *Trichoderma* is a fungal genus that is able to induce the expression of genes involved in plant defense response and also to promote plant growth, root development and nutrient uptake. In this article, a strategy that combines *in silico* analysis and real time PCR to detect additional bean defense-related genes, regulated by the presence of *Trichoderma velutinum* and/or *R. solani* has been applied. Based in this strategy, from the 48 bean genes initially analyzed, 14 were selected, and only *WRKY33*, *CH5b* and *hGS* showed an up-regulatory response in the presence of *T. velutinum*. The other genes were or not affected (*OSM34*) or down-regulated by the presence of this fungus. *R. solani* infection resulted in a down-regulation of most of the genes analyzed, except *PR1*, *OSM34* and *CNGC2* that were not affected, and the presence of both, *T. velutinum* and *R. solani*, up-regulates *hGS* and down-regulates all the other genes analyzed, except *CH5b* which was not significantly affected. As conclusion, the strategy described in the present work has been shown to be effective to detect genes involved in plant defense, which respond to the presence of a BCA or to a pathogen and also to the presence of both. The selected genes show significant homology with previously described plant defense genes and they are expressed in bean leaves of plants treated with *T. velutinum* and/or infected with *R. solani*.

Keywords: biotic stress, systemic acquired resistance, induced systemic resistance, hypersensitive response, defense genes, biocontrol agent, *Phaseolus vulgaris*

INTRODUCTION

The common bean (*Phaseolus vulgaris* L.) is the most important food legume crop worldwide. Bean production is often affected by biotic and abiotic factors (Guerrero-González et al., 2011) by microorganisms, humidity, temperature... that are detected as signals for the activation of plant response mechanisms. This crop is affected by a wide diversity of fungal pathogens (*Sclerotinia* spp., *Fusarium* spp., *Phytium* spp., *Botrytis* spp.,...) among them *Rhizoctonia solani* JG Kühn [Teleomorph: *Thanatephorus cucumeris* (AB Frank) Donk] has a remarkable importance as responsible of important economic losses in this crop (Valenciano et al., 2006). *R. solani* is a necrotrophic pathogen responsible for the root and hypocotyl diseases. Plant infection occurs through wounds or by the direct action of the fungal mycelium, which tears the cuticle and penetrates the epidermis (Guerrero-González et al., 2011).

As a strategy to control bean infectious diseases, mainly those caused by fungi, the use of biocontrol agents (BCA) can reduce the negative effects of plant pathogens and they also can promote positive responses in the plant (Shoresh et al., 2010). The genera *Trichoderma, Gliocladium, Rhizobium, Pseudomonas,* are beneficial organisms that have shown good efficiency as BCAs against pathogenic microorganisms. *Trichoderma* (Teleomorph: *Hypocrea*) is a fungal genus that is found in the soil, and it is a secondary fast growing opportunistic invasive. In addition, *Trichoderma* biocontrol strains are able to induce the expression of genes involved in defense response and also to promote plant growth, root development, and nutrient uptake (Hermosa et al., 2012).

The relationships established between plant and micro-organisms are very diverse. When a plant is exposed to a pathogenic microorganism, the production of molecules associated to salicylic acid is increased, being this a systemic acquired resistance (SAR) response. The response of plants against non-pathogenic microorganisms is different, resulting in activation of signaling cascades that are dependent on jasmonic acid and ethylene, such as hydroperoxide lyase, peroxidase, and phenylalanine ammonia lyase, all of which belong to an induced systemic resistance (ISR) response (Druzhinina et al., 2011). Other responses result in a rapid cell death in infected tissues, then plants activate the hypersensitive response that involves the accumulation of salicylic acid, reactive oxygen species and an increased the influx of Ca^{2+} (Guerrero-González et al., 2011).

In the tripartite interaction of bean plants with the pathogen *R. solani* and a biocontrol *Trichoderma* species, several changes are produced in the plant, such as the increase in phenolic acid and lignin, accumulation of phytoalexins (Guerrero-González et al., 2011), and down- or up-regulation of defense-related genes expression (Mayo et al., 2015). Different categories of defense-related genes whose expression is modulated by biotic stresses have been described in bean plant interacting with pathogen and non-pathogenic microorganisms (Mayo et al., 2015).

Our hypothesis is that the combination of real time PCR with "*in silico*" analysis is a valid strategy to identify bean defense-related genes regulated by BCAs and/or plant pathogens. The aim is develop a systematic strategy to detect bean defense-related genes regulated by the presence of *Trichoderma velutinum* and/or *R. solani*. Finally, the procedure has been validated by the analysis of expression of the selected genes in the presence or absence of these two fungi.

MATERIALS AND METHODS

Trichoderma and *Rhizoctonia solani* Isolates and Culture Collections

Trichoderma velutinum T028, was collected from the bean traditional production area (Protected Geographical Indication, PGI), called "Alubia La Bañeza - León" (EC Reg. n.256/2010 published on March 26th, 2010, OJEU L880/17), from a High Quality variety of beans (**Figure 1**) without any genetic manipulation. It was isolated from soil plot bean in the Astorga region (León, Spain). This isolate gave percentages of inhibition greater that 60% in membrane assays and 40% in direct confrontation assays with *R. solani*, and that was able to sporulate on potato-dextrose-agar (PDA) medium.

Rhizoctonia solani R43 was isolated from bean plants of the same PGI and selected based on its high virulence. The isolated strains were stored in the collection "Pathogens and Antagonists of the Laboratory Diagnosis of Pests and Diseases" (PALDPD, University of León, León, Spain).

Isolates were inoculated on PDA (Becton Dickinson, Germany) medium and grown at 25°C in the dark for 1 week. After this incubation time *T. velutinum* T028 was exposed to light in order to induce the spore's formation.

FIGURE 1 | Bean seeds of "Canela" variety of the Protected Geographical Indication "Alubia La Bañeza – León" (Spain).

Plant Materials and Growth Conditions

Sixty bean seeds (Canela landrace, PGI "Alubia de la Bañeza - León," **Figure 1**) per treatment were germinated and cultured in presence or absence of the fungi in four conditions according to the procedure previously described by Mayo et al. (2015): (i) *T. velutinum* (T028) isolate plus *R. solani* (R43) (RT028); (ii) *T. velutinum* isolate (T028) without pathogen (C = control) (CT028); (iii) control (without *T. velutinum*) with *R. solani* (RC) and (iv) control without fungi (CC). The culture was carried out in climatic chamber and growth conditions were performed as previously described (Mayo et al., 2015). Six bean leaves from 45 day-old plants of each treatment were randomly collected and stored at −80°C until use.

RNA Extraction and Purification

The procedures for RNA extraction were performed as described previously (Reid et al., 2006). Bean leaves were lyophilized and were ground to a fine powder in liquid nitrogen using a mortar and pestle. The powder was mixed with 20 ml of extraction buffer/g of sample (extraction buffer: 0.1% SDS, 100 mM LiCl, 10 mM EDTA, 100 mM Tris-HCl, pH9) pre-warmed at 65°C, and 20 ml/g of phenol-chloroform-isoamyl alcohol 25:24:1 (Sigma-Aldrich, St. Louis, MO, USA). Then, the mixtures, in eppendorf tubes, were centrifuged at 13,000 rpm for 10 min at 4°C. The aqueous layer was transferred to a new tube. This step was repeated twice. Nucleic acids were precipitated with 1 volume of LiCl 4 M, mixed and kept overnight at 4°C. Tubes were then centrifuged at 13,000 rpm for 30 min at 4°C, and the resulting pellets were washed with ice cold ethanol 70%-DEPC, centrifuged again at 13,000 rpm for 10 min at 4°C and air dried. Finally, the pellets were dissolved in 50–200 μl H_2O-DEPC and stored at −20°C until use.

RNA concentrations and its purity were estimated from the A260/280 absorbance ratio with a NanoDrop (Thermo Scientific, Wilmington, DE, USA), considering the ideal absorbance ratio ($1.8 \leq A260/280 \leq 2.0$) and 1% agarose gel was run to visualize the integrity of the RNA.

cDNA Synthesis

Approximately 5 μg of RNA were treated with DNase using the TURBO DNAfree™ Kit (Applied Biosystems, Foster City, CA, USA), according to the manufacturer instructions. cDNA was synthesized using High-Capacity cDNA Reverse Transcription kit (Applied-Biosystems, Foster City, CA, USA) according to the manufacture's manual.

qPCR Conditions and Analysis

qPCR reactions were performed with 7300 System (Applied Biosystems, Foster City, CA, USA) using SYBR® Green. Each reaction was performed in 20 μl containing 10 μl of 2 X Power SYBR Green PCR Master Mix (Life Technologies), 0.2–0.3 μM primers and cDNA samples diluted 1:20. Each qPCR reaction was performed in triplicate. Reactions were run using the cycling parameter described previously (Reid et al., 2006) and the qPCR data were analyzed by the $2^{-\Delta\Delta Ct}$ method (Pfaffl, 2001). In order to analyze the qPCR data, *Act11* gene was used as housekeeping to determine the relative expression level of the other genes analyzed in this work (Borges et al., 2012). *T. velutinum* T028 strain was selected as reference strain in this study based on its positive effects on bean phenotype with and without *R. solani* infection (data no published). For the determination of qPCR efficiency of each primer pairs, a standard curve was performed using the following cDNA dilutions: 1:4, 1:16, 1:64; 1:256 and 1:1024. Every measurement was made in triplicate. The corresponding qPCR efficiencies (E) were calculated for every primer pair with the software 7300 System SDS software (Applied Biosystems, Foster City, CA, USA) according to the equation $E = (10^{-1/slope} - 1) \times 100$ (Rutledge and Stewart, 2008).

The significance of the differences between the gene expressions levels were compared by the Student's *t*-test using SAS (SAS Institute Inc., 2004, Cary, NC, USA).

RESULTS

Selection of Putative Bean Defense-Related Genes

Following an exhaustive and systematic analysis, summarized in the **Figure 2**, several bean genes were selected for their expression analysis in leaves from bean plants grown in interaction with *T. velutinum* and infected or not with *Rhizoctonia solani*. Thus, as result of the search in the literature, 48 genes were firstly found, showing stress and/or defense response (**Table 1**). Only those genes that resulted to be expressed in *P. vulgaris* leaves, based on transcriptomic data reported in the Phytozome database[1], were considered for qPCR expression analysis in leaves. The genes for which we confirmed expression in leaves were considered for further analyses.

As result, from the 48 genes selected for their involvement in bean stress and/or defense responses, only 19 were selected which showed a detectable level of expression in bean leaves.

The selected genes can be included in nine different groups (**Table 2**): (i) involved in the regulation of the balance between necrotrophic and biotrophic pathogen responses: *WRKY33* (WRKY transcription factor) (NM129404.3) (Bakshi and Oelmüller, 2014); (ii) pathogenesis related genes: *PR1* (pathogenesis related 1) (HO864272) (Guerrero-González et al., 2011), *PR2* (β 1-3 endoglucanase) (HO864270) (Guerrero-González et al., 2011), *PR3* (chitinase class I) (TC18606) (Pereira et al., 2014), *PR4* (pathogenesis related 4) (HO864354) (Guerrero-González et al., 2011), *PR16a* (germin.like protein 8) (CB540239) (Borges et al., 2012), *IPER* (basic peroxidase) (AF007211) (Upchurch and Ramirez, 2010), *PPO* (polyphenol oxidase) (EF158428) (Upchurch and Ramirez, 2010); (iii) related with the ethylene signaling pathway: *ERF1* (ethylene-responsive transcription factor 1) (AF076277) (Lorenzo et al., 2003), *ERF5* (ethylene-responsive transcription factor 5) (At5g47230) (Moffat et al., 2012), and *CH5b* (endochitinase precursor) (FE897014.1) (Vellicce et al., 2006); (iv) involved in phytoalexin

[1] http://phytozome.jgi.doe.gov/pz/portal.html

biosynthesis: *PAL1* (phenylalanine and histidine ammonia-lyase) (KF279696) (Kim and Hwang, 2014); (v) related in osmotin biosynthesis: *OSM34* (osmitin-like protein) (At4g11650) (Sharma et al., 2013); (vi) involved in Ca^{2+} signaling: *CNGC2* (cyclic nucleotide-gated ion channel 2) (CB542582) (Borges et al., 2012); (vii) needed for antimicrobials and oxylipins (defense signaling molecules): *HPL* (hydroperoxide lyase) (AW733791) (Upchurch and Ramirez, 2010), *Lox2* (lipoxygenase 2) (D13949) (Upchurch and Ramirez, 2010), *Lox7* (lipoxygenase 2) (Upchurch and Ramirez, 2010); (viii) *GSTa* (2,4-D inducible glutathione *S*-transferase) (HO864392) (Guerrero-González et al., 2011); and (ix) *hGS* (homoglutathione synthetase) (HO864377) both related with oxidative stress (Guerrero-González et al., 2011).

However, only 14 genes were selected to the study of the expression genes because *PR16a, IPER, PPO, Lox2,* and *Lox7,* showing negative qPCR results, were finally discarded.

Selection of a *Trichoderma* Strain to Validate the Gene Selection Strategy

Trichoderma velutinum T028 was the selected isolate, based on its positive effect on bean growth. Thus, plants inoculated with this strain showed a significant increase in dry weight of both aerial parts and root system, including when *R. solani* was present in the substrate (**Figure 3**). Thus, when bean plants were treated with *T. velutinum* T028 they increased respect to control plants (CC) 4.75% their diameter of hypocotyl, 10.75% their length of root system, 4.27 and 5.51% in dry weight of aerial parts and root system, respectively. When plants were infected with *R. solani*, the action of *T. velutinum* T028 caused an increased respect to the control plant with the pathogen (RC) of the diameter of hypocotyl in 8.76, 21.15% in the length of root system, and 11.05 and 3.43% in dry weight of aerial parts and root system respectively.

Based on these results, this isolate was used for further studies. In addition, this is the first report in which the effects of this strain on bean phenotype and plant gene regulation are studied.

Effect of *R. solani* Infection on Expression of the Selected Genes. Validation of the Procedure Used to Select Bean Genes Involved in Defense Responses (Strategy Validation I)

A significant down-regulation of expression of *PR2, PR3, PR4, ERF1, ERF5, PAL1, HPL,* and *GTSa* genes with ratios of expression ranging from 0.149 fold for *PAL1* and 0.763 fold for *PR3* was observed in bean plants grown in the presence of *R. solani* (RC) compared to control plants (CC). Conversely, expression of *PR1, OSM34, CNGC2,* and *hGS* genes was up-regulated, but with non-statistically significant differences with a ratios between 1.289 and 1.193 for *PR1* and *hGS*, respectively (**Figure 4**).

Effect of *Trichoderma* on Expression of the Selected Genes (Strategy Validation II)

Trichoderma treatment also down-regulates expression of most of the bean defense-related genes, but at a lower level than *R. solani.* Thus, when *T. velutinum* T028 was in the substrate (CT028), *PR2, PR3, PR4, ERF1, ERF5, PAL1, CNGC2, HPL,* and *GSTa*

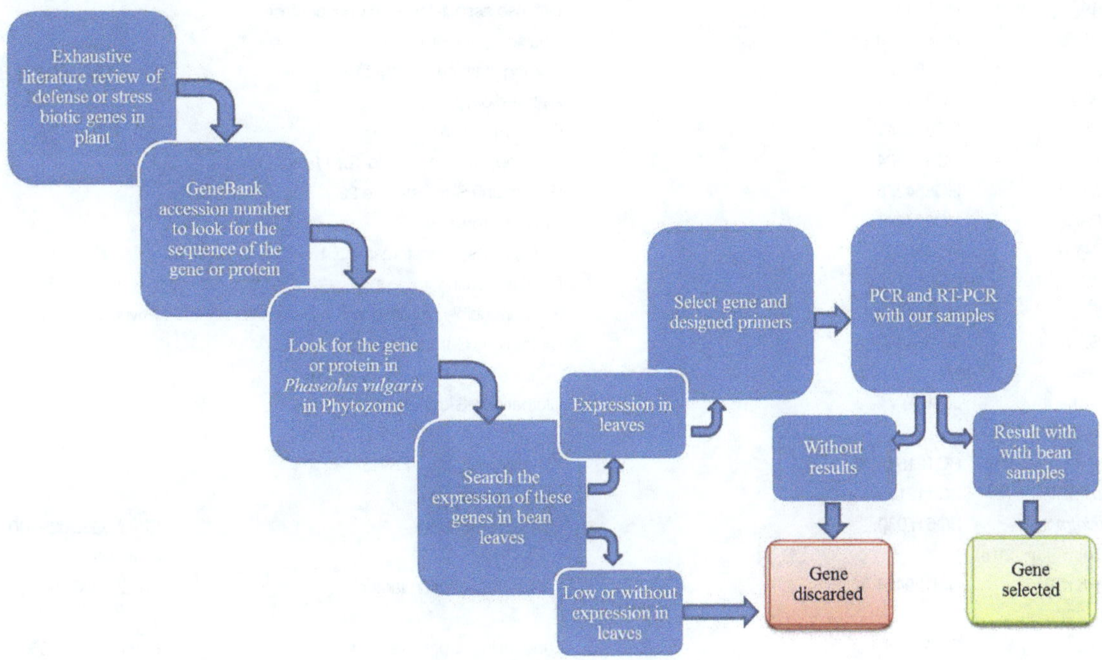

FIGURE 2 | Schematic representation of the work flow set up in the present work to select bean genes involved in plant defense.

TABLE 1 | Genes selected for stress and/or defense response and their empirical expression in *Phaseolus vulgaris* leaves.

Id	Gene	Accession number	Functional annotation	NCBI Phytozome
Pereira et al., 2014				
1	*Chit*	AY357300.2	Chitanase	
2	*Glu1*	DQ093563.1	β-1,3-glucanase	
3	*Pod3*	AF485265.1	Peroxidase	
4	*PR3*	TC18606	Chitinase class I	Phvul.009G116600
5	*Lox1*	U76687.2	Lipoxygenase	
Upchurch and Ramirez, 2010				
6	*PPO*	EF158428	Polyphenol oxidase	
7	*PR10*	AJ289155	Stress-induced ribonuclease-like protein	
8	*PR12*	BU964598	Defensin precursor	
9	*MMP2*	AY057902	Matrix metalloproteinase 2	
10	*CHS*	X53958	Chalcone synthase	
11	*AOS*	DQ288260	Allene oxide synthase	
12	*HPL*	AW733791	Hydroperoxide lyase	Phvul.005G116800
13	*LOX2*	D13949	Lipoxygenase 2	Phvul.005G156700
14	*LOX7*	U36191	Lipoxygenase 2	Phvul.005G156900
15	*IPER*	AF007211	Basic peroxidase	Phvul.009G215000
Borges et al., 2012				
16	*PR16a*	CB540239	Germin-like protein 8	Phvul.010G129900
17	*PGla*	CB542106	Polygalacturonase-inhibitor-like protein	
18	*MAPKK*	CB543156	MEK map kinase kinase	
19	*PROF*	CB543496	Profilin	
20	*CNGC2*	CB542582	Cyclin nucleotide-gated ion channel 2	Phvul.008G036200
Guerrero-González et al., 2011				
21	*PR1*	HO864272	Pathogenesis related protein 1	Phvul.003G109100
22	*PR2*	HO864270	Pathogenesis related protein 2	Phvul.003G109200
23	*PR4*	HO864354	Pathogenesis related protein 4	Phvul.006G102300
24	*PR10*	HO864271	Pathogenesis related protein 10)	
25	*LTP2*	HO864366	Lipid-transfer protein 2	
26	*SIP*	HO864290	Syringolide-induced protein B13-1-9	
27	*DAAP*	HO864358	Defense associated acid phosphatase	
28	*CHI*	HO864289	Chalcone isomerase	
29	*hGS*	HO864377	Homoglutathione synthetase	Phvul.006G094500
30	*aDO1*	HO864351	Alpha- dioxygenase 1	
31	*CPRD14*	HO864341	CPRD14 protein	
32	*OPR5*	HO864304	12-oxophytodienoic acid 10, 11-reductase	
33	*GST22*	HO864275	Glutathione S-transferase 22	
34	*CPRD8*	HO864396	CPRD8 protein	
35	*UDPGT*	HO864301	UDP-glucosyl transferase 72E1	
36	*ERD15*	HO864375	ERD15 protein	
37	*GTSa*	HO864392	2,4-D inducible glutathione S-transferase	Phvul.002G241400
38	*GST15*	HO864369	Glutathione S-transferase 15	
Gallou et al., 2009				
39	*GST1*	J03679	Gluthatione-S-transferase 1	
Lehtonen et al., 2008				
40	*TSI-1*	BQ121547	TSI-1 protein	
41	*Lip*	BQ112158	Lipase-like protein	
42	*Amintransf2*	BQ517030	Aminotransferase 2	Phvul.006G029100
Bakshi and Oelmüller, 2014				
43	*WRKY33*	NM129404.3	WRKY transcription factors	Phvul.008G090300
Vellicce et al., 2006				
44	*CH5b*	FE897014.1	Endochitinase precursor	Phvul.009G116500
Lorenzo et al., 2003				

(Continued)

TABLE 1 | Continued

Id	Gene	Accession number	Functional annotation	NCBI Phytozome
45	*ERF1*	AF076277	Ethylene-Responsive Transcription Factor 1	Phvul.007G127800
Moffat et al., 2012				
46	*ERF5*	At5g47230	Ethylene-Responsive Transcription Factor 5	Phvul.002G055700
Kim and Hwang, 2014				
47	*PAL1*	KF279696	Phenylalanine and histidine ammonia-lyase	Phvul.001G177800
Sharma et al., 2013				
48	*OSM34*	At4g11650	Osmotin-like protein	Phvul.002G155500

were significantly down-regulated with expression ratios ranging from 0.168 for *PR4* to 0.754 for *ERF1*. However, *WRKY 33*, *CH5b*, and *hGS* were up-regulated when compared with the levels of expression in control plants, with relative expression levels between 2.462 for *CH5b* and 1.576 for *hGS* (**Figure 5**). *OSM34* was slightly but not significantly up-regulated.

Effect of Interaction of *T. velutinum* and *R. solani* on Expression of the Selected Genes (Strategy Validation III)

When *T. velutinum* T028 and *R. solani* (RT028) were in the substrate, the genes *WRKY33, PR2, PR3, PR4, ERF1, ERF5, PAL1, OSM34, HPL and GSTa* were significantly down-regulated with values between 0.179 for *PAL1* and 0.631 for *WRKY33*. In the case of *PR1* and *CNGC2*, they were also down-regulated but not significantly respect to control plant (C). Conversely, *hGS* was up-regulated with a significant ratio of 1.589 respect to control plants, while *CH5b* was not significantly up-regulated with a ratio of 1.613 (**Figure 6**).

DISCUSSION

Plants have developed some defensive strategies to perceive pathogen attack and to translate this perception into an appropriate adaptive response. During attack, plants are able to enhance their resistance (induced, acquired, hypersensitive) (Lodha and Basak, 2012). Contact with pathogenic and non-pathogenic microorganisms triggers two mechanisms: (i) SAR that is usually triggered by local infections, it provides long-term systemic resistance to pathogen attack and requires the involvement of the signal molecule salicylic acid (Durrant and Dong, 2004), and (ii) ISR that is known to result from colonization of roots by certain non-pathogenic microorganisms and is dependent on components of the jasmonic acid and ethylene signaling pathways (Shoresh et al., 2010). Then, the combination of both types of induced resistance response can protect the plant against pathogens and can even result in additive level of induced protection against pathogens through both the jasmonic acid/ethylene and salicylic acid pathways (Verhagen et al., 2006).

In the present work we developed a strategy to select genes involved in bean defense response, which would belong to those pathways, but also genes that can contribute to plant defense by other mechanisms. In this sense several previous works have described genes involved in bean defense response (Guerrero-González et al., 2011; Mayo et al., 2015). However, in the present work, by a systematic approach, 48 genes were initially considered, and 14 finally selected, which match with the criteria set up in this work: (i) they showed significant homology with previously described plant defense genes, and (ii) were expressed in bean leaves of plants treated with *Trichoderma* and/or infected with *R. solani*.

The expression of *P. vulgaris* defense-related genes was analyzed in leaves, although the interaction with *Trichoderma* and/or *R. solani* is initially produced at the root level, to determine if the signals generated in roots as result of this interaction are able to systematically stimulate the bean defense along long distance from roots to the leaves. The isolate *T. velutinum* T028 was selected following a similar strategy to that previously described (Mayo et al., 2015), and based on its positive effect on bean growth. In this work, to select a *Trichoderma* isolate, the results of the *in vitro* membrane assays and direct confrontation assays against *R. solani* were analyzed. Isolate *Trichoderma* T019 was then selected, showing a percentage of inhibition higher than 40% in the membrane assays, and/or 20% in the direct confrontation assays. This isolate also showed the best positive effects on plant phenotype among all the analyzed isolates.

WRKY transcription factors have been involved in the regulation of plant defense gene expression (Rushton and Somssich, 1998; Singh et al., 2002). Thus, WRKY33 has a role in biotic stress defense, where it regulates the balance between necrotrophic and biotrophic pathogen responses (Lippok et al., 2007; Pandey and Somssich, 2009; Birkenbihl et al., 2012). Previous studies have pointed out the involvement of *Arabidopsis* WRKY transcription factors in regulating the expression of *PR* genes by direct binding (Chen et al., 2002; Kim et al., 2006). A rapid pathogen-induced *WKRY33* expression did not require salicylic acid signaling but a downregulation of this gene involved a direct activation of jasmonic acid (Bakshi and Oelmüller, 2014). In the present case, when bean plants were in contact *T. velutinum* T028 without pathogen, the *WRKY33* gene expression was significantly up-regulated while the *PR* genes expression (*PR2, PR3* and *PR4*) was significantly down-regulated compared to expression levels in plants without *Trichoderma* treatment. In the present work, when *R. solani* was added to the substrate, expression of *WRKY33* was significantly down-regulated in plants with *Trichoderma* inoculation, while *PR2, PR3* and *PR4* were down-regulated. In the study by Mayo et al. (2015),

TABLE 2 | Common bean sequences used for primer design for RT-PCR analysis.

Gene	Functional annotation	NCBI Phytozome	Forward/Reverse	Efficiency Reference
Reference genes				
Act11	Actin-11	Phvul.008G011000	TGCATACGTTGGTGATGAGG AGCCTTGGGGTTAAGAGGAG	1.084
Ukn1	Unknown	Phvul.011G023200	ATTCCCATCATGCAGCAAAG AGATCCCTCCAGGTCAATCC	0.937
Balance between necrotrophic and biotrophic pathogen responses				
WRKY33	WRKY transcription factors	Phvul.008G090300	TTTCACAGGACAGGTTCCAGC CCTTTGACAGAAATGACTGAAGGA	0.938
Pathogenesis related genes				
PR1	Pathogenesis Related 1	Phvul.003G109100	TGGTCCTAACGGAGGATCAC TGGCTTTTCCAGCTTTGAGT	1.094 Mayo et al., 2015
PR2	Beta 1-3 Endoglucanase	Phvul.003G109200	GTGAAGGACGCCGATAACAT ACTGAGTTTGGGGTCGATTG	1.048 Mayo et al., 2015
PR3	Chitinase class I	Phvul.009G116600	TGGAGTTGGTTATGGCAACAA ATTCTGATGGGATGGCAGTGT	1.034
PR4	Pathogenesis-related 4	Phvul.006G102300	CGCAGTGAGTGCATATTGCT TGTTTGTCACCCTCAAGCAC	0.922 Mayo et al., 2015
PR16a	Germin-like protein 8	Phvul.010G129900	GGCAGTCTCATGGTTATGGTTT GCATGCTCAAGTCTCAACACAT	–
IPER	Peroxidase precursor	Phvul.009G215000	GGCAAGCATTATATGGTTGAAA GATGGCAACATCCATCACTTTA	–
PPO	Polyphenol oxidase	Phvul.008G073200	GAAGACGATGATTTGCTGGTTA AAGAAACATTTTCCTTTGTGAAA	–
Ethylene signaling pathway				
ERF1	Ethylene-Responsive Transcription Factor 1	Phvul.007G127800	CGCTCTCAAGAGGAAACACTCC TGAATCAGAAGGAGGAGGGAAT	0.937
ERF5	Ethylene-Responsive Transcription Factor 5	Phvul.002G055700	GGCTCCAAGTGGATTGAGAAC TCAGAATCAGATAACTACAAAGCACAA	0.932
CH5b	Endochitinase precursor	Phvul.009G116500	CAGCCAAAGGCTTCTACACC TTGTTTCGTGAGACGTTTGC	0.883 Mayo et al., 2015
Phytoalexins biosynthesis				
PAL1	Phenylalanine and histidine ammonia-lyase	Phvul.001G177800	TGAGAGAGGAGTTGGGCACT TTCCACTCTCCAAGGCATTC	1.034
Osmotin biosynthesis				
OSM34	Osmotin-like protein	Phvul.002G155500	GAACGGAGGGTGTCACAAAATC CGTAGTGGGTCCACAAGTTCCT	0.927
Involved in Ca^{2+} signaling				
CNGC2	Cyclic nucleotide-gated ion channel 2	Phvul.008G036200	ATTCAATTTGCTTGGAGACGTT ACAGTTTTATTGAAGGCCAGGA	0.98
Antimicrobials and oxylipins (defense signaling molecules)				
HPL	Hydroperoxide lyase	Phvul.005G116800	TCAAGGCTACATTTGTATTTCCA TGGTGCACATTTCTTAGTAGCAA	0.984
Lox2	Lipoxygenase 2	Phvul.005G156700	ATGCAAGGCTAAAGAGATCCAA ATGGTGACAGGAGCTAAACACA	–
Lox7	Lipoxygenase 2	Phvul.005G156900	GAAGGCTTGACTTTCAGAGGAA AACACACGAGAAGATTCAACCA	–
Oxidative stress				
GSTa	2.4-D inducible glutathione S-transferase	Phvul.002G241400	AGGGAGTCACACTGGCTATGTT ATGTGCCATTTGCATTTTAGTG	1.013
hGS	Homoglutathione synthetase	Phvul.006G094500	GTGGCTATATGGTGCGTACAAA GAAACAAGAATGCATCTCCTCA	1.023
Amintransf2	Aminotransferase 2	Phvul.006G029100	TTCTTCCTTTTCTGCTCTTTCAA AGATGACAAGATGCAATGATTTTT	–

(–) Genes that empirically showing expression but showing negative qPCR results.

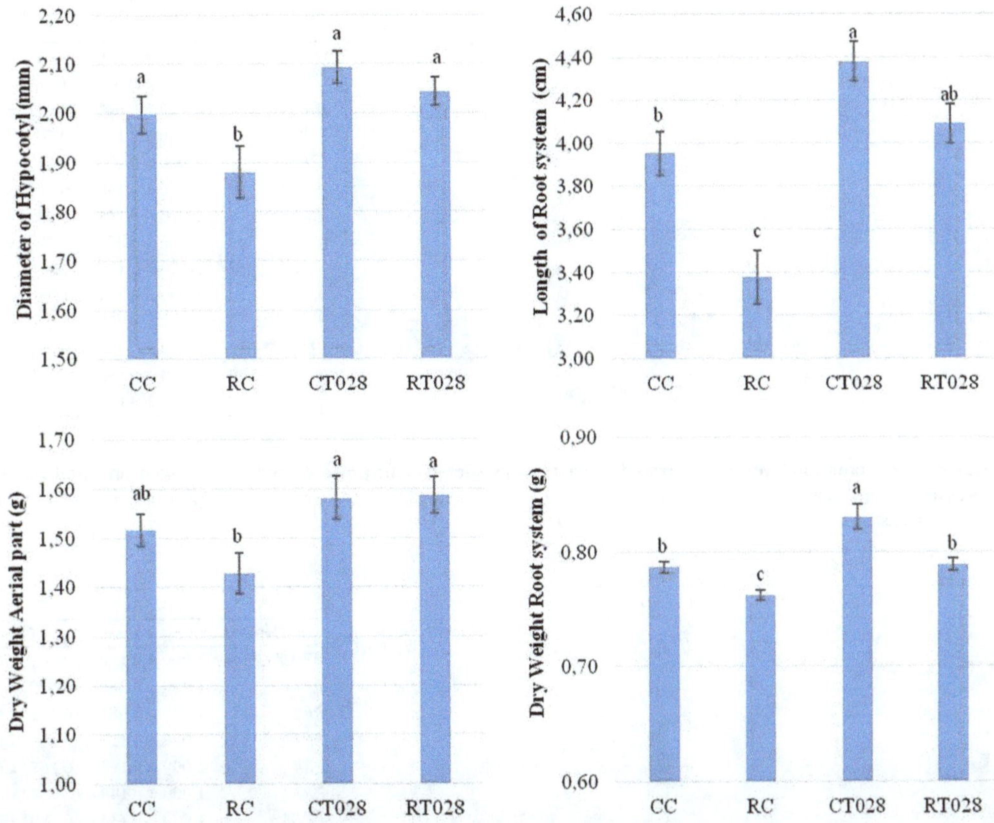

FIGURE 3 | Evaluation of the diameter of the hypocotyl (above left), length of root system (above right) dry weight of the aerial part (below left) and root system (below right) of bean plants grown during 45 days after sowing. [*Trichoderma velutinum* T028 without pathogen (CT028), *T. velutinum* T028 with *Rhizoctonia solani* (RT028), *R. solani* control (RC) and control without fungus (CC)]. Differences statistically significant respect to control plants ($p < 0.05$) are indicated with different letters.

the expression of *PR1*, *PR2*, *PR3*, and *PR4* was down-regulated when beans were inoculated with *R. solani*.

WRKY33 is also involved in the regulation of the expression of genes modulated by components of the ethylene signaling pathway. In this work, expression of the *ERF1* and *ERF5* reached similar significant values either with or without *Trichoderma* and or *R. solani* in the substrate. This result contrasts with previous reports showing that *ERF5* was up-regulated and *WRKY33* was down-regulated in *Arabidopsis* infected with *Alternaria brassicicola* (Son et al., 2012). *WRKY33* would act as a represor of *ERF1* and *ERF5* expression. Thus, when the expression of *WRKY33* is increased, expression of *ERF1* and *ERF5* is down-regulated.

CH5b encodes an endochitinase precursor and it is also related with the ethylene signaling pathway. In previous works, it has been shown that, when this gene was over-expressed the *R. solani* symptoms were reduced in crops like *Nicotiana tabacum* and *Brassica napus* (Broglie et al., 1991). However, in this study, when bean plants were in contact with *R. solani*, the expression of this gene was down-regulated but not significantly, while treatment of these infected plants with *T. velutinum* resulted in its significant up-regulation. These results are in agreement with previous data, showing that the pathogen represses its expression, and the presence of *Trichoderma* induced it (Mayo et al., 2015).

PAL plays an important role in plant defense; it is involved in the biosynthesis of salicylic acid, which is related to plant systemic resistance (Nugroho et al., 2002; Chaman et al., 2003). *PAL* gene expression is also regulated in response to pathogen infection. In this work, the presence of *T. velutinum* and *R. solani* in the soil resulted in a significant down-regulation of this gene compared with control plants.

Osmotins have plant protective effects against pathogen infection (Narasimhan et al., 2009). In this study, when *T. velutinum* or *R. solani* were present in the soil, the expression of *OSM34* was not significantly up-regulated respect to control plants, but when both fungi were in the soil at the same time, *OSM34* was slightly but significantly down-regulated.

The *CNGC* genes can be related to early plant defense responses due to changes in ion flux, including H^+ and Ca^{2+} influx and K^+ and Cl^- efflux (Atkinson et al., 1996). The up-regulation of *CNGC2* can confirm the importance of ion channels for the plant resistance response (Borges et al., 2012). In this work, this gene was up-regulated when *R. solani* was present in the soil not significant. Conversely, *CNGC2* was down-regulated

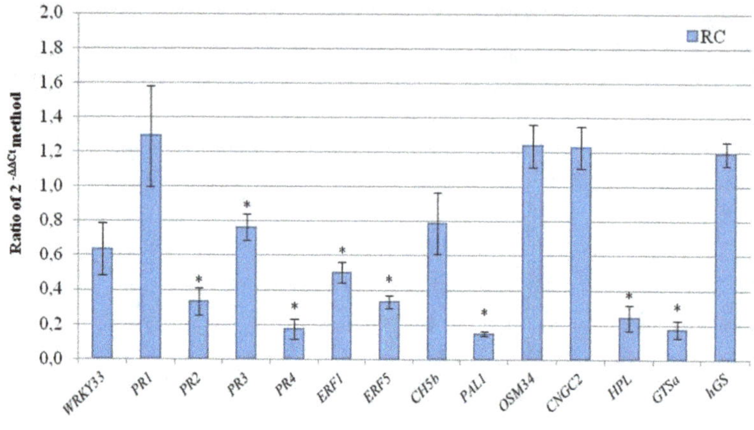

Gene	CC	RC
WRKY33	1.000 ± 0.190	0.634 ± 0.152
PR1	1.000 ± 0.266	1.289 ± 0.292
PR2	1.000 ± 0.085	0.331* ± 0.079
PR3	1.000 ± 0.127	0.763* ± 0.077
PR4	1.000 ± 0.197	0.176* ± 0.057
ERF1	1.000 ± 0.149	0.502* ± 0.061
ERF5	1.000 ± 0.134	0.333* ± 0.037
CH5b	1.000 ± 0.272	0.790 ± 0.179
PAL1	1.000 ± 0.116	0.149* ± 0.013
OSM34	1.000 ± 0.264	1.237 ± 0.125
CNGC2	1.000 ± 0.235	1.228 ± 0.122
HPL	1.000 ± 0.279	0.245* ± 0.073
GTSa	1.000 ± 0.158	0.177* ± 0.051
hGS	1.000 ± 0.115	1.193 ± 0.068

FIGURE 4 | Analysis of relative expression levels of the bean defense genes selected in the present work in bean plants infected with *R. solani versus* their levels of expression in control plants. The data were analyzed by the $2^{-\Delta\Delta Ct}$ method. The differences statistically significant respect to control plants ($p < 0.05$) are indicated with an asterisk.

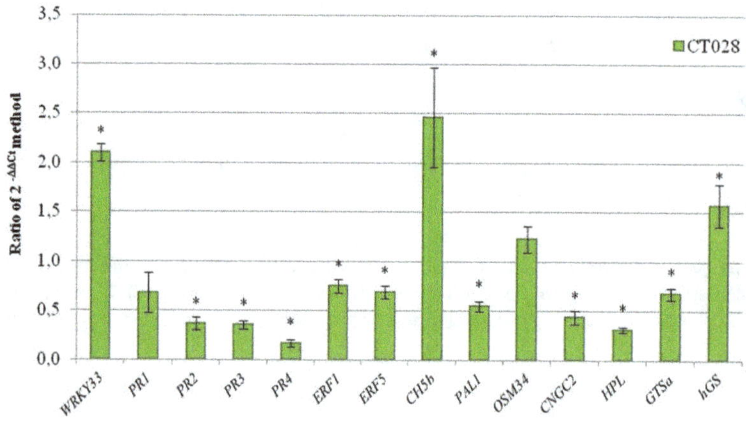

Gene	CC	CT028
WRKY33	1.000 ± 0.190	2.102* ± 0.086
PR1	1.000 ± 0.266	0.679 ± 0.201
PR2	1.000 ± 0.085	0.368* ± 0.065
PR3	1.000 ± 0.127	0.357* ± 0.041
PR4	1.000 ± 0.197	0.168* ± 0.038
ERF1	1.000 ± 0.149	0.754* ± 0.067
ERF5	1.000 ± 0.134	0.689* ± 0.064
CH5b	1.000 ± 0.272	2.462* ± 0.503
PAL1	1.000 ± 0.116	0.550* ± 0.052
OSM34	1.000 ± 0.264	1.230 ± 0.132
CNGC2	1.000 ± 0.235	0.441* ± 0.069
HPL	1.000 ± 0.279	0.316* ± 0.026
GTSa	1.000 ± 0.158	0.679* ± 0.059
hGS	1.000 ± 0.115	1.576* ± 0.215

FIGURE 5 | Analysis of relative expression levels of the bean defense genes selected in the present work in bean plants treated with *T. velutinum versus* their levels of expression in control plants. The data were analyzed as indicated in the legend to the **Figure 4**.

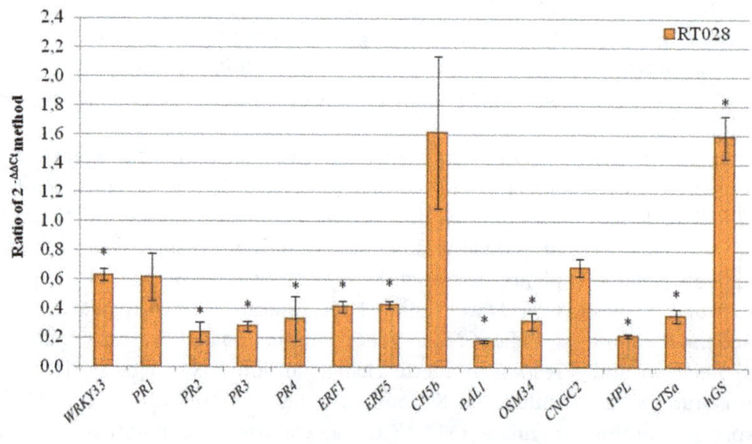

Gene	CC	RT028
WRKY33	1.000 ± 0.190	0.631* ± 0.042
PR1	1.000 ± 0.266	0.617 ± 0.164
PR2	1.000 ± 0.085	0.238* ± 0.068
PR3	1.000 ± 0.127	0.278* ± 0.037
PR4	1.000 ± 0.197	0.330* ± 0.156
ERF1	1.000 ± 0.149	0.416* ± 0.041
ERF5	1.000 ± 0.134	0.429* ± 0.026
CH5b	1.000 ± 0.272	1.613 ± 0.522
PAL1	1.000 ± 0.116	0.179* ± 0.011
OSM34	1.000 ± 0.264	0.316* ± 0.060
CNGC2	1.000 ± 0.235	0.689 ± 0.060
HPL	1.000 ± 0.279	0.219* ± 0.014
GTSa	1.000 ± 0.158	0.357* ± 0.046
hGS	1.000 ± 0.115	1.589* ± 0.147

FIGURE 6 | Analysis of relative expression levels of the bean defense genes selected in the present work in bean plants infected with *R. solani* and treated with *T. velutinum versus* their levels of expression in control plants. The data were analyzed as indicated in the legend to the **Figure 4**.

in plants treated with *T. velutinum*. Then, the pathogen would induce an activation of hypersensitive defense mechanisms.

Hydroperoxide lyase (*HPL*) is involved in the production of antimicrobial and defense signaling oxylipins (Noordermeer et al., 2001; Huang et al., 2010). In this study, the presence of *T. velutinum* and *R. solani*, resulted in a down-regulation of this gene expression respect to control plants. In previous works, when tomato plants were in contact with *Botrytis cinerea*, *HPL* expression increased 24 h after gray mold infection, but after that time the expression of this gene decreased gradually (Wan et al., 2013). In the present case, after 45 days in contact with the fungus *T. velutinum* and/or *R. solani*, its expression was down-regulated, indicating that the plant identifies *Trichoderma* and *Rhizoctonia* as two invader organisms, and some of the mechanisms activated against the presence of both are similar, independently of the final response specifically activated in the plant by each one.

GSTa (2,4-D inducible glutathione *S*-transferase) expression also responds to pathogen attack (Mauch and Dudler, 1993) and can be induced by molecules such as salicylic acid, methyl jasmonate, abscisic acid and H_2O_2 (Dixon et al., 2002; Moons, 2005). In *Gossypium arboretum*, GST provides resistance to fungal pathogens and oxidative stress (Barthelson et al., 2010). *GST* expression was up-regulated during fungal infection in barley, *Arabidopsis*, and cotton (Dowd et al., 2004; Durrant and Dong, 2004; Lu et al., 2005). However, in banana *GST* was down-regulated following *Fusarium oxysporum* f *specialis* (f. sp.) *cubense* infection (Wang et al., 2013), which is in agreement with the present case, where the expression of *GSTa* was down-regulated when *T. velutinum* and/or *R. solani* were present in the soil.

hGS encodes a homoglutathione synthetase that is involved in response to oxidative stress. There is not much information about the behavior of this gene in the plant. In the present study, when bean plants were in contact with *T. velutinum* and/or *R. solani*, expression of this gene was significantly up-regulated compared to control plants. In other studies, treatment of *Medicago truncatula* plants with compounds that release nitric oxide, a key signaling molecule in plants, induced expression of *GST* but not *hGS* in roots (Innocenti et al., 2007). Similarly, common bean plants treated with H_2O_2 showed up-regulation of *hGS* in nodules, whereas treatments with cadmium, sodium chloride, or jasmonic acid had no effect (Loscos et al., 2008).

CONCLUSION

From 48 genes initially analyzed, 14 bean genes were selected in the present work and only *WRKY33*, *CH5b* and *hGS* showed an up-regulatory response in the presence of *T. velutinum*, the other genes were or not affected (*OSM34*) or down-regulated by the presence of this fungus. *R. solani* infection resulted in a down-regulation of most of the genes analyzed, except *PR1*, *OSM34* and *CNGC2* that were not affected, and the presence of both, *T. velutinum* and *R. solani*, up-regulates *hGS* and down-regulates all the other genes analyzed, except *CH5b* which was not significantly affected.

As conclusion, the strategy described in the present work has been shown to be effective to detect genes involved in plant defense, which respond to the presence of a BCA or to a pathogen and also to the presence of both. The selected genes showed significant homology with described plant defense genes and they are expressed in bean leaves of plants treated with *T. velutinum* and/or infected with *R. solani*. The proposed strategy will be very useful in studies about the interaction of bean with pathogens and biocontrol fungi.

AUTHOR CONTRIBUTIONS

PC and SG conceived the research. SM, OG-L, and AR-G designed the research. SM, OG-L, and AR-G conducted the experiments. SM, PC, and SG analyzed data. SM, PC, SG, EC, and FS interpreted the data. SM, PC, SG, EC, and FS wrote the manuscript. All authors were agreed to be accountable for all aspects of the work in ensuring that questions related to the accuracy or integrity of any part of the work are appropriately investigated and resolved. All authors critically revised the manuscript. All authors approved the final version to be published.

ACKNOWLEDGMENTS

The grant awarded to Sara Mayo Prieto (FPU12/00677) by the Ministry of Education, Culture and Sport (Spain) according the Resolution of April 25, 2012 (BOE of 10 May), as amended by Resolution of May 18, 2012 (BOE of 22 May), and by Resolution of October 31, 2012 (BOE of 12 November) and granted in the Resolution of December 20, 2012. Ministry of Economy and Competitiveness for National project "Farnesol as self-regulatory molecule in *Trichoderma*. Tyrosol and farnesol signaling in *Trichoderma*-bean interaction" (AGL2012-40041-C02-02). Junta de Castilla y León, Consejería de Educación for the project "Effect of terpenes and physiologically related compounds produced by *Trichoderma parareesei* in the development of common bean (*Phaseolus vulgaris* L.) and in defensive responses of this plant" (LE228U14).

REFERENCES

Atkinson, M. M., Midland, S. L., Sims, J. J., and Keen, N. T. (1996). Syringolide 1 triggers Ca2 influx, K efflux, and extracellular alkalization in soybean cells carrying the disease-resistance gene Rpg4. *Plant Physiol.* 112, 297–302. doi: 10.1104/pp.112.1.297

Bakshi, M., and Oelmüller, R. (2014). WRKY transcription factors. *Plant Signal. Behav.* 9:e27700. doi: 10.4161/psb.27700

Barthelson, R. A., Qaisar, U., and Galbraith, D. W. (2010). Functional analysis of the *Gossypium arboreum* genome. *Plant Mol. Biol. Rep.* 28, 334–343. doi: 10.1007/s11105-009-0157-5

Birkenbihl, R. P., Diezel, C., and Somssich, I. E. (2012). *Arabidopsis* WRKY33 is a key transcriptional regulator of hormonal and metabolic responses toward *Botrytis cinerea* infection. *Plant Physiol.* 159, 266–285. doi: 10.1104/pp.111.192641

Borges, A., Melotto, M., Tsai, S. M., and Caldas, D. G. G. (2012). Changes in spatial and temporal gene expression during incompatible interaction between common bean and Anthracnose pathogen. *J. Plant Physiol.* 169, 1216–1220. doi: 10.1016/j.jplph.2012.04.003

Broglie, K., Chet, I., Holliday, M., Cressman, R., Biddle, P., Knowlton, S., et al. (1991). Transgenic plants with enhanced resistance to the fungal pathogen *Rhizoctonia solani*. *Science* 254, 1194–1197. doi: 10.1126/science.254.5035.1194

Chaman, M. E., Copaja, S. V., and Argandoña, V. H. (2003). Relationships between salicylic acid content, Phenylalanine Ammonia-Lyase (PAL) activity, and resistance of barley to aphid infestation. *J. Agric. Food Chem.* 51, 2227–2231. doi: 10.1021/jf020953b

Chen, W., Provart, N. J., Glazebrook, J., Katagiri, F., Chang, H. S., Eulgem, T., et al. (2002). Expression profile matrix of *Arabidopsis* transcription factor genes suggests their putative functions in response to environmental stresses. *Plant Cell* 14, 559–574. doi: 10.1105/tpc.010410

Dixon, D. P., Lapthorn, A., and Edwards, R. (2002). Plant glutathione transferases. *Genome Biol.* 3, 3004.1–3004.10. doi: 10.1186/gb-2002-3-3-reviews3004

Dowd, C., Wilson, I. W., and McFadden, H. (2004). Gene expression profile changes in cotton root and hypocotyl tissues in response to infection with *Fusarium oxysporum* f. sp. vasinfectum. *Mol. Plant-Microbe Interact.* 17, 654–667. doi: 10.1094/MPMI.2004.17.6.654

Druzhinina, I. S., Seidl-Seiboth, V., Herrera-Estrella, A., Horwitz, B. A., Kenerley, C. M., Monte, E., et al. (2011). Trichoderma: the genomics of opportunistic success. *Nat. Rev. Microbiol.* 9, 749–759. doi: 10.1038/nrmicro2637

Durrant, W. E., and Dong, X. (2004). Systemic acquired resistance. *Annu. Rev. Phytopathol.* 42, 185–209. doi: 10.1146/annurev.phyto.42.040803.140421

Gallou, A., Cranenbrouck, S., and Declerck, S. (2009). Trichoderma harzianum elicits defence response genes in roots of potato plantlets challenged by *Rhizoctonia solani*. *Eur. J. Plant Pathol.* 124, 219–230. doi: 10.1007/s10658-008-9407-x

Guerrero-González, M. L., Rodríguez-Kessler, M., Rodríguez-Guerra, R., González-Chavira, M., Simpson, J., Sanchez, F., et al. (2011). Differential expression of *Phaseolus vulgaris* genes induced during the interaction with *Rhizoctonia solani*. *Plant Cell Rep.* 30, 1465–1473. doi: 10.1007/s00299-011-1055-5

Hermosa, R., Viterbo, A., Chet, I., and Monte, E. (2012). Plant-beneficial effects of *Trichoderma* and of its genes. *Microbiology* 158, 17–25. doi: 10.1099/mic.0.052274-0

Huang, F. C., Studart Witkowski, C., and Schwab, W. (2010). Overexpression of hydroperoxide lyase gene in *Nicotiana benthamiana* using a viral vector system. *Plant Biotechnol. J.* 8, 783–795. doi: 10.1111/j.1467-7652.2010.00508.x

Innocenti, G., Pucciariello, C., Le Gleuher, M., Hopkins, J., de Stefano, M., Delledonne, M., et al. (2007). Glutathione synthesis is regulated by nitric oxide in *Medicago truncatula* roots. *Planta* 225, 1597–1602. doi: 10.1007/s00425-006-0461-3

Kim, D. S., and Hwang, B. K. (2014). An important role of the pepper phenylalanine ammonia-lyase gene (PAL1) in salicylic acid-dependent signalling of the defence response to microbial pathogens. *J. Exp. Bot.* 65, 2295–2306. doi: 10.1093/jxb/eru109

Kim, K. C., Fan, B., and Chen, Z. (2006). Pathogen-induced *Arabidopsis* WRKY7 is a transcriptional repressor and enhances plant susceptibility to *Pseudomonas syringae*. *Plant Physiol.* 142, 1180–1192. doi: 10.1104/pp.106.082487

Lehtonen, M. J., Somervuo, P., and Valkonen, J. P. T. (2008). Infection with *Rhizoctonia solani* induces defense genes and systemic resistance in potato sprouts grown without light. *Phytopathology* 98, 1190–1198. doi: 10.1094/PHYTO-98-11-1190

Lippok, B., Birkenbihl, R. P., Rivory, G., Brümmer, J., Schmelzer, E., Logemann, E., et al. (2007). Expression of AtWRKY33 encoding a pathogen-or PAMP-responsive WRKY transcription factor is regulated by a composite DNA motif containing W box elements. *Mol. Plant-Microbe Interact.* 20, 420–429. doi: 10.1094/MPMI-20-4-0420

Lodha, T. D., and Basak, J. (2012). Plant-pathogen interactions: what microarray tells about it? *Mol. Biotechnol.* 50, 87–97. doi: 10.1007/s12033-011-9418-2

Lorenzo, O., Piqueras, R., Sánchez-Serrano, J. J., and Solano, R. (2003). Ethylene response factor 1 integrates signals from ethylene and jasmonate pathways in plant defense. *Plant Cell* 15, 165–178. doi: 10.1105/tpc.007468

Loscos, J., Matamoros, M. A., and Becana, M. (2008). Ascorbate and homoglutathione metabolism in common bean nodules under stress conditions and during natural senescence. *Plant Physiol.* 146, 1282–1292. doi: 10.1104/pp.107.114066

Lu, Z.-X., Gaudet, D. A., Frick, M., Puchalski, B., Genswein, B., and Laroche, A. (2005). Identification and characterization of genes differentially expressed in the resistance reaction in wheat infected with *Tilletia tritici*, the common bunt pathogen. *BMB Rep.* 38, 420–431. doi: 10.5483/BMBRep.2005.38.4.420

Mauch, F., and Dudler, R. (1993). Differential induction of distinct glutathione-S-transferases of wheat by xenobiotics and by pathogen attack. *Plant Physiol.* 102, 1193–1201. doi: 10.1104/pp.102.4.1193

Mayo, S., Gutiérrez, S., Malmierca, M. G., Lorenzana, A., Campelo, M. P., Hermosa, R., et al. (2015). Influence of *Rhizoctonia solani* and *Trichoderma* spp. in growth of bean (*Phaseolus vulgaris* L.) and in the induction of plant defense-related genes. *Front. Plant Sci.* 6:685. doi: 10.3389/fpls.2015.00685

Moffat, C. S., Ingle, R. A., Wathugala, D. L., Saunders, N. J., Knight, H., and Knight, M. R. (2012). ERF5 and ERF6 play redundant roles as positive regulators of JA/Et-mediated defense against *Botrytis cinerea* in *Arabidopsis*. *PLoS ONE* 7:e35995. doi: 10.1371/journal.pone.0035995

Moons, A. (2005). Regulatory and functional interactions of plant growth regulators and plant glutathione S-transferases (GSTs). *Vitam. Horm.* 72, 155–202. doi: 10.1016/S0083-6729(05)72005-7

Narasimhan, M. L., Bressan, R. A., D'Urzo, M. P., Jenks, M. A., and Mengiste, T. (2009). Unexpected turns and twists in structure/function of PR-proteins that connect energy metabolism and immunity. *Adv. Bot. Res.* 51, 439–489. doi: 10.1016/S0065-2296(09)51011-7

Noordermeer, M. A., Veldink, G. A., and Vliegenthart, J. F. G. (2001). Fatty acid hydroperoxide lyase: a plant cytochrome P450 enzyme involved in wound healing and pest resistance. *Chembiochem* 2, 494–504. doi: 10.1002/1439-7633(20010803)2:7/8<494::AID-CBIC494>3.0.CO;2-1

Nugroho, L. H., Verberne, M. C., and Verpoorte, R. (2002). Activities of enzymes involved in the phenylpropanoid pathway in constitutively salicylic acid-producing tobacco plants. *Plant Physiol. Biochem.* 40, 755–760. doi: 10.1016/S0981-9428(02)01437-7

Pandey, S. P., and Somssich, I. E. (2009). The role of WRKY transcription factors in plant immunity. *Plant Physiol.* 150, 1648–1655. doi: 10.1104/pp.109.138990

Pereira, J. L., Queiroz, R. M. L., Charneau, S. O., Felix, C. R., Ricart, C. A. O., Lopes Da Silva, F., et al. (2014). Analysis of *Phaseolus vulgaris* response to its association with Trichoderma harzianum (ALL-42) in the presence or absence of the phytopathogenic fungi *Rhizoctonia solani* and *Fusarium solani*. *PLoS ONE* 9:e98234. doi: 10.1371/journal.pone.0098234

Pfaffl, M. W. (2001). A new mathematical model for relative quantification in real-time RT-PCR. *Nucleic Acids Res.* 29, e45–e45. doi: 10.1093/nar/29.9.e45

Reid, K. E., Olsson, N., Schlosser, J., Peng, F., and Lund, S. T. (2006). An optimized grapevine RNA isolation procedure and statistical determination of reference genes for real-time RT-PCR during berry development. *BMC Plant Biol.* 6:27. doi: 10.1186/1471-2229-6-27

Rushton, P. J., and Somssich, I. E. (1998). Transcriptional control of plant genes responsive to pathogens. *Curr. Opin. Plant Biol.* 1, 311–315. doi: 10.1016/1369-5266(88)80052-9

Rutledge, R. G., and Stewart, D. (2008). Critical evaluation of methods used to determine amplification efficiency refutes the exponential character of real-time PCR. *BMC Mol. Biol.* 9:96. doi: 10.1186/1471-2199-9-96

Sharma, S., Lin, W., Villamor, J. G., and Verslues, P. E. (2013). Divergent low water potential response in *Arabidopsis* thaliana accessions *Landsberg erecta* and Shahdara. *Plant. Cell Environ.* 36, 994–1008. doi: 10.1111/pce.12032

Shoresh, M., Harman, G. E., and Mastouri, F. (2010). Induced systemic resistance and plant responses to fungal biocontrol agents. *Annu. Rev. Phytopathol.* 48, 21–43. doi: 10.1146/annurev-phyto-073009-114450

Singh, K. B., Foley, R. C., and Oñate-Sánchez, L. (2002). Transcription factors in plant defense and stress responses. *Curr. Opin. Plant Biol.* 5, 430–436. doi: 10.1016/S1369-5266(02)00289-3

Son, G. H., Wan, J., Kim, H. J., Nguyen, X. C., Chung, W. S., Hong, J. C., et al. (2012). Ethylene-responsive element-binding factor 5, ERF5, is involved in chitin-induced innate immunity response. *Mol. Plant-Microbe Interact.* 25, 48–60. doi: 10.1094/MPMI-06-11-0165

Upchurch, R. G., and Ramirez, M. E. (2010). Defense-related gene expression in soybean leaves and seeds inoculated with *Cercospora kikuchii* and *Diaporthe phaseolorum* var. meridionalis. *Physiol. Mol. Plant Pathol.* 75, 64–70. doi: 10.1016/j.pmpp.2010.08.007

Valenciano, J. B., Casquero, P. A., Boto, J. A., and Marcelo, V. (2006). Evaluation of the occurrence of root rots on bean plants (*Phaseolus vulgaris*) using different sowing methods and with different techniques of pesticide application. *New Zeal. J. Crop Hortic. Sci.* 34, 291–298. doi: 10.1080/01140671.2006.9514419

Vellicce, G. R., Ricci, J. C. D., Hernández, L., and Castagnaro, A. P. (2006). Enhanced resistance to *Botrytis cinerea* mediated by the transgenic expression of the chitinase gene CH5b in strawberry. *Transgenic Res.* 15, 57–68. doi: 10.1007/s11248-005-2543-6

Verhagen, B. W. M., Van Loon, L. C., and Pieterse, C. M. J. (2006). "Induced disease resistance signaling in plants," in *Floriculture, Ornamental and Plant Biotechnology*, ed. J. A. Teixeira da Silva (Utrecht: Global Science Books), 334–343.

Wan, X.-H., Chen, S.-X., Wang, C.-Y., Zhang, R.-R., Cheng, S.-Q., Meng, H.-W., et al. (2013). Isolation, expression, and characterization of a hydroperoxide lyase gene from cucumber. *Int. J. Mol. Sci.* 14, 22082–22101. doi: 10.3390/ijms141122082

Wang, Z., Huang, S., Jia, C., Liu, J., Zhang, J., Xu, B., et al. (2013). Molecular cloning and expression of five glutathione S-transferase (GST) genes from Banana (*Musa acuminata* L. AAA group, cv. cavendish). *Plant Cell Rep.* 32, 1373–1380. doi: 10.1007/s00299-013-1449-7

Conflict of Interest Statement: The authors declare that the research was conducted in the absence of any commercial or financial relationships that could be construed as a potential conflict of interest.

The reviewer EVG declared a past co-authorship with one of the authors SG to the handling Editor, who ensured that the process met the standards of a fair and objective review.

19

Transcriptional Profiling of *Medicago truncatula* during *Erysiphe pisi* Infection

Edited by:
Antonio M. De Ron,
National Spanish Research Council,
Spain

Reviewed by:
Ana M. Gonzalez,
Misión Biológica de Galicia-Spanish National Research Council, Spain
Tom Warkentin,
University of Saskatchewan, Canada

***Correspondence:**
Miguel Curto,
Department of Plant Breeding,
Institute for Sustainable Agriculture,
Spanish National Research Council,
Apdo. 4084, E-14080 Córdoba, Spain
b72curum@uco.es

Miguel Curto [1], Franziska Krajinski [2], Armin Schlereth [2] and Diego Rubiales [1]*

[1] Department of Plant Breeding, Institute for Sustainable Agriculture, Spanish National Research Council, Córdoba, Spain,
[2] Department of Plant-Microbe Interactions, Max Planck Institute of Molecular Plant Physiology, Potsdam, Germany

Resistance to powdery mildew has been studied in a number of plant species, yet the molecular mechanisms remain largely unknown. Transcription factors (TFs) play a critical role in the plant defense response by regulating the transcriptional machinery which coordinates the expression of a large group of genes involved in plant defense. Using high-throughput quantitative real-time PCR (qPCR) technology more than 1000 *Medicago truncatula* TFs were screened in a pair of susceptible and resistant genotypes of *M. truncatula* after 4 h of *Erysiphe pisi* infection. Seventy nine TF genes, belonging to 33 families showed a significant transcriptional change in response to *E. pisi* infection. Forty eight TF genes were differentially expressed in the resistant genotypes compared to the susceptible one in response to *E. pisi* infection, including pathogenesis-related transcriptional factors, AP2/EREBP (APETALA2/ETHYLENE-RESPONSIVE ELEMENT BINDING FACTORS), WRKY (highly conserved WRKYGQK amino-acid sequence), MYB (Myeloblastoma), homeodomain (HD) and zinc finger C_2C_2 (CYS_2-CYS_2), C_2H_2, (CYS_2-HIS_2), LIM (*L*in-11, *I*sl-1, *M*ec-3) gene families, which are involved in known defense responses. Our results suggest that these TF genes are among the *E. pisi* responsive genes in resistant *M. truncatula* that may constitute a regulatory network which controls the transcriptional changes in defense genes involved in resistance to *E. pisi*.

Keywords: *Erysiphe pisi*, legumes, *Medicago truncatula*, transcription factors, qPCR

Introduction

Plants grown in the natural environment are confronted by a variety of pathogens. Remaining healthy depends on their ability to recognize pathogens and to activate defense mechanisms against them. The plant defense responses are regulated by a broad number of signaling pathways. Transcription factors (TFs) control the transfer of genetic information from DNA to RNA by activation or repression of transcription, playing important roles in plant development and defense by regulating different signaling pathways (Singh et al., 2002; Udvardi et al., 2007). Data from several plant genome projects suggest that more than five percent of the plant genome encodes TF sequences (around 2000 TFs) (Riechmann and Ratcliffe, 2000). Therefore, many biologic processes, including responses to pathogens, are controlled by multiple genes managed by TFs (Singh et al., 2002). Several analyses have shown their differential expression in plants as responses to interactions with biotic and abiotic effectors (Udvardi et al., 2007). In spite of the importance of legumes as sources of protein and oil, and in the symbiotic nitrogen fixation, less than one percent of the transcript-specific regulation roles of TFs have been characterized in legumes (Udvardi et al., 2007). Among legumes, *Medicago truncatula* is a model with key attributes such as self-fertility, rapid generation time, and a small diploid genome (Singh et al., 2007) that have facilitated the use of molecular and genetic tools (Rose, 2008).

Powdery mildews are biotrophic plant pathogens that seriously constrain crop production worldwide (Bélanger et al., 2002). *Erysiphe* spp. cause considerable losses in various important legume crops (Sillero et al., 2006). This fungus has been classified into three physiologically specialized forms, f.sp. *pisi* specialized on *Pisum*, f.sp. *medicaginis* specialized on *Medicago*, and f.sp. *vicia sativa*, specialized on *Vicia* (Falloon and Viljanen-Rollinson, 2001). Breeding for powdery mildew resistance is the most desirable strategy to control this disease by means of resistant cultivars (Fondevilla and Rubiales, 2012; Rubiales et al., 2015). Consequently, several genes involved in resistance to powdery mildew have been reported in different plant species (Fondevilla et al., 2007; Yang et al., 2013; Barilli et al., 2014; Curto et al., 2015; Iglesias-García et al., 2015). High-throughput methods have resulted in identification of genes potentially associated with specific processes and characterization of the regulatory networks that control their expression (Czechowski et al., 2004; Caldana et al., 2007). Among them, DNA microarrays have been used successfully to characterize global gene expression patterns in *M. truncatula* (Foster-Hartnett et al., 2007; Samac et al., 2011; Zhang et al., 2014; Curto et al., 2015; Song et al., 2015) providing detailed information of metabolic pathways involved in the analyzed systems. Previous studies have analyzed the *E. pisi/M. truncatula* pathosystem (Foster-Hartnett et al., 2007; Samac et al., 2011; Curto et al., 2015) using different genotypes and microarray platforms, such as Mt16kOLI1, Mt16kOLI1plus (Küster et al., 2004, 2007), and Affymetrix GeneChip® (http://www.affymetrix.com). These studies have increased the knowledge of mechanisms involved in *E. pisi* resistance in *M. truncatula*, which are agreement that a wide variety of mechanisms and pathways are involved in *E. pisi* resistance including pathogenesis-related genes (i.e., *PR10, Pprg2*), as well as other genes involved in signal transduction, cell wall metabolism (i.e., *Glucan endo-1,3-beta-D-glucosidase, Pectinase*) and abiotic stress, such as *Heat shock protein 17.7, UVB-resistance protein BudCAR5,* and *Dehydration-responsive protein (RD22)*. Although DNA microarrays have been shown to be five times less sensitive than qPCR (Czechowski et al., 2004), due to its high cost qPCR remains a technique used for low- to middle-scale studies. Several large-scale TF profiling approaches have employed the *M. truncatula* qPCR-based platform available (Kakar et al., 2008) in various studies (Verdier et al., 2008; Gao et al., 2010; Madrid et al., 2010; Villegas-Fernández et al., 2014). In this study, we screened the TF transcriptome of *M. truncatula* for altered expression during *E. pisi* infection using qPCR. Previous histological assessments showed that the resistance mechanisms carried out by the resistant genotype SA1306 is mainly related to hampering spore germination and further colony establishment by epidermal cell death as a hypersensitive response to *E. pisi* germlings that develop appressoria (Curto et al., 2015). Several mechanisms capable of monitoring changes in the plant cell wall are carried out by cellular signaling responses (Ringli, 2010; Cheung and Wu, 2011). The present study has allowed us to identify the transcription factor-encoding genes involved in the *E. pisi/M. truncatula* pathosystem, which are candidates for further functional studies. In addition, this approach provides a model for the regulatory network controlling the expression of TF genes in this pathosystem.

Materials and Methods

Plant Material, Growth Conditions, and Inoculation

The study was performed through an analysis of two genotypes of *M. truncatula*, the commercial cultivar *M. truncatula* Gaertn. v. Parabinga and the accession SA1306, shown to be susceptible and resistant, respectively to *E. pisi* f.sp. *Medicaginis* (Curto et al., 2015).

The seeds of *M. truncatula* were pre-soaked in filter paper, kept in dark conditions at 4°C for 24 h, and germinated in the dark for 48 h in a growth chamber at 65% relative humidity and 20°C. The seedlings were placed in pots (125 ml) containing a 1:1 mixture of perlite and sand substrate, fertilized with half-strength Hoagland's solution (Hoagland and Arnon, 1950) 3 times a week, and grown (25°C, 12 h photoperiod, 250 μmol/m^2 light intensity, 80% relative humidity) for 4 weeks before pathogen inoculation.

As pathogen, we used a monosporic isolate of *E. pisi* f.sp. *medicaginis* strain CO05, derived from a mildew population collected on *M. truncatula* plants at Córdoba (Prats et al., 2007), which was maintained and propagated by infecting Parabinga plants. One day before inoculation the highly infected plants were shaken to remove old conidia in order to produce an inoculum with vigorous young spores. *M. truncatula* plants of both lines were inoculated when the fourth trifoliate leaf was completely expanded (4-week-old-plant). Inoculation was carried out using a setting tower to give an inoculum density of 5 conidia mm^{-2} (Prats et al., 2007). Five plants of each *M. truncatula* genotype were inoculated per triplicate, at the same time keeping five non-infected plants as a control, in total 60 *M. truncatula* plants. Thus, three independent biological replicates, five plants per condition (control and infected) and per genotype (Parabinga and SA1306), were performed with leaflets of control and *E. pisi* infected plants of both *M. truncatula* genotypes that were harvested 4 h after *E. pisi* inoculation. The samples were immediately washed with water, blot dried with filter paper, frozen in liquid nitrogen, and stored at −80°C until RNA extraction.

RNA Extraction, cDNA Synthesis, and qPCR Assays

RNA was purified from collected samples using the Nucleospin RNA II kit (MACHEREY-NAGEL, Bethlehem, PA) following the manufacturer's procedure. The integrity of total RNA was assessed on 1% agarose gels (samples were denatured in formaldehyde/formamide buffer), as well as for quantity and purity by using a NanoDrop ND-100 spectrophotometer (NanoDrop Technologies, Wilmington, DE) to measure the optical density. RNA samples were digested with RNase-free DNasel (Ambion Inc., Houston, TX), according to the manufacturer's protocol. The absence of genomic DNA was checked by PCR analysis using primers designed on the *M. truncatula* ubiquitin gene intron sequence (Kakar et al., 2008).

Synthesis of first-strand cDNA was carried out with oligo-dT12–18 (Qiagen, Hilden, Germany) using SuperScript III reverse transcriptase (Invitrogen GmbH, Karlsruhe, Germany). The efficiency of cDNA synthesis was evaluated by real-time quantitative PCR (qPCR) amplification of 5′ and 3′ regions of two reference genes, GAPDH (*Glyceraldehyde 3 phosphate dehydrogenase*) and Ubiquitin (Kakar et al., 2008). A single peak in the dissociation curve at the end of the PCR reaction allowed confirmation of the specificity of the amplified products.

A *M. truncatula* transcription factor platform composed of more than 1000 *M. truncatula* TFs gene-specific primers was used to carry out the qPCR experiments (Kakar et al., 2008). The qPCR reactions were carried out in triplicate in an optical 384-well plate with an ABI PRISM® 7900 HT Sequence Detection System (Applied Biosystems, Foster City, CA) as described previously (Kakar et al., 2008).

Normalization and Data Analysis

SDS software ver. 2.3 (Applied Biosystems) was used to analyze fluorescent signals and calculate the quantification cycle (Cq) (Bustin et al., 2009). The baseline data were collected from the fluorescence signal between cycles 3 and 15, and used to correct the fluorescence signal of the samples. The PCR efficiencies (E) and correlation coefficients (R^2) from linear regression analysis were calculated for each performed PCR reaction by the software LinRegPCR ver. 7.5 as described previously (Kakar et al., 2008) (**Table S1**). The amplification reactions with $R^2 < 0.99$ that showed efficiencies lower than 1.8 were excluded for further analysis (24.4% of reactions). TF genes were considered detected if they were expressed in at least two biological replicates with a $Cq < 40$.

Eight reference genes encoding Pentatricopeptide repeat protein (PPRrep; TC96273), Protein phosphatase 2A subunit A3 (PDF 2; TC107161), Polypyrimidine tract-binding protein homolog (PTB; TC111751), Helicase (CB892427), Ubiquitin (TC102473), Ubiquitin-protein ligase 7 (UPL7; TC111218), Ubiquitin-conjugating enzyme E2 (UBC; AW686873), and Ubiquitin-conjugating enzyme E2 9 (UBC9; TC106312) (Kakar et al., 2008) were studied in order to determine the best suited reference genes for transcript normalization. The expression stability of the eight reference genes was analyzed by the geNorm software (Vandesompele et al., 2002; Hellemans et al., 2007) for each cDNA sample under study. In addition, pair-wise comparison analysis allowed determining the optimal number of reference genes in this assay (Vandesompele et al., 2002).

Expression values were calculated from E^{Cq} of each individual plot. To normalize the gene expression of each PCR reaction, ratios of the geometric mean of the selected reference genes to the different biological conditions were used using the Equation (1):

$$\left(\frac{E_{ref}^{(Cq\ ref)}}{E_{gene}^{(Cq\ gene)}}\right) \quad (1)$$

The relative induction/repression of TFs from *E. pisi* infected samples compared to untreated samples was calculated using the Equation (2):

$$\left(\frac{\left(\frac{E_{ref}^{(Cq\ ref)}}{E_{gene}^{(Cq\ gene)}}\right)\text{infected condition}}{\left(\frac{E_{ref}^{(Cq\ ref)}}{E_{gene}^{(Cq\ gene)}}\right)\text{untreated condition}}\right) \quad (2)$$

Non-parametric Levene's test and Spearman's correlation coefficient were used to verify the equality of variances in the samples and to study the similarity between TF gene expression profiles, respectively. TF genes showing statistically significant differences ($P < 0.05$) were clustered using a hierarchical cluster analysis by complete linkage. A model for the regulatory network controlling the expression of regulated genes induced by *E. pisi* in both *M. truncatula* genotypes studied was built using NodeXL (http://nodexl.codeplex.com).

Results

Evaluation of Resistance in *Medicago truncatula* Genotypes

Differences in the response to *E. pisi* between the two genotypes were not yet visible at the time the leaves were sampled for RNA extraction. Powdery mildew infection was macroscopically visible on remaining leaflets 2 weeks after inoculation, with profuse sporulation in the susceptible Parabinga genotype and absence of symptoms in the resistant SA1306 genotype (**Figure S1**). Thus, previous study described that colony formation was much higher in Parabinga than in SA1306, as well as the hypersensitive response associated with epidermal cell death was negligible in Parabinga, but marked in SA1306 (Curto et al., 2015). Former studies unveiled that at early *E. pisi* infection times, such as 4 h (Curto et al., 2015) and 12 h (Samac et al., 2011), *M. truncatula* plants induce a highly number of metabolic pathways in response to *E. pisi* infection. Hence, we choose to analyze the TF transcriptome of *M. truncatula* during *E. pisi* infection at 4 h after pathogen infection.

Selection of Reference Genes

Eight reference genes were studied to determine those best suited for transcript normalization. Transcripts levels of all reference genes were calculated, in each cDNA sample, using the average expression stability (M) calculated by geNorm software (**Figure 1A**). All reference genes showed high average expression stability ($M < 0.66$) among them the *UBC9*, *Helicase*, *PTB*, and *UPL7* reference genes showed the lower average expression stability (M) indicating a greater transcript stability (**Figure 1A**). Pair-wise variation (V) was also calculated as described by Vandesompele et al. (2002) allowing determining the optimal number of stable reference genes. The results indicated that the inclusion of a third gene ($V_{3/4}$) or more genes ($V_{4/5}$, $V_{5/6}$, $V_{6/7}$, and $V_{7/8}$) has no significant effect (**Figure 1B**). Therefore, we selected *UBC9* and *Helicase* as the best reference genes for this experiment, which were used for transcript normalization of the analyzed TF genes.

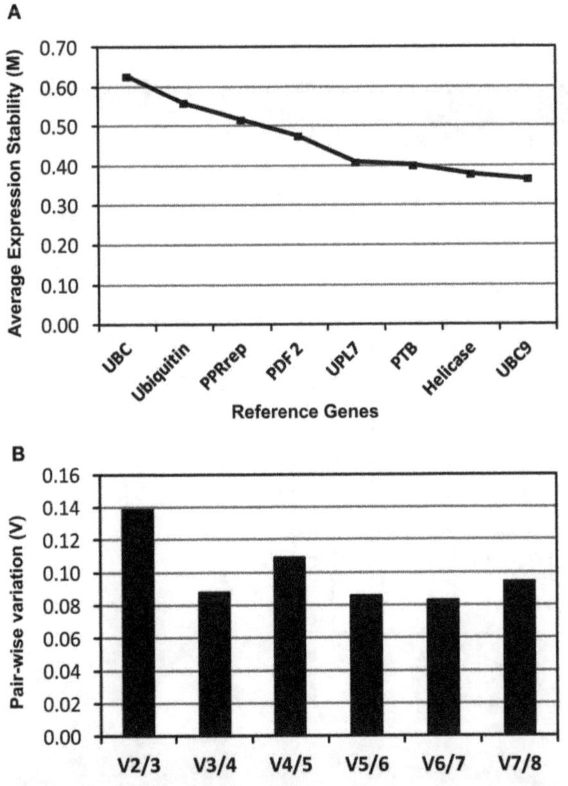

FIGURE 1 | Evaluation of candidate reference genes analyzed using geNorm software. Expression stability **(A)** and pair-wise variation **(B)** plots for the eight reference genes studied. A lower M value indicates a more stable expression. The pair-wise variation (V) values indicate the optimal number of reference genes.

Expression Patterns of TF Genes in *M. truncatula* Following *E. pisi* Infection

We analyzed and compared the expression patterns of TF genes in the susceptible cv. Parabinga and the resistant SA1306 genotypes at 4 h after *E. pisi* inoculation (**Figure S1**). A total of 623 genes of the qPCR TF platform (59.6%) were considered detected ($Cq < 40$; $n \geq 2$) and 95 showed statistically significant differences ($P < 0.05$) upon *E. pisi* infection in SA1306 and Parabinga genotypes. The relative gene expression ratios (m), $\log_2$expression ratios inoculated/control after *E. pisi* inoculation, were calculated for all TF genes. TF genes were considered to be differentially up- or down-regulated in response to *E. pisi* infection, if they met the prerequisites $p \leq 0.05$ and $m \geq 0.7$ or $m \leq -0.7$, respectively. We studied the expression pattern of these genes that showed statistically significant differences through a hierarchical clustering analysis based on gene expression profiles (**Figure 2, Table 1**).

Transcription Factor Regulatory Network Induced by *E. pisi* Infection in *M. truncatula*

The qPCR platform allowed identification of TF genes differentially expressed in the two *M. truncatula* genotypes in response to *E. pisi* infection. Genes regulated in response to *E. pisi* infection ($p \leq 0.05$ and $m \geq 0.7$ or $m \leq -0.7$) were clustered into 10 groups with different expression patterns (**Figure 3**). The first group (GI) includes 15 genes that were down-regulated in SA1306. In Parabinga eight of the genes were differentially expressed, one was up-regulated and seven TFs were repressed. These genes belong to 12 TF families, including Zn-Finger members (C_2H_2 (CYS_2-HIS_2), TTF-type (THYROID TRANSCRIPTION FACTOR), LIM (*Lin*-11, *Isl*-1, *Mec*-3), AP2/EREBP (APETALA2/ETHYLENE-RESPONSIVE ELEMENT BINDING FACTORS), AUX/IAA (AUXIN/INDOLE ACETIC ACID), bHLH (BASIC-HELIX-LOOP-HELIX), BTB/POZ (BROAD COMPLEX, TRAMTRACK, BRIC-A-BRAC/POX VIRUS, AND ZINC FINGER), E2F (E2 FACTOR), HMG (HIGH-MOBILITY GROUP), MYB/HD-like (MYELOBLASTOSIS/ HOMEODOMAIN-LIKE), NAC (NAM/ATAF/CUC), and SBP (SQUAMOSA-PROMOTER BINDING PROTEIN) (**Table 1**). The second group (GII) includes 13 genes which were down-regulated in Parabinga, except for one gene that was not differentially expressed. In the resistant SA1306 genotype, three were up-regulated and other three repressed belong to Zn-finger families, such as C_2C_2 (CYS_2-CYS_2) and C_3H-type I (CYS_3-HIS_1) family. The remaining genes belong to AP2/EREBP, DDT (DNA BINDING HOMEOBOX, AND DIFFERENT TRANSCRIPTION FACTOR), HD, HD-like, HSF (HEAT STRESS TRANSCRIPTION FACTOR), NAC, RR (RESPONSE REGULATOR RECEIVER), and WRKY TF families.

Groups III, IV, and V include genes that were induced in both genotypes. Group III contains six genes that showed lower transcription levels in SA1306 than in Parabinga. These genes belong to AP2/EREBP, HD, HD-like, MYB, and MYB/HD-like TF families. Group IV includes nine genes that showed stronger transcription activation in the resistant SA1306 genotype. These genes belong to bHLH, C_2H_2 (Zn), HD, HD-like, MADS (MADS box), MYB, and PHD (PLANT HOMEODOMAIN MOTIF) TF families. Group V contains 10 genes that showed similar up-regulation expression patterns in both susceptible and resistant genotypes. Genes of this fifth group encode proteins belonging to AP2/EREBP, ARF (AUXIN-RESPONSE FACTOR), bHLH, GRAS (GAI, RGA, SCR), HD, HTH (HELIX-TURN-HELIX), NAC, RR, and DHHC (ASP-HIS-HIS-CYS) (Zn) TF families.

Genes clustered in groups VI and VII were mainly not expressed differentially in either genotype. None of the genes included in group VI were regulated whereas seven of the 11 genes of group VII were regulated. Three genes were induced in both genotypes and they are members of HD-like, PHD, and bZIP (BASIC LEUCINE ZIPPER) TF families, respectively. The remaining four regulated genes of group VII were specifically up-regulated in Parabinga and SA1306, which were included in HD, bZIP, and MYB/HD-like TF families.

Group VIII is comprised of six genes which were induced in Parabinga and not differentially expressed in SA1306. These genes belong to bHLH, FHA (FORKHEAD-ASSOCIATED), MYB/HD-like and three Zn-finger TF families (C_2C_2, CCHC (CYS-X8-CYS-X5-CYS-X3-HIS) and U1-type). The eight genes included in group IX were induced in SA1306, whereas in Parabinga half of them were up-regulated and the remaining genes were not differentially expressed. Genes of group IX are members of ARID (AT-RICH INTERACTION DOMAIN), C_2H_2

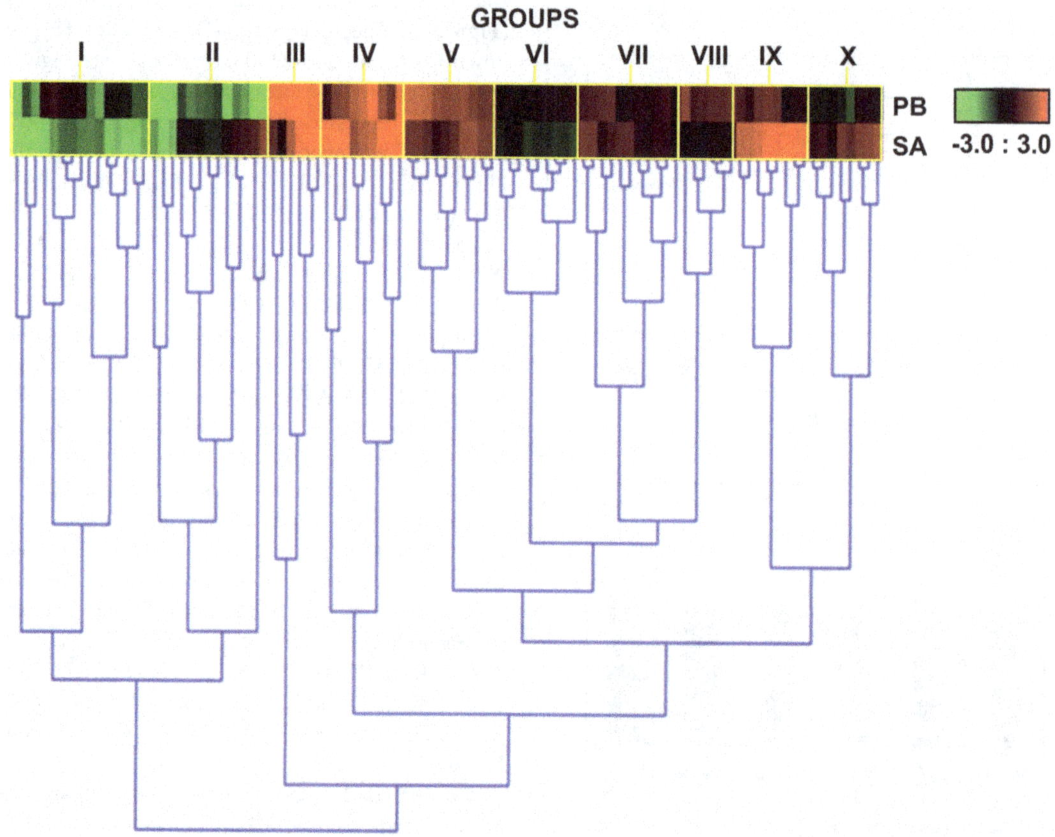

FIGURE 2 | Heat map expression profiles of TF genes. Heat map showing expression profiles of 95 genes that were differentially expressed in Parabinga (PB) and SA1306 (SA) *M. truncatula* genotypes in response to *E. Pisi* infection. Genes were considered differentially expressed if they met the prerequisites $p \leq 0.05$ and $m \leq -0.7$ or $m \geq 0.7$. Up-regulation ($m \geq 0.7$) is indicated in red; down-regulation ($m \leq -0.7$) in green; black indicates no differential expression ($-0.7 \leq m \leq 0.7$). The heat map expression profiles are grouped by yellow rectangles (I–X). Additional information is available in **Table 1**.

(Zn), HD, HD-like, NAC, and SBP TF families. Finally, group X includes eight genes which were mostly up-regulated in SA1306 and not differentially expressed in Parabinga. All genes belonging to this group were induced in the resistant genotype, except for two genes that were not differentially expressed. Only one repressed gene was detected in Parabinga. The genes of this last cluster belong to Zn finger families C_2C_2, CCHC, HD, HD-like, JUMONJI (JmjC domain), MYB/HD-like, and SBP TF families.

Around 80% of TF genes (79/95 genes) that showed statistically significant differences ($P < 0.05$) had at least a 1.6-fold change in transcript accumulation ($-0.7 \geq m \geq 0.7$) (**Table S2**). To study the regulatory network controlling the expression and interactions of these 79 genes during *E. pisi* infection, we further analyzed their expression in the susceptible Parabinga and the resistant SA1306 genotypes (**Figure 4**).

Our analysis revealed that 16 and 18 of the 79 TF genes were specifically regulated in Parabinga and in SA1306, respectively. The remaining 45 genes were regulated in both genotypes (**Figure 4**, **Table S2**). In the susceptible Parabinga genotype 10 of the 16 specifically regulated genes were induced and the remaining six genes were repressed. Most of the genes specifically induced in Parabinga are members of MYB/HD-like, HD, C_2C_2, CCHC, RR, FHA, bHLH, and U1-type Zn finger families. Meanwhile, the six genes specifically down-regulated in Parabinga are included in RR, AP2/EREBP, NAC, HD-family, C_3H- type 1 (Zn), and DDT TF families. Moreover, the resistant SA1306 genotype showed 11 and seven genes specifically up- and down-regulated, respectively. Genes specifically induced in SA1306 are members of HD/HD-like, MYB/HD-like, SBP, C_2C_2 (Zn), C_2H_2 (Zn), bZIP, ARID, and JUMONJI TF families. The genes down-regulated in SA1306 belong to LIM, C_2H_2 (Zn), TTF-type (Zn), MYB/HD-like, bHLH, E2F, and AUX/IAA TF families.

On the other hand, the genes regulated in both genotypes were mainly up-regulated; 30 induced and 10 repressed genes. Most of these 30 induced genes are in the HD-like, HD, and MYB TF families (**Figure 4**, **Table S2**). Moreover, the 10 repressed genes are included mainly in Zn finger families, C_2C_2 and C_2H_2. Interestingly, four of five common regulated genes were induced in SA1306 and repressed in Parabinga, and belong to the HD-like (TF626), CCHC (Zn) (TF660), WRKY (TF913), and AP2/EREBP

TABLE 1 | Details of regulated TF genes clustered in groups among *Medicago truncatula* genotypes analyzed, SA1306 (SA), and Parabinga (PB), in response to *Erysiphe pisi* infection.

ID[a]	TC[b]	TF family; subfamily[c]	Relative gene expression ratios (m)[d]		F[e]	Group[f]
			PB	SA		
TF700	CX528154	BTB/POZ	−2.31*	−3.55*	−1.24	I
TF479	TC110178	NAC	−1.54*	−1.12*	0.42	I
TF823	–	C2H2 (Zn)	−1.46*	−3.36*	−1.89	I
TF436	TC98230	SBP	−1.16*	−1.24*	−0.08	I
TF855	–	AP2/EREBP	−0.87*	−1.49*	−0.62	I
TF303	–	AP2/EREBP	−0.86*	−1.83*	−0.97	I
TF485	TC109384	bHLH	−0.83*	−3.51*	−2.68	I
TF425	–	bHLH	−0.36	−1.98*	−1.62	I
TF101	TC103429	C2H2 (Zn)	−0.28	−1.62*	−1.34	I
TF8	TC109476	LIM	−0.28	−1.75*	−1.47	I
TF388	BG448434	TTF-type (Zn)	0.61	−2.13*	−2.74	I
TF244	TC108091	MYB/HD-like	0.54	−1.04*	−1.58	I
TF780	–	AUX/IAA	0.49	−1.21*	−1.70	I
TF691	BG450549	E2F	0.49	−1.00*	−1.49	I
TF233	–	HMG	0.72*	−1.43*	−2.15	I
TF962	–	AP2/EREBP	−5.08*	1.56*	6.64	II
TF626	–	HD-like	−3.62*	0.91*	4.52	II
TF913	TC97332	WRKY family; WRKY	−1.48*	0.77*	2.25	II
TF393	TC95605	C2C2 (Zn); DOF	−2.65*	−2.19*	0.45	II
TF265	–	C2C2 (Zn); DOF	−4.31*	−1.42*	2.89	II
TF716	AL382911	C2C2 (Zn); GATA	−2.46*	−1.60*	0.86	II
TF767	–	C3H; C3H-type 1(Zn)	−2.42*	0.44	2.86	II
TF822	–	DDT	−1.24*	0.69	1.92	II
TF448	TC97611	RR	−1.23*	0.06	1.29	II
TF546	AL366881	AP2/EREBP	−0.71*	0.26	0.97	II
TF618	–	HD family; HD	−1.03*	−0.27	0.76	II
TF598	TC95256	NAC	−0.92*	−0.47	0.45	II
TF631	TC96049	HSF	−0.69	−0.47	0.22	II
TF837	–	MYB	5.96*	1.87*	−4.09	III
TF996	–	HD-like	4.24*	3.23*	−1.01	III
TF129	TC102127	MYB/HD-like	4.45*	2.37*	−2.08	III
TF549	TC107542	HD family; HD-ZIP	3.89*	3.02*	−0.87	III
TF3	–	AP2/EREBP	3.11*	1.02*	−2.08	III
TF934	TC109302	MYB/HD-like	4.22*	0.39	−3.82	III
TF428	–	C2H2 (Zn)	2.88*	2.17*	−0.71	IV
TF879	CB066652	PHD	2.42*	3.53*	1.11	IV
TF333	–	bHLH	2.18*	1.99*	−0.19	IV
TF322	–	MADS	1.98*	1.70*	−0.28	IV
TF63	–	MYB	1.76*	2.80*	1.04	IV
TF270	–	C2H2 (Zn)	1.59*	4.49*	2.90	IV

(Continued)

TABLE 1 | Continued

ID[a]	TC[b]	TF family; subfamily[c]	Relative gene expression ratios (m)[d]		F[e]	Group[f]
			PB	SA		
TF258	–	HD family; HD	1.41*	5.26*	3.85	IV
TF563	–	MADS	1.25*	3.13*	1.88	IV
TF230	TC109855	HD-like	0.71*	5.66*	4.95	IV
TF600	TC103296	bHLH	1.44*	1.72*	0.29	V
TF87	TC96308	ZF DHHC	1.20*	1.40*	0.20	V
TF639	TC100932	bHLH	1.32*	1.40*	0.08	V
TF296	–	GRAS	1.68*	1.66*	−0.02	V
TF565	–	HTH; FIS	1.57*	1.09*	−0.48	V
TF386	TC110943	ARF	1.55*	0.79*	−0.75	V
TF473	TC107897	AP2/EREBP	2.15*	1.22*	−0.93	V
TF136	TC96243	NAC	2.17*	1.21*	−0.96	V
TF497	BF636434	HD family; HD	2.08*	1.01*	−1.07	V
TF276	BE249457	RR	1.71*	0.65	−1.06	V
TF364	TC96319	CCAAT; CCAAT-HAP3	−0.20	−0.15	0.05	VI
TF196	TC106782	EIL	−0.05	−0.12	−0.07	VI
TF351	TC103599	bHLH	−0.21	−0.42	−0.21	VI
TF543	TC101251	BD	0.25	−0.44	−0.70	VI
TF449	–	CCHC (Zn)	0.21	−0.59	−0.80	VI
TF438	–	PHD	0.38	−0.43	−0.81	VI
TF899	–	TCP	0.39	−0.51	−0.90	VI
TF441	–	JUMONJI	0.41	−0.54	−0.95	VI
TF959	–	C2H2 (Zn)	0.60	−0.67	−1.28	VI
TF537	–	bZIP	1.09*	0.80*	−0.29	VII
TF429	TC111833	MYB/HD-like	0.99*	0.59	−0.40	VII
TF140	TC107912	PHD	0.86*	1.22*	0.37	VII
TF97	CX534602	HD-like	0.77*	1.28*	0.51	VII
TF588	TC106806	HD family; HD	0.73*	0.63	−0.10	VII
TF523	TC102139	MYB/HD-like	0.63	0.67	0.04	VII
TF199	–	NAC	0.58	0.55	−0.03	VII
TF464	TC109097	bZIP	0.48	1.05*	0.57	VII
TF437	–	CCHC (Zn)	0.47	0.44	−0.03	VII
TF816	TC96871	MYB/HD-like	0.25	0.29	0.04	VII
TF372	TC96859	HD family; HD	0.15	0.97*	0.82	VII
TF389	TC104194	C2C2 (Zn); DOF	1.59*	−0.06	−1.66	VIII
TF1007	–	bHLH	1.11*	0.32	−0.79	VIII
TF349	TC98775	CCHC (Zn)	1.00*	0.30	−0.70	VIII
TF179	CX533076	U1-type (Zn)	0.99*	0.32	−0.67	VIII
TF481	TC112164	FHA	0.88*	−0.03	−0.91	VIII
TF143	–	MYB/HD-like	0.80*	0.21	−0.59	VIII
TF797	–	HD-like	1.26*	2.47*	1.21	IX
TF200	TC109833	NAC	1.10*	2.23*	1.13	IX
TF814	BG451025	HD-like	1.11*	2.45*	1.34	IX
TF552	TC107542	HD family; HD-ZIP	0.87*	2.09*	1.22	IX
TF666	CX541503	HD-like	0.65	2.13*	1.48	IX
TF982	TC96831	C2H2 (Zn)	0.52	2.77*	2.25	IX

(Continued)

TABLE 1 | Continued

ID[a]	TC[b]	TF family; subfamily[c]	Relative gene expression ratios (m)[d]		F[e]	Group[f]
			PB	SA		
TF901	–	SBP	0.48	2.71*	2.23	IX
TF27	–	ARID	0.05	3.00*	2.95	IX
TF1009	–	JUMONJI	0.08	1.73*	1.65	X
TF726	AL375449	MYB/HD-like	0.45	1.52*	1.07	X
TF158	–	HD-like	0.44	1.43*	0.99	X
TF540	–	SBP	−0.38	1.31*	1.69	X
TF660	–	CCHC (Zn)	−0.83*	1.04*	1.88	X
TF308	–	C2C2 (Zn); DOF	−0.33	0.75*	1.08	X
TF295	–	HD family; HD	−0.35	0.65	1.00	X
TF562	TC98196	HD family; HD	−0.08	0.48	0.55	X

Relative gene expression ratios (m) are listed and sorted by cluster (Group).

[a] *TF gene identification number. Additional information is given in* ***Table S1***.

[b] *Identifier in the TIGR M. truncatula Gene index (MtGI 7).*

[c] *TF families; sub-families are showed as described (Kakar et al., 2008).*

[d] *Relative gene expression ratios values (m) were calculated for Parabinga (PB) and SA1306 (SA) genotypes,* *using the following equation:* $\log_2$ *differential expression ratio* $(\frac{\text{inoculated}}{\text{control}})$.

[e] *Fold change expression ratio (F) in SA1306 compared to Parabinga were calculated using the equation:* $\log_2$ *expression ratio* $(\frac{\text{SA1306}}{\text{Parabinga}})$.

[f] *Heat map expression profiles clustered. Additional information is given in* ***Figures 2, 3***.

* *Genes were considered differentially expressed in response to E. pisi infection if to meet the prerequisites* $p \leq 0.05$ *and* $m \leq -0.7$ *or* $m \geq 0.7$.

(TF962) TF families. The fifth gene, TF233 (HMG), was up-regulated in Parabinga and repressed in SA1306.

Interestingly, 48 genes were differentially expressed in SA1306 compared to Parabinga ($P < 0.05$; $-0.7 > F > 0.7$) in response to *E. pisi* infection, and they belong to 25 TF families (**Figure 5**, **Table S3**). Among them, the most represented TF families are HD-like, C_2H_2 (Zn), AP2/EREBP, MYB/HD-like, and HD/HD, which comprise approximately half of the differentially expressed genes. The bHLH, C_2C_2/DOF (CYS_2-CYS_2/DNA-BINDING ONE ZINC FINGER) (Zn), HD-ZIP (HD-ZINC-REGULATED TRANSPORTER), MYB, NAC, and SBP TF families were also well represented (≈24%) and the rest of TF families were poorly represented (**Figure 5**). Eleven of 25 families include genes with greater transcript accumulation ($F > 0.7$) in SA1306 compared to Parabinga (**Figure 5**). Among them, the genes belonging to the ARID, WRKY family/WRKY, C_2C_2 (Zn)/DOF, SBP, HD-like, MADS, and CCHC (Zn) families were up-regulated around two fold in SA1306 compared to Parabinga. Meanwhile only three families, TTF-type (Zn), bHLH, and HMG, include genes that were down-regulated ($F \leq -2$) in SA1306 compared to Parabinga (**Figure 5**). Generally, families HD-ZIP and NAC include genes that were induced in both genotypes. However, the AP2/ERBP family was lightly up-regulated in SA1306 and almost not regulated in Parabinga (**Table S3**).

Discussion

Thanks to the high-throughput methods genes, potentially associated with specific processes and characterization of the regulatory networks that control their expression, have been identified (Czechowski et al., 2004; Caldana et al., 2007). DNA microarrays have been successfully applied to characterize global gene expression patterns in *M. truncatula* (Foster-Hartnett et al., 2007; Samac et al., 2011; Zhang et al., 2014; Curto et al., 2015; Song et al., 2015). Previous DNA microarray studies have analyzed the *E. pisi/M. truncatula* pathosystem (Foster-Hartnett et al., 2007; Samac et al., 2011; Curto et al., 2015) using different genotypes and microarray platforms, such as Mt16kOLI1, Mt16kOLI1plus, and Affymetrix GeneChip® (http://www.affymetrix.com), which have increased the knowledge of mechanisms involved in *E. pisi* resistance in *M. truncatula*. Several large-scale TF profiling approaches have employed the *M. truncatula* qPCR-based platform available (Kakar et al., 2008) in various studies (Verdier et al., 2008; Gao et al., 2010; Madrid et al., 2010; Villegas-Fernández et al., 2014; Noguero et al., 2015). In spite of the progress in characterizing TFs, those involved in the expression of stress-related genes in plants remain undiscovered (Singh et al., 2002). Particularly, the TFs involved in the defense mechanisms against *E. pisi* need to be clarified in order to completely understand the mechanisms involved in the plant's defense against this pathogen.

In our study we found that 95 of the TF genes analyzed (15%) were expressed differentially. These results agree with similar approaches carried out in response to infection by *Uromyces striatus* (≈13%) (Madrid et al., 2010) and *Botrytis* spp. (≈20%) (Villegas-Fernández et al., 2014). A subset of these genes belong to 25 TF families (**Figure 5**, **Table S3**), including AUX/IAA, bHLH, E2F, HD, JUMONJI, MYB, SBP, and zinc finger families (C_2C_2, C_2H_2, LIM), were specifically regulated in the resistant SA1306 genotype suggesting that they act as major regulators of transcription throughout *E. pisi* defense responses.

Zinc finger and HD families represent the most of these genes specifically regulated in the resistant SA1306 genotype, and are members of the C_2H_2 (Zn), C_2C_2/DOF, LIM, HD-like, HD/HD, and HD-ZIP TF families, which are agree with recent studies that used the same qPCR TF platform (Villegas-Fernández et al., 2014). The C_2H_2 (Zn) family playing a critical role as key transcriptional repressors involved in the defense response of plants to stress (Brayer and Segal, 2008; Ciftci-Yilmaz and Mittler, 2008; Kiełbowicz-Matuk, 2012) and *M. truncatula* to biotrophic and necrotrophic pathogens. Nevertheless, previous studies have reported the relationship between SBP genes and plant disease resistance, such as programmed cell death in *Arabidopsis* (Stone et al., 2005) and these studies are in agreement that programmed cell death, frequently associated to host cell death, is a common plant defense mechanisms against *E. pisi* (Curto et al., 2006; Barilli et al., 2014). Thus, a member of the SPB TF family (*VpSBP5*) has been reported to be induced by powdery mildew (*E. necator*) (Hou et al., 2013), suggesting that this TF is involved in the resistance to powdery mildew by inducing salicylic acid and methyl jasmonate molecular signals. The C_2C_2 (Zn)/DOF family member have divergent physiological roles (Yanagisawa, 2002) including defense gene expression in response to salicylic acid and oxidative stress signals (Chen et al., 1996; Yanagisawa, 2002) and phytohormone-regulated expression (De Paolis et al., 1996; Yanagisawa, 2002). Recent studies have shown that *E. pisi* infection induces several enzymes, such as *psCHS1* and

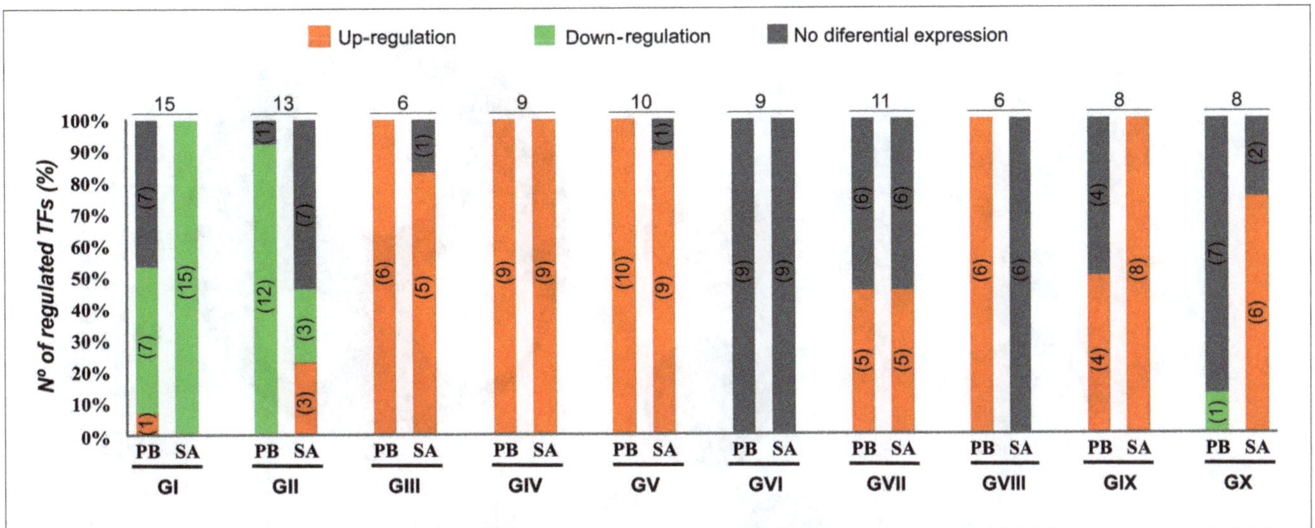

FIGURE 3 | Expression profiles of TF genes. Percentages and numbers (into the brackets) of up-regulated (in red), down-regulated (in green), and no differential expressed (in gray) genes detected in Parabinga (PB) and SA1306 (SA) for each of the 10 clusters (groups GI–GX). Top on each cluster the total number of TF regulated genes is indicated. Additional information is available in **Table 1**.

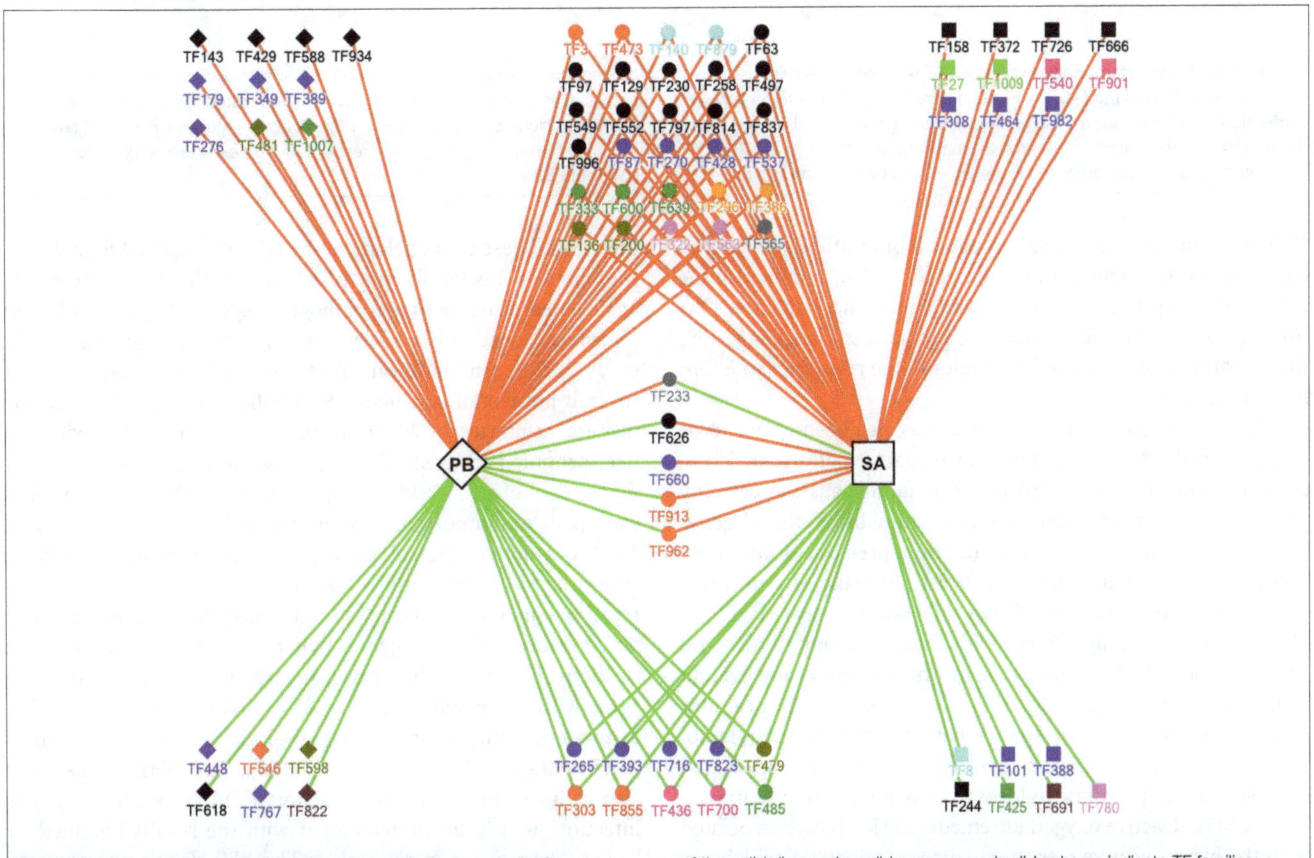

FIGURE 4 | A model for the regulatory network that controls the expression of *E. Pisi*-induced TF genes ($p = 0.05$; $-0.7 \geq m \geq 0.7$) in Parabinga (PB; diamond) and SA1306 (SA; square) *M. truncatula* genotypes. Solid diamonds and solid squares indicate the TF genes regulated in Parabinga and SA1306, respectively. The TF genes that were regulated in both genotypes are indicated by solid spheres. Up- and down-regulation are indicated by red and green lines, respectively. The colors of the solid diamonds, solid squares, and solid spheres indicate TF families: Green (bHLH); black (HD family; HD-Like; MYB; MYB/HD-like); orange (ARF; GRAS); red (AP2/ERBP; WRKY); blue (Zn-fingers TF families; bZIP); olive green (FHA; NAC); violet (RR); pink (SBP; BTB/POZ); gray (HMG; HTH); brown (E2F; DDT); sky-blue (LIM; PHD); pea green (JUMONJI; ARID); dark pink (AUX/IAA; MADS). A detailed description of these genes is shown in **Table S2**.

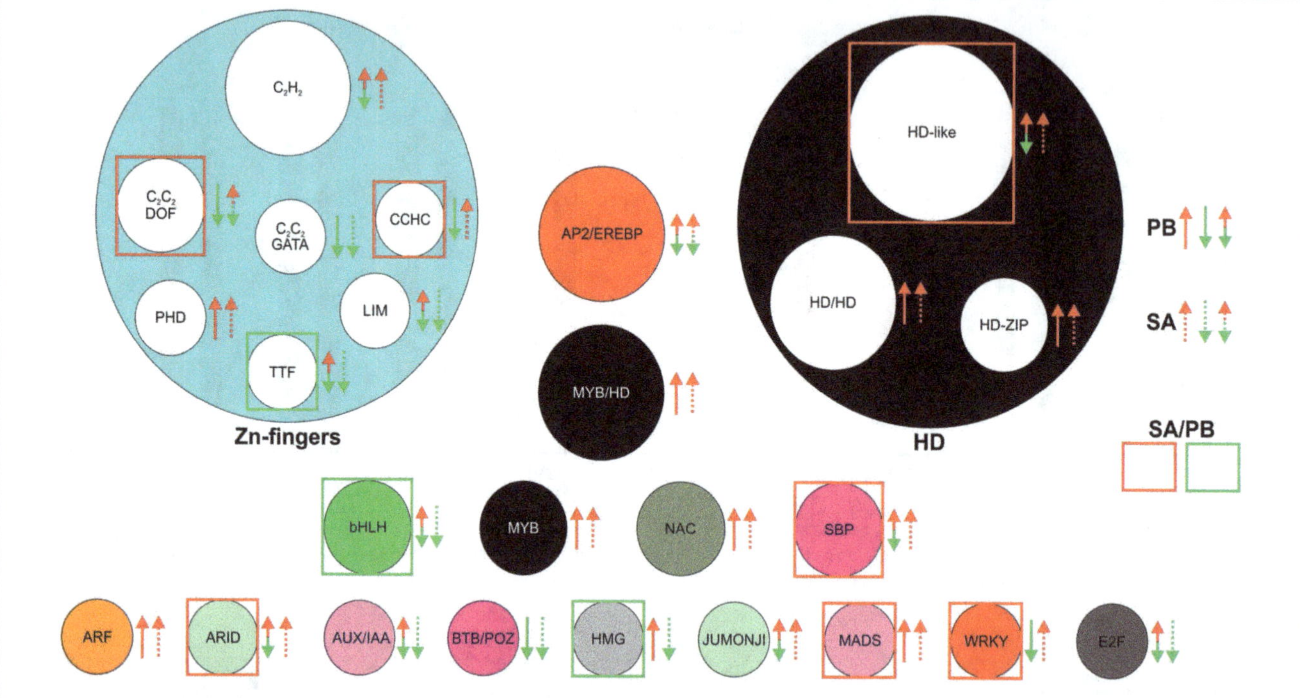

FIGURE 5 | Most differentially expressed TF families in SA1306 compared to Parabinga (−0.7 > *F* > 0.7) in response to *Erysiphe pisi* infection. The TF families are represented by circles whose size is proportional to the number of genes in the family. Up-regulation ($m \geq 0.7$) is indicated by red arrows; down-regulation ($m \leq -0.7$) by green arrows; no differential expression ($-0.7 \leq m \leq 0.7$) are indicate by double red/green arrows; solid lines represent Parabinga (PB), and dotted lines SA1306 (SA). TF families highly up-regulated ($F \geq 1.8$) and down-regulated ($F \leq -1.8$) are indicated by red and green boxes, respectively. For expanded information, see **Table S3**.

PEAPAL2, involved in phenylpropanoid biosynthesis leading to biosynthesis of phytoalexin (Barilli et al., 2014), reinforcing the role of phenylpropanoid pathway in the elicited defense. Our findings are in agreement with these previous studies, suggesting the important role of these TF families in the resistance to *E. pisi* in *M. truncatula*.

Moreover, the LIM zinc TF family has also been found to be involved in the resistance mechanism against pathogens. Thus, several studies have revealed that this family has the capacity to regulate the expression of some lignin biosynthetic genes (Rogers and Campbell, 2004). In addition, previous studies have reported the role of lignification processes into the *M. truncatula* defense responses against to *E. pisi* infection (Prats et al., 2007). Our findings are in agreement with these previous studies and recent studies (Barilli et al., 2014), in which enzymes involved in ROIs stress, such as peroxidase *Prx7*, are regulated in pea after *E. pisi* inoculation. Other studies have found that monolignol biosynthesis plays a critical role in cell wall apposition mediated defense against powdery mildew fungus penetration (Bhuiyan et al., 2009). Reactive oxygen intermediates (ROIs) are associated with the hypersensitive response (Torres et al., 2005), which has been related to programmed cell death (hypersensitive response) that plays a critical role in resistance to *E. pisi* (Barilli et al., 2014).

Our results also reveal a subset of TF genes that encode TFs with a HD (homeodomain) protein domain, which showed a different expression pattern in the resistant SA1306 compared to the susceptible Parabinga genotype in response to *E. pisi* infection. These genes belong to the HD-like and HD family/HD. Members of this family may play a role in the defense response against necrotrophic fungal pathogens regulated by jasmonic acid (Korfhage et al., 1994; Coego et al., 2005; Villegas-Fernández et al., 2014). In addition, a recent study has revealed that several homeodomain-like TF families are involved in the defense responses in *M. truncatula* when confronted with the necrotrophic pathogens, *Botrytis fabae,* and *B. cinerea* (Villegas-Fernández et al., 2014), supporting that these TF families may play an important role in the defensive mechanism of *M. truncatula* to fungal pathogens. Nevertheless, our results show that the Auxin/Indole-3-acetic acid (AUX/IAA) family was repressed in the resistant SA1306 genotype in response to *E. pisi* infection, which suggests that this family may play a role in the resistance mechanism against *E. pisi.* Earlier studies are agree with our results and have described that down regulation of auxin signaling contribute to plant induced immune responses in *Arabidopsis* (Navarro et al., 2006). The bHLH TFs were also induced in the resistant genotype in response to *E. pisi* infection, which are in agreement with the results obtained by Villegas-Fernández et al. (2014). The bHLH TFs up-regulated by *E. pisi* key TF regulating the expressions of jasmonic acid responsive genes (Fernández-Calvo et al., 2011), which mediate the transcriptional reprogramming associated with the plant immune response.

Our findings also indicate that E2F and JUMONJI TF families are involved in the defense response against *E. pisi.*

The E2F transcription factor family has been found to induce transcription of genes required for cell cycle progression and DNA replication (Vandepoele et al., 2005). In addition, more than 100 E2F target genes have been identified, including genes involved in several defense responses and signaling (Ramirez-Parra et al., 2003). The members if the family JUMONJI play a role in the histone methylation process (Li et al., 2013). Dimethylated or trimethylated histone H3 lysine 27 (H3K27me2/3) marks silent or repressed genes involved in stress responses in plants. Li et al. (2013) studied the *jumonji C* protein gene *JMJ705* that is induced by stress signals during pathogen infection, and is involved in methyl jasmonate–induced dynamic removal of H3K27me3 and in gene activation increasing their basal and induced expression during pathogen infection (Balciunas and Ronne, 2000; Li et al., 2013). Our results are in agreement with the results obtained by Villegas-Fernández et al. (2014), which It is that have reported that the JUMONJI TF family may be involved in the defense response to a fungal pathogen, reinforcing the role of this TF family in defense responses against fungal pathogens.

Moreover, an important number of TF genes were regulated in both genotypes, which mainly belonged to AP2/EREBP (Singh et al., 2002; Gutterson and Reuber, 2004; Dietz et al., 2010; Villegas-Fernández et al., 2014), C_2H_2 (Zn) (Ciftci-Yilmaz and Mittler, 2008; Villegas-Fernández et al., 2014), MYB (Singh et al., 2002; Villegas-Fernández et al., 2014), HD (Coego et al., 2005; Villegas-Fernández et al., 2014), MYB/HD-like (Singh et al., 2002; Coego et al., 2005), NAC (Dangl and Jones, 2001; Villegas-Fernández et al., 2014), and PHD (Libault et al., 2007; Villegas-Fernández et al., 2014) TF families. Interestingly, the susceptible and resistant genotypes showed different expression patterns in five of these TF genes in response to *E. pisi* infection (**Figure 4**). Four of these five genes (TF626, TF660, TF913, and TF962) encode TFs belonging to known defense system pathways, which were induced in SA1306 and repressed in Parabinga, respectively. TF962 encodes a member of the AP2/EREBP family known to be linked to response to abiotic and biotic stresses (Dietz et al., 2010), as well as involved in response to a chitin elicitor and in metabolism of the plant hormone methyl jasmonate (McGrath et al., 2005; Libault et al., 2007). Results obtained were similar to those generated in of a study of *M. truncatula* of responses to *Botrytis* infection, in which AP2/EREBP was shown to be a key regulator of defense responses (Villegas-Fernández et al., 2014). TF913 encodes a WRKY protein, whose family members have been shown to be key components in the regulation of plant disease resistance (Eulgem and Somssich, 2007). Previous studies have shown that WRKY TF family members are involved in the regulation of *R* gene-mediated disease resistance as well as in the regulation of transcriptional reprogramming associated with plant immune responses (Eulgem and Somssich, 2007; Buscaill and Rivas, 2014). Several genes encoding WRKY proteins (*AtWRKY18*, *AtWRKY40*) have been identified that confer resistance against powdery mildew (Shen et al., 2007). In addition, *AtWRKY18* has been characterized to act as a positive regulator required for full SAR (Wang et al., 2006), whose transcriptional expression may be is linked to *AtWRKY70*, which modulates the cross-talk between signaling pathways regulating salicylic acid (SA)-dependent and jasmonic acid-dependent responses (Eulgem and Somssich, 2007). These results are in accord with results of recent studies in *M. truncatula* that WRKY TFs involved in the defensive reaction of *M. truncatula* to *Uromyces* and *Botrytis* (Madrid et al., 2010; Villegas-Fernández et al., 2014), supporting the critical role of this TF family in the plant defense responses against fungal pathogens.

The Zn finger family CCHC (TF660) was also detected as induced in SA1306 and repressed in Parabinga. Previous studies (Mangeon et al., 2010; Villegas-Fernández et al., 2014) are in agreement with our results, suggesting the important role of this TF family, A member of HD-like (TF626) was also induced in the resistant genotype and highly repressed in Parabinga. This family has been found to be involved in defense responses against fungal necrotroph pathogens regulated by jasmonic acid (Korfhage et al., 1994; Coego et al., 2005; Villegas-Fernández et al., 2014). We suggest that the 10 TFs that were repressed in both genotypes may play a role suppressing genes involved in photosynthetic metabolism leading to a reduction in the photosynthetic rate, as previously suggested (Swarbrick et al., 2006; Bolton, 2009).

Conclusion

We have screened more than 1000 TFs genes of *M. truncatula* for altered expression during *E. pisi* infection using qPCR platform. Forty eight TFs from them showed significant differences in the resistant SA1306 genotype compared to the susceptible Parabinga. These TF genes belong mainly to AP2/EREBP, WRKY, MYB, HD, and zinc finger families (C2C2, C2H2, LIM) gene families, which are involved in known defense responses. In addition, we suggest a regulatory network that controls the expression in *M. truncatula* of genes involved in resistance to *E. pisi*. These results will help to systematically decipher the functional roles of TF genes and to develop new strategies against powdery mildew.

Acknowledgments

This work was supported by the European Commission FP6 Framework Programme Grain Legume Integrated Project (FOOD-CT-2004-506223) and by Spanish project AGL2014-52871-R co-financed by FEDER. We are grateful to Udvardi research Group (The Samuel Roberts Noble Foundation, Okla, USA) for advice and support and to Carolina Johnstone for grammatical review. This research work is lovingly dedicated to my mother, who has fought cancer two times.

Supplementary Material

The Supplementary Material for this article can be found online at: http://journal.frontiersin.org/article/10.3389/fpls.2015.00517

Figure S1 | Macroscopic disease symptoms in the susceptible cultivar Parabinga (A) and in the resistant SA1306 accession (B).

Table S1 | Complete list of TF genes. The accession numbers (AC), the TIGR *Medicago* gene accession numbers (TC) if available, as well as the transcription factor family and the subfamily names are shown as described Kakar et al. (Kakar et al., 2008). The quantification cycle (Cq), PCR efficiencies (PCReff) and correlation coefficients for the qPCR reactions performed on both *Medicago truncatula* genotypes, Parabinga and SA1306, in three biological replicates are indicated.

Table S2 | Details of regulated TF genes among *Medicago truncatula* genotypes analyzed, SA1306 (SA) and Parabinga (PB), in response to *Erysiphe pisi* infection ($p \leq 0.05$; $-0.7 \geq m \geq 0.7$). Relative gene expression ratios (m) are listed and sorted by "Class" and "m" values. ID: Identification number of TF gene. TF family: Transcription factor family. TF subfamily: Subfamily names. m: Average log_2 differential expression ratios (inoculated/control) for Parabinga (PB) and SA1306 (SA) genotypes. F: Fold change expression ratios of differentially expressed genes in SA1306 compared to Parabinga, log_2 expression ratio SA1306/Parabinga. Class: Genes that were specifically regulated in SA1306 (SA), Parabinga (PB) and both genotypes (Common) (**Figure 4**).

Table S3 | Most differentially expressed TF families in SA1306 compared to Parabinga ($-0.7>F>0.7$) in response to *Erysiphe pisi* infection.

References

Balciunas, D., and Ronne, H. (2000). Evidence of domain swapping within the jumonji family of transcription factors. *Trends Biochem. Sci.* 25, 274–276. doi: 10.1016/S0968-0004(00)01593-0

Barilli, E., Rubiales, D., Gjetting, T., and Lyngkjaer, M. F. (2014). Differential gene transcript accumulation in peas in response to powdery mildew (*Erysiphe pisi*) attack. *Euphytica* 195, 1–16. doi: 10.1007/s10681-014-1062-z

Bélanger, R. R., Bushnell, W. R., Dik, A. J., and Carver, T. L. W. (2002). *The Powdery Mildews: A Comprehensive Treatise.* St. Paul, MN: APS Press.

Bhuiyan, N. H., Selvaraj, G., Wei, Y., and King, J. (2009). Role of lignification in plant defense. *Plant. Signal. Behav.* 4, 158–159. doi: 10.4161/psb.4.2.7688

Bolton, M. D. (2009). Primary metabolism and plant defense-fuel for the fire. *Mol. Plant Microbe Interact.* 22, 487–497. doi: 10.1094/MPMI-22-5-0487

Brayer, K. J., and Segal, D. J. (2008). Keep your fingers off my DNA: protein-protein interactions mediated by C2H2 zinc finger domains. *Cell Biochem. Biophys.* 50, 111–131. doi: 10.1007/s12013-008-9008-5

Buscaill, P., and Rivas, S. (2014). Transcriptional control of plant defence responses. *Curr. Opin. Plant Biol.* 20, 35–46. doi: 10.1016/j.pbi.2014.04.004

Bustin, S. A., Benes, V., Garson, J. A., Hellemans, J., Huggett, J., Kubista, M., et al. (2009). The MIQE guidelines: minimum information for publication of quantitative real-time PCR experiments. *Clin. Chem.* 55, 611–622. doi: 10.1373/clinchem.2008.112797

Caldana, C., Scheible, W.-R., Mueller-Roeber, B., and Ruzicic, S. (2007). A quantitative RT-PCR platform for high-throughput expression profiling of 2500 rice transcription factors. *Plant Methods* 3:7. doi: 10.1186/1746-4811-3-7

Chen, W., Chao, G., and Singh, K. B. (1996). The promoter of a H_2O_2-inducible, *Arabidopsis* glutathione S-transferase gene contains closely linked OBF- and OBP1-binding sites. *Plant J.* 10, 955–966. doi: 10.1046/j.1365-313X.1996.10060955.x

Cheung, A. Y., and Wu, H.-M. (2011). THESEUS 1, FERONIA and relatives: a family of cell wall-sensing receptor kinases? *Curr. Opin. Plant Biol.* 14, 632–641. doi: 10.1016/j.pbi.2011.09.001

Ciftci-Yilmaz, S., and Mittler, R. (2008). The zinc finger network of plants. *Cell Mol. Life Sci.* 65, 1150–1160. doi: 10.1007/s00018-007-7473-4

Coego, A., Ramirez, V., Gil, M. J., Flors, V., Mauch-Mani, B., and Vera, P. (2005). An *Arabidopsis* homeodomain transcription factor, *OVEREXPRESSOR OF CATIONIC PEROXIDASE* 3, mediates resistance to infection by necrotrophic pathogens. *Plant Cell* 17, 2123–2137. doi: 10.1105/tpc.105.032375

Curto, M., Camafeita, E., Lopez, J. A., Maldonado, A. M., Rubiales, D., and Jorrín, J. V. (2006). A proteomic approach to study pea (*Pisum sativum*) responses to powdery mildew (*Erysiphe pisi*). *Proteomics* 6, S163–S174. doi: 10.1002/pmic.200500396

Curto, M., Krajinski, F., Küster, H., and Rubiales, D. (2015). Plant defense responses in *Medicago truncatula* unveiled by microarray analysis. *Plant Mol. Biol. Rep.* 33, 569–583. doi: 10.1007/s11105-014-0770-9

Czechowski, T., Bari, R. P., Stitt, M., Scheible, W.-R., and Udvardi, M. K. (2004). Real-time RT-PCR profiling of over 1400 *Arabidopsis* transcription factors: unprecedented sensitivity reveals novel root- and shoot-specific genes. *Plant J.* 38, 366–379. doi: 10.1111/j.1365-313X.2004.02051.x

Dangl, J. L., and Jones, J. D. G. (2001). Plant pathogens and integrated defence responses to infection. *Nature* 411, 826–833. doi: 10.1038/35081161

De Paolis, A., Sabatini, S., De Pascalis, L., Costantino, P., and Capone, I. (1996). A *rolB* regulatory factor belongs to a new class of single zinc finger plant proteins. *Plant J.* 10, 215–223. doi: 10.1046/j.1365-313X.1996.10020215.x

Dietz, K.-J., Vogel, M., and Viehhauser, A. (2010). AP2/EREBP transcription factors are part of gene regulatory networks and integrate metabolic, hormonal and environmental signals in stress acclimation and retrograde signalling. *Protoplasma* 245, 3–14. doi: 10.1007/s00709-010-0142-8

Eulgem, T., and Somssich, I. E. (2007). Networks of WRKY transcription factors in defense signaling. *Curr. Opin. Plant Biol.* 10, 366–371. doi: 10.1016/j.pbi.2007.04.020

Falloon, R. E., and Viljanen-Rollinson, S. L. H. (2001). "Powdery mildew," in *Compendium of Pea Diseases and Pests, 2nd Edn.*, eds J. M. Karaft and F. L. Pfleger (St. Paul, MN: APS Press), 28–29.

Fernández-Calvo, P., Chini, A., Fernández-Barbero, G., Chico, J.-M., Gimenez-Ibanez, S., Geerinck, J., et al. (2011). The *Arabidopsis* bHLH transcription factors MYC3 and MYC4 are targets of JAZ repressors and act additively with MYC2 in the activation of jasmonate responses. *Plant Cell* 23, 701–715. doi: 10.1105/tpc.110.080788

Fondevilla, S., and Rubiales, D. (2012). Powdery mildew control in pea. A review. *Agron. Sustain. Dev.* 32, 401–409. doi: 10.1007/s13593-011-0033-1

Fondevilla, S., Torres, A. M., Moreno, M. T., and Rubiales, D. (2007). Identification of a new gene for resistance to powdery mildew in *Pisum fulvum*, a wild relative of pea. *Breed. Sci.* 57, 181–184. doi: 10.1270/jsbbs.57.181

Foster-Hartnett, D., Danesh, D., Peñuela, S., Sharopova, N., Endre, G., Vandenbosch, K. A., et al. (2007). Molecular and cytological responses of *Medicago truncatula* to *Erysiphe pisi*. *Mol. Plant Pathol.* 8, 307–319. doi: 10.1111/j.1364-3703.2007.00395.x

Gao, L.-L., Kamphuis, L. G., Kakar, K., Edwards, O. R., Udvardi, M. K., and Singh, K. B. (2010). Identification of potential early regulators of aphid resistance in *Medicago truncatula* via transcription factor expression profiling. *New Phytol.* 186, 980–994. doi: 10.1111/j.1469-8137.2010.03229.x

Gutterson, N., and Reuber, T. L. (2004). Regulation of disease resistance pathways by AP2/ERF transcription factors. *Curr. Opin. Plant Biol.* 7, 465–471. doi: 10.1016/j.pbi.2004.04.007

Hellemans, J., Mortier, G., De Paepe, A., Speleman, F., and Vandesompele, J. (2007). qBase relative quantification framework and software for management and automated analysis of real-time quantitative PCR data. *Genome Biol.* 8:R19. doi: 10.1186/gb-2007-8-2-r19

Hoagland, D. R., and Arnon, D. I. (1950). The water-culture method of growing plants without soil. *Calif. Agric. Exp. Stn. Circ.* 347, 1–32.

Hou, H., Yan, Q., Wang, X., and Xu, H. (2013). A SBP-Box gene *VpSBP5* from chinese wild *Vitis* Species Responds to *Erysiphe necator* and defense signaling molecules. *Plant Mol. Biol. Rep.* 31, 1261–1270. doi: 10.1007/s11105-013-0591-2

Iglesias-García, R., Rubiales, D., and Fondevilla, S. (2015). Penetration resistance to *Erysiphe pisi* in pea mediated by *er1* gene is associated with protein cross-linking but not with callose apposition or hypersensitive response. *Euphytica* 201, 381–387. doi: 10.1007/s10681-014-1221-2

Kakar, K., Wandrey, M., Czechowski, T., Gaertner, T., Scheible, W.-R., Stitt, M., et al. (2008). A community resource for high-throughput quantitative RT-PCR analysis of transcription factor gene expression in *Medicago truncatula*. *Plant Methods* 4:18. doi: 10.1186/1746-4811-4-18

Kiełbowicz-Matuk, A. (2012). Involvement of plant C2H2-type zinc finger transcription factors in stress responses. *Plant Sci.* 185–186, 78–85. doi: 10.1016/j.plantsci.2011.11.015

Korfhage, U., Trezzini, G. F., Meier, I., Hahlbrock, K., and Somssich, I. E. (1994). Plant homeodomain protein involved in transcriptional regulation of a pathogen defense-related gene. *Plant Cell* 6, 695–708. doi: 10.1105/tpc.6.5.695

Küster, H., Becker, A., Firnhaber, C., Hohnjec, N., Manthey, K., Perlick, A. M., et al. (2007). Development of bioinformatic tools to support EST-sequencing, *in silico*- and microarray-based transcriptome profiling in mycorrhizal symbioses. *Phytochemistry* 68, 19–32. doi: 10.1016/j.phytochem.2006.09.026

Küster, H., Hohnjec, N., Krajinski, F., El Yahyaoui, F., Manthey, K., Gouzy, J., et al. (2004). Construction and validation of cDNA-based Mt6k-RIT macro- and microarrays to explore root endosymbioses in the model legume *Medicago truncatula*. *J. Biotechnol.* 108, 95–113. doi: 10.1016/j.jbiotec.2003.11.011

Li, T., Chen, X., Zhong, X., Zhao, Y., Liu, X., Zhou, S., et al. (2013). Jumonji C domain protein *JMJ705*-mediated removal of histone H3 lysine 27 trimethylation is involved in defense-related gene activation in rice. *Plant Cell* 25, 4725–4736. doi: 10.1105/tpc.113.118802

Libault, M., Wan, J., Czechowski, T., Udvardi, M., and Stacey, G. (2007). Identification of 118 *Arabidopsis* transcription factor and 30 ubiquitin-ligase genes responding to chitin, a plant-defense elicitor. *Mol. Plant Microbe Interact.* 20, 900–911. doi: 10.1094/MPMI-20-8-0900

Madrid, E., Gil, J., Rubiales, D., Krajinski, F., Schlereth, A., and Millán, T. (2010). Transcription factor profiling leading to the identification of putative transcription factors involved in the *Medicago truncatula–Uromyces striatus* interaction. *Theor. Appl. Genet.* 121, 1311–1321. doi: 10.1007/s00122-01 0-1390-x

Mangeon, A., Junqueira, R. M., and Sachetto-Martins, G. (2010). Functional diversity of the plant glycine-rich proteins superfamily. *Plant Signal. Behav.* 5, 99–104. doi: 10.4161/psb.5.2.10336

McGrath, K. C., Dombrecht, B., Manners, J. M., Schenk, P. M., Edgar, C. I., Maclean, D. J., et al. (2005). Repressor- and activator-type ethylene response factors functioning in jasmonate signaling and disease resistance identified via a genome-wide screen of *Arabidopsis* transcription factor gene expression. *Plant Physiol.* 139, 949–959. doi: 10.1104/pp.105.068544

Navarro, L., Dunoyer, P., Jay, F., Arnold, B., Dharmasiri, N., Estelle, M., et al. (2006). A plant miRNA contributes to antibacterial resistance by repressing auxin signaling. *Science* 312, 436–439. doi: 10.1126/science.1126088

Noguero, M., Le Signor, C., Vernoud, V., Bandyopadhyay, K., Sanchez, M., Fu, C., et al. (2015). *DASH* transcription factor impacts *Medicago truncatula* seed size by its action on embryo morphogenesis and auxin homeostasis. *Plant J.* 81, 453–466. doi: 10.1111/tpj.12742

Prats, E., Llamas, M. J., and Rubiales, D. (2007). Characterization of resistance mechanisms to *Erysiphe pisi* in *Medicago truncatula*. *Phytopathology* 97, 1049–1053. doi: 10.1094/PHYTO-97-9-1049

Ramirez-Parra, E., Fründt, C., and Gutierrez, C. (2003). A genome-wide identification of E2F-regulated genes in *Arabidopsis*. *Plant J.* 33, 801–811. doi: 10.1046/j.1365-313X.2003.01662.x

Riechmann, J. L., and Ratcliffe, O. J. (2000). A genomic perspective on plant transcription factors. *Curr. Opin. Plant Biol.* 3, 423–434. doi: 10.1016/S1369-5266(00)00107-2

Ringli, C. (2010). Monitoring the outside: cell wall-sensing mechanisms. *Plant Physiol.* 153, 1445–1452. doi: 10.1104/pp.110.154518

Rogers, L. A., and Campbell, M. M. (2004). The genetic control of lignin deposition during plant growth and development. *New Phytol.* 164, 17–30. doi: 10.1111/j.1469-8137.2004.01143.x

Rose, R. J. (2008). *Medicago truncatula* as a model for understanding plant interactions with other organisms, plant development and stress biology: past, present and future. *Funct. Plant Biol.* 35, 253–264. doi: 10.1071/FP07297

Rubiales, D., Fondevilla, S., Chen, W., Gentzbittel, L., Higgins, T. J. V., Castillejo, M. A., et al. (2015). Achievements and challenges in legume breeding for pest and disease resistance. *Crit. Rev. Plant Sci.* 34, 195–236. doi: 10.1080/07352689.2014.898445

Samac, D. A., Peñuela, S., Schnurr, J. A., Hunt, E. N., Foster-Hartnett, D., Vandenbosch, K. A., et al. (2011). Expression of coordinately regulated defence response genes and analysis of their role in disease resistance in *Medicago truncatula*. *Mol. Plant Pathol.* 12, 786–798. doi: 10.1111/j.1364-3703.2011.00712.x

Shen, Q.-H., Saijo, Y., Mauch, S., Biskup, C., Bieri, S., Keller, B., et al. (2007). Nuclear activity of MLA immune receptors links isolate-specific and basal disease-resistance responses. *Science* 315, 1098–1103. doi: 10.1126/science.1136372

Sillero, J., Fondevilla, S., Davidson, J., Patto, M., Warkentin, T., Thomas, J., et al. (2006). Screening techniques and sources of resistance to rusts and mildews in grain legumes. *Euphytica* 147, 255–272. doi: 10.1007/s10681-006-6544-1

Singh, K. B., Foley, R. C., and Oñate-Sánchez, L. (2002). Transcription factors in plant defense and stress responses. *Curr. Opin. Plant Biol.* 5, 430–436. doi: 10.1016/S1369-5266(02)00289-3

Singh, R. J., Chung, G. H., and Nelson, R. L. (2007). Landmark research in legumes. *Genome* 50, 525–537. doi: 10.1139/G07-037

Song, J., Wang, Y., Li, H., Li, B., Zhou, Z., Gao, S., et al. (2015). The F-box family genes as key elements in response to salt, heavy mental, and drought stresses in *Medicago truncatula*. *Funct. Integr. Genomics* 1–13. doi: 10.1007/s10142-015-0438-z

Stone, J. M., Liang, X., Nekl, E. R., and Stiers, J. J. (2005). *Arabidopsis AtSPL14*, a plant-specific SBP-domain transcription factor, participates in plant development and sensitivity to fumonisin B1. *Plant J.* 41, 744–754. doi: 10.1111/j.1365-313X.2005.02334.x

Swarbrick, P. J., Schulze-Lefert, P., and Scholes, J. D. (2006). Metabolic consequences of susceptibility and resistance (race-specific and broad-spectrum) in barley leaves challenged with powdery mildew. *Plant Cell Environ.* 29, 1061–1076. doi: 10.1111/j.1365-3040.2005.01472.x

Torres, M. A., Jones, J. D. G., and Dangl, J. L. (2005). Pathogen-induced, NADPH oxidase-derived reactive oxygen intermediates suppress spread of cell death in *Arabidopsis thaliana* *Nat. Genet.* 37, 1130–1134. doi: 10.1038/ng1639

Udvardi, M. K., Kakar, K., Wandrey, M., Montanari, O., Murray, J., Andriankaja, A., et al. (2007). Legume transcription factors: global regulators of plant development and response to the environment. *Plant Physiol.* 144, 538–549. doi: 10.1104/pp.107.098061

Vandepoele, K., Vlieghe, K., Florquin, K., Hennig, L., Beemster, G. T. S., Gruissem, W., et al. (2005). Genome-wide identification of potential plant E2F target genes. *Plant Physiol.* 139, 316–328. doi: 10.1104/pp.105.066290

Vandesompele, J., De Preter, K., Pattyn, F., Poppe, B., Van Roy, N., De Paepe, A., et al. (2002). Accurate normalization of real-time quantitative RT-PCR data by geometric averaging of multiple internal control genes. *Genome Biol.* 3: research0034.0031. doi: 10.1186/gb-2002-3-7-research0034

Verdier, J., Kakar, K., Gallardo, K., Le Signor, C., Aubert, G., Schlereth, A., et al. (2008). Gene expression profiling of *M.* truncatula transcription factors identifies putative regulators of grain legume seed filling. *Plant Mol. Biol.* 67, 567–580. doi: 10.1007/s11103-008-9320-x

Villegas-Fernández, A. M., Krajinski, F., Schlereth, A., Madrid, E., and Rubiales, D. (2014). Characterization of transcription factor following expression profiling of *Medicago truncatula – Botrytis spp* interactions. *Plant Mol. Biol. Rep.* 32, 1030–1040. doi: 10.1007/s11105-014-0710-8

Wang, D., Amornsiripanitch, N., and Dong, X. (2006). A genomic approach to identify regulatory nodes in the transcriptional network of systemic acquired resistance in plants. *PLoS Pathog.* 2:e123. doi: 10.1371/journal.ppat.0020123

Yanagisawa, S. (2002). The Dof family of plant transcription factors. *Trends Plant. Sci.* 7, 555–560. doi: 10.1016/S1360-1385(02)02362-2

Yang, S., Tang, F., Caixetab, E. T., and Zhu, H. (2013). Epigenetic regulation of a powdery mildew resistance gene in *Medicago truncatula*. *Mol. Plant.* 6, 2000–2003. doi: 10.1093/mp/sst106

Zhang, J.-Y., Cruz De Carvalho, M. H., Torres-Jerez, I., Kang, Y. U. N., Allen, S. N., Huhman, D. V., et al. (2014). Global reprogramming of transcription and metabolism in *Medicago truncatula* during progressive drought and after rewatering. *Plant Cell Environ.* 37, 2553–2576. doi: 10.1111/pce.12328

Conflict of Interest Statement: The authors declare that the research was conducted in the absence of any commercial or financial relationships that could be construed as a potential conflict of interest.

20

Trade-Offs between Economic and Environmental Impacts of Introducing Legumes into Cropping Systems

Edited by:
Antonio M. De Ron,
Spanish National Research Council - Misión Biológica de Galicia, Spain

Reviewed by:
Salvador Nadal Moyano,
Junta de Andalucía, Spain
Francisco Javier Ibarra Perez,
Instituto Nacional de Investigaciones Forestales Agrícolas y Pecuarias, Mexico

***Correspondence:**
Moritz Reckling
moritz.reckling@zalf.de

†Present Address:
Frederick L. Stoddard,
Department of Food and Environmental Sciences, University of Helsinki, Helsinki, Finland

Moritz Reckling[1,2], Göran Bergkvist[2], Christine A. Watson[2,3], Frederick L. Stoddard[4†], Peter M. Zander[5], Robin L. Walker[3], Aurelio Pristeri[6], Ion Toncea[7] and Johann Bachinger[1]*

[1] Institute of Land Use Systems, Leibniz Centre for Agricultural Landscape Research, Müncheberg, Germany, [2] Department of Crop Production Ecology, Swedish University of Agricultural Sciences, Uppsala, Sweden, [3] Crop and Soil Systems, Scotland's Rural College, Aberdeen, UK, [4] Department of Agricultural Sciences, University of Helsinki, Helsinki, Finland, [5] Institute of Socio-Economics, Leibniz Centre for Agricultural Landscape Research, Müncheberg, Germany, [6] Department of Agricultural Science, Mediterranean University of Reggio Calabria, Reggio Calabria, Italy, [7] National Agricultural Research and Development Institute, Fundulea, Romania

Europe's agriculture is highly specialized, dependent on external inputs and responsible for negative environmental impacts. Legume crops are grown on less than 2% of the arable land and more than 70% of the demand for protein feed supplement is imported from overseas. The integration of legumes into cropping systems has the potential to contribute to the transition to a more resource-efficient agriculture and reduce the current protein deficit. Legume crops influence the production of other crops in the rotation making it difficult to evaluate the overall agronomic effects of legumes in cropping systems. A novel assessment framework was developed and applied in five case study regions across Europe with the objective of evaluating trade-offs between economic and environmental effects of integrating legumes into cropping systems. Legumes resulted in positive and negative impacts when integrated into various cropping systems across the case studies. On average, cropping systems with legumes reduced nitrous oxide emissions by 18 and 33% and N fertilizer use by 24 and 38% in arable and forage systems, respectively, compared to systems without legumes. Nitrate leaching was similar with and without legumes in arable systems and reduced by 22% in forage systems. However, grain legumes reduced gross margins in 3 of 5 regions. Forage legumes increased gross margins in 3 of 3 regions. Among the cropping systems with legumes, systems could be identified that had both relatively high economic returns and positive environmental impacts. Thus, increasing the cultivation of legumes could lead to economic competitive cropping systems and positive environmental impacts, but achieving this aim requires the development of novel management strategies informed by the involvement of advisors and farmers.

Keywords: crop rotation, framework, land use and impacts, multi-criteria assessment, protein crops, resource-efficiency, rotation generator

INTRODUCTION

Crop production in Europe is highly specialized and currently relies on a very small number of crop species, raising questions about the sustainability of farming (Tilman et al., 2002). Furthermore, less than 30% of the demand for protein feed supplement is produced within Europe (Bouxin, 2014). The reintegration of legumes into European agriculture could reduce the current protein deficit and contribute to the transition to more sustainable agricultural systems (Voisin et al., 2013).

The use of legumes also affects the performance of cropping systems and their environmental impacts, including (i) nitrogen supply via biological nitrogen fixation (BNF), reducing the demand for external nitrogen fertilizers, (ii) positive pre-crop benefits through a combination of residual nitrogen and break-crop effects (Angus et al., 2015; Preissel et al., 2015), (iii) reduced fossil energy consumption in crop production (Jensen et al., 2011), and (iv) increased crop diversification and biodiversity (Köpke and Nemecek, 2010).

Despite these benefits, there are good reasons why European farmers grew grain legumes on only 1.6% of the arable land in 2014 (FAOstat, 2015). These include specialization in a few crops (Zander et al., 2016), low and unstable yields of grain legumes (Von Richthofen et al., 2006; Cernay et al., 2015; Reckling et al., 2015), low and unpredictable policy support (Bues et al., 2013), and lack of awareness of, and inability to evaluate the benefits of including legumes in the cropping system (Preissel et al., 2015). As a consequence, the area under grain legumes declined from 4.7% of arable land in 1961 to currently below 2% (FAOstat, 2015). Organic farms used 6.8% of their arable land for production of grain legumes in 2014 (Eurostat, 2015). Forage legume areas also declined because of a transformation of the ruminant feeding system toward maize silage and soybean oil cakes, and geographical separation between cereal and cattle-rearing areas (Voisin et al., 2013).

By introducing legumes into the cropping system, farmers also introduce more complexity into their planning. To maximize the benefits of legumes while minimizing potential threats, a number of issues need to be taken into account. The preceding crop in a rotation should have a high ability to take up available nitrogen in order to reduce residual soil N levels and allow BNF of the legumes to be optimized. Legume crops need to be kept several years apart in the rotation because of their susceptibility to soil-borne diseases. The following crop should take up the residual N of the legume quickly, in order to minimize opportunities for nitrate leaching and nitrous oxide emission. Legume crops, their management and their effects on other crops in the rotation are specific to different climatic and edaphic conditions (Döring, 2015), e.g., narrow-leafed lupin (*Lupinus angustifolius* L.) is considered suitable for sandy soils with low soil pH, whereas faba bean (*Vicia faba* L.) is typically grown on clay soils with sufficient rainfall, soybean (*Glycine max* (L.) Merr.) requires appropriate daylength to flower and sufficient growing degree-days to mature, and pea (*Pisum sativum* L.) is suitable for many different soil types. Therefore, the design of cropping systems needs to be site-specific and based on regional expert knowledge.

Although the positive effects of legumes on subsequent crops, including yield benefits and reductions in nitrogen fertilizer demand, soil tillage and biocide applications are well known (Angus et al., 2015; Preissel et al., 2015) from field experiments, these effects are difficult to formalize and are rarely considered in modeling (Bergez et al., 2010). Life cycle assessments comparing rotations with and without legumes have been carried out (e.g., Nemecek et al., 2008; Köpke and Nemecek, 2010) but these work with a limited set of rotations and do not include the detailed levels of field management required to evaluate all rotational effects.

In order to focus research and policy intervention, a method is needed to assess the economic and environmental effects of changes to cropping system design, such as the integration of legumes. To identify novel practices that achieve high economic and environmental benefits the method should provide a range of crop combinations and evaluate management options beyond current farming practices.

We earlier (Reckling et al., 2016) formulated an assessment framework that allows a systematic analysis of a large range of cropping options. It operates on the cropping system scale and takes rotational effects of crops such as legumes into account.

The objective of this study is to assess the economic and environmental effects of integrating legumes into cropping systems, in order to identify potentials and limitations of increasing legume cultivation in Europe. For the analysis we apply the cropping system framework by Reckling et al. (2016) in case studies in Germany, Italy, the United Kingdom, Romania, and Sweden. We hypothesize that the integration of legumes increases positive environmental impacts and at the same time reduces economic returns, thus leading to greater trade-offs between economic and environmental benefits.

METHODS

The static and rule-based framework developed by Reckling et al. (2016) was used to assess cropping systems in five case studies following three steps: (i) *generate crop rotations* (using a rule-based rotation generator), (ii) *calculate the impact of crop production activities* (using environmental and economic indicators), and (iii) *assess and compare cropping systems* with and without legumes. Following Bergez et al. (2010), we used a multi-criteria analysis for the identification of trade-offs.

Case Studies

Five contrasting regions were selected across Europe (**Figure 1**): Brandenburg (BB) in north-eastern Germany, Calabria (CB) in southern Italy, eastern Scotland (ES) in the United Kingdom, Sud-Muntenia (SM) in Romania, and Västra Götaland (VG) in western Sweden. The regions are characterized by contrasting climatic conditions and cropping systems, and were selected to represent a broad range of bio-physical and socio-economic conditions and possible roles of legume production. In all regions, legumes occupy less than 2.5% of the arable land. Within case studies, land capability types were defined by regional classifications and differences in soil texture (**Table 1**).

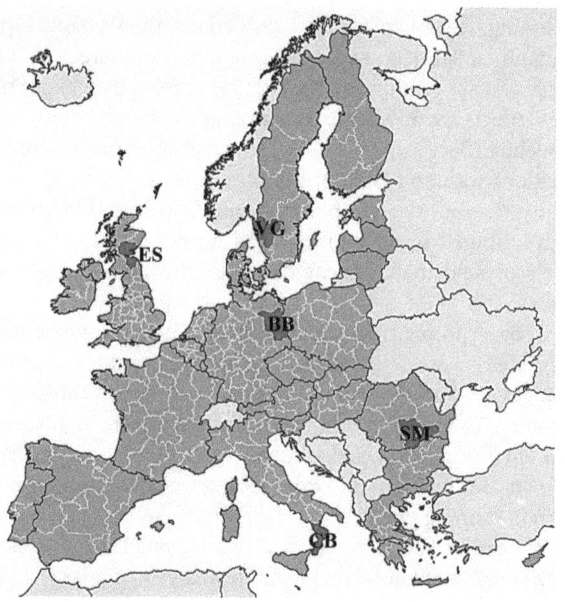

FIGURE 1 | Case study regions across Europe, Brandenburg (BB) in north-eastern Germany, Calabria (CB) in southern Italy, eastern Scotland (ES) in the United Kingdom, Sud-Muntenia (SM) in Romania, and Västra Götaland (VG) in western Sweden (adapted from Eurostat, 2010).

These determine the choice of the crops, their management and performance.

In Brandenburg, soil types are classified according to LELF (2010). Arable framing concentrates on Type 1–2 with higher yield potential and forage oriented farming on Type 3–4 with a lower yield potential. Soil Type 2 was selected for arable and Type 3 for forage systems. In Calabria, the assessment focused on arable rainfed systems in the lowlands, where pea and faba bean are the main annual legume crops. In Eastern Scotland, mixed cropping is common across soil Types 1–3, and forage production on Types 3 and 4 (classified according to Bibby et al., 1982). For the assessment, Type 1 and 2 were selected for arable and Type 3 for forage systems. Cereals occupy 23% and grassland 73% of the agricultural land, while faba bean and pea together account for only 0.3% (Scottish-Agricultural-Census, 2010). Arable farmers in the lowlands grow large shares of cereals and relatively few break crops, mainly winter and spring oilseed rape, potato, and swede. In the hilly areas, farms are dominated by grassland with low shares of clover. In Sud-Muntenia, Chernozem soils with high humus contents were selected, as they underlie the major cropping areas with medium- and large-scale farms. The study considered only arable farms, because information on forage grass production was limited. Legumes already grown include pea and soybean for feed and common bean (*Phaseolus vulgaris* L.) for the food market. In Västra Götaland, soils with high clay contents were selected for this study, because they represent the major cropping areas with both arable and mixed farms. Arable farmers use 80–100% of their land for cereals, and legumes and other break crops play minor roles (1.7 and 1.5% of the arable land). In forage-oriented farms, perennial grasses with a relatively low share of legumes (<20%) are common.

Across the case studies, pea, faba bean and clover (*Trifolium* L.) (sole cropped or in mixture with grasses) were the most widely tested legumes. Wheat, barley and oilseed rape were the most common non-legume crops (**Table 1**).

Generate Crop Rotations

Crop rotations were generated with the previously described static and rule-based rotation generator (Reckling et al., 2016) that produces fixed and cyclical "agronomically sound rotations" for arable fields, following a fixed set of site-specific agronomic rules. In each region, agronomists defined suitable crops and then the restriction values for crop sequence, including suitability of crop-crop combinations, the minimum cultivation break and frequency restrictions (% of a crop and crop type in a rotation) to allow the site-specific generation of rotations. In addition, a rotation of *current farming* ("business as usual") was based on the most common crops grown.

The rotation generator combined crops to produce all possible 3- to 6-year sequences following the criteria (i) crop sequence suitability (e.g., pea-rye), (ii) maximum frequency of a crop in the rotation (e.g., 20% for pea), (iii) minimum break between the same crop (4 years for pea), and (iv) maximum frequency of crops of the same crop type (e.g., 25% grain legumes). Duplicates and multiples were removed and rotations were filtered by soil type. Agronomists checked the plausibility of the results in comparison to the existing sequences.

Calculate Impact Indicators

The crop production activities (CPA) covered the crop, preceding crop and site-specific crop management. CPA were designed by collecting management data in a structured survey among 2–4 experienced agronomists in each case study in 2013. The data included inputs (seed, N-P-K fertilizer and pesticides), outputs (grain yield, forage yield, and straw yield) and management characteristics (fertilizer intensity, machinery use, and harvesting method) for each crop. In order to consider rotational effects, pre-crop types were defined according to their ability to influence the yield of the following crop through N from residue decomposition. Besides the effect on yield, differences in fertilization and agro-chemical applications were considered for each pre-crop type. The following types were separated, in the order of increasing N residue levels and positive effect on yield of subsequent crops: (i) spring cereals, winter cereals and maize; (ii) grain legumes, rapeseed and grass (<10% legumes); (iii) forage legumes in pure stands and in mixtures with grass (>30% legumes).

We selected five indicators to assess the economic and environmental performance of cropping systems. Nitrogen is not only important in crop production, it also contributes to environmental pollution of the atmosphere and water bodies, so nitrous oxide emission and nitrate leaching were selected as indicators. Nitrogen fertilizer use and nitrogen fertilizer efficiency were selected as indicators of resource use efficiency. Gross margins provided the standard indicator for economic

TABLE 1 | Cropping characteristics of selected regions and land capability types.

Region	Site	% sand	% clay	% of agric. area	Non-legume crops	Legume crops
ARABLE CROPPING SYSTEMS						
BB	Type 2	50	25	22	WW, WR, SB, WB, WT, WOR, SM, SO	FP, FB
CB	Type 3	47	23	36	WB, DW, SO, WOR, WW, WT	FP, FB
ES	Type 1/2	78	14	21	SB, WB, SO, WO, SOR, WOR, SW, WW	FP, FB
SM	Type 1	65	18	42	WW, WOR, GM, SF, WB	FP, CB, SY
VG	Type 1	17	33	50	SB, SO, SW, WW, SOR, WOR, WR, WT	FP, FB
FORAGE CROPPING SYSTEMS						
BB	Type 3	70	20	37	WW, WR, WOR, SM, SO, SB	FP, LU, GC, AF
ES	Type 3	68	17	10	SB, WB, SOR, WOR, SW, WW, SO, WO, GR	GC
VG	Type 1	17	33	50	SM, SB, SO, SW, WW, SOR, WOR, WR, WT, GR	FP, FB, FP-SO, GC

BB, Brandenburg; CB, Calabria; ES, Eastern Scotland; SM, Sud-Muntenia; and VG, Västra Götaland. Types represent different land capabilities. AF, Alfalfa; CB, common bean; DW, durum wheat; FB, faba bean; FP, field pea; GC, grass-clover; GM, grain maize; GR, grass; LU, lupin; SB, spring barley; SF, sunflower; SM, silage maize; SO, spring oat; SOR, spring oilseed rape; SW, spring wheat; SY, soybean; WB, winter barley; WO, winter oat; WOR, winter oilseed rape; WR, winter rye; WT, winter triticale; WW, winter wheat.

performance of crops. All impacts were calculated on a per-hectare and per-year basis.

Calculation of Nitrate Leaching

Nitrate leaching is calculated as nitrate-N based on the soil type, preceding crop and crop management as a function of the soil leaching probability and the nitrogen surplus. Adapted from Gäth and Wohlrab (1992) and Bachinger and Zander (2007), nitrate leaching is calculated as:

$$N_{leaching} = N_{surplus} * L_P$$

where L_P is the leaching probability during the winter (mean winter precipitation divided by the water holding capacity at rooting depth and a crop-specific leaching coefficient) and the $N_{surplus}$ is calculated as:

$$N_{surplus} = \left(N_{manureP} + N_{fertilizer} + N_{mineralization} - N_{dfs}\right)$$

where $N_{manureP}$ is the plant-available N content in solid and liquid manure and $N_{mineralization}$ is calculated as a function of the total organic N content (typical contents per soil type) and a region-specific mean annual mineralization rate modified by the pre-crop specific N supply level. N_{dfs} is the nitrogen derived from soil and is calculated as:

$$N_{dfs} = N_{uptake} - N_{fixation}$$

Where N_{uptake} is the nitrogen accumulated by the crop and $N_{fixation}$ is BNF of grain and forage legumes calculated as a function of the yield, the N content, the crop-specific ratio of above and below ground plant N, the percentage of N derived from the atmosphere (%Ndfa) depending on soil mineral N content using minimum and maximum %Ndfa values (Peoples et al., 2009), the percentage of legumes in crop mixtures (grass-clover and cereal-legume intercropping), and the ratio of fixed N transferred to grass in grass-clover mixtures. The soil mineral N content is estimated taking into account possible N mineralization from preceding crop residues in spring, along with N inputs from plant-available N in manure and mineral N fertilizer (Reckling et al., 2016). Atmospheric nitrogen deposition and non-symbiotic nitrogen fixation were not considered because reliable data was not available.

Calculation of Nitrogen Fertilizer Efficiency and Use

Nitrogen fertilizer efficiency was calculated as the ratio of the N output in harvested grain or biomass to the N input from mineral and organic fertilizer and N in seed. N fertilizer use was calculated as the average N applied from organic and mineral N fertilizer.

Calculation of Nitrous Oxide Emissions

Soil-based nitrous oxide emission from crop cultivation was calculated with the IPCC 2006 Tier 1 methodology, including direct and indirect emission from applied fertilizer, manure and crop residues using standard factors and parameters, and assuming no emission from nitrogen fixation (IPCC, 2006).

Calculation of Gross Margins

Gross margins are calculated by subtracting variable costs (inputs, variable costs of machinery and services) from the revenues (yield multiplied by the product price). Labor costs, subsidies, interest and insurance were not taken into account. Data for the economic analysis were obtained from regional statistics (KTBL, 2010; LELF, 2010; Scottish-Agricultural-Census, 2010; SJV, 2011; ISTAT, 2013; MADR, 2013; SAC, 2013; Eurostat, 2015).

Analysis of Cropping Systems

Cropping systems were generated for each land capability and evaluated using the impact indicators. The evaluated systems were grouped as either *arable*, including only grain crops, or *forage*, including at least one forage crop, and the groups were further divided into those *with legumes* or *without legumes*. The dataset was analyzed by plotting environmental impacts against economic impacts of all single cropping systems and by calculating the mean impact and difference between arable and forage systems with and without legumes.

Trade-offs were analyzed between economic and environmental impacts using multi-criteria analysis based on five indicators: (i) Gross margin, (ii) nitrate leaching,

(iii) nitrous oxide emission, (iv) N fertilizer use, and (v) N fertilizer efficiency. To allow relative comparisons, impacts were normalized for each land capability and arable and forage systems separately by dividing the impact of each single cropping system by the overall mean. Trade-offs were analyzed for selected cropping systems:

1) *current farming* without legumes
2) *economic-environmental optimized* systems without legumes
3) *economic-environmental optimized* systems with legumes

Current farming without legumes represents the most common cropping system per land capability based on the crop proportions and expert knowledge. *Economic-environmental optimized* cropping systems with and without legumes are selected from the large range of generated systems according to the criteria:

- Gross margin is equal to or 50€ lower than the system with the highest gross margin
- Nitrate leaching is equal to or lower than the system with the highest gross margin
- Nitrous oxide emission is equal to or lower than system with the highest gross margin

The selected cropping systems were compared in radar charts to identify trade-offs and to evaluate the performance from a systems perspective.

Model Evaluation

Generated rotations were compared by agronomists and advisors with current and plausible cropping patterns. Algorithms for nitrate leaching were already validated by Bachinger and Zander (2007) using HERMES, a dynamic model that simulates water and soil nitrogen dynamics (Kersebaum, 1995). Model outputs for each land capability type were compared against published data and by experts in nitrate leaching. The IPCC method to estimate nitrous oxide emissions is acknowledged to be simple and static, so is open to criticism (Philibert et al., 2012). Nevertheless, it is widely used in GHG inventories and emission models (Lokupitiya and Paustian, 2006; Berdanier and Conant, 2012) because it can be relatively easily applied with only a few input parameters. The model outputs were checked by a specialist in nitrous oxide emission measurements and models (B. Rees, pers. communication). Calculations of impacts at the cropping systems scale were evaluated by a panel of 12 agronomists who discussed the generated systems and compared the results with current management practices in their regions.

RESULTS

Integration of legumes affected the economic and environmental performance of cropping systems. The environmental performance was evaluated in terms of reduced leaching and nitrous oxide loss and lower fertilizer use, and economic performance was evaluated in terms of higher gross margin. On average across all regions, arable cropping systems with legumes had 18% less nitrous oxide emission, 24% less N fertilizer use, no increase in nitrate leaching and no decrease in gross margin (**Table 2**). In forage cropping systems, average effects were stronger, namely 33% less nitrous oxide emission, 38% less N fertilizer use, 22% less nitrate leaching and 21% higher gross margin (**Table 2**). Effects varied between the case studies due to the different crop management, bio-physical and socio-economic conditions, and between arable and forage systems.

TABLE 2 | Average economic and environmental impacts for cropping systems with and without legumes.

Region	Site	Legume	Generated systems [no.]	Gross margin [€/ha]	NO_3-N leaching [kg/ha]	N fertilizer use [kg/ha]	N_2O emissions [kg/ha]
ARABLE CROPPING SYSTEMS							
BB	Type 2	+legume	249	14	21	88	3.0
		-legume	68	51	23	114	3.6
CB	Type 1	+legume	328	195	24	32	1.3
		-legume	12	263	23	53	1.9
ES	Type 1/2	+legume	1237	637	23	107	4.1
		-legume	227	600	30	132	4.6
SM	Type 1	+legume	220	476	13	86	3.0
		-legume	20	369	13	108	3.6
VG	Type 1	+legume	10,127	420	30	100	3.5
		-legume	1756	452	31	121	4.0
FORAGE CROPPING SYSTEMS							
BB	Type 3	+legume	102	130	18	53	2.2
		–legume	89	80	37	126	4.7
ES	Type 3	+legume	18	733	24	220	7.4
		-legume	23	715	30	311	9.7
VG	Type 1	+legume	146	482	15	146	4.7
		-legume	108	483	14	201	6.1

BB, Brandenburg; CB, Calabria; ES, Eastern Scotland; SM, Sud-Muntenia; and VG, Västra Götaland. Types represent different land capabilities.

Effects of Integrating Legumes into Arable Cropping Systems

The inclusion of grain legumes, such as pea and faba bean, in arable cropping systems reduced N fertilizer use by 17–40% and nitrous oxide emission by 12–30% (**Table 2**). The smallest effects were found in VG for N fertilizer use and ES for nitrous oxide emission and the greatest in CB for N fertilizer use and nitrous oxide emission. Gross margins were 6 and 29% higher in ES and SM, and 7, 26, and 73% lower in VG, CB and BB, respectively (**Table 2**). Nitrate leaching was similar or slightly reduced in arable systems. Cropping systems with high gross margin tended to have higher nitrous oxide emissions than cropping systems with low gross margin. No relationship was observed between gross margin and nitrate leaching (**Figure 2**).

The *current farming* scenario without legumes achieved high economic returns in BB, CB and ES and resulted in relatively high nitrate leaching and nitrous oxide emission in all regions (**Figure 2**). In VG and SM the *current farming* system achieved a low economic and environmental performance although it is practiced by a large share of farmers according to crop share statistics. The inclusion of oilseed rape improved gross margins in BB and VG and environmental benefits in cropping systems with and without legumes. The *economic-environmental optimized* system without legumes performed economically better than *current farming* in ES, SM and VG and environmentally better in all regions (**Table 3**).

The *optimized* system with legumes achieved 12 and 97% higher gross margins in ES and SM than the *optimized* system without legumes, because of the high yield and high market value (for food use) of pea and common bean in the two regions (**Table 3**). Besides, adding a legume as another break crop to the cropping system led to an increased yield of the subsequent crop and hence to a higher gross margin. In VG, CB, and BB the gross margin was 4–21% lower in the *optimized* system with legumes than in the *optimized* system without legumes, mainly due to low yields and low prices for pea and faba bean (**Table 3**). The *optimized* system with legumes had 12–25% lower N fertilizer use, 9–28% lower nitrate leaching (except in CB) and 9–23% lower nitrous-oxide emissions than that without legumes (**Table 3**). The major contributory factors to the improved environmental effects were that no N fertilizer was applied to the legume crop, the N fertilizer dose to the subsequent crop was reduced and substitution of winter wheat with crops that received smaller amounts of fertilizer. In CB, the *optimized* system with and without legumes reduced nitrate leaching by 72–81% compared to common practice, because higher leaching was calculated for winter oat than for winter oilseed rape (**Table 3**). In BB and CB, a proportion of legume-supported systems resulted in negative gross margins because they did not include a sufficient proportion of profitable crops such as oilseed rape (**Figure 2**).

Effects of Integrating Legumes into Forage Cropping Systems

The inclusion of forage legumes such as grass-clover and alfalfa and grain legumes in forage cropping systems reduced N fertilizer use by 27–58% and nitrous oxide emission by 23–52% (**Table 2**). The greatest reduction in N fertilizer use and nitrous oxide loss was found in BB where highly fertilized silage maize was replaced by unfertilized grass-clover. The effect was smaller in VG and ES where perennial grasses with low legume content were replaced by grass-clover including 30% clover in the crop biomass. Gross margins were similar between systems with and without legumes in VG and ES and 62% higher with legumes in BB (**Table 2**). Nitrate leaching was similar in both systems in VG and was reduced by the introduction of legumes in ES and BB. Similar to arable systems, forage cropping systems with high gross margins tended to have high nitrous oxide emissions, but no relationship was observed between gross margin and nitrate leaching (**Figure 3**).

The current cropping systems without legumes achieved high economic returns in BB and VG and resulted in relatively high nitrate leaching and nitrous oxide emission in all regions (**Figure 3**). In ES the *current farming* system with an additional cereal crop in the cropping system did not perform as well as the *optimized* systems economically or environmentally (**Table 3**), although it is practiced by a large proportion of farmers. Silage maize dominates *current farming* in BB, and has the best economic performance. Maize yields relatively well on these soils and the economic value of silage maize for feed and biogas is high. The *optimized* system without legumes performed better economically than *current farming* in ES and VG and better environmentally in all regions (**Table 3**). In BB, the *optimized* system with a lower share of silage maize resulted in improved environmental performance than the *current* system, but with lower gross margin.

The *optimized* system with legumes achieved 3–4% higher gross margins than without legumes in VG and ES because of the similar yield and price of grass-clover compared to grass, lower costs of fertilization, and a better pre-crop effect that led to a yield increase in the subsequent crop (**Table 3**). In BB, the integration of grass-clover resulted in 40% higher gross margin due to the high price of grass-clover forage, lower fertilizer costs and a greater pre-crop effect compared to silage maize, and despite the higher costs of harvesting 3 cuts of grass-clover rather than one crop of silage maize (**Table 3**). Across the regions, the *optimized* system with legumes had 23–54% lower N fertilizer use, 21–58% lower nitrous-oxide emissions, and, in BB, 70% lower nitrate leaching than that without legumes (**Table 3**). The major reasons for this difference include that no N was applied to the legume crop, the reduced N fertilizer doses to the subsequent crop and changes in the crop sequence.

Trade-Offs between Economic and Environmental Impacts

In all arable cropping systems, trade-offs were found between gross margin and environmental impacts (**Figure 4**). In BB, CB, and ES *current farming* had a high, above-average, economic and a low, below-average, environmental performance. The *optimized* system without legumes performed similarly in environmental terms to *current farming* in BB and ES. In CB, SM, and VG the *optimized* systems without legumes performed better on at least

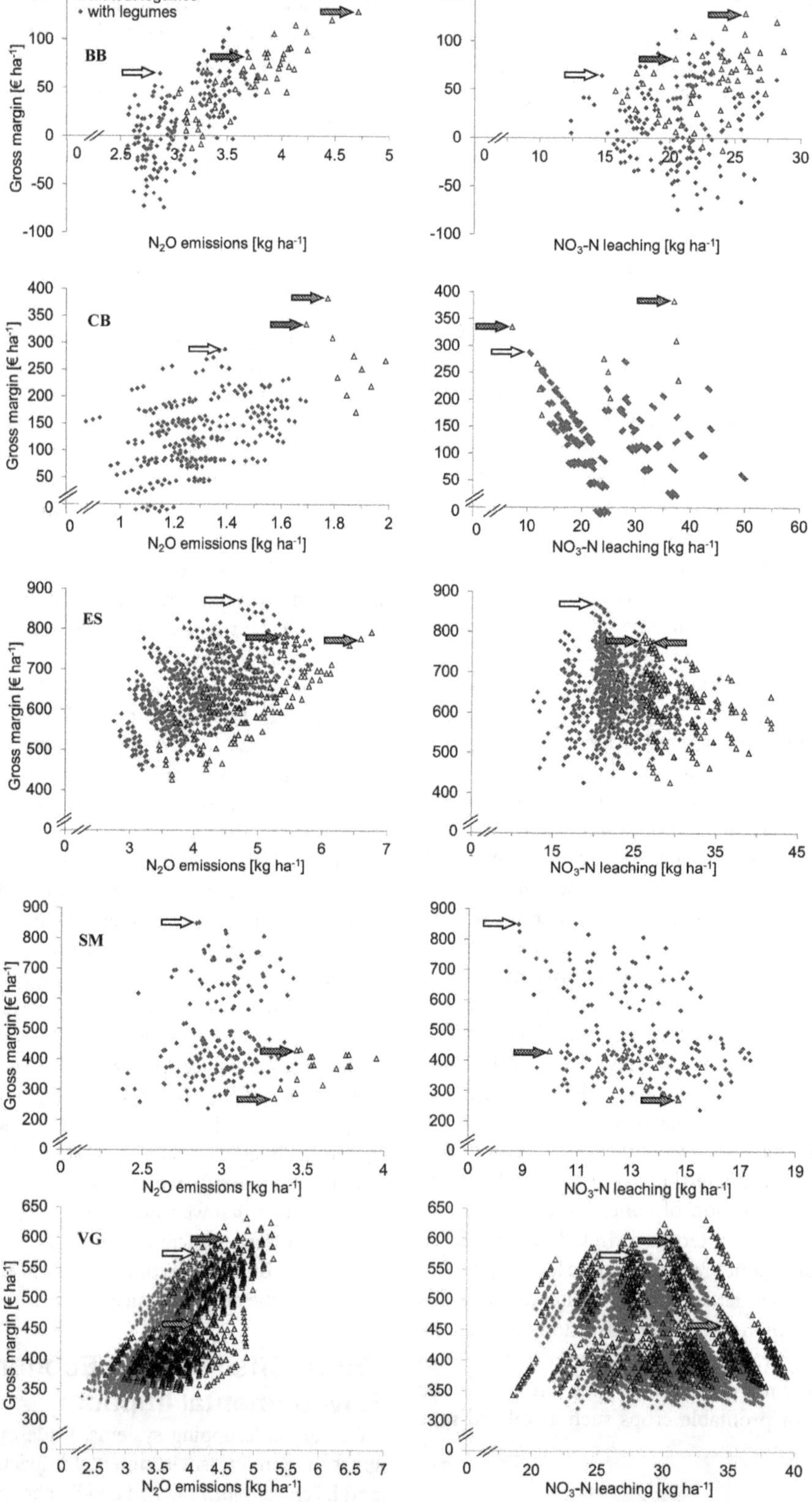

FIGURE 2 | Gross margin plotted against N_2O emissions and NO_3-N leaching for arable cropping systems with and without legumes in Brandenburg (BB), Calabria (CB), Eastern Scotland (ES), Sud-Muntenia (SM), and Västra Götaland (VG). Blank and dotted arrows indicate economic-environmental optimized systems with and without legumes and dashed arrows indicate current farming.

TABLE 3 | Economic and environmental impacts for current and optimized cropping systems with and without legumes.

Region	Site	System	Legume	Crop 1	Crop 2	Crop 3	Crop 4	Crop 5	Crop 6	Gross margin [€/ha]	NO_3-N leaching [kg/ha]	N fertilizer use [kg/ha]	N_2O emissions [kg/ha]
ARABLE CROPPING SYSTEMS													
BB	Type 2	Current	−legume	WW	WB	OR				128	26	139	4.7
		Econ.−env. optimized	−legume	WR	WR	WR	SB	OR		82	20	125	3.7
		Econ.−env. optimized	+legume	WW	WR	WR	WR	**FP**		64	15	94	2.9
CB	Type 1	Current	−legume	WB	WO	WB	WO			383	37	49	1.8
		Econ.−env. optimized	−legume	WB	OR	WB	OR			335	7	46	1.7
		Econ.−env. optimized	+legume	WB	OR	WB	**FB**	OR		287	10	34	1.4
ES	Type 1/2	Current	−legume	WB	OR	WW	WW	WB		776	27	192	6.6
		Econ.−env. optimized	−legume	WB	OR	WB	WO	SB		780	26	156	5.4
		Econ.−env. optimized	+legume	WB	OR	WB	WO	**FP**		869	20	130	4.7
SM	Type 1	Current	−legume	GM	WW	SF				272	15	102	3.3
		Econ.−env. optimized	−legume	GM	WB	OR				430	10	103	3.5
		Econ.−env. optimized	+legume	GM	WB	OR	**CB**			848	9	80	2.8
VG	Type 1	Current	−legume	WW	WW	SO				459	35	124	4.1
		Econ.−env. Optimized	−legume	WW	SB	OR	WR	SB		598	31	143	4.6
		Econ.−env. optimized	+legume	WW	SB	OR	WR	**FB**		573	28	126	4.1
FORAGE CROPPING SYSTEMS													
BB	Type 3	Current	−legume	SM	SM	SM	WR			262	40	166	6.9
		Econ.−env. optimized	−legume	SM	WR	SM	SO	WR		131	34	131	4.8
		Econ.−env. optimized	+legume	**GC**	**GC**	WR	WR	WR	SB	184	10	60	2.0
ES	Type 3	Current	−legume	GR	GR	GR	SB	SO		664	29	293	9.1
		Econ.−env. optimized	−legume	GR	GR	GR	SB			767	18	340	10.5
		Econ.−env. optimized	+legume	**GC**	**GC**	**GC**	SB			795	12	235	8.0
VG	Type 1	Current	−legume	GR	GR	GR	WW	SO		535	12	222	6.6
		Econ.−env. optimized	−legume	GR	GR	GR	OR	WT	SO	551	12	212	6.3
		Econ.−env. optimized	+legume	**GC**	**GC**	**GC**	OR	WR	SO	567	13	163	5.0

BB, Brandenburg; CB, Calabria; ES, Eastern Scotland; SM, Sud-Muntenia; and VG, Västra Götaland. Types represent different land capabilities. CB, Common bean; FB, faba bean; FP, field pea; GC, grass-clover; GM, grain maize; GR, grassland; OR, winter oilseed rape; SB, spring barley; SF, sunflower; SM, silage maize; SO, spring oat; WB, winter barley; WO, winter oat; WR, winter rye; WW, winter wheat; legumes are highlighted in bold.

one of the environmental measures, particularly nitrate leaching. The *optimized* systems with legumes performed economically well in all regions, except in BB, and reached above-average scores for all environmental indicators (**Figure 4**).

Trade-offs between gross margin and environmental impacts were found for forage cropping systems without legumes (**Figure 5**). All systems without legumes performed below average, except regarding gross margin and nitrate leaching. The legume-supported systems reached above-average scores on all indicators, including gross margin. The environmental performance was generally good, especially considering nitrate leaching (**Figure 5**).

DISCUSSION

Environmental and Economic Impacts

The reduction in the use of N fertilizer when legumes were integrated into cropping systems, 17–40% in arable and 27–58% in forage systems (**Table 2**), is mainly attributable to the nitrogen added through the BNF of legume crops (Peoples et al., 2009). The greatest savings were made in forage systems, because the perennial legumes included (clover and alfalfa) could fix more than 350 kg ha^{-1} (Carlsson and Huss-Danell, 2003), which is much more than the corresponding figures for the annual legumes, pea and faba bean, that fix about 130 kg ha^{-1} (Peoples et al., 2009). In BB, the N fertilizer savings were particularly large (58%), because unfertilized grass-clover replaced silage maize that received high doses of N fertilizer. In VG and ES, less N fertilizer was saved (27 and 29%, respectively) (**Table 2**), because the grass-clover was fertilized, even if the doses were quite low. The different management of grass-clover represents common farming practice in the respective regions. The potential of BNF in grass-clover could be exploited more effectively than is common practice in VG and ES if N fertilizer doses to grass-clover were reduced and/or distributed in time in a way that allows a sufficient share of legumes, as well as reducing losses (Eriksen et al., 2015). Nitrogen fertilizer savings in subsequent crops depend on the economic trade-off between securing maximum yields and maximizing N savings (Preissel et al., 2015). In our study, we used N fertilizer prices that were relatively

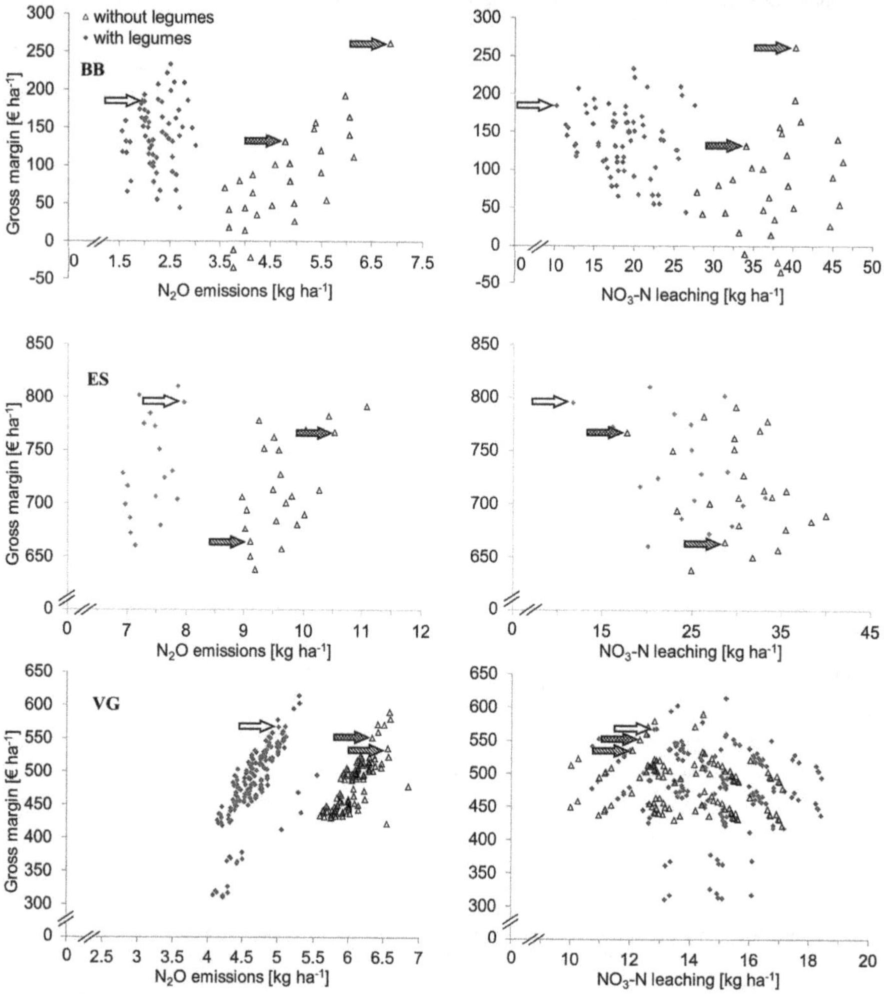

FIGURE 3 | Gross margin plotted against N_2O emissions and NO_3-N leaching for forage cropping systems with and without legumes in Brandenburg (BB), Eastern Scotland (ES), and Västra Götaland (VG). Blank and dotted arrows indicate economic-environmental optimized systems with and without legumes and dashed arrows indicate current farming.

low in relation to product prices. Therefore maintaining high yields came out as more important than saving N fertilizer, which means that we estimated small fertilizer savings, but high yield increases in subsequent crops. However, the framework allows new applications, such as to maximize N fertilizer savings, and the results of this analysis would have been different with this aim.

The average nitrous oxide emissions in cropping systems with legumes were 18 and 30% lower in arable and forage systems, respectively (**Table 2**). The differences among regions were greater for forage systems than for arable systems. The difference was greatest in BB (52%) due to the lower amounts of N applied to grass-clover than to silage maize. There is a direct relationship between N fertilizer input and nitrous oxide emissions, as shown by several studies compiled by Buckingham et al. (2014). However, we have used the assumption that 1% of each kg of N fertilizer is released as field-based nitrous oxide emission (IPCC, 2006), and this is currently under consideration. Philibert et al. (2012) calculated lower emission factors (EF) when the amount of N applied was below 160 kg N ha^{-1}, Hinton et al. (2015) estimated EF to be between 0.28 and 1.35% of applied N depending on the N input, and Rees et al. (2012) concluded from a meta-analysis using measured data from Europe that annual emissions from arable sites were significantly greater than predicted by IPCC. Besides N fertilizer inputs, field-based nitrous oxide emissions are influenced by soil, management and environmental conditions (Ball et al., 2014). It seems clear that the EF used in the present study is approximate, but there is no consensus on how to modify it to better represent actual emissions.

The risk of nitrous oxide emission is generally considered to be lower from legumes than from cereals, oilseed rape and grassland (Stehfest and Bouwman, 2006; Schwenke et al., 2015). In a meta-analysis, Jensen et al. (2011) calculated that emissions from legume crops were around 40% lower than those from N-fertilized crops, and similar to unplanted soils or crops that were not N fertilized. The reason is that legume N inputs through BNF

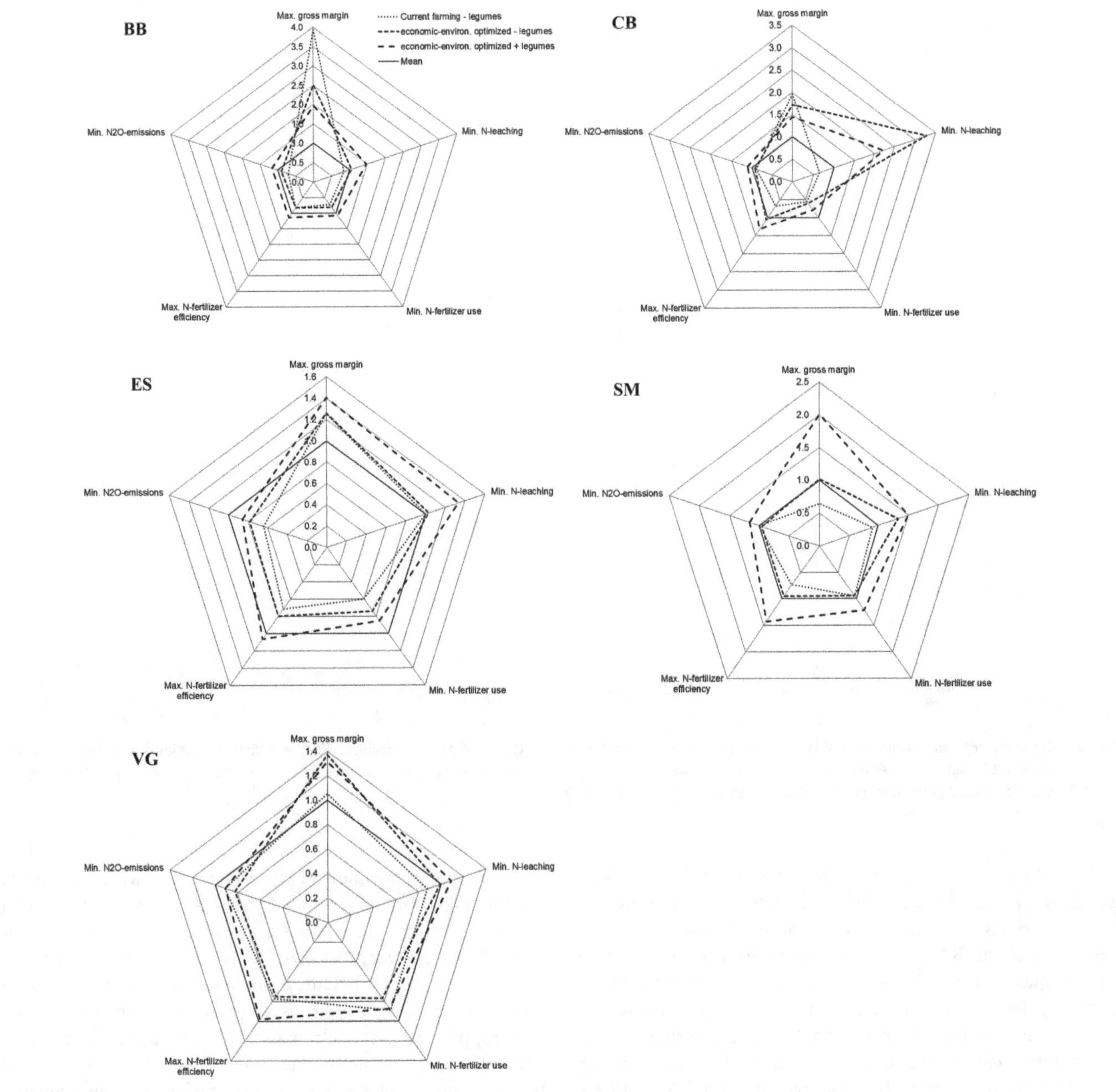

FIGURE 4 | Multi-criteria assessment of arable cropping systems in Brandenburg (BB), Calabria (CB), Eastern Scotland (ES), Sud-Muntenia (SM), and Västra Götaland (VG), for scenarios with and without legumes. Values are the ratio of the single impact relative to the average impact calculated for that indicator across all cropping systems per region (outside values represent positive impacts). Absolute values are shown in **Figure 2**.

supplement the uptake of soil mineral N to meet crop N demand. The preferential use of soil mineral N leads to low availability of nitrate N for potential denitrification losses (Schwenke et al., 2015).

Nitrate leaching was similar with and without legumes in arable cropping systems (**Table 2**). Similarly to nitrous oxide emissions, nitrate leaching is soil-dependent and influenced by tillage, which did not differ between cropping systems with and without legumes. Tillage operations could in some circumstances be reduced before and after grain legumes (Luetke-Entrup et al., 2003; López-Bellido et al., 2003, 2004), which could reduce losses of nitrogen and increase gross margins of legume-supported systems (Preissel et al., 2015). The lower leaching in CB in the *optimized* cropping systems than in the current system was a result of the substitution of winter oat by winter oilseed rape (**Table 3**). The latter can take up more N before winter and thus reduce the risk of nitrate leaching (Sieling and Kage, 2012).

In forage systems, management factors such as soil tillage, nitrogen fertilization and the management of legumes in grass-clover mixtures influence nitrate leaching significantly (Eriksen

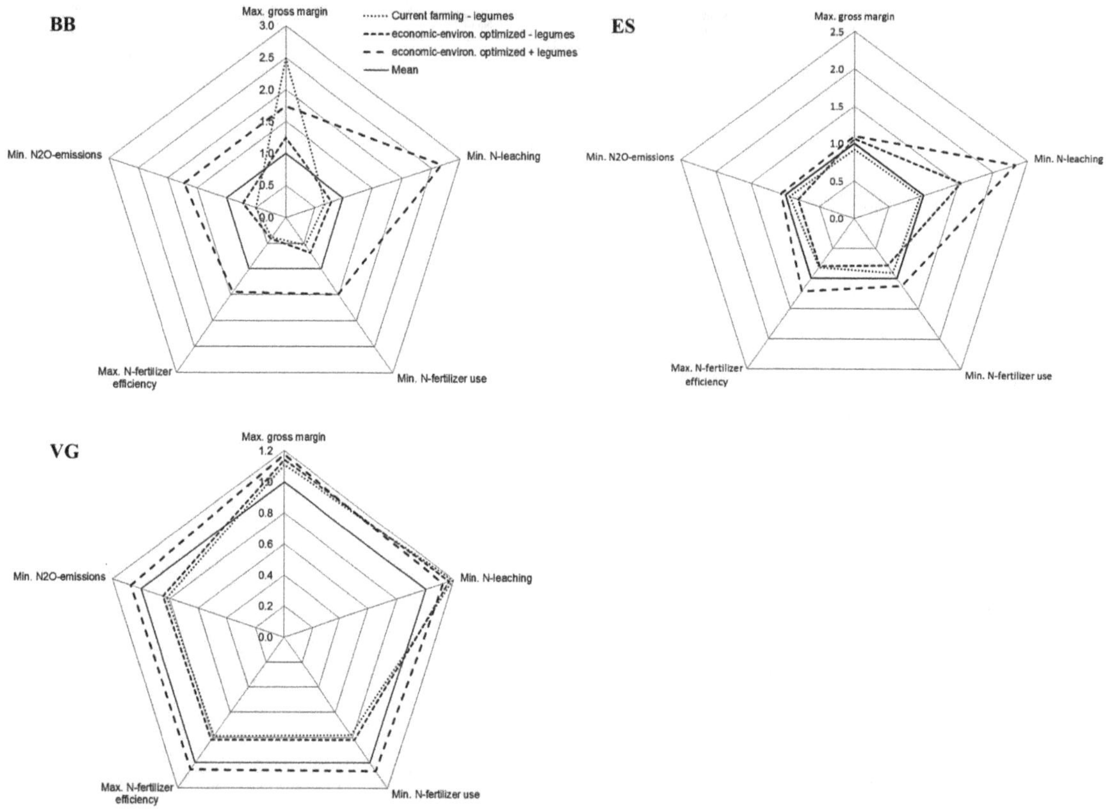

FIGURE 5 | Multi-criteria assessment of forage cropping systems in Brandenburg (BB), Eastern Scotland (ES), and Västra Götaland (VG), for scenarios with and without legumes. Values are the ratio of the single impact relative to the average impact calculated for that indicator across all cropping systems per region (outside values represent positive impacts). Absolute values are shown in **Figure 3**.

et al., 2015). In our study, nitrate leaching was similar in both *optimized* systems in VG, slightly reduced with legumes in ES and strongly reduced in BB (**Table 3**). Reduced leaching with legumes in BB can be attributed to continuous crop cover in grass-clover compared to winter fallow between maize crops and high N fertilization of the maize crop compared to no N fertilization of grass-clover, in combination with a humid winter climate that causes most of the leaching to occur from October to March. In ES, lower doses of N fertilizer to grass-clover than to pure grass explain the slightly lower leaching.

The so-called break-crop effect is another aspect that increases the efficiency of utilizing N after a legume. Roots of a given crop are healthier after an unrelated crop has been grown, because pathogen populations are reduced (Angus et al., 2015), allowing more N to be taken up by the crop, and reducing the availability of N for leaching. There is a risk of large leaching losses after incorporation of N-rich crop residues, such as legume residues, so proper management is required to avoid losses (Eriksen et al., 2015). In our study, the risk of losses was reduced through the generation of optimized cropping systems where pre-crop effects were used effectively. Cover cropping is another option to reduce nitrate leaching after grain legumes (Plaza-Bonilla et al., 2015), but was not tested in this study.

Economic performance is regarded a key driver responsible for low adoption of legumes in cropping systems by farmers (Von Richthofen et al., 2006). As individual crops, legumes generally have lower gross margins than cereals and oilseed crops (Preissel et al., 2015). Our study revealed that the economic performance of legume crops was improved when assessed at the cropping system scale that allows consideration of break-crop effects on fertilizer use, other inputs, and yields of the following crops. The rotational effects of legumes are generally not considered in the economic evaluations of cropping systems (Zander et al., 2016). When cropping system effects are included, the competitiveness of legumes improved and the difference in gross margin between systems with and without legumes ranged from −67 to 106€ in arable and from 0 to 50€ in forage systems across all regions (**Figures 2, 3**). The high economic performance of legume systems in SM depends on stable market prices for common bean.

Trade-Offs between Economic and Environmental Impacts

Cropping systems with a high economic performance tended to have high nitrous oxide emissions due to high input use. Such systems could be relevant for the intensification of agricultural systems, and concentrate the negative environmental impacts onto less arable land according to the land-sparing paradigm (Lamb et al., 2016).

The integration of grain legumes in arable systems had environmental benefits in all regions, but cropping systems with legumes also performed well economically in VG, SM, and ES (**Figure 4**). Thus the introduction of legumes in these regions added services without associated economic cost or losses in productivity, so they contributed to the ecological intensification of cropping systems (Doré et al., 2011). In CB and BB, the gross margin was lower in legume-supported systems than in those without legumes because of the low yields and prices of grain legumes.

Forage systems with legumes achieved both economic and environmental benefits in all regions because of relatively high yields, similar prices and better residual effects on subsequent crops considering yield and N fertilizer use (**Figure 5**).

Thus, adding a legume to cropping systems reduced trade-offs compared to systems without legumes resulting either in (a) win-win situations where legume cultivation is economically attractive and increases environmental benefits, or (b) situations where legumes are not economically attractive but increase environmental benefits.

Opportunities and Constraints for Integrating Legumes into Cropping Systems

In the case of win-win situations, legumes are already economically attractive and provide environmental benefits. The opportunities for forage legumes are easier to utilize because of (i) their generally better economic and environmental performance compared to annual grain legumes, (ii) their high quality feed value for livestock (relatively high prices), and (iii) their relatively simple integration into existing temporary grassland. The perennial nature of forage legumes offers opportunities for climate change mitigation and adaptation (Lüscher et al., 2014), improved biodiversity (Stein-Bachinger and Fuchs, 2012) and soil organic carbon content (Jensen et al., 2011), and lowered risk of both soil erosion (Jensen et al., 2011) and weed infestation (Håkansson, 2003). The deep rooting of some species improves subsoil accessibility for subsequent crops and increases the water infiltration rate (Fischer et al., 2014). However, the adoption of forage legumes is restricted to mixed farms with crop and livestock production, farm collaborations with livestock farmers, and farms that deliver green biomass to biogas plants (Tidåker et al., 2014). In mixed farms, nutrient management through the integration of crops and livestock remains a challenge, because of the high risk of nitrogen losses in manure management (Watson et al., 2005). Important agronomic constraints in legume production include crop establishment and to maintain about one third of legumes in grass-clover mixtures to achieve the maximum benefits from the clover (Suter et al., 2015). When feeding clovers and alfalfa to livestock, there is also an increased risk that ruminants suffer problems with digestion, i.e., bloat (Majak et al., 1995; Dewhurst et al., 2009). The specialization of farming and associated spatial decoupling of livestock and crop production is the major reason for the low proportion of forage legumes in Europe (Lemaire et al., 2015).

There are situations in arable systems were the introduction of legumes has both economic and environmental advantages, especially were grain legumes achieve high prices as a human food product (SM and ES) and where grain yields are relatively high. However, in most situations, grain legumes are still not competitive and there are multiple reasons why European farmers do not grow these crops. The major drivers are economic forces favoring cost-effective production systems and gains from international trade over the economic benefits of diversified production systems (Zander et al., 2016). Agronomic risks include lower yields and yield stability of grain legumes compared to cereals (Cernay et al., 2015), especially under conditions with low water availability such as in BB (Reckling et al., 2015) and in CB. The magnitude and causes of yield variability requires further research (Döring, 2015) and is attributable partly to the much lower investment in legume breeding than in cereal breeding (Meynard et al., 2013). Management strategies to reduce yield stability beyond good farming practices are limited. Other limitations mentioned by experts from the case study regions were increased weed infestation, root rot caused by *Aphanomyces euteiches* in pea, susceptibility to soil compaction, and lack of adapted cultivars (Stoddard, 2013). These constraints require that the legume crop be carefully selected to fit the local conditions and that the crop rotation be well designed with sufficient breaks between legumes that are affected by similar pathogens.

Marketing constraints for grain legumes are mainly due to low prices and few marketing channels. Prices are currently often lower than the actual feed value, and EU-produced grain legumes have difficulties competing with other protein crops on the world market (Voisin et al., 2013). The value chains for EU-produced grain legumes for feed and food are poorly developed in many European countries (Meynard et al., 2013). Regional supply chains for the feed and food market could increase competitiveness of legumes in Europe (Voisin et al., 2013) e.g., a GMO-free protein supply. Furthermore, ecosystem services, including reduced nitrous oxide emissions and nitrate leaching, and increased biodiversity are currently not rewarded through payments, so they are not considered in farmers' economic calculations and could justify policy support (Zander et al., 2016).

Farmers and advisors seldom consider the long-term benefits, focusing instead on single years. This leads to an underestimation of the services provided by legumes. The valuation of such services requires an assessment at the cropping-system scale (Reckling et al., 2016). Appreciation of these services could lead to wider adoption of legumes by farmers, but would also require that practical information on their rotational effects and proper management is available (Stoddard, 2013).

Framework Evaluation

The framework was evaluated using (i) design validation, (ii) plausibility checking of outputs and (iii) end-user validation, as proposed by Bockstaller and Girardin (2003) and described in detail by Reckling et al. (2016). The plausibility checking, especially through experts, provided sufficient information to adapt the model for the specific application in each of the contrasting case studies. Several iterations were needed with agronomists to define the input data and to evaluate the outputs. Thus the quality of the model outputs was dependent to a large extent on the quality of the expert knowledge, as is usually the case for rule-based models (Bachinger and Zander, 2007; Naudin et al., 2015).

The major sensitivity of the framework was the quantification of pre-crop effects that were estimated by experts and supported by information from field experiments (Reckling et al., 2016). Data from field experiments and dynamic models alone do not provide all the required data. Field experiments show varying effects between years and sites, and dynamic models have difficulties in handling pre-crop effects (Lorenz et al., 2013; Kollas et al., 2015).

The static calculation of nitrate leaching and nitrous oxide emission quantifies effects that do not consider the variability of weather, yields and environmental impacts (as is the case in process-based models), and are used here to make only relative comparisons between cropping systems. Nevertheless, static approaches are often used for nitrous oxide emission (Lokupitiya and Paustian, 2006; Berdanier and Conant, 2012) and nitrate leaching (Bachinger and Zander, 2007; Stein-Bachinger et al., 2015) when numerous systems are compared and measurements are not sufficiently available. The equation for nitrate leaching is based on the N surplus at the end of the growing season of each crop and might not sufficiently take into account all losses from mineralizing crop residues. This is especially the case when the N surplus is low due to high N uptake, e.g., the combination of high yields and low N input. Potential N losses through mineralization of N-rich residues from legumes after harvest until the end of the growing season might be underestimated. The indicators used for the evaluation were, however, found to be suitable to account for agronomic, environmental and economic aspects. Additional indicators such as erosion, soil organic matter, diseases and weed infestation, and the water foot print of crops could be added to the sustainability assessment in future.

The assessment framework generated and assessed large numbers of rotations that provide challenges to users, but the large number and diversity of systems provided the opportunity to explore cropping strategies outside existing system configurations and boundaries. This ability to explore a large number of options can also be used to complement information derived from life cycle assessments e.g., Nemecek et al. (2008). Through applying the selection criteria, we were able to identify novel *economic-environmental optimized* systems with legumes that could in the future be tested as prototypes by farmers (Vereijken, 1997).

A particular strength of this assessment framework is that it involves stakeholders from research and practical farming in the process of redesigning and assessing cropping systems that fulfill both economic and environmental aims, and that this can be done without having to test a wide range of systems on farms or experimental stations.

Outlook

To utilize the multiple services of legumes, their production needs to increase significantly. Where legumes already lead to win-win situations with high economic and environmental benefits, such effects need to be communicated to advisors and farmers, and regional supply chains need to be developed for the feed and human food market.

In situations where legumes are not (yet) competitive but their environmental impact is considered desirable, constraints of low yield stability, weed and disease infestation, require innovative solutions from on-farm and on-station experimentation, use of crop management tools and advances in plant breeding. In such situations, policy support could be justified.

AUTHOR CONTRIBUTIONS

All authors contributed to manuscript writing and read and approved the final manuscript. MR, GB, FS, CW, PZ, and JB designed the study. MR, PZ, and JB performed the modeling. RW, AP, IT, GB, MR, and JB provided data for the case studies and were involved in the validation of the results and interpretation.

FUNDING

The work was financed by the German Federal Ministry of Food and Agriculture; the Brandenburg Ministry of Sciences, Research and Cultural Affairs; the Scottish Government RESAS Strategic Research Programme; the EU FP7 project "Legume Futures" (Grant 245216 CP-FP), and the FACCE-ERA-NET+ project Climate-CAFE (Grant PTJ-031A544). The publication of this article was funded by the Open Access fund of the Leibniz Association.

ACKNOWLEDGMENTS

We thank Nicole Schläfke, Renate Wille, Kairsty Topp, Vera Eory, Juliette Maire, Bodil Frankow-Lindberg, Birgitta Båth, Marioara Rusu, and Michele Monti for providing and managing data, and for advice.

REFERENCES

Angus, J. F., Kirkegaard, J. A., Hunt, J. R., Ryan, M. H., Ohlander, L., and Peoples, M. B. (2015). Break crops and rotations for wheat. *Crop Pasture Sci.* 66, 523–552. doi: 10.1071/CP14252

Bachinger, J., and Zander, P. (2007). ROTOR, a tool for generating and evaluating crop rotations for organic farming systems. *Eur. J. Agr.* 26, 130–143. doi: 10.1016/j.eja.2006.09.002

Ball, B. C., Griffiths, B. S., Topp, C. F. E., Wheatley, R., Walker, R. L., Rees, R. M., et al. (2014). Seasonal nitrous oxide emissions from field soils under reduced tillage, compost application or organic farming. *Agric. Ecosyst. Environ.* 189, 171–180. doi: 10.1016/j.agee.2014.03.038

Berdanier, A. B., and Conant, R. T. (2012). Regionally differentiated estimates of cropland N_2O emissions reduce uncertainty in global calculations. *Glob. Chang. Biol.* 18, 928–935. doi: 10.1111/j.1365-2486.2011.02554.x

Bergez, J. E., Colbach, N., Crespo, O., Garcia, F., Jeuffroy, M. H., Justes, E., et al. (2010). Designing crop management systems by simulation. *Eur. J. Agr.* 32, 3–9. doi: 10.1016/j.eja.2009.06.001

Bibby, J., Douglas, K. H., and Thomasson, A. J. S. R. (1982). *Land Capability Classification for Agriculture. Soil Survey of Scotland Monograph*. Aberdeen: The Macaulay Institute for Soil Research.

Bockstaller, C., and Girardin, P. (2003). How to validate environmental indicators. *Agric. Syst.* 76, 639–653. doi: 10.1016/S0308-521X(02)00053-7

Bouxin, A. (2014). *Food and Feed Statistical Yearbook 2013*. Brussels: FEFAC, European Feed Manufacturers' Association.

Buckingham, S., Anthony, S., Bellamy, P. H., Cardenas, L. M., Higgins, S., McGeough, K., et al. (2014). Review and analysis of global agricultural N_2O emissions relevant to the UK. *Sci. Total Environ.* 487, 164–172. doi: 10.1016/j.scitotenv.2014.02.122

Bues, A., Preissel, S., Reckling, M., Zander, P., Kuhlmann, T., Topp, K., et al. (2013). *The Environmental Role of Protein Crops in the New Common Agricultural Policy. European Parliament, Directorate General for Internal Policies, Policy Department B: Structural and Cohesion Policies, Agricultural and Rural Development IP/B/AGRI/IC/2012-067*. 112. Available online at: www.europarl.europa.eu/studies [29.10.2015].

Carlsson, G., and Huss-Danell, K. (2003). Nitrogen fixation in perennial forage legumes in the field. *Plant Soil* 253, 353–372. doi: 10.1023/A:1024847017371

Cernay, C., Ben-Ari, T., Pelzer, E., Meynard, J.-M., and Makowski, D. (2015). Estimating variability in grain legume yields across Europe and the Americas. *Sci. Rep.* 5:11171. doi: 10.1038/srep11171

Dewhurst, R. J., Delaby, L., Moloney, A., Boland, T., and Lewis, E. (2009). Nutritive value of forage legumes used for grazing and silage. *Irish J. Agri. Food Res.* 48, 167–187. Available online at: http://www.jstor.org/stable/20720367

Doré, T., Makowski, D., Malézieux, E., Munier-Jolain, N., Tchamitchian, M., and Tittonell, P. (2011). Facing up to the paradigm of ecological intensification in agronomy: Revisiting methods, concepts and knowledge. *Eur. J. Agr.* 34, 197–210. doi: 10.1016/j.eja.2011.02.006

Döring, T. F. (2015). "Grain legume cropping systems in temperate climates," in *Grain Legumes. Handbook of Plant Breeding*, ed A. M. De Ron (New York, NY: Springer Science+Business Media), 401–434.

Eriksen, J., Askegaard, M., Rasmussen, J., and Søegaard, K. (2015). Nitrate leaching and residual effect in dairy crop rotations with grass–clover leys as influenced by sward age, grazing, cutting and fertilizer regimes. *Agric. Ecosyst. Environ.* 212, 75–84. doi: 10.1016/j.agee.2015.07.001

Eurostat (2010). Available online at: http://ec.europa.eu/eurostat/ [15.3.2012].

Eurostat (2015). Available online at: http://ec.europa.eu/eurostat/web/agriculture/ [29.10.2015]

FAOstat (2015). *Production—Crops. Grain Legume Production Areas EU Countries and Switzerland*. Available online at: http://faostat.fao.org [29.10.2015]

Fischer, C., Roscher, C., Jensen, B., Eisenhauer, N., Baade, J., Attinger, S., et al. (2014). How do earthworms, soil texture and plant composition affect infiltration along an experimental plant diversity gradient in grassland? *PLoS ONE* 9:e98987. doi: 10.1371/journal.pone.0098987

Gäth, S., and Wohlrab, B. (1992). *Strategien zur Reduzierung Standort- und Nutzungsbedingter Belastungen des Grundwassers mit Nitrat*. Oldenburg: Deutsche Bodenkundliche Gesellschaft—AG Bodennutzung in Wasserschutz- und -schongebieten.

Håkansson, S. (2003). *Weeds and Weed Management on Arable Land: An Ecological Approach*. Wallingford: CABI Publishing. doi: 10.1079/9780851996516.0000

Hinton, N. J., Cloy, J. M., Bell, M. J., Chadwick, D. R., Topp, C. F. E., and Rees, R. M. (2015). Managing fertiliser nitrogen to reduce nitrous oxide emissions and emission intensities from a cultivated Cambisol in Scotland. *Geod. Reg.* 4, 55–65. doi: 10.1016/j.geodrs.2014.12.002

IPCC (2006). "2006 IPCC guidelines for national greenhouse gas inventories," in *Prepared by the National Greenhouse Gas Inventories Programme*, eds H. S. Eggleston, L. Buendia, K. Miwa, T. Ngara, and K. Tanabe (Institute for Global Environmental Strategies (IGES)).

ISTAT (2013). *National Institute of Statistics (Italy)*. Available online at: www.istat.it [29.10.2015]

Jensen, E. S., Peoples, M. B., Boddey, R. M., Gresshoff, P. M., Hauggaard-Nielsen, H., Alves, B. J. R., et al. (2011). Legumes for mitigation of climate change and the provision of feedstock for biofuels and biorefineries. A review. *Agron. Sust. Dev.* 32, 329–364. doi: 10.1007/s13593-011-0056-7

Kersebaum, K. C. (1995). Application of a simple management model to simulate water and nitrogen dynamics. *Ecol. Modell.* 81, 145–156. doi: 10.1016/0304-3800(94)00167-G

Kollas, C., Kersebaum, K. C., Nendel, C., Manevski, K., Müller, C., Palosuo, T., et al. (2015). Crop rotation modelling – A European model intercomparison. *Eur. J. Agron.* 70, 98–111. doi: 10.1016/j.eja.2015.06.007

Köpke, U., and Nemecek, T. (2010). Ecological services of faba bean. *Field Crops Res.* 115, 217–233. doi: 10.1016/j.fcr.2009.10.012

KTBL (2010). *Betriebsplanung Landwirtschaft, 22. Auflage*. Darmstadt: Kuratorium für Technik und Bauwesen in der Landwirtschaft.

Lamb, A., Green, R., Bateman, I., Broadmeadow, M., Bruce, T., Burney, J., et al. (2016). The potential for land sparing to offset greenhouse gas emissions from agriculture. *Nat. Clim. Chang.* 6, 488–492. doi: 10.1038/nclimate2910

LELF (2010). *Datensammlung für die Betriebsplanung und die Betriebswirtschaftliche Bewertung landwirtschaftlicher Produktionsverfahren im Land Brandenburg, 7.* Auflage. Landesamt für Ländliche Entwicklung, Landwirtschaft und Flurneuordnung, Frankfurt (Oder). 131.

Lemaire, G., Gastal, F., Franzluebbers, A., and Chabbi, A. (2015). Grassland-cropping rotations: an avenue for agricultural diversification to reconcile high production with environmental quality. *Environ. Manage.* 56, 1065–1077. doi: 10.1007/s00267-015-0561-6

Lokupitiya, E., and Paustian, K. (2006). Agricultural soil greenhouse gas emissions. *J. Environ. Qual.* 35, 1413–1427. doi: 10.2134/jeq2005.0157

López-Bellido, L., López-Bellido, R. J., Castillo, J. E., and López-Bellido, F. J. (2004). Chickpea response to tillage and soil residual nitrogen in a continuous rotation with wheat: I. Biomass and seed yield. *Field Crops Res.* 88, 191–200. doi: 10.1016/j.fcr.2004.01.011

López-Bellido, R. J., López-Bellido, L., López-Bellido, F. J., and Castillo, J. E. (2003). Faba bean (*Vicia faba* l.) response to tillage and soil residual nitrogen in a continuous rotation with wheat (*Triticum aestivum* l.) under rainfed Mediterranean conditions. *Agron. J.* 95, 1253–1261. doi: 10.2134/agronj2003.1253

Lorenz, M., Furst, C., and Thiel, E. (2013). A methodological approach for deriving regional crop rotations as basis for the assessment of the impact of agricultural strategies using soil erosion as example. *J. Environ. Manage.* 127, S37–S47. doi: 10.1016/j.jenvman.2013.04.050

Luetke-Entrup, N., Schneider, M., and Braun, J. (2003). Bearbeitungsgänge sparen, aber mit Augenmaß. *Neue Land. Sond. Konserv. Bodenbearb.* 9–14.

Lüscher, A., Mueller-Harvey, I., Soussana, J. F., Rees, R. M., and Peyraud, J. L. (2014). Potential of legume-based grassland–livestock systems in Europe: a review. *Grass Forage Sci.* 69, 206–228. doi: 10.1111/gfs.12124

MADR (2013). *Project no. 6.2.3. "Technical-Economic Foundation of Production Technologies, Costs Determination and Estimation of the Profitability Level of Vegetal and Animal Products"*.

Majak, W., Hall, J. W., and McCaughey, W. P. (1995). Pasture management strategies for reducing the risk of legume bloat in cattle. *J. Anim. Sci.* 73, 1493–1498.

Meynard, J. M., Messéan, A., Charlier, A., Charrier, F., Farès, M., Le Bail, M., et al. (2013). *RE: Crop Diversification: Obstacles and Levers. Study of Farms and Supply Chains. Synopsis of the Study Report, INRA*. Available online at: www6.paris.inra.fr/depe/content/download/3736/35824/version/1/file/Version+Anglaise+Diversification-8pages.pdf [29.10.2015].

Naudin, K., Husson, O., Scopel, E., Auzoux, S., Giner, S., and Giller, K. E. (2015). PRACT (Prototyping Rotation and Association with Cover crop and no Till) – a tool for designing conservation agriculture systems. *Eur. J. Agron.* 69, 21–31. doi: 10.1016/j.eja.2015.05.003

Nemecek, T., von Richthofen, J.-S., Dubois, G., Casta, P., Charles, R., and Pahl, H. (2008). Environmental impacts of introducing grain legumes into European crop rotations. *Eur. J. Agron.* 28, 380–393. doi: 10.1016/j.eja.2007.11.004

Peoples, M. B., Brockwell, J., Herridge, D. F., Rochester, I. J., Alves, B. J. R., Urquiaga, S., et al. (2009). The contributions of nitrogen-fixing crop legumes to the productivity of agricultural systems. *Symbiosis* 48, 1–17. doi: 10.1007/BF03179980

Philibert, A., Loyce, C., and Makowski, D. (2012). Quantifying uncertainties in N_2O emission due to N fertilizer application in cultivated areas. *PLoS ONE* 7:e50950. doi: 10.1371/journal.pone.0050950

Plaza-Bonilla, D., Nolot, J.-M., Raffaillac, D., and Justes, E. (2015). Cover crops mitigate nitrate leaching in cropping systems including grain legumes: field evidence and model simulations. *Agric. Ecosyst. Environ.* 212, 1–12. doi: 10.1016/j.agee.2015.06.014

Preissel, S., Reckling, M., Schläfke, N., and Zander, P. (2015). Magnitude and farm-economic value of grain legume pre-crop benefits in Europe: a review. *Field Crops Res.* 175, 64–79. doi: 10.1016/j.fcr.2015.01.012

Reckling, M., Döring, T., Stein-Bachinger, K., Bloch, R., and Bachinger, J. (2015). Yield stability of grain legumes in an organically managed monitoring experiment. *Aspects Appl. Biol.* 128, 57–62. doi: 10.13140/RG.2.1.1122.0966

Reckling, M., Hecker, J.-M., Bergkvist, G., Watson, C., Zander, P., Stoddard, F., et al. (2016). A cropping system assessment framework—evaluating effects of introducing legumes into crop rotations. *Eur. J. Agron.* 76, 186–197. doi: 10.1016/j.eja.2015.11.005

Rees, R. M., Augustin, J., Alberti, G., Ball, B. C., Boeckx, P., Cantarel, A., et al. (2012). Nitrous oxide emissions from European agriculture; an analysis of variability and drivers of emissions from field experiments. *Biogeosci. Dis.* 9, 9259–9288. doi: 10.5194/bgd-9-9259-2012

SAC (2013). *Farm Management Handbook.* Available online at: http://www.sruc.ac.uk/info/120376/farm_management_handbook [10.10.2014]

Schwenke, G. D., Herridge, D. F., Scheer, C., Rowlings, D. W., Haigh, B. M., and McMullen, K. G. (2015). Soil N_2O emissions under N_2-fixing legumes and N-fertilised canola: a reappraisal of emissions factor calculations. *Agric. Ecosyst. Environ.* 202, 232–242. doi: 10.1016/j.agee.2015.01.017

Scottish-Agricultural-Census (2010). *Scottish-Agricultural-Census.* Available online at: http://www.gov.scot/Topics/Statistics/Browse/Agriculture-Fisheries/PubFinalResultsJuneCensus [10.10.2014]

Sieling, K., and Kage, H. (2012). Efficient N management using winter oilseed rape. A review. *Agron. Sust. Dev.* 30, 271–279. doi: 10.1051/agro/2009036

SJV (2011). *Yearbook of Agricultural Statistics 2011.* Available online at: www.jordbruksverket.se/swedishboardofagriculture/statistics/agriculturalstatistics/2011.4.13af540213600c8df7a8000399.html [29.10.2015]

Stehfest, E., and Bouwman, L. (2006). N_2O and NO emission from agricultural fields and soils under natural vegetation: summarizing available measurement data and modeling of global annual emissions. *Nut. Cycl. Agroecosyst.* 74, 207–228. doi: 10.1007/s10705-006-9000-7

Stein-Bachinger, K., and Fuchs, S. (2012). Protection strategies for farmland birds in legume–grass leys as trade-offs between nature conservation and farmers' needs. *Organ. Agric.* 2, 145–162. doi: 10.1007/s13165-012-0029-4

Stein-Bachinger, K., Reckling, M., Bachinger, J., Hufnagel, J., Koker, W., and Granstedt, A. (2015). Ecological recycling agriculture to enhance agro-ecosystem services in the baltic sea region: guidelines for implementation. *Land* 4, 737–753. doi: 10.3390/land4030737

Stoddard, F. L. (2013). Agronomic case studies in Legume Futures. *Leg. Fut. Rep.* 1.2., 83. Available online at: www.legumefutures.de/images/Legume_Futures_Report_1.2.pdf

Suter, M., Finn, J. A., Connolly, J., Loges, R., and Lüscher, A. (2015). Gain in nitrogen yield from grass-legume mixtures is robust over a wide range of legume proportions and environmental conditions. *Proc. Environm. Sci.* 29, 187–188. doi: 10.1016/j.proenv.2015.07.253

Tidåker, P., Sundberg, C., Öborn, I., Kätterer, T., and Bergkvist, G. (2014). Rotational grass/clover for biogas integrated with grain production—A life cycle perspective. *Agric. Syst.* 129, 133–141. doi: 10.1016/j.agsy.2014.05.015

Tilman, D., Cassman, K. G., Matson, P. A., Naylor, R., and Polasky, S. (2002). Agricultural sustainability and intensive production practices. *Nature* 418, 671–677. doi: 10.1038/nature01014

Vereijken, P. (1997). A methodical way of prototyping integrated and ecological arable farming systems (I/EAFS) in interaction with pilot farms. *Eur. J. Agron.* 7, 235–250. doi: 10.1016/S1161-0301(97)00039-7

Voisin, A.-S., Guéguen, J., Huyghe, C., Jeuffroy, M.-H., Magrini, M.-B., Meynard, J.-M., et al. (2013). Legumes for feed, food, biomaterials and bioenergy in Europe: a review. *Agron. Sust. Dev.* 34, 361–380. doi: 10.1007/s13593-013-0189-y

Von Richthofen, J.-S., Pahl, H., Bouttet, D., Casta, P., Cartrysse, C., Charles, R., et al. (2006). What do European farmers think about grain legumes. *Grain Legumes* 45, 14–15. Available online at: www.ias.csic.es/grainlegumesmagazine/Grain_Legumes_issue_45.pdf

Watson, C. A., Öborn, I., Eriksen, J., and Edwards, A. C. (2005). Perspectives on nutrient management in mixed farming systems. *Soil Use Managem.* 21, 132–140. doi: 10.1079/SUM2005304

Zander, P., Amjath-Babu, T. S., Preissel, S., Reckling, M., Bues, A., Schläfke, N., et al. (2016). Grain legume decline and potential recovery in European agriculture: a review. *Agron. Sust. Dev.* 36, 1–20. doi: 10.1007/s13593-016-0365-y

Conflict of Interest Statement: The authors declare that the research was conducted in the absence of any commercial or financial relationships that could be construed as a potential conflict of interest.

21

Quinoa for Marginal Environments: Toward Future Food and Nutritional Security in MENA and Central Asia Regions

Edited by:
Didier Bazile,
CIRAD, France

Reviewed by:
Abu Hena Mostafa Kamal,
University of Texas at Arlington, USA
Hector Daniel Bertero,
University of Buenos Aires, Argentina

***Correspondence:**
Redouane Choukr-Allah
r.choukrallah@biosaline.org.ae

Redouane Choukr-Allah*, Nanduri K. Rao, Abdelaziz Hirich, Mohammad Shahid, Abdullah Alshankiti, Kristina Toderich, Shagufta Gill and Khalil Ur Rahman Butt

Research and Innovation Division, International Center for Biosaline Agriculture, Dubai, UAE

Quinoa is recognized as a crop of great value in terms of tolerance to abiotic stresses and there is growing interest to introduce it in marginal agricultural production systems worldwide. Also, quinoa is one of the most nutritious food crops currently known and the nutritive properties of the crop are seen as a mean to fight malnutrition globally. Various quinoa cultivars have been screened for tolerance to salinity, water-use efficiency and nutritional quality and the positive attributes found in them have created even wider global interest in its cultivation. This paper summarizes 15 years of studies on assessing the potential for introducing the crop in a few countries of the Middle East and North Africa (MENA) and Central Asia regions and describes the key constraints for scaling-up the production under marginal growing conditions in the newly introduced countries.

Keywords: irrigation, salinity, yield, sowing dates, food security

INTRODUCTION

Quinoa has been recognized as a climate resilient crop of great value and there is an increasing effort to introduce it in different marginal agriculture production systems of the world. Various quinoa cultivars have been screened for tolerance to abiotic stresses, especially salinity, drought, and frost and the positive attributes of the crop have created wider global interest in its cultivation (Jacobsen, 2003; Jacobsen et al., 2003). Also, quinoa is one of the most nutritious food crops currently known. The seeds contains high quality protein, which has all of the essential amino acids including lysine, methionine and threonine that are scarce in cereals and legumes (Repo-Carrasco et al., 2003). In view of its exceptional nutritional quality and ability to grow under marginal environments, the Food and Agriculture Organization of the United Nations (FAO) has identified quinoa as one of the crops that will play an important role in ensuring future food security and designated the year 2013 as the "Year of Quinoa" (Bazile et al., 2015). Worldwide, the demand for quinoa is growing, especially in the health food segment, but current supplies are unable to match it. Besides the use for human consumption, quinoa seed has other uses as livestock and poultry feed. The whole plant can be used as green fodder and harvest residues can be fed to the animals. In the context of the Middle East and North Africa (MENA) and Central Asia regions, where soil and water salinity is increasingly becoming a constraint to agricultural production, quinoa is seen as an alternative crop with significant potential to have a central role in sustaining farm productivity.

The International Center for Biosaline Agriculture (ICBA) in Dubai, United Arab Emirates (UAE) has been working since 2007 on quinoa in partnership with the Ministry of Environment and Water (MOEW) of the UAE, Abu Dhabi Farmer's Service Center (ADFSC) and its Peruvian partners [Instituto Nacional de Innovacion Agraria (INIA) and Universidad National Agraria La Molina (UNALM)] to evaluate the performance of several quinoa cultivars for their productivity when grown in marginal conditions. ICBA has identified and developed five high yielding salt- and heat-tolerant lines that are now ready to be tested in other agro-ecological zones. The FAO has also been in the forefront of quinoa research, keeping in view its unique nutritional characteristics and adaptability to a wide range of agro-ecological conditions, especially in the marginal areas and the organization has initiated a Regional Project TCP/RAB/3403 for the introduction, appropriation and institutionalization of its production in Algeria, Egypt, Iran, Iraq, Lebanon, Mauritania, Sudan, and Yemen. ICBA considers quinoa as an essential crop in its efforts to sustain agricultural productivity amidst growing threat of salinity and water scarcity, as well as to tackle hunger, malnutrition and poverty in the MENA and Central Asia regions. To fully exploit the potential of the crop for marginal environments, identification of new and high-yielding quinoa genotypes with good local adaptation and high nutritional quality are crucial, which requires intensified screening and adaption research.

This paper summarizes 15 years of studies in assessing the potential of quinoa for introducing it in some countries of the MENA and Central Asia Regions. In view of quinoa's potential as a stress-tolerant, climate resilient future-proof crop, ICBA is developing best practices for its production and management in marginal environments. The Center is also focusing on utilization and marketing of the produce and on fostering knowledge transfer, strengthening competencies and exchange of experiences among the various stakeholders within the MENA and Central Asia regions.

SCREENING FOR LOCAL ADAPTATION AND YIELD POTENTIAL

ICBA acquired 121 germplasm accessions from the United States Department of Agriculture (USDA) and evaluated them for growth performance and yield at its research station (25°05′49″ N and 55°23′25″E) during the cropping season November 2006–March 2007. The soil at the experimental site was fine sand and moderately alkaline (pH 8.2) with very low organic matter (<0.5%). The seeds were sown in the first week of November 2006 and each accession was planted in three 3 m-rows with spacing of 50 cm between the rows and one meter between two accessions. The distance between plants within each row was maintained at 25 cm. Before sowing, the fertility of the soil was improved by incorporating organic fertilizer (compost) at the rate of 40 t ha^{-1} and during crop growth two split doses of NPK (20:20:20) at the rate of 50 kg ha^{-1}, were applied by banding alongside the rows. The amount of N added to the soil by the combination of compost and NPK was estimated to be 570 kg ha^{-1}. The plants were irrigated with low-salinity water with an electrical conductivity (EC_w) of 2-3 dS m^{-1} using the drip system. Water was applied once every day for 20 min at a flow rate of 4 l h^{-1} per plant and in addition, about 60 mm of rainfall was received during the growing season. Of the 121 accessions planted, seeds were harvested from 73 accessions that survived through the growing season and based on the seed yield data, the top 20 performers were selected for subsequent evaluation, carried out during the cropping seasons 2007-08 and 2008-09 (Rao and Shahid, 2012). The agronomic practices such as irrigation and applications of fertilizers were the same as described previously. The data on grain yield of the 20 accessions from the two cropping seasons (2007-08 and 2008-09) are presented in **Table 1**. The seed yield among the accessions varied between 53.86 and 359.86 g m^{-2} in 2007-08, and between 3.32 and 258.42 g m^{-2} in 2008-09. Averaged over the two seasons, the yield ranged between 34.06 g m^{-2} and 238.99 g m^{-2} among the accessions. The studies showed that quinoa has good adaptation and can be successfully cultivated in the Arabian Peninsula. In a further study conducted at the ICBA Research Station in Dubai during 2009-10, the performance of five top-yielding accessions selected on the basis of average seed yield from previous studies were further evaluated using low-salinity irrigation water (EC_w 2.8 dS m^{-1}) (Rao and Shahid, 2012). Analysis of the data revealed significant differences for plant height, number of primary branches and number of inflorescences per plant. However, differences among accessions for all other traits including inflorescence length, seed yield per plant, fresh and dry biomass were found to be marginal. The seed yield among the five accessions ranged between 374.4 g m^{-2} (Ames 13757) and 533.6 g m^{-2} (Ames 13761) with an average of 456.6 g m^{-2} over accessions. The dry matter yield, averaged over the accessions was 1464 g m^{-2}, with accession Ames 13742 producing the maximum yield of 1624 g m^{-2} (Rao and Shahid, 2012). During course of the yield trials over the years, mass selection to eliminate inferior plant types and improve the yield potential of the selected accessions has resulted in five improved lines (Q1–Q5) which were used in further studies described below.

AGRONOMIC EVALUATION AND PERFORMANCE OF SELECTED LINES

The results from the preliminary trials undertaken at ICBA Research Station gave sufficient indication of the potential of quinoa as a salt-tolerant alternative crop. However, further investigations were needed to study the performance under a range of biophysical environments especially at various soil and water salinities to introduce the crop to the farmers. Hence field trials were conducted under a range of agro-ecological conditions—at Ghayathi in the Western Region of Abu Dhabi Emirate in 2012-13 with two of the five selected lines (Q3, Q5) and NSL 106399; and subsequently in five locations—three in the Northern Emirates (Dibba, Hamraniah, and Al Dhaid) and two in the Western Region of Abu Dhabi (Madinat Zayed and Ghayathi) in 2013-14, involving four of the five selected quinoa lines (Q1, Q2, Q3, and Q5) (**Table 2**). While the two locations planted in the Western Region were agricultural farms leased for the trials, those in the Northern Emirates were the federal government's regional research stations. Each set of lines was

TABLE 1 | Grain yields in 20 selected quinoa accessions grown in winter 2007-08 and 2008-09 (source: Rao and Shahid, 2012).

Accession no.	Source	Yield (g m^{-2})		
		2007-08	2008-09	Mean
Ames 13220	Bolivia	84.57	100.95	92.76
Ames 13719	USA	74.61	87.05	80.83
Ames 13723	USA	80.09	134.42	107.25
Ames 13724	USA	132.32	62.42	97.37
Ames 13727	USA	111.39	133.02	122.2
Ames 13736	USA	57.50	109.07	83.29
Ames 13742	USA	359.86	118.12	238.99
Ames 13749	USA	211.06	79.25	145.16
Ames 13757	USA	187.26	120.32	153.79
Ames 13758	USA	64.81	3.32	34.06
Ames 13761	USA	60.50	258.42	159.46
Ames 21931	Bolivia	84.19	39.88	62.04
Ames 22154	Chile	50.85	45.13	47.99
Ames 22155	Chile	147.34	91.43	119.39
Ames 22157	Chile	118.58	133.33	125.95
NSL 106395	USA	105.77	119.25	112.51
NSL 106398	USA	54.00	193.92	123.96
NSL 106399	USA	167.10	148.65	157.87
NSL 86649	USA	53.86	143.1	98.48
PI 478410	Bolivia	58.06	13.27	35.67
Mean		113.19	106.72	109.95
SE		17.79	14.40	16.12

TABLE 2 | Quinoa yield trails in the UAE: Soil and water characteristics at the locations and details of accessions evaluated (Rao, 2016).

Location	Growing season	Soil texture	Water salinity (dS m^{-1})	Identity
ICBA Reserach Station	2009-10	Sand	2.8	Ames 13742, Ames 13749, Ames 13757, Ames 13761, NSL 106399
Ghayathi	2012-13	Sand	15.1	ICBA-Q3, ICBA-Q5, NSL 106399
Ghayathi	2013-14	Sand	16.3	ICBA-Q1, Q3, Q4, Q5
Madinat Zayed	2013-14	Sandy loam	18.9	ICBA-Q1, Q3, Q4, Q5
Dibba	2013-14	Sand	6.1	ICBA-Q1, Q3, Q4, Q5
Hamraniah	2013-14	Sandy loam	4.5	ICBA-Q1, Q3, Q4, Q5
Al Dhaid	2013-14	Loamy sand	2.3	ICBA-Q1, Q3, Q4, Q5

sown in a randomized complete block design (RCBD) with three replications. In all cases, the distance between rows was 50 cm and between plants within the row was 25 cm. Spacing between two adjacent plots was maintained at 1 m. The amount of water and fertilizers applied were the same for all the five locations and as described in the previous section.

In 2012-13, at Ghayathi, despite high salinity of irrigation water (EC_w14-15 dS m^{-1}), the mean seed yield obtained of the three lines was 750 g m^{-2} which was on par with the highest yields reported from the non-saline traditional quinoa growing areas. The green biomass yield was also high, the mean of the three cultivars being 4.3 kg m^{-2}, indicating the potential of quinoa as an alternative forage crop for saline areas (Rao et al., 2013). In 2013-14 trials, the salinity of irrigation water in the five locations varied between 2.3 and 18.9 dS m^{-1} (**Table 2**). Seed yield from trials in the Northern Emirates correlated with the salinity of irrigation water (EC_w), which was 6.1 dS m^{-1} in Dibba, 4.5 dS m^{-1} in Hamraniah and 2.3 dS m^{-1} in Al Dhaid. Thus, averaged over lines, the seed yield was highest in Al Dhaid (541 g m^2), followed by Hamraniah (398 g m^{-2}) and Dibba (190 g m^2) (Rao, 2016; **Figure 1**). In the western region despite higher salinity of the irrigation water, significantly higher seed yields were recorded than those from the Northern Emirates. In Ghayathi, the mean seed yields of the four lines was 1050 g m^{-2}, much higher than the yields recorded in the previous year (750 g m^{-2}) from the same location. In Madinat Zayed the mean seed yield of the four lines was 700 g m^{-2}, which was similar to the yield obtained from Ghayathi in the previous year (Rao et al., 2013; Rao, 2016; **Figure 1**). The results from yield trials showed high degree of variability in the performance of the different cultivars across locations (**Supplementary Figure 1**). While no major differences existed among the different locations in terms of radiation, mean maximum and minimum temperatures or precipitation during the growing period, the differences in yields could only be attributed to differences in physicochemical properties of the soils and water (other than the EC and pH which were measured), for which more detailed analyses are warranted. The two locations planted in the Western Region being agricultural soils, possibly had a more favorable microenvironment compared to the sandy soils with poorly developed profiles at the research stations in Northern Emirates. Nevertheless, the exceptionally higher yields obtained with highly saline irrigation water in the Western Region showed that quinoa, which is a facultative halophyte, not only endures salinity but also some of its cultivars prosper under saline conditions, therefore has huge potential as an alternative food and feed crop when growing traditional crops becomes uneconomical due to increased groundwater salinity. The results besides confirming quinoa's suitability to withstand high salinity in water and soils, also demonstrated that it has good adaptation to the hyper-arid desert environments and is an excellent candidate for crop diversification in the UAE as well as other countries with similar climatic conditions.

In addition to the UAE, ICBA has recently undertook pilot trials in six other MENA and Central Asian countries and the selected quinoa lines proved to be very productive even under very poor soils (sandy soils) and under conditions of deficit irrigation and high salinity. For instance, in 2015, five quinoa lines (Q1–Q5) were evaluated for seed and forage yields in three countries of Central Asia, namely: Uzbekistan, Tajikistan and Kyrgyzstan under different eco-agroclimatic zones that significantly differ in soil characteristics and agricultural practices (ICBA, 2015). Results from Kyrgyzstan were variable due to late planning but in Uzbekistan the seed yields among the lines ranged between 294 g m^{-2} (Q3) and 557 g m^{-2} (Q5) with a mean

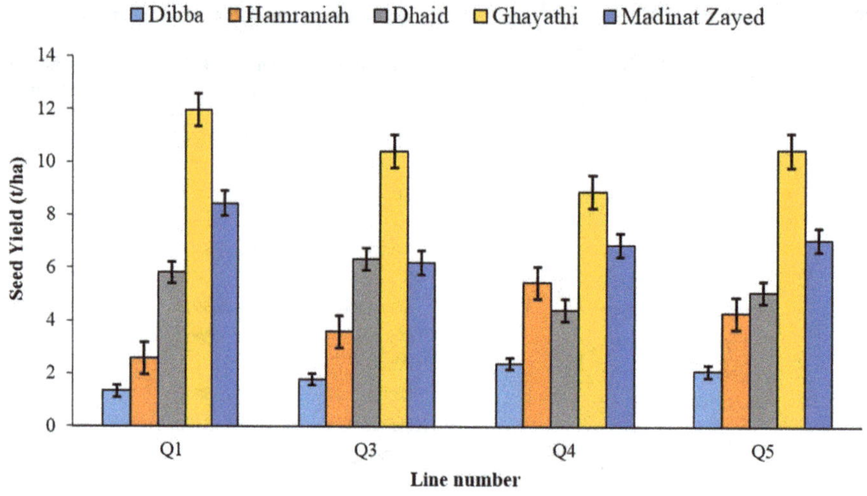

FIGURE 1 | The mean seed yields of four quinoa lines grown at five locations in the UAE during 2013-14. The error bars represent LSD of the means ($p < 0.05$) (Rao, 2016).

of 436 g m^{-2} and in Tajikistan they ranged from 147 g m^{-2} (Q2) to 336 g m^{-2} (Q1) with a mean of 227 g m^{-2} (**Figure 2**).

The high yields obtained under marginal agro-climatic conditions in different locations are indicative of the potential and further investigations are needed to study the performance of a much wider range of genetically diverse accessions at various soil and water salinities to fully exploit the available genetic diversity within the crop. Quinoa is a predominantly self-pollinating species and considerable variation exists between cultivars for many of the desired characters. Therefore, it should be possible to select better adapted genotypes with high yields and nutritional quality combined with salt-and drought-tolerance. Identification of desirable genotypes needs to be followed by work on optimization of cultural practices to maximize productivity under the local conditions.

Quinoa was evaluated and introduced in Morocco for the first time in 2000-01 growing season in two locations, Khenifra (mountain region) and Rabat (coastal region). Obtained results are presented in **Figure 3**. Among the tested lines, G205-95DK performed the best under high altitude conditions due to its origin in the high altitudes of Altiplano (Andean Plateau), while most of the other lines showed adaptation to the coastal region probably because of provenance effect (Benlhabib et al., 2015).

In Morocco, in a field experiment conducted in Rhamna region, four selected quinoa lines (L11, L119, L123, L142, L143) along with two commercial varieties Titicaca and Puno were evaluated for phenological and agro-morphological characteristics (Filali, 2011). Titicaca and L143 recorded the highest seed yield (1.5 t ha^{-1}), while L11 showed the lowest seed yield (0.46 t ha^{-1}). The same lines when cultivated in the experimental farm of the Agronomic and Veterinary Medicine Hassan II Institute, Horticultural complex, Agadir (IAV-CHA) and tested under four different irrigation levels (100, 75, 50, and 25% of full irrigation), responded differently to water stress (**Figure 4**). The line L143 was most affected by water stress, L123 and L11 were less affected as the reduction in yield compared to the fully irrigated treatment was less (**Supplementary Figures 5, 6**).

NUTRITIONAL QUALITY

Quinoa seeds are an exceptionally nutritious food source, owing to their high protein content with all essential amino acids, lack of gluten, and high content of several minerals such as Ca, Mg, Fe, and it is also rich in vitamins. Abiotic stress is known to induce considerable changes in the composition and quality of cereal grains such as wheat (Ashraf, 2014). Despite quinoa's growing popularity, the knowledge about the relationship between the growth conditions and the nutritional profile is still limited. Characterization of the nutritional and anti-nutritional properties of quinoa and evaluation of the effects of extreme growing conditions on the nutritional quality will provide strategic information for the introduction and promotion of quinoa in new environments besides aiding in the selection of nutritionally high and stable cultivars. Results from analysis of the chemical composition of the seeds harvested from three locations (Al Dahid, Ghayathi and Madinat Zayed) with varying levels of irrigation water salinity (2.3, 16.3, and 18.9 dS m^{-1}, respectively) showed that salt stress has marginal effect on the proximate composition (Rao, 2016). However, significant differences were found in the mineral content of the seeds. Especially, seeds harvested from Ghayathi and Madina zayed had higher Na content, but lower levels of Ca and Fe compared to those harvested from Al Dhaid (Rao, 2016). These findings have significant implications, especially for introduction of quinoa in highly saline areas, therefore detailed investigations are recommended for confirmation.

AGRONOMIC TRIALS AND CULTIVATION

In the FAO's Regional Project TCP/RAB/3403, 11 quinoa cultivars were evaluated for seed yield potential in seven of

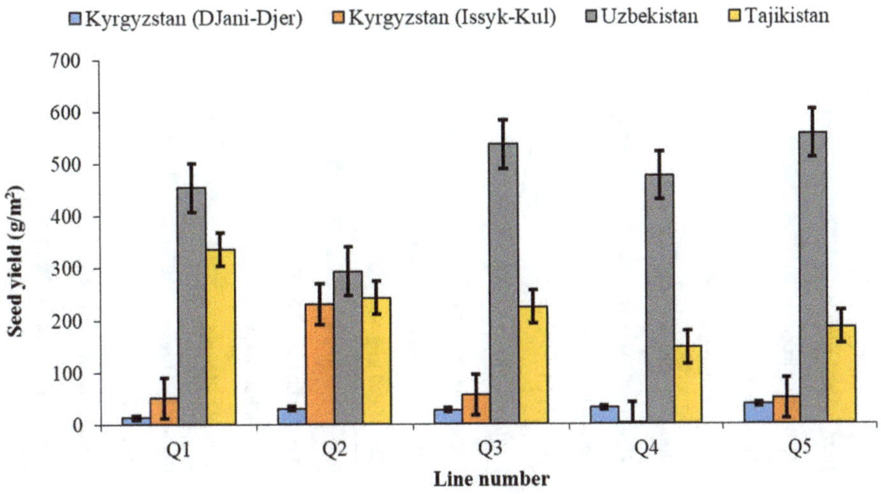

FIGURE 2 | The mean seed yields of five quinoa lines grown at four locations in Central Asia. The error bars represent standard errors of the means (ICBA, 2015).

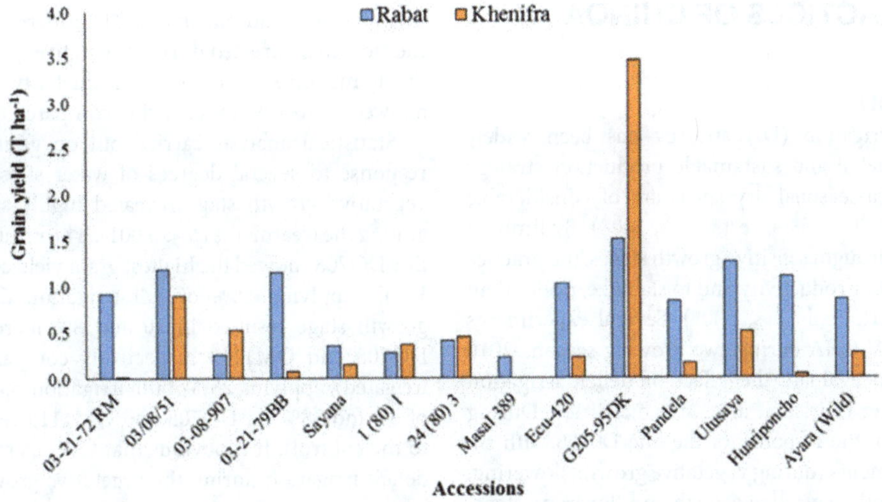

FIGURE 3 | Quinoa grain yields (T ha^{-1}) under high altitude (Khenifra) and coastal (Rabat) conditions in 2000-01 season (Benlhabib et al., 2015).

the eight participating countries (Egypt, Iraq, Iran, Lebanon, Mauritania, Sudan and Yemen) during January-June 2014. Large yield variations per hectare were observed across locations within and between countries, the mean seed yields ranging from 0.11–0.96 t ha^{-1} in Iraq, 0.24–1.90 t ha^{-1} in Yemen, 0.41–3.87 t ha^{-1} in Egypt, 1.50–7.50 t ha^{-1} in Lebanon, 0.16–1.56 t ha^{-1} in Iran, and 0.03–0.23 t ha^{-1} in Mauritania. The highest yield (7.50 t ha^{-1}) was recorded in Lebanon, followed by Egypt (3.87 t ha^{-1}), while the lowest yield was recorded in Mauritania (0.23 t ha^{-1}) (Dost, 2015). A comparison of the cultivars across the countries/locations revealed very interesting and consistent patterns of performance that resulted in the identification of several potential/suitable varieties for the countries. The cultivars Q12 and Q21 exhibited outstanding performance in almost all the countries of evaluation, while Q29, Q18, Q19, Q22, Q26, and Q27 produced superior yields as compared to other cultivars at all locations/countries except Iran. Convinced by quinoa's performance, one private agriculture company in Egypt and a farmer in Lebanon planted quinoa on a large-scale (1–2 hectares) with a plan to further extend the production together with improvement in capacity of farmers and generating market demand. However, private sector involvement in quinoa production will mainly depend on the conducive policies of the governments, allocation of financial and human resources for research and development of new varieties along with the improvement in the production technologies, processing and utilization within the region.

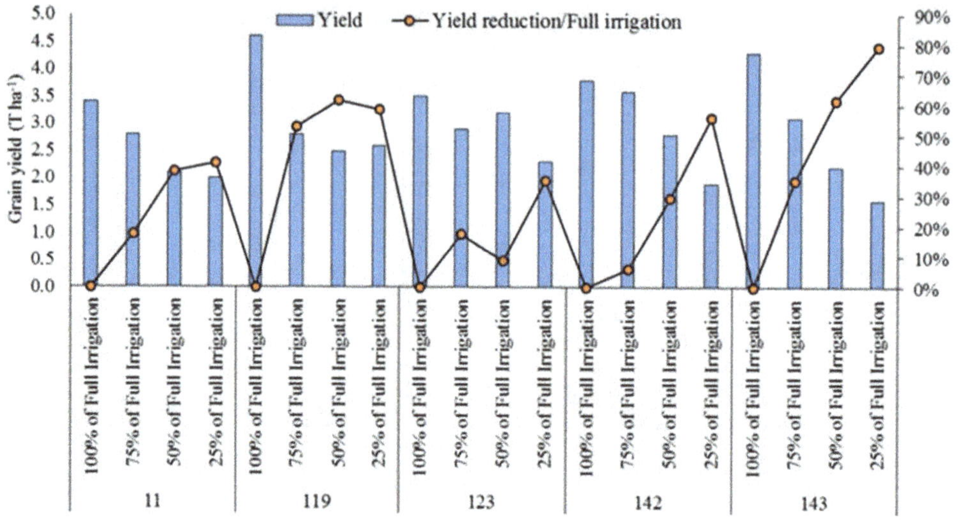

FIGURE 4 | Responses of 5 quinoa lines to different irrigation levels in terms of yield (Filali, 2011).

CROPPING PRACTICES OF QUINOA PRODUCTION

Deficit Irrigation

In quinoa, deficit irrigation (DI) strategy has been widely investigated as a valuable and sustainable production strategy in regions where intra-seasonal dry spells are of considerable importance (Garcia et al., 2003; Geerts et al., 2008). By limiting water applications to drought sensitive growth stages, the practice aims to maximize water productivity and to stabilize, rather than maximize yields (Geerts and Raes, 2009). Several experiments were conducted at IAV-CHA during two growing seasons 2010 and 2011 in order to evaluate the effect of deficit irrigation with treated wastewater (Hirich et al., 2012a,b,c, 2013). During the first season (2010), the response of the line D0708 with six deficit irrigation treatments (during vegetative growth, flowering, grain filling, during both vegetative growth and flowering stage, alternating with 100% of full irrigation as non-stress condition, and 50% of full irrigation as water deficit condition) were studied. During the second season (2011), two lines (D0708 and QM1113) were studied with six other irrigation treatments (rainfed, 0, 25, 50, 75, and 100% of full irrigation) only during the vegetative growth stage, while in the remaining stages, full irrigation was provided except for the rainfed treatment. The results clearly showed that the highest yields were obtained in the fully irrigated treatment followed by the treatment stressed during vegetative growth stage. When subjected to deficit irrigation during flowering, grain filling, during both vegetative growth and flowering, and during all crop stages, a yield reduction of 32, 37, 46, and 50% was recorded respectively, in comparison with the control (full irrigation). The finding indicates that applying deficit irrigation during vegetative growth stage results in increased root to shoot ratio. Early water deficit induced root development and full irrigation applied during flowering and grain filling allowed the plants to produce more biomass due to the well-developed root system, possibly leading to higher levels of water and nutrient uptake. The increased yield observed in the treatment stressed during vegetative growth stage is probably due to maximized crop water productivity (CWP), which further allowed up to 19% water saving compared to full irrigation.

Statistical analysis carried out on grain yield of quinoa in response to several degrees of water stress applied during the vegetative growth stage revealed highly significant differences among the treatments ($p < 0.001$). The results clearly showed that line DO708 showed the highest grain yield compared to QM1113. While applying 50% of full irrigation during the vegetative growth stage resulted in 10 and 8% increase in seed yield in DO708 and QM1113 respectively compared to control (fully irrigated), applying 25% of full irrigation caused a yield reduction of 31 and 38% (in DO708 and QM1113 respectively, compared to the control). It is obvious that CWP is maximized under 50% deficit irrigation during the vegetative growth stage in both the lines.

Results from field trials conducted in sandy soil during two growing seasons (2011 and 2012) at the Experimental Station of Cadi Ayyad University located in 70 km south West Marrakech (Morocco) and also under rainfed condition in a farmer's field at Tnin Bouchan, about 250 km away from Agadir indicated that deficit irrigation (50 and 33% of full irrigation) affected quinoa performance and resulted in seed yield reduction of 15.8, 30.1% respectively in the first season (2011) and by 15.2, 41.5% respectively, in the second season (2012) in comparison with full irrigation. Under rainfed conditions, seed yields were reduced by 62.1 and 59.3% in the 2011 and 2012 growing seasons, respectively compared to full irrigation. CWP was maximized in the treatment receiving 50% of full irrigation (Fghire et al., 2013).

Organic Amendment

One of the most important factors which limits crop growth and production is water scarcity. As already discussed, studies show that deficit irrigation is a judicious strategy to increase water use efficiency and CWP in water-scarce environments. Organic amendments, on the other hand, improve the water holding capacity of soils and combing them with deficit irrigation can be a practical solution for sustainable yields as the negative effect of water scarcity can be compensated through the increased water holding capacity of the soils and there by improved availability of water and nutrients to the plants. In line with this, a study was conducted in at IAV-CHA to evaluate the combined effect of organic amendment (compost) and deficit irrigation using treated wastewater on productivity of the quinoa line (D0708) during October 2011 to March 2012 (Hirich et al., 2013, 2014a). Three levels of organic amendments (0, 5, and 10 t ha^{-1}) combined with two levels of deficit irrigation (50 and 100% of full irrigation) were studied. While the highest seed yield (66.3 g plant^{-1}) was recorded when quinoa was subjected to full irrigation and received 10 t ha^{-1} of compost, yield was low under water deficit conditions without organic amendment (**Figure 5**). The results indicated that organic amendment of 10 and 5 t ha^{-1} increased grain yield by 16 and 3% respectively, under full irrigation conditions, and by 18 and 13% under deficit irrigation (respectively). It can be concluded that organic amendment improve quinoa yields significantly under deficit irrigation conditions by improving the water holding capacity of the soils and the access to water and nutrients by the plants. Combining deficit irrigation and organic amendment also allowed maximizing the CWP.

Quinoa Fertilization

The combined effect of water stress and nitrogen application was studied on a quinoa line (D0708) in a field trial laid out in split-plot design at IAV-CHA, during the year 2013 (**Supplementary Figures 2, 4**). The treatments included four levels of water stress (25, 50, 75, and 100% of full irrigation) and seven different rates of nitrogen (0, 40, 80, 120, 160, 200, and 240 kg ha^{-1}) application. The results suggested that seed yield increased with increasing nitrogen supply, though the response varied with the level of water stress (**Figure 6**). The yield was highest in the 50% of full irrigation treatment with 240 kg ha^{-1} of nitrogen. CWP increased with higher supply of nitrogen and the degree of water stress, the value being highest in the most stressed treatment (25% of full irrigation) and 240 kg ha^{-1} of nitrogen and lowest with the full irrigation without nitrogen supply (Hirich, 2014).

In a field trial conducted in Wadi El-Natroon region, Beheira Governorate, Egypt during 2008/2009 and 2009/2010 winter seasons, Shams (2011) studied different rates of nitrogen fertilization (0, 90, 180, 270, and 360 Kg N ha^{-1}) for improvement of growth and yield in sandy soils. High nitrogen fertilizer rate significantly increased yield during both the seasons. However, nitrogen use efficiency has reduced with increased rate of nitrogen application. In a pot experiment with two quinoa lines (Quinoa-52 and Quinoa-37) and two commercial varieties (Titicaca and Puno). Lavini et al. (2014) studied the effect of five rates of nitrogen application (0, 50, 100, 150, and 200 mg kg^{-1} of soil). The results showed that both the lines responded similarly to the application of nitrogen and yield has improved significantly with increased nitrogen rate.

In studies conducted by Gomaa (2013) in Egypt, quinoa plants were fertilized with ammonium nitrate (34% N) at 0, 120, 238, 357 kg ha^{-1} in combination with nitrobin (biofertilizer) or calcium super phosphate (15.5% P_2O_5) at 0, 120, 238, 357 kg ha^{-1} in combination of phosphorin (biofertilizer). The plants performed the best in the treatment receiving 238 kg ha^{-1} of ammonium nitrate in combination with nitrobin.

Use of Saline Water

A pot experiment was carried out at the IAV-CHA to evaluate the response of quinoa (line D0708) to different levels of irrigation water salinity (1, 10, 20, and 30 dS m^{-1}) (Hirich et al., 2014c,d; Lavini et al., 2014). Irrigation with water at 10 dS m^{-1} salinity has not affected seed yield significantly and yield reduction was only 9% compared to the fresh water control (1 dS m^{-1}). Increasing salinity to 20 dS m^{-1} and then to 30 dS m^{-1} resulted in a significant reduction in seed yield (24 and 34%, respectively, compared to the control. Contrarily, saline water irrigation had positive effect on CWP—the most efficient treatment being the most salt-stressed treatment (30 dS m^{-1}), which produced 0.8 kg for 1 m^3 of water and the lowest being the control (0.6 kg m^{-1}) ($p < 0.001$). This difference in terms of CWP can be explained in terms of the difference in water uptake which was higher in the fresh water treatment and lower in the saline treatments.

Three lines of quinoa (QM1113, QS0938, D0708) cultivated in open field were irrigated with treated wastewater with different levels of salinity in order to assess the impact on yield (El Youssfi et al., 2012). Significant differences were found in the performance of the lines—QM1113 being the most productive with an average grain yield of 6.92 t ha^{-1} at 6 dS m^{-1} salinity followed by D0708 with an average yield of 5.65 t ha^{-1} at 3 dS m^{-1}. The findings indicated that quinoa yields increase with increasing salinity of up to 6 dS m^{-1} and being a facultative halophyte, salinity promotes growth but up to a certain threshold, beyond which growth and productivity start to be negatively affected. Results from these studies differ with those of Rao (2016) from on-farm trials where seed yields were maximum despite irrigation with highly saline water (15–18 dS m^{-1}); and those from field experiments in Turkey where salinity stress of up to 40 dS m^{-1} did not interfere with seed and biomass yields significantly (Yazar et al., 2015); or of Razzaghi et al. (2011) who showed an yield reduction of 50% with saline water of 40 dS m^{-1} compared with fresh water.

Simple seed germination tests were conducted at ICBA, Dubai to study the role of Bontera™ (Biofertilizer and humic acid) in enhancing seed germination and in mitigating the negative impact of salinity (**Figure 7**). Different levels of sodium chloride of up to 50% seawater salinity (1.2% NaCl) were tested in combination with different dilutions of Bontera™ (ranging from 1:20 to 1:500). The results of these tests showed that salinity in the growth medium inhibited seed germination to the extent that at 1.6% NaCl seed germination did not occur. At lower concentrations of Bontera™ (dilutions of 1:500 and above) significantly enhanced seed germination and seedling growth, while higher concentrations (1:300) were inhibitory (Gill et al., 2016). Under field conditions, foliar application of Bontera™ consistently showed up to 28% improvement in grain yield. It was concluded that application of Bontera™ at 1–2 l ha^{-1} could reduce the chemical fertilizers application by 25%. Foliar application of Bontera™ in combination with two irrigation

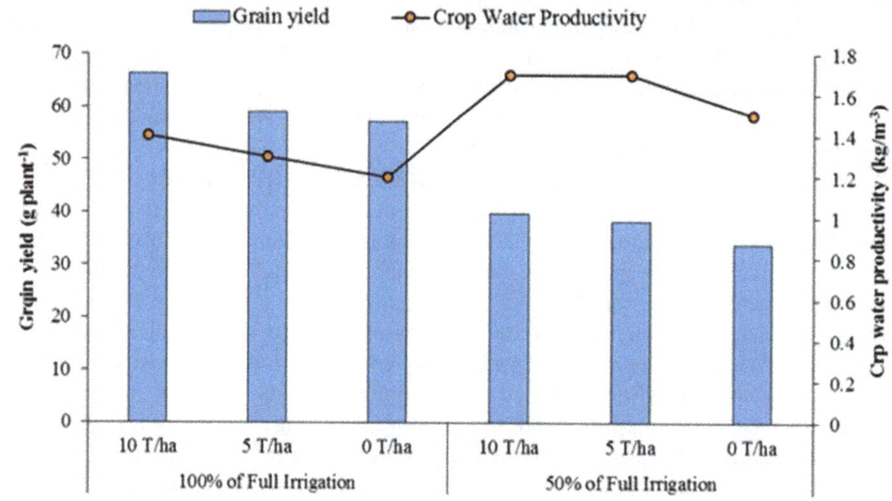

FIGURE 5 | Combined effect of organic amendment and deficit irrigation on quinoa yield and water productivity (Hirich et al., 2014a).

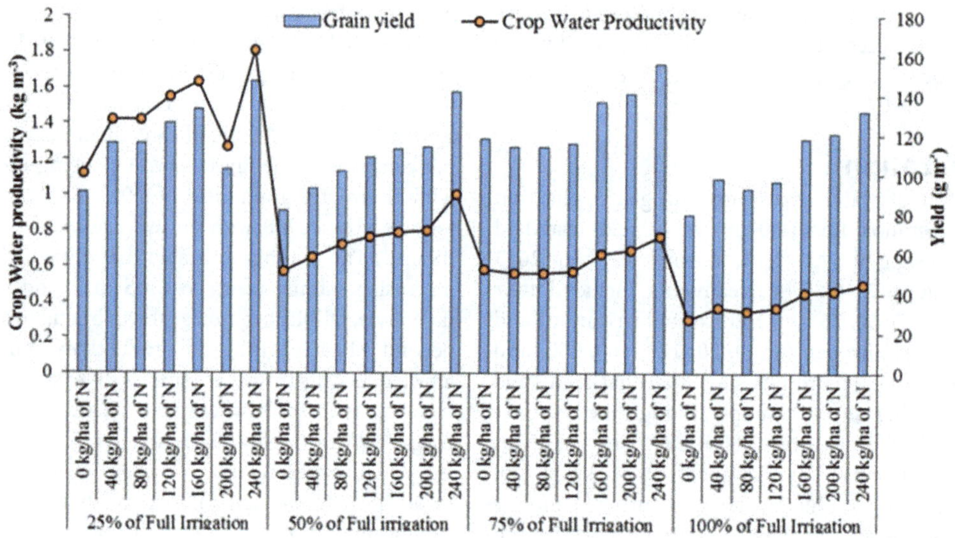

FIGURE 6 | Combined effect of water stress and N fertilization on quinoa yield and water productivity (Hirich, 2014).

regimes (100% ET and 50% ET) showed greater impact at 50% ET suggesting a mitigating effect for stresses like water deficiency and drought (Gill et al., 2016).

Sowing Dates

Hirich et al. (2014b) carried out a series of field trials at IAV-CHA (**Supplementary Figure 3**), to investigate the effects of 10 sowing dates—from 1st November to 15th March with 15-day intervals, on quinoa performance. The study showed that sowing dates affected growth and productivity obviously due to differences in temperature, precipitation and radiation over time (**Figure 8**). Seed and dry matter yields were highest when quinoa was sown in November and early December. Reduced yield during the late sowing dates was explained by delay in germination due to prevailing low temperatures and occurrence of downy mildew during February and March triggered by high air humidity. The length of growing period increased from November to January and decreased from January to March. The longest growing period was when sown on 1st of January and the shortest was when sown on 15th of March.

CONSTRAINTS FOR SCALING UP

Poor germination and crop establishment are particular problems likely to be encountered with quinoa especially in saline or other marginal environments. In the trials conducted in the UAE and the FAO project countries, significantly high variability was observed in the yields of the lines/varieties. Therefore, understanding of the influence of edaphic and climatic factors on productivity in different locations is important. There

FIGURE 7 | Seed germination and seedling growth after 2 days- under different NaCl & Bontera solution Concentrations (Gill et al., 2016).

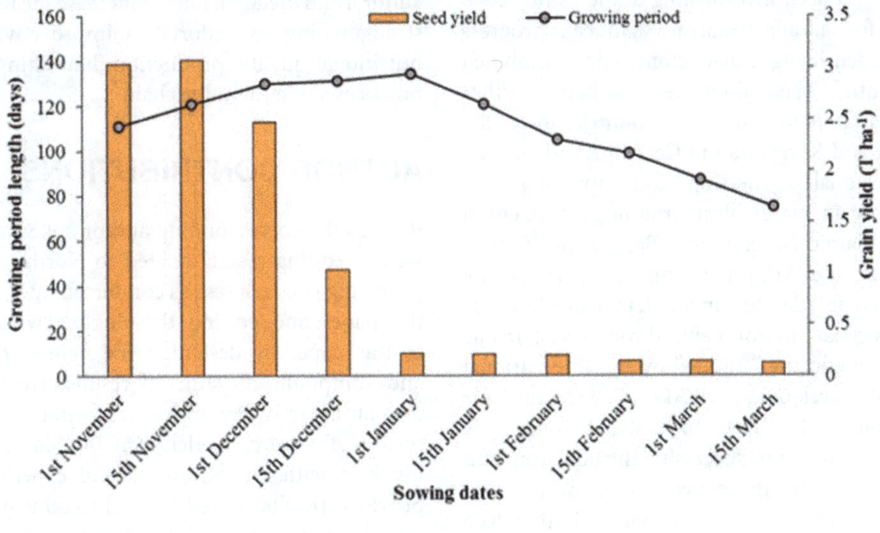

FIGURE 8 | Effect of sowing dates on quinoa yield and growing period length (Hirich et al., 2014b).

is also considerable knowledge gap regarding quinoa pests and diseases, particularly in areas outside its traditional growing regions. Some of the wild relatives such as *Chonopodium murale* and C. album are common weeds in several of the countries where quinoa is introduced and they possibly harbor a number of biotic stressors that can easily migrate onto cultivated species. Introduction and scaling-up of novel crops such as quinoa to non-traditional environments also requires the study and establishment of the entire chain and building capacity of the researchers and the farmers in the basics of production and harvesting, storage and processing technologies (Rao, 2016).

CONCLUSIONS AND RECOMMENDATIONS

In recent years, quinoa is receiving significant attention as a nutritively rich multi-purpose agro-industrial crop that can thrive in extreme soil and climatic conditions. Faced with the challenge of increasing the production of high-quality food to feed the growing population, quinoa offers an excellent alternative to ensure food and nutrition security in marginal environments in the MENA as well as other regions across the globe. In the MENA region, while many countries have recently initiated work on quinoa, Morocco and the UAE have made significant advances toward introducing the crop in the local production systems. In Morocco, IAV HASSAN II Institute has been studying quinoa since 2000 as a drought and salinity-tolerant alternative crop to contribute to the food and nutrition security of people in the mountainous areas. As part of this program, quinoa has already been introduced to a dozen sites across the country and each year, the number of quinoa growers' is increasing all over the regions, especially Oujda, Benslimane, Fes, Boulmane, and Marrakech. However, quinoa yields over the years remained unpredictable and very low, averaging between 1.2 and 1.4 t ha^{-1}, while the maximum attainable yield can be up to 8–10 t ha^{-1}. A range of factors such as the choice of cultivars, optimal sowing date and nutrient management were suggested to affect the production. Constraints which still need to be overcome include crop stand establishment, sensitivity to high temperatures and salinity, weed control and saponin removal. Furthermore, there is also a need to design a product marketing strategy and raise awareness among farmers and the government agencies about quinoa's potential as a stress-tolerant alternative crop for marginal environments.

In the UAE, ICBA has been investigating quinoa since 2006 as an alternative crop for salt-affected areas. Significant progress has been made in identifying salt-tolerant lines combined with high yield potential. These lines are now being further evaluated for their yield potential in several countries including Uzbekistan, Tajikistan and Kyrgyzstan in Central Asia (through the project "Cross-regional partnerships for improving food and nutritional security in marginal environments of Central Asia" funded by the Islamic Development Bank); and Yemen, Jordan, and Egypt in the MENA region [as part of the collaborative project "Adaptation to climate change in West Asia and North Africa marginal environments through sustainable crop and livestock diversification" funded by the International Fund for Agricultural Development (IFAD), OPEC Fund for International Development (OFID)]. As with any other new crop, one of the key factors for successful introduction and establishment of quinoa under the novel climates will be the identification of appropriate planting material. It is therefore important to study the adaptation and yield potential of several quinoa genotypes from different provenances to select the most promising genotypes suitable for the local agro-climatic conditions. Information on these aspects as well as the economic assessment of the profitability of quinoa cultivation is essential, especially when planted under sub-optimal growth conditions with low-quality water in the targeted countries. ICBA's future research on quinoa focuses on developing best practices in relation to the production, management under diverse farming systems and agro-ecological regions, and on utilization and marketing of the produce, besides fostering knowledge transfer, strengthening competencies and exchange of experiences among quinoa researchers within the MENA and Central Asia regions.

It is well known that climate change is rapidly degrading the conditions of crop production. Salinization and aridity are forecasted to increase in most parts of the world, especially in the MENA region. As a consequence, new stress-tolerant genotypes of the existing crops or new and alternative crops or species must be identified and used for future food security. Stress-tolerant crops such as quinoa offer major opportunities due to their comparative advantage over the staple food crops in terms of tolerance to harsh growing conditions. The results presented in this paper have indeed shown that quinoa maintains productivity in rather poor soils and under water stress conditions and high salinity. Moreover, quinoa seeds are an exceptionally nutritious food source, owing to their high protein content with all essential amino acids, lack of gluten, high mineral content (e.g., Ca, Mg, Fe), and health-promoting compounds such as flavonoids. Higher yield do not guarantee quinoa's success in the region and to be successful it must fit in the current cropping patterns, farming systems and prove its worth in rigorous and properly designed on-farm trials in marginal lands affected by salinity and alkalinity and in the areas where the majority of the food crops could not be produced economically. For scaling up and large scale adoption by the farmers in the region, creating opportunities for marketing the produce are also vital. Quinoa being a new crop to the region, more efforts will be needed to create awareness for its incorporation in the social, cultural and dietary habits to increase the market demand besides strengthening the efforts to improve yields combined with nutritional quality, production, harvesting and post-harvesting practices along the value chain.

AUTHOR CONTRIBUTIONS

RC was the corresponding author collecting data and results as well as editing results related to North Africa region. NR was providing research results conducted by ICBA. He was reviewing the paper and editing the English writing. AH contributed to this paper by designing the figures, commenting the data and compiling all gathered results. He was also revising the formatting according to Journal Author guideline. KT provided results of research undertaken in Central Asia. Butt has been involved with Dr. NR in several experiment on quinoa. SG provided results related to seed treatment in order to improve crop tolerance to salinity. AA provided results related to seed treatment in order to improve crop tolerance to salinity.

SUPPLEMENTARY MATERIAL

The Supplementary Material for this article can be found online at: http://journal.frontiersin.org/article/10.3389/fpls.2016.00346

Supplementary Figure 1 | Quinoa panicle variability in terms of color and panicle volume.

Supplementary Figure 2 | Trial on combined impacts of water stress and nitrogen application carried out in Agadir, Morocco.

Supplementary Figure 3 | Sowing date trial carried out in Agadir, Morocco.

Supplementary Figure 4 | Quinoa harvest within field trial on combined impacts of water stress and nitrogen application carried out in Agadir, Morocco.

Supplementary Figure 5 | Field trial about responses of seven quinoa accessions to several irrigation levels.

Supplementary Figure 6 | Genetic and morphological variability of several quinoa accessions.

REFERENCES

Ashraf, M. (2014). Stress-induced changes in wheat grain composition and quality. *Crc. Rev. Food. Sci. Nutr.* 54, 1576–1583. doi: 10.1080/10408398.2011.644354

Bazile, D., Bertero, D., and Nieto, C. (2015). *State of the Art Report on Quinoa around the World in 2013.* Rome: Food and Agriculture Organization of the United Nations (FAO) & CIRAD (Centre de coopération internationale en recherche agronomique pour le développement).

Benlhabib, O., Jacobsen, S. E., Jellen, E. N., Maughan, P. J., and Choukr-Allah, R. (2015). "Status of quinoa production and research in Morocco," in *State of the Art Report on Quinoa around the World in 2013,* eds D. Bazile, D. Bertero, and C. Nieto (Rome: Food and Agriculture Organization of the United Nations), 178–491.

Dost, M. (2015). "Field evaluation results across locations and identification of suitable QUINOA varieties," in *Wrap up Workshop of Regional Quinoa Project (TCP/RAB/3403–FAO)* (Rome: Food and Agriculture Organization of the United Nations).

El Youssfi, L., Choukr-Allah, R., Zaafrani, M., Mediouni, T., Ba Samba, M., and Hirich, A. (2012). Effect of domestic treated wastewater use on three varieties of Quinoa (*Chenopodium quinoa*) under semi arid conditions. *World Acad. Sci. Eng. Technol.* 68, 306–309.

Fghire, R., Oudou, I. A., Filali, K., Benlhabib, O., and Wahbi, S. (2013). "Deficit irrigation and fertilization impact on quinoa water and yield productions ," in *International Conference on Sustainable Water Use for Securing Food Production in the Mediterranean Region under Changing Climate,* ed R. Choukr-Allah. (Agadir: SWUP-MED project), 409–416.

Filali, K. (2011). *Caractérisation et Évaluation du Rendement de Lignées de Quinoa (Chenopodium quinoa) dans la Région de Rhamna.* Rabat: Diplôme de Master en Biotechnologies et Amélioration Génétique des Productions Agricoles, Institut Agronomique et Vétérinaire Hassan II, Morocco.

Garcia, M., Raes, D., and Jacobsen, S.-E. (2003). Evapotranspiration analysis and irrigation requirements of quinoa (*Chenopodium quinoa*) in the Bolivian highlands. *Agric. Water Manage.* 60, 119–134. doi: 10.1016/S0378-3774(02)00162-2

Geerts, S., and Raes, D. (2009). Deficit irrigation as an on-farm strategy to maximize crop water productivity in dry areas. *Agric. Water Manage.* 96, 1275–1284. doi: 10.1016/j.agwat.2009.04.009

Geerts, S., Raes, D., Garcia, M., Vacher, J., Mamani, R., Mendoza, J., et al. (2008). Introducing deficit irrigation to stabilize yields of quinoa (*Chenopodium quinoa* Willd.). *Eur. J. Agron.* 28, 427–436. doi: 10.1016/j.eja.2007.11.008

Gill, S., Alshankiti, A., and Shahid, S. A. (2016). "Evaluation of BonteraTM microbial soil enhancer for improving crop production and water saving in sandy soils," in *Joint Publication of ICBA and Flozyme,* ed S. Gill (Dubai: International Center for Biosaline Agriculture).

Gomaa, E. F. (2013). Effect of Nitrogen, Phosphorus and biofertilizers on quinoa plant (*Chenopodium quinoa*). *J. Appl. Sci. Res.* 9, 5210–5222.

Hirich, A. (2013). Using deficit irrigation with treated wastewater to improve crop water productivity of sweet corn, chickpea, faba bean and quinoa. *Moroccan Rev. Agron. Vet. Sci.* 2, 15–22.

Hirich, A. (2014). *Effects of Deficit Irrigation using Treated Wastewater and Irrigation with Saline Water on Legumes, Corn and Quinoa Crops.* Ph.D., thesis, Hassan II Institue of Agronomy and Veterinary Medicine, Morocco.

Hirich, A., Choukr-Allah, R., and Jacobsen, S. E. (2014a). Deficit irrigation and organic compost improve growth and yield of quinoa and pea. *J. Agron. Crop. Sci.* 200, 390–398. doi: 10.1111/jac.12073

Hirich, A., Choukr-Allah, R., and Jacobsen, S. E. (2014b). Quinoa in Morocco – effect of sowing dates on development and yield. *J. Agron. Crop. Sci.* 200, 371–377. doi: 10.1111/jac.12071

Hirich, A., Choukr-Allah, R., and Jacobsen, S.-E. (2013). The combined effect of deficit irrigation by treated wastewater and organic amendment on quinoa (*Chenopodium quinoa* Willd.) productivity. *Desal Water Treat* 52, 2208–2213. doi: 10.1080/19443994.2013.777944

Hirich, A., Choukr-Allah, R., Jacobsen, S.-E., and Benlhabib, O. (2012a). Could quinoa be an alternative crop of wheat in the Mediterranean region: case of Morocco? *Les Notes d'Alerte du CIHEAM* 86, 1–8.

Hirich, A., Choukr-Allah, R., Jacobsen, S.-E., El Yousfi, L., and El Omari, H. (2012b). Using deficit irrigation with treated wastewater in the production of quinoa (*Chenopodium quinoa* Willd.) in Morocco. *Rev. Científica UDO Agric.* 12, 570–583.

Hirich, A., Choukr-Allah, R., Jelloul, A., and Jacobsen, S.-E. (2014c). Quinoa (*Chenopodium quinoa* Willd.) seedling, water uptake and yield responses to irrigation water salinity. *Acta Hort.* 1054, 145–152. doi: 10.17660/ActaHortic.2014.1054.16

Hirich, A., Choukr-Allah, R., Ragab, R., Jacobsen, S.-E., El Youssfi, L., and El Omari, H. (2012c). The SALTMED model calibration and validation using field data from Morocco. *J. Mat. Environ. Sci.* 3, 342–359.

Hirich, A., Jelloul, A., Choukr-Allah, R., and Jacobsen, S. E. (2014d). Saline water irrigation of quinoa and chickpea: seedling rate, stomatal conductance and yield responses. *J. Agron. Crop. Sci.* 200, 378–389. doi: 10.1111/jac.12072

ICBA (2015). *ICBA Annual Report 2015.* Dubai: International Center for Biosaline Agriculture.

Jacobsen, S. E. (2003). The worldwide potential for quinoa (*Chenopodium quinoa* Willd.). *Food Rev. Int.* 19, 167–177. doi: 10.1081/FRI-120018883

Jacobsen, S. E., Mujica, A., and Jensen, C. R. (2003). The resistance of Quinoa (*Chenopodium quinoa* Willd.) to adverse abiotic factors. *Food Rev. Int.* 19, 99–109. doi: 10.1081/FRI-120018872

Lavini, A., Pulvento, C., D'Andria, R., Riccardi, M., Choukr-Allah, R., Belhabib, O., et al. (2014). Quinoa's potential in the Mediterranean region. *J. Agron. Crop Sci.* 200, 344–360. doi: 10.1111/jac.12069

Rao, N. K. (2016). "Quinoa: a future-proof crop for climate smart agriculture," in *Global Forum for Innovations in Agriculture-2016,* ed R. Choukr-Allah (Abu Dhabi).

Rao, N. K., Rahman, K. U., and Ismail, S. (2013). "Quinoa: prospects as an alternative crop for salt-affected areas," in *Book of Abstracts, 3rd International Conference on Neglected and Under-utilized Species (NUS) for Food Secure Africa,* eds R. A. Hall, P. Rudebjer, and S. Padulosi (Rome: Bioversity International), 95.

Rao, N. K., and Shahid, M. (2012). Quinoa–a promising new crop for the Arabian Peninsula. *American-Eurasian J. Agric. Environ. Sci.* 12, 1350–1355.

Razzaghi, F., Ahmadi, S. H., Adolf, V. I., Jensen, C. R., Jacobsen, S. E., and Andersen, M. N. (2011). Water relations and transpiration of Quinoa (*Chenopodium quinoa* Willd.) under salinity and soil drying. *J. Agron. Crop Sci.* 197, 348–360. doi: 10.1111/j.1439-037X.2011.00473.x

Repo-Carrasco, R., Espinoza, C., and Jacobsen, S.-E. (2003). Nutritional value and use of the andean crops quinoa (*Chenopodium quinoa*) and Kañiwa (*Chenopodium pallidicaule*). *Food Rev. Int.* 19, 179–189. doi: 10.1081/FRI-120018884

Shams, A. (2011). Response of quinoa to nitrogen fertilizer rates under sandy soil conditions. *Int J. Water Resour. Arid Environ.* 1, 318–325.

Yazar, A., Incekaya, Ç., Sezen, S. M., and Jacobsen, S.-E. (2015). Saline water irrigation of quinoa (*Chenopodium quinoa*) under Mediterranean conditions. *Crop Pasture Sci.* 66, 993–1002. doi: 10.1071/CP14243

Conflict of Interest Statement: The authors declare that the research was conducted in the absence of any commercial or financial relationships that could be construed as a potential conflict of interest.

Permissions

All chapters in this book were first published in PCFFTF, by Frontiers Media SA; hereby published with permission under the Creative Commons Attribution License or equivalent. Every chapter published in this book has been scrutinized by our experts. Their significance has been extensively debated. The topics covered herein carry significant findings which will fuel the growth of the discipline. They may even be implemented as practical applications or may be referred to as a beginning point for another development.

The contributors of this book come from diverse backgrounds, making this book a truly international effort. This book will bring forth new frontiers with its revolutionizing research information and detailed analysis of the nascent developments around the world.

We would like to thank all the contributing authors for lending their expertise to make the book truly unique. They have played a crucial role in the development of this book. Without their invaluable contributions this book wouldn't have been possible. They have made vital efforts to compile up to date information on the varied aspects of this subject to make this book a valuable addition to the collection of many professionals and students.

This book was conceptualized with the vision of imparting up-to-date information and advanced data in this field. To ensure the same, a matchless editorial board was set up. Every individual on the board went through rigorous rounds of assessment to prove their worth. After which they invested a large part of their time researching and compiling the most relevant data for our readers.

The editorial board has been involved in producing this book since its inception. They have spent rigorous hours researching and exploring the diverse topics which have resulted in the successful publishing of this book. They have passed on their knowledge of decades through this book. To expedite this challenging task, the publisher supported the team at every step. A small team of assistant editors was also appointed to further simplify the editing procedure and attain best results for the readers.

Apart from the editorial board, the designing team has also invested a significant amount of their time in understanding the subject and creating the most relevant covers. They scrutinized every image to scout for the most suitable representation of the subject and create an appropriate cover for the book.

The publishing team has been an ardent support to the editorial, designing and production team. Their endless efforts to recruit the best for this project, has resulted in the accomplishment of this book. They are a veteran in the field of academics and their pool of knowledge is as vast as their experience in printing. Their expertise and guidance has proved useful at every step. Their uncompromising quality standards have made this book an exceptional effort. Their encouragement from time to time has been an inspiration for everyone.

The publisher and the editorial board hope that this book will prove to be a valuable piece of knowledge for researchers, students, practitioners and scholars across the globe.

List of Contributors

Antonio M. De Ron
Biology of Agrosystems, Misión Biológica de Galicia, National Spanish Research Council, Pontevedra, Spain
Sistemas Agroforestales, Unidad Asociada a la Misión Biológica de Galicia (CSIC), Pontevedra, Spain

P. Rodiño, Marta Santalla and Ana M. González
Biology of Agrosystems, Misión Biológica de Galicia, National Spanish Research Council, Pontevedra, Spain

María J. Lema
Sistemas Agroforestales, Unidad Asociada a la Misión Biológica de Galicia (CSIC), Pontevedra, Spain
Phytopathological Station do Areeiro, Provincial Chamber, Pontevedra, Spain

Isaura Martín
National Center for Plant Genetic Resources, National Institute for Agricultural and Food Research and Technology, Alcalá de Henares, Spain

Jaime Kigel
The Robert H. Smith Faculty of Agriculture, Food and Environment, Hebrew University of Jerusalem, Rehovot, Israel

María J. Suso
Department of Plant Breeding, Institute for Sustainable Agriculture, Spanish National Research Council, Córdoba, Spain

Penelope J. Bebeli
Laboratory of Plant Breeding and Biometry, Department of Crop Science, Agricultural University of Athens, Athens, Greece

Stefanie Christmann
International Center for Agricultural Research in the Dry Areas, Rabat, Morocco

Célia Mateus
Agrarian and Forestry Systems and Plant Protection Unit, National Institute for Agrarian and Veterinarian Research/Linking Landscape, Environment, Agriculture and Food, Oeiras, Portugal

Valeria Negri and Renzo Torricelli
Department of Agricultural, Food and Environmental Sciences, University of Perugia, Perugia, Italy

Miguel A. A. Pinheiro de Carvalho
ISOPlexis Genebank, University of Madeira, Funchal, Portugal
Mediterranean Institute of Environmental and Agricultural Sciences, University of Evora, Evora, Portugal

Maria M. Veloso
Biotechnology and Genetic Resources Unit, National Institute for Agrarian and Veterinarian Research/ Linking Landscape, Environment, Agriculture and Food, Oeiras, Portugal

M. Mercedes Lucas and José J. Pueyo
Institute of Agricultural Sciences, ICA-CSIC, Madrid, Spain

Frederick L. Stoddard
Department of Food and Environmental Sciences, University of Helsinki, Helsinki, Finland

Paolo Annicchiarico
Research Centre for Fodder Crops and Dairy Productions, CRA, Lodi, Italy

Juana Frías and Cristina Martínez-Villaluenga
Institute of Food Science Technology and Nutrition, ICTAN-CSIC, Madrid, Spain

Daniela Sussmann
Fraunhofer-Institute for Process Engineering and Packaging, Freising, Germany

Marcello Duranti
Department of Food Environment and Nutritional Sciences, Università degli Studi di Milano, Milan, Italy

Alice Seger
Terrena Lup'Ingredients, Martigne-Ferchaud, France

Peter M. Zander
Leibniz Centre for Agricultural Landscape Research, ZALF, Müncheberg, Germany

Eliot M. Herman and Monica A. Schmidt
School of Plant Sciences, University of Arizona, Tucson, AZ, USA

Didier Bazile
UPR Green, Department of Environment and Societies, French Agricultural Research and International Cooperation Organization, CIRAD, Montpellier, France

Sven-Erik Jacobsen
Department of Plant and Environmental Sciences, Faculty of Science, University of Copenhagen, Taastrup, Denmark

Alexis Verniau
Ecole Supérieure d'Agricultures, Angers, France

Incoronata Galasso, Roberto Russo, Sergio Mapelli, Elena Ponzoni, Ida M. Brambilla and Remo Reggiani
Institute of Agricultural Biology and Biotechnology, CNR, Milano, Italy

Giovanna Battelli
Institute of Science of Food Production, CNR, Milano, Italy

David Correa-Galeote, Eulogio J. Bedmar, Antonio J. Fernández-González and Manuel Fernández-López
Department of Soil Microbiology and Symbiotic Systems, Estación Experimental del Zaidín, Agencia Estatal Consejo Superior de Investigaciones Científicas, Granada, Spain

Gregorio J. Arone
Department of Agricultural Sciences, National University of Huancavelica, Huancavelica, Peru

Salliana R. Stetina and Rickie B. Turley
Crop Genetics Research Unit, Plant Physiology, United States Department of Agriculture, Agricultural Research Service, Stoneville, MS, USA

Ouafae Benlhabib and Noura Boujartani
Departement de Production, Protection et Biotechnologies Vegetales, Agronomic and Veterinary Institute Hassan II, Rabat, Morocco

Peter J. Maughan and Eric N. Jellen
Department of Plant and Wildlife Sciences, Brigham Young University, Provo, UT, USA

Sven E. Jacobsen
Department of Plant and Environmental Sciences, Faculty of Life Sciences, University of Copenhagen, Copenhagen, Denmark

Emanuela Pedrazzini, Davide Mainieri, Claudia A. Marrano and Alessandro Vitale
Istituto di Biologia e Biotecnologia Agraria, Consiglio Nazionale delle Ricerche, Milano, Italy

Francesca Sparvoli, Eleonora Cominelli, Roberto Bollini and Maria G. Daminati
CNR, Institute of Agricultural Biology and Biotechnology, Milan, Italy

Monica Laureati and Ella Pagliarini
Department of Food, Environmental and Nutritional Sciences, University of Milan, Milan, Italy

Roberto Pilu and Ivan Toschi
Department of Agricultural and Environmental Sciences - Production, Landscape, Agroenergy, University of Milan, Milan, Italy

Gianluca Giuberti and Paola Fortunati
Alimentari e Ambientali, Facoltà di Scienze Agrarie, Istituto di Scienze degli Alimenti e della Nutrizione, Università Cattolica del Sacro Cuore, Piacenza, Italy

Hamid Khazaei, Carolyn T. Caron, Michael Fedoruk, Albert Vandenberg and Kirstin E. Bett
Department of Plant Sciences, University of Saskatchewan, Saskatoon, SK, Canada

Clarice J. Coyne and Rebecca McGee
USDA-ARS, Pullman, WA

Marwan Diapari
Department of Plant Sciences, University of Saskatchewan, Saskatoon, SK, Canada
London Research and Development Centre, Agriculture and Agri-Food Canada, London, ON, Canada

Imade Y. Nsa
Department of Microbiology, University of Lagos, Lagos, Nigeria

Kehinde T. Kareem
Institute of Agricultural Research and Training, Moor Plantation, Obafemi Awolowo University, Ibadan, Nigeria

Sara Mayo, Alicia Lorenzana, M. Piedad Campelo and Pedro A. Casquero
Research Group of Engineering and Sustainable Agriculture, Natural Resources Institute, University of León, León, Spain

Santiago Gutiérrez and Monica G. Malmierca
Area of Microbiology, University School of Agricultural Engineers, University of León, Ponferrada, Spain

Rosa Hermosa
Department of Microbiology and Genetics, Spanish-Portuguese Centre for Agricultural Research, University of Salamanca, Salamanca, Spain

Abirami Ramalingam, Himabindu Kudapa1 and Lekha T. Pazhamala1
International Crops Research Institute for the Semi-Arid Tropics (ICRISAT), Hyderabad, India

Wolfram Weckwerth
Department of Ecogenomics and Systems Biology, University of Vienna, Vienna, Austria

Rajeev K. Varshney
International Crops Research Institute for the Semi-Arid Tropics (ICRISAT), Hyderabad, India
School of Plant Biology and Institute of Agriculture, The University of Western Australia, Crawley, WA, Australia

Longjing Zhou, Qin Zhou, Jian Cai, Xiao Wang, Tingbo Dai, Weixing Cao and Dong Jiang
National Engineering and Technology Center for Information Agriculture/Key Laboratory of Crop Physiology and Ecology in Southern China, Ministry of Agriculture, Nanjing Agricultural University, Nanjing, China

Fulai Liu
Department of Plant and Environmental Sciences, Faculty of Science, University of Copenhagen, Taastrup, Denmark

Xiangnan Li
National Engineering and Technology Center for Information Agriculture/Key Laboratory of Crop Physiology and Ecology in Southern China, Ministry of Agriculture, Nanjing Agricultural University, Nanjing, China
Department of Plant and Environmental Sciences, Faculty of Science, University of Copenhagen, Taastrup, Denmark

Markus G. Stetter, Leo Zeitler, Adrian Steinhaus, Karoline Kroener, Michelle Biljecki and Karl J. Schmid
Institute of Plant Breeding, Seed Science and Population Genetics, University of Hohenheim, Stuttgart, Germany

Sara Mayo, Oscar González-López, Alvaro Rodríguez-González, and Pedro A. Casquer
Research Group of Engineering and Sustainable Agriculture, Department of Agrarian Engineering and Sciences, Natural Resources Institute, University of León, León, Spain

Eleonora Cominelli and Francesca Sparvoli
Institute of Agricultural Biology and Biotechnology, Consiglio Nazionale delle Ricerche, Milan, Italy

Santiago Gutiérrez
Area of Microbiology, University School of Agricultural Engineers, University of León, Ponferrada, Spain

Miguel Curto and Diego Rubiales
Department of Plant Breeding, Institute for Sustainable Agriculture, Spanish National Research Council, Córdoba, Spain

Franziska Krajinski and Armin Schlereth
Department of Plant-Microbe Interactions, Max Planck Institute of Molecular Plant Physiology, Potsdam, Germany

Moritz Reckling
Institute of Land Use Systems, Leibniz Centre for Agricultural Landscape Research, Müncheberg, Germany
Department of Crop Production Ecology, Swedish University of Agricultural Sciences, Uppsala, Sweden

Göran Bergkvist
Department of Crop Production Ecology, Swedish University of Agricultural Sciences, Uppsala, Sweden

Christine A. Watson
Department of Crop Production Ecology, Swedish University of Agricultural Sciences, Uppsala, Sweden
Crop and Soil Systems, Scotland's Rural College, Aberdeen, UK

Frederick L. Stoddard
Department of Agricultural Sciences, University of Helsinki, Helsinki, Finland

Peter M. Zander
Institute of Socio-Economics, Leibniz Centre for Agricultural Landscape Research, Müncheberg, Germany

Robin L. Walker
Crop and Soil Systems, Scotland's Rural College, Aberdeen, UK

Aurelio Pristeri
Department of Agricultural Science, Mediterranean University of Reggio Calabria, Reggio Calabria, Italy

Ion Toncea
National Agricultural Research and Development Institute, Fundulea, Romania

Johann Bachinger
Institute of Land Use Systems, Leibniz Centre for Agricultural Landscape Research, Müncheberg, Germany

Redouane Choukr-Allah, Nanduri K. Rao, Abdelaziz Hirich, Mohammad Shahid, Abdullah Alshankiti, Kristina Toderich, Shagufta Gill and Khalil Ur Rahman Butt
Research and Innovation Division, International Center for Biosaline Agriculture, Dubai, UAE

Index